The colonisation of land
Origins and adaptations of terrestrial animals

COLIN LITTLE

Lecturer in Zoology, University of Bristol

CAMBRIDGE UNIVERSITY PRESS

Cambridge

London   New York   New Rochelle

Melbourne   Sydney

CAMBRIDGE UNIVERSITY PRESS
Cambridge, New York, Melbourne, Madrid, Cape Town, Singapore, São Paulo, Delhi

Cambridge University Press
The Edinburgh Building, Cambridge CB2 8RU, UK

Published in the United States of America by Cambridge University Press, New York

www.cambridge.org
Information on this title: www.cambridge.org/9780521106832

© Cambridge University Press 1983

First published 1983
This digitally printed version 2009

A catalogue record for this publication is available from the British Library

Library of Congress Catalogue Card Number: 83–1787

ISBN 978-0-521-25218-8 hardback
ISBN 978-0-521-10683-2 paperback

The colonisation of land
Origins and adaptations of terrestrial animals

Contents

Contents

Preface

The response of one of my colleagues, when told that I was writing a book about the colonisation of land, summed up the problem: he replied that it sounded 'an ambitious and open-ended project'. After more than 10 years work on the book, I have to agree that the scope of the subject is far beyond the capacity of one individual. However, it may be that the single-author approach has some justification. I have tried to make comparisons both between and within animal groups, and to infer general principles where appropriate. I have also tried to maintain some kind of similarity in coverage when discussing different animal groups, in terms of morphological, physiological and behavioural adaptations. In many cases, however, research emphasis has been towards a few topics in particular, and I have allowed this to influence my accounts so that different chapters do have different slants.

With a topic so vast I have not attempted to produce complete coverage, or to quote all the relevant literature. I hope, however, that the works quoted do give a fair picture of the situation as it stands. In trying to represent authors' views correctly, I am indebted to a large number of experts who have allowed themselves to be coerced into commenting upon individual chapters. Very many people have been helpful, including those who, when asked for permission to use their material in illustrations, kindly brought me up to date on the subject. For their comments, and without holding them in any way responsible for the final text, I would particularly like to thank: Drs J. Moore and R. Gibson for their comments on nemertines; Dr A. E. Dorey, on flatworms; Dr E. B. Andrews and Professor A. Graham, on molluscs; Dr R. S. K. Barnes and Dr S. P. Hopkin, on crustaceans; Dr R. S. Wilson, on chelicerates; Dr J. G. E. Lewis, on myriapods; Professor O. Dahl, on diplurans; Dr R. G. Davies and Dr G. Vannier, on insects; and Dr A. L. Panchen, on vertebrates. Above all, I would like to thank Dr R. S. K. Barnes, who has encouraged me from the start, and has waded through the entirety of the first draft, plying his correcting pen and helping to reduce the errors and idiocies in the final account.

The figures are almost all adaptations of published work, and I am grateful to the authors and publishers concerned for permission to use them here. Acknowledgement of the source is given in each case. They have all been redrawn, and I hope that the present versions will meet with the approval of their originators. The photographs have all been taken especially for the book, and here I have been helped by grants from the Royal Society, the British Museum and the British Council, which have allowed me to visit Papua New Guinea and New Zealand. Wherever possible the photographs have been taken in the field, and not in simulated natural conditions in the laboratory. I realise that this has not helped in photographic terms, but hope that because of it the animals are in natural poses. For most of the photography I have been lucky enough to be advised and helped by my wife, P. E. Stirling, who has also encouraged me enormously in the preparation of the whole book. To her I am particularly grateful for help and forbearance while 'the book' has been my major preoccupation.

The technical problems involved in producing a final copy have been greatly reduced for me by Mrs J. Cochran, who typed the initial draft, and by Mrs J. Ablett, who undertook the herculean task of sorting out the reference list and also produced an excellent index. At the same time, the seemingly endless process of hunting for appropriate literature has been made almost a pleasure by the helpfulness of successive librarians in the Botany–Zoology Library at Bristol. I would like to thank all these people for making the book possible.

Lastly, to all those slighted authors whom I have either misquoted, or not quoted at all, I extend my apologies. I hope that they will feel able to write and let me know of my errors.

Colin Little
Bristol, August 1982

Acknowledgements

The source of each figure is given in the legend. I am grateful to the following for permission to reproduce material: Academic Press Inc. (London) Ltd (Fig. 8.14*a*, from *Transport of ions and water in animals*, copyright 1977; Fig. 9.7, from *Basic structure and evolution of vertebrates*, copyright 1980). Academic Press Inc. (New York) (Fig. 8.13). Alan R. Liss Inc. (Figs. 6.7*b*, 8.15). American Midland Naturalist (Fig. 5.6). American Society of Zoologists (Figs. 5.13, 5.26, 5.27, 5.29, 5.30). *Archives italiennes de Biologie* (Fig. 6.9*a*). *Arkiv för Zoologi* (Fig. 9.4). A. Asher & Co. (Fig. 4.18*a,c*). *Biological Bulletin* (Figs. 2.3*b*, 5.18, 5.19). The Company of Biologists Ltd (Figs. 2.3, 3.7, 3.8, 4.19, 4.20, 5.8, 5.10, 5.31, 6.4, 6.10, 7.4, 8.12, 9.5, 9.9). CSIRO Editorial and Publications Service (Fig. 9.2*a*). Elsevier Biomedical Press BV (Figs. 5.21, 9.11*b*, 10.8). *Entomologica experimentalis et applicata* (Fig. 5.7). VEB Gustav Fischer Verlag (Figs. 4.23, 7.5*d*, 7.6*a*, 7.7*b*, 8.7). Hutchinson & Ross Inc. (Fig. 8.9*a*, from *Environmental physiology of desert organisms*, ed. N. F. Hadley, copyright 1975 by Dowden, Hutchinson & Ross Inc., Stroudsburg, PA). The Linnean Society of London (Figs. 7.6*b,c*, 7.10, by permission of the Council). Macmillan Journals Ltd (Fig. 6.2, from *Nature*, **251**, copyright 1974). The Malacological Society of London (Figs. 4.8, 10.3). Masson, S. A., Paris (Fig. 7.3, from *Archives d'Anatomie microscopique*, **46**, copyright 1957). *Monitore Zoologico Italiano* (Fig. 7.7*a*). New York Zoological Society (Fig. 6.11). *Oikos* (Fig. 6.1). Verlag Paul Parey (Fig. 8.3). Pergamon Press Ltd (Figs. 4.15, 5.23, 5.25, 6.3, 8.10, 8.17, 8.18, 9.3). The Royal Society (Figs. 8.16, 10.7). Royal Entomological Society of London (Fig. 8.11). Royal Meteorological Society (Fig. 8.9*b*). The Royal Society of Edinburgh (Figs. 3.4, 10.10). Society for the Study of Evolution (Fig. 6.5). Springer Verlag (Figs. 3.1, 4.20*b*, 5.12, 5.28, 6.7*a*, 7.5*a*, 7.5*a,b,c*). The University of Chicago Press (Fig. 5.9, from *Physiological Zoölogy*, **38**, copyright 1965). The University of Michigan, Museum of Zoology (Fig. 10,11). The University of Texas Press (Fig. 10.4, from *Coevolution of animals and plants*, ed. L. E. Gilbert & P. H. Raven, copyright 1975). *Vie et Milieu* (Fig. 7.2). The Zoological Society of London (Figs. 4.9, 4.14, 9.6, 10.1). Zoological Society of Southern Africa (Figs. 5.22, 8.2).

1

Introduction

'It was different from the universe to which he had been accustomed. For one thing, the heaven or sky above him was now a perfect circle. The horizon had closed to this. In order to imagine yourself into the Wart's position, you would have to picture a round horizon, a few inches above your head, instead of the flat horizon which you usually see. Under this horizon of air you would have to imagine another horizon of under water, spherical and practically upside down – for the surface of the water acted partly as a mirror to what was below it. It is difficult to imagine.'

From *The once and future king* by T. H. White, 1958, Collins, London.

There are in existence today approximately 30 animal phyla, whose distribution can be summarised as follows (see Table 1.1). Three are parasitic for most of their life cycle, and of the remainder 11 are confined entirely to the sea. Sixteen have representatives in the sea and in fresh water, and of these six also have terrestrial members. Three phyla that are basically aquatic contain some members able to resist drying in the state known as cryptobiosis. No phylum is restricted to fresh water or to land, except the Onychophora, which some would call a subphylum of the phylum Arthropoda. This distribution, together with considerations such as the composition of animal body fluids, has led the majority of authors to consider that life originated in the sea, and that freshwater and terrestrial forms evolved later (e.g. Pantin, 1931a; Pearse, 1936; Baldwin, 1964). Another hypothesis is that complex organic compounds, and therefore primitive organisms, were more likely to be formed on land (Hinton & Blum, 1965). Whether or not life itself originated in the sea, it seems likely that most of the animal phyla developed there, and have only later moved on to land. This change of habitat provides the subject matter of the following pages.

The various terrestrial groups are considered individually in later chapters, but here we may anticipate these by briefly referring to the phyla that have terrestrial representatives. Perhaps surprisingly, both the flatworms and nemertines have produced species which live on land. The flatworms, usually known by their small freshwater representatives, also include large, tropical land dwellers that are often spectacular

Table 1.1. *The distribution of the various animal phyla*

| Phylum | Presence or absence in various environments | | | |
	Marine	Fresh water (with small soil forms)	Parasitic	Terrestrial
Protozoa	x	x	x	–
Porifera	x	x	–	–
Mesozoa	–	–	x	–
Coelenterata	x	x	–	–
Platyhelminthes	x	x	x	x (few)
Ctenophora	x	–	–	–
Nemertinea	x	x (few)	–	x (few)
Acanthocephala	–	–	x	–
Rotifera	x (few)	x	–	(Cryptobiotic species)
Gastrotricha	x	x	–	–
Kinorhyncha	x	–	–	–
Nematoda	x	x	x	(Cryptobiotic species)
Nematomorpha	x (few)	x	x	–
Entoprocta	x	x (few)	··	
Annelida	x	x	x	x
Mollusca	x	x	x (few)	x
Phoronidea	x	–	–	–
Bryozoa	x	x	–	–
Brachiopoda	x	–	–	–
Sipunculoidea	x	–	–	–
Echiuroidea	x	–	–	–
Priapulida	x	–	–	–
Pentastomida	–	–	x	–
Tardigrada	x (few)	x	–	(Cryptobiotic species)
Onychophora	–	–	–	x
Arthropoda	x	x	x	x
Echinodermata	x	–	–	–
Chaetognatha	x	–	–	–
Pogonophora	x	–	–	–
Hemichordata	x	–	–	–
Chordata	x	x	–	x

'x' denotes presence in any particular environment.

in shape and colour. The nemertines, mainly marine and containing such familiar forms as species of *Lineus*, also have terrestrial species, but these are not usually as large as terrestrial flatworms.

Three phyla have cryptobiotic species, as mentioned above: the rotifers, nematodes and tardigrades. There are also some cryptobiotic insects. These species are all very small, and live in soil, in mosses and other places which although often wet may also dry up completely. During the dry season the animals become desiccated, but they recover when rehydrated.

The remainder of the land animals are more widely known. The molluscs have only produced one terrestrial class, the gastropods; while the annelids are represented mainly by the oligochaetes, although polychaetes and leeches are found on land in the tropics. The arthropods are the best-known terrestrial group apart from the vertebrates. Their land dwellers include insects, arachnids, myriapods and their more primitive relatives the onychophorans, and crustaceans. Among the latter the isopods are commonest and best known, but various members of the decapods and amphipods are found on land in the tropics. Lastly, the vertebrates form the terrestrial representatives of the phylum Chordata, and show increasing adaptations to life on land in the sequence fish, amphibians, reptiles, birds and mammals.

Of this large assemblage of terrestrial animals, past authors have mainly concerned themselves with the vertebrates (e.g. Carter, 1967). Little attention has been paid to invertebrate groups, other than at a general level (e.g. Goin & Goin, 1974). Early comparative physiologists and biochemists were most interested in the adaptations shown by animals living in various types of habitat. Baldwin, for example, devoted much of his book, *An introduction to comparative biochemistry*, first published in 1937, to the colonisation of land habitats, and the attendant biochemical and physiological problems. Of necessity the biochemical aspects were at that time mainly limited to the vertebrates. Pearse (1936) in a book entitled *The migrations of animals from sea to land* dealt with the routes that animals have taken, the causes of their migration and the changes that have occurred in the animals that have moved on to land. Pearse was able to add much to previous knowledge, which had mainly referred to the vertebrates, by his studies on various crustaceans and their adaptations to life in the littoral zone and out of water; but inevitably he was restricted by the few detailed studies available at the time. In 1947, Gislén gave a stimulating and wide-ranging address entitled 'Conquering terra firma', in which he commented upon a great variety of terrestrial invertebrates, and the routes by which their ancestors had emerged from an aquatic life. He concluded that most animal groups became terrestrial not directly by movement across the marine littoral zone, but by indirect movement first into brackish or fresh water.

Recently, the approach of comparative physiologists has been more to study in detail the adaptations of terrestrial animals than to theorise about the routes and causes of their movement from water to land. Such pioneering work as that of Edney (1957, 1960, 1977) on the terrestrial adaptations of arthropods has provided the pattern of much recent investigation. However, the immense amount of information accumulated has not been brought together, nor has it been considered in the context of the origins of terrestrial animals. The present volume is an attempt to examine terrestrial animal groups from the point of view of their origins as well as in terms of their adaptations.

Before discussing the information that is available concerning the various phyla and classes of animals that have representatives on land, some general points must be made. These will recur with the discussion of every animal group, but it is essential first to consider them in a general way.

1.1 The terrestrial environment

Superficially, the differences between dry land and water, as environments for the various animals that live in them, are easy to define. That is to say, when a typical aquatic environment is compared with a typical terrestrial environment, the following differences are found (see Table 1.2). Water is about 1000 times as dense as air, and therefore produces greater support, but since it is about 50 times as viscous, it also impedes movement. The partial pressure of oxygen may be the same in water as in air, but the maximum amount of oxygen per unit volume is over 30 times greater in air. In contrast, the solubility of carbon dioxide in water is greater than that of oxygen by a factor of 25. Various properties concerning the availability of water and salts are vastly different in water and on land: in water, salts may be readily available in large quantities as in the sea, or in smaller quantities as in estuaries and fresh water, whereas on land, salts are not available in solution; however, differences in salt concentration between an animal and its aquatic environment will produce tendencies for salts to diffuse across the body wall, while in air this problem does not arise. Although the availability of water may appear to be greater in aqueous media, individual animals may find it difficult to obtain water because of differences in osmotic pressure between the external water and their body fluid. In general, though, water loss is a greater problem on land where the water vapour pressure may at times be low and is usually variable. Variability is, in fact, greater for most factors on land than in water; for example, because of the low thermal capacity of air, air temperature may change rapidly over short time periods or over short distances. The result of this greater general variability is that more rapid changes of environment with time and distance are found on land than in water.

This brief comparison of aquatic and terrestrial environments avoids the crucial point that the two environments intergrade. In a sense the subject of this book could not exist if they did not, as it would be extremely unlikely for animals to make the sudden jump between the two environments described. Not only do they intergrade, but the intergradations occur in many different series. For example, marine environments merge into littoral areas, which may merge into rocky or sandy shores, marshes or mangrove swamps, or directly into forests or deserts. Some of these intergradations are obviously more gradual than others, and provide more gentle gradients in most physical factors, and therefore possibly easier routes for colonisation by animals.

The various intergradations between environments will be further discussed when we come to an examination of the routes by which animal groups that are primitively aquatic have moved on to land. Meanwhile, there is another set of environments which pose more of an immediate problem if we wish to define what is meant by terrestrial life: such habitats as leaf litter, soil, temporary pools and so on. Here, there is a continuous gradient between areas which are very constant in their physical conditions, often have liquid water available, and are always 100% water-saturated, and areas which are very variable indeed and often dry up completely. It then becomes exceedingly difficult to define which areas are truly terrestrial and which are not. However, it is perhaps slightly easier to examine the situations as they affect the animals living in them. From this point of view the important factor is the relation of the animals to the available water: if they are effectively covered by a layer of water then they are living as aquatic animals; if they are not so covered, which often means just that they are bigger, as earthworms are usually bigger than soil-dwelling nematodes, then they can be said to be terrestrial. This definition results in such forms as soil-dwelling nematodes, mites, rotifers and tardigrades being called aquatic, while the burrowing amphibians and earthworms are called terrestrial; and such a division at least has the merit that it fits the known facts concerning terrestrial adaptations in soil dwellers. Thus the nematodes have no obvious adaptations for terrestrial life, whereas the earthworms can

reproduce out of water and obtain their oxygen from the air; although this is not to say that many of the problems faced by earthworms are not faced by freshwater animals as well.

The situation in temporary pools is somewhat simpler to define, since the animals inhabiting them are only active when liquid water is present, and must therefore be considered aquatic. Of course, there will be borderline cases, difficult to fit into either category; but this merely reflects the fact that within aquatic and terrestrial environments there are great variations in physical conditions.

It is occasionally convenient to have a classification of the relationships of these terrestrial and semi-terrestrial animals to their environment, and to give a rough guideline to the approach taken in this book a scheme is outlined in Table 1.3. It is not suggested that this be used as a rigid method of classifying animals, but rather to indicate some of the ways in which animals have responded to the rigours of the terrestrial environment.

1.2 · The nature of the evidence

This book sets out to discuss the available evidence about which groups of animals have colonised land, which routes they have used, and what modifications or adaptations have been involved, together with some comment on the time taken by these colonists, and on the degree of 'terrestrialisation' reached by each group. The evidence on these various counts falls under four headings.

1.2.1 Geological evidence

It is unfortunate that this line of approach, which could theoretically tell us so much about the history of animal invasions of land, has so far provided very little information, especially for the invertebrates. This is due in the main to some of the mechanisms of fossil formation, including the lack of preservation of soft-bodied

Table 1.2. *Some physical properties of air and water*

Property	Sea water	Pure water	Air
Density (g/ml at 20°C and 76 cmHg)	1.025	1.000	0.0012
Viscosity (centipoises at 20°C)	1.09	1.00	0.02
Oxygen content (ml/100 ml, in equilibrium with air at 20°C)	0.53	0.66	20.95
Oxygen solubility (i.e. content in equilibrium with oxygen at 76 cmHg and 20°C, ml/100 ml)	2.36	2.94	—
Carbon dioxide content (ml/100 ml, in equilibrium with air at 20°C)	0.02*	0.03	0.033
Carbon dioxide solubility (i.e. content in equilibrium with CO_2 at 76 cmHg and 20°C, ml/100 ml)	76.2*	88.3	—
Thermal capacity per unit volume (cal ml^{-1} $(°C)^{-1}$ at 20°C)	0.93	1.00	0.0003
Refractive index (at 20°C)	1.34	1.33	1.00
Velocity of sound (m/s at 20°C and 76 cmHg)	1519	1486	343
Surface tension (dynes/cm in contact with air at 20°C)	73.2	72.8	—

* The figures quoted are for free carbon dioxide plus carbonic acid. In pure water the majority of carbon dioxide is present in these forms, but in sea water the majority exists as HCO_3^-, and the total carbon dioxide of normal sea water including HCO_3^- is 4.6 ml/100 ml.

Figures taken from Chemical Rubber Co. (1966), Cox (1965), Geiger (1961), Richards (1965), Skirrow (1965), Sverdrup, Johnson & Fleming (1942).

animals and soft parts in general, and the scarcity of fossils formed by terrestrial animals. Various exceptional circumstances, such as the trapping of insects in amber, have provided occasional shafts of illumination into the otherwise dark record of terrestrial invertebrate palaeontology. The record for terrestrial vertebrates is in a much happier state, but still is poor compared to our knowledge of aquatic faunas. From the scattered records available, and from the better-documented aquatic records, we can at least see at what times in geological history many of the main groups have evolved, and in some cases we can make a fairly well-educated guess as to when particular groups moved on to land. Often, too, we can see whether or not particular groups moved into freshwater as well as marine habitats, and this is important when we consider the routes by which aquatic animals migrated to land.

The sequence of geological periods is given in Table 1.4 with approximate figures for the age of each period.

1.2.2 Evidence from present-day distribution

Together with the palaeontological evidence, where this is available, the present-day distribution of various groups suggests likely routes of colonisation of land. For example, it is found that modern gastropod molluscs of the superfamily Littorinacea are found in marine and brackish habitats, and on land; but there are no freshwater members of the group. This must strongly suggest that movement on to land has not taken place via fresh water, but more directly from the sea or from areas of brackish water.

1.2.3 Evidence from comparative anatomy and physiology

It appears that nearly every facet of life has been altered with a change from water to land, and these alterations can be viewed from two aspects. In the first place they can be approached as examples of the ways in which various species are adapted to particular environments, and this is the approach of the comparative physiologist as emphasized by Edney (1957). Additionally, the information gained by comparative physiologists and anatomists can be used to suggest possible mechanisms by which animals have, in the past, been able to colonise land from aquatic habitats. This second procedure is certainly a much more dangerous and speculative one than the first, and should probably never be carried to the limit by assuming that mechanisms present in living species evolved directly from those of other living species. It can, however, indicate general trends which may have occurred in evolution. In the present book both these approaches will be used to some extent.

Since this evidence from comparative studies constitutes by far the greatest weight of available information, a brief list of some of the changes which are thought to have taken place during the movement of animals on to land, and which have been discussed by such authors as Baldwin (1964), Carter (1951), Pearse (1936) and Young (1950), will be given here, to anticipate the detailed cases given later.

Table 1.3. *Animal responses to water supply on land*

A. *Aquatic animals*

Besides those conventionally regarded as aquatic, this category includes small soil animals that are only active when covered by a water film, e.g. protozoans, ostracods, nematodes

B. *Cryptozoic animals*

These require constant high humidity and are intolerant of desiccation. They include a variety of forms that mostly maintain themselves in humid places by behavioural adaptations

(i)   Soil dwellers, e.g. earthworms, burrowing amphibians, many insect larvae
(ii) Leaf litter animals, also sometimes found under stones etc., e.g. many woodlice, centipedes, millipedes
(iii) Tropical forest dwellers, e.g. flatworms, nemertines, leeches

C. *Hygrophilic animals*

These require high humidity or water for activity, but are tolerant of desiccation. They often aestivate seasonally, e.g. some amphibious and many land snails. Cryptobiotic species might be included here

D. *Xerophilic animals*

These are active in dry conditions. This category includes many vertebrates – reptiles, birds and mammals – and many arthropods. Of the latter most are insects, but many arachnids and some isopod crustaceans are active even in desert conditions

Table 1.4. *The geological time scale*

Era	Period	Approximate age from start of period in millions of years
Cainozoic	Quaternary	1.8
	Tertiary	65
Mesozoic	Cretaceous	135
	Jurassic	192
	Triassic	230
Palaeozoic	Permian	290
	Carboniferous	350
	Devonian	410
	Silurian	435
	Ordovician	485
	Cambrian	560
Precambrian		?3750

Figures from Cohee, Glaessner & Hedberg (1978) and Dott & Batten (1981).

Salts and water

Differences in environment appear to have led many times to the production of mechanisms for conserving water, be they physiological, structural, behavioural, or, more often, a combination of all these three. The external surface of terrestrial animals is often relatively impermeable to water or, if not, the animals remain in areas of high humidity. Loss of water in the urine and in faeces is cut to a minimum, and salts are reabsorbed from the urine. Little is known about the mechanisms by which salts are obtained by land animals, but presumably they are taken up from the food. A comparison of the ionic composition of the body fluids of terrestrial animals with that of their aquatic relatives may often suggest whether the terrestrial forms are more closely related to marine or to freshwater ancestors; and this in turn may indicate that a particular group has colonised land from fresh water or from the sea. In a similar way conclusions may be drawn from many of the other physiological attributes discussed below.

Respiration

Mechanisms for obtaining oxygen and for eliminating carbon dioxide which are efficient in water often do not function well in air; for example, gills tend to collapse in air, causing a change from a very large surface area to only a small effective surface area for gas exchange, while 'lungs' or vascularised air-filled cavities may function well. On the other hand, all respiratory surfaces have to be kept moist so that oxygen can dissolve before passing across the respiratory epithelium.

Other changes in the chain of respiratory processes must follow, especially in the mechanisms by which oxygen and carbon dioxide are carried by the circulatory fluid: with different partial pressures of carbon dioxide, the whole acid–base balance of the fluid will be modified, and this will in turn affect the position with regard to osmotic and ionic regulation and kidney function.

Nitrogenous waste

In water the nitrogenous waste product is often ammonia; this is highly toxic, but also very soluble, and is usually carried away in large volumes of water. On land, when there is no such water supply, the nitrogenous waste products are usually eliminated as somewhat less toxic substances such as urea (slightly toxic but very soluble) or uric acid (insoluble). There are many other substances which may be produced, and many exceptions to the generalisations given, but the production of non-toxic and/or insoluble compounds is often regarded as the norm.

Reproduction

In many marine animals, eggs and sperm are shed into the external sea water, fertilisation occurs there, and when the larvae hatch they become planktonic for some period. This process is obviously not feasible for a truly terrestrial animal, and in general fertilisation has become internal, and the young are either placed in a moist environment to hatch or are carried round with the parent in some type of brood pouch. One notable exception appears to be the decapod Crustacea, which must almost all return to the water to breed.

Sense organs

In aquatic animals most of the chemical receptors can be described as contact receptors, since they respond to dilute solutions of various substances. Terrestrial animals often maintain taste receptors, but many also possess olfactory organs which are distance receptors, capable in some cases of detecting minute quantities of specific chemicals transported by air currents.

The sense of hearing in aquatic invertebrates is either primitive or non-existent. The organs of hearing in animals are usually linked to the organs of balance, since statocysts are often employed in both. Aquatic invertebrates often possess statocysts, but they have so far only been shown to receive low-frequency vibrations. Land invertebrates can, in contrast, often perceive high-frequency vibrations, and this capacity is coupled with a greater capacity for sound production. Mechanisms [1] of light reception and image formation probably differ little in the sea and on land; but the difference between the refractive index of water and that of air ensures some differences in detail. In aquatic eyes there is little refraction at the outer surface or cornea, and the convergence of the light rays must all be due to the lens; whereas in eyes functioning in air much refraction occurs at the cornea so that the lens does not need to be so powerful. This factor of refractive index may partly explain why although cephalopod eyes are very similar to mammalian eyes they are even closer in structure to those of fish. For animals in transition from water to land the difference in refractive index must therefore pose a problem; just how important this problem is can be seen from the complex structure of insects such as those in the genus *Dytiscus* and fish such as those in the genus *Anableps* where one half of the eye is adapted for vision above the water surface and the other half for vision below.

Movement and behaviour

It has been pointed out that air is much less dense than water, and also less viscous. These differences have meant that although the typical methods of moving in water have had to be abandoned in air, movement on land has in many cases become much speeded up; while the adaptations produced by birds and insects have allowed utilisation of the air without contact with the ground, and the consequent advantages of even more rapid motion. With this capacity for greater speed and manoeuvrability, and with sense organs becoming adapted to the land habitat, changes in behaviour patterns have often been extreme. Behaviour has in general become more complex, as exemplified by the exceedingly complicated courtship displays of many terrestrial, as opposed to most aquatic, animals. Behavioural complexity has reached its peak in social animals such as insects and vertebrates.

1.2.4 Negative evidence

Some further discussion of the factors which have allowed the colonisation of land arises from the types of marine animals which have no terrestrial representatives. This 'negative' evidence must. however, be treated cautiously, as it is only too easy to suggest that a particular aquatic type could never have become terrestrial because of a particular structural form or physiological mechanism, and then to find that there are in fact land relatives with features very similar to those of the aquatic ancestor. For instance, it might be argued that the planarians are basically unsuited to terrestrial life because they have a permeable skin and would be immediately subject to desiccation. Yet planarians have been successful in invading certain terrestrial habitats such as tropical forests, where the environment may become relatively dry during the day but is always humid at night. By restricting their activity to the night time, planarians have in fact avoided those characteristics of the environment that they cannot tolerate. In other words, a behavioural adaptation has allowed colonisation of a particular habitat which would otherwise have been impossible. We shall see that this type of adaptation is common in terrestrial invertebrates.

Despite the inadequacies of the negative argument given above, some consideration of those types which are represented only in the sea may be useful in suggesting factors that have prevented some animals from moving on to land; but we shall be in a better position to investigate these forms when we have examined the groups that do have terrestrial representatives, and so discussion of these marine groups will be largely postponed to the final chapter.

With this brief summary of the types of evidence that are available to us in any examination of the mechanisms and routes of invasion of land, we can turn to an examination of the various animal groups that have terrestrial representatives.

2

Flatworms, nemertines and nematodes

'The existence of a division of the genus *Planaria*, which inhabits the dry land, interested me much ... Numerous species inhabit both salt and fresh water; but these to which I allude were found, even in the drier parts of the forest, beneath logs of rotten wood, on which I believe they feed.'

From *Journal of researches into the natural history and geology of the countries visited during the voyage of II.M.S. 'Beagle' round the world, under the command of Captain FitzRoy, R.N.* by Charles Darwin, 1845, John Murray, London.

The three groups that will be considered in this chapter, the flatworms (phylum Platyhelminthes), the nemertines (phylum Nemertinea or Rhynchocoela) and the nematodes (phylum Nematoda or class Nematoda of the phylum Aschelminthes) are all worm-like forms. All possess a relatively lowly grade of organisation, and all have, in many senses, avoided the problems of terrestrial life, although they all have representatives that live on land. Yet these terrestrial forms show some differences from their aquatic relatives, and these differences are instructive in indicating some of the changes necessary for any transition from aquatic to terrestrial life. The characteristics of these forms also highlight some of the difficulties involved in deciding by which routes particular groups have moved on to land, because although some species resemble littoral marine types, others resemble those found in fresh water.

The three groups are first considered individually, and then some general comments are made.

2.1 The flatworms

Of the three classes of the phylum Platyhelminthes, both the Trematoda (flukes) and the Cestoda (tapeworms) are parasitic, and hence the environment in which they live is decided mainly by the characteristics of the host, and not by those of the environment in which the host lives. While it is true that both flukes and tapeworms may pass part of the life cycle as cysts which can withstand desiccation and extremes of temperature, liquid water must be available for any activity

to occur. This discussion will therefore be limited to the free-living forms, the class Turbellaria.

Within the Turbellaria, terrestrial forms which inhabit for the most part damp forest habitats are found in the rhabdocoels, 'alloeocoels' and triclads (Hyman, 1951a). Unfortunately very little is known about terrestrial representatives of the first two groups, in which all the animals are very small; but the triclads (Fig. 2.1), which are often brightly coloured and may reach lengths of 60 cm, have attracted some attention. Triclads are found in the sea, in brackish water, in fresh water and on land, but are regarded as being marine in origin (Ball, 1981). Their division into suborders follows these habitat differences, and the three suborders are on this account called the Maricola, Paludicola and Terricola. Although Hyman (1951a) assumed that the Turbellaria have spread from the sea to fresh water, and only thence into terrestrial habitats, other authors have suggested that the Terricola and Maricola may be closely related (see Ball, 1981). The origin of the Terricola is thus obscure, but some discussion of the origin of osmoregulatory abilities within the suborder is helpful in assessing possible routes on to land.

At the present time few triclad species are found in the marine littoral zone, although many rhabdocoels are common in saltmarshes (den Hartog, 1974). The existence of these rhabdocoels and of a few genera of brackish water triclads such as *Uteriporus* and *Procerodes* demonstrates the ability of turbellarian physiological systems to cope with large changes in the availability of salts and water. The only species in which details of the mechanisms of response to salinity changes are well documented is *Procerodes littoralis*. This species lives in a specialised habitat where freshwater streams run over shingle substrates into the sea (Pantin, 1931b). Here the salt concentration changes rapidly over the tidal cycle from that of full-strength sea water to that of fresh water. When sudden dilution occurs, water moves through the epidermis into the parenchyma, and this water is then taken up by the cells of the gut (Beadle, 1934). Epidermal permeability to water is then lowered, by a mechanism involving the presence of calcium, and the gut cells remain swollen while the animal stays in fresh water. According to this explanation, the proto-nephridial system, with its large number of flame cells, is not involved in the operation. In freshwater planarians this protonephridial excretory system is well developed, although few details are available about its function. Observations on the rhabdocoel *Gyratrix hermaphroditus* suggest that the system is involved in water balance, because while those specimens which live in fresh water have a complicated system of flame cells, tubules, ampullae and an excretory bladder, specimens living in sea water have no tubules, ampullae or bladders (Kromhout, 1943). Terrestrial planarians also possess well-developed protonephridial systems, and these have been implicated in salt and water balance, mostly by circumstantial evidence. Their fine structure is like that of freshwater planarians (Silveira & Corinna, 1976). Some of the details of flame cell function will be considered in the section dealing with nemertines. For the moment it is sufficient to conclude that the osmoregulatory system in *P. littoralis* is designed to cope with a very extreme situation, and that it is probably not typical of marine triclads. The development of turbellarian protonephridia as osmoregulatory organs is more likely to have been associated with later movement into fresh water, and the similarity of the protonephridia in Terricola to those in Paludicola supports the idea that movement on to land occurred from fresh water, and not from the sea or from brackish water.

Terrestrial triclads are restricted to damp habitats, and the problems that they encounter are likely to be similar to those of freshwater forms, although they cannot withstand submersion. They are unusual among invertebrates in converting much of their waste nitrogen

to urea: in *Bipalium kewense*, for instance, up to 73% of excreted nitrogen may be in the form of urea (Campbell *et al.*, 1972). This feature is not, however, a unique development of Terricola, but a phenomenon widespread in the flatworms (Campbell, 1965; Simmons, 1970). Terricola possess a moist epidermis without a cuticle, provided with many microvilli, and ciliated in parts, like that of the aquatic forms (Bedini & Papi, 1974; Bautz, 1977). They are dependent upon the use of mucus for locomotion, and are not, consequently, very resistant to desiccation. Their behavioural responses tend to prevent them from being exposed to low humi-dities, however, since many show a strong negative phototaxis, and are thigmotactic. The phototaxis is mediated through the eyes, and some of these eyes have a more complex structure than those of fresh-water triclads. Indeed, it has been suggested that some land planarians have image-forming eyes, although there is no experimental evidence available (Hyman, 1951*a*). Because of these behavioural tendencies, terrestrial flatworms remain hidden during the day and are usually only active at night, when they come out to feed on small invertebrates such as earthworms, nematodes and snails. They themselves seem to have few predators, probably because they appear distasteful to otherwise likely pursuers. It seems probable that the bright colour patterns so often found in the tropical forms may serve as warning coloration (Fig. 2.1).

Almost all turbellarians are hermaphrodites, but in general copulation occurs, with internal fertilisation (Hendelberg, 1974). This preadaptation for life on land is accompanied by two other factors in development: the eggs have a hard and resistant shell and the young hatch directly as young worms, not as larvae. There seems, in fact, to be little difference in these respects between terrestrial and aquatic forms.

The terrestrial triclads consist of three major families, the Geoplanidae, the Rhynchodemidae and the

Fig. 2.1. Terrestrial flatworm from Papua New Guinea. This species has a bright colour pattern (yellow and black), typical of many terrestrial flatworms and thought to be aposematic (warning coloration). Length *c.* 2 cm.

Bipaliidae, all of which are in the main restricted to the floor of tropical and sub-tropical rain forest. The Geoplanidae are mostly South American. They have a dorso-ventrally flattened body and creep upon their entire ventral surface. Some species have been introduced into California, where they have become established in gardens. The Rhynchodemidae are widespread in the tropics, but are also found in some humid situations in the temperate zones. The most widespread endemic species in Europe is *Microplana terrestris*, which is found within the rectangle bounded by Greece, southern France, Ireland and southern Sweden (Gislén, 1944). In Britain this species is accompanied by others, although the validity of some of these is uncertain (Pantin, 1944; Reynoldson, 1974; Jones, 1978*a*). In general the Rhynchodemidae have a more cylindrical cross-section than the Geoplanidae, and they have a specialised creeping sole, so that much less of the surface area of the body is in contact with the substratum. Although Pantin (1950) believed that muscular waves were involved in this locomotion, Jones (1978*a*) found that the cilia were responsible for propulsion. Under some circumstances, *Microplana terrestris* and *Rhynchodemus bilineatus* further reduce the area of contact with the ground by producing stationary muscular waves along the body, and contacting the substratum only at a series of points. This phenomenon is shown in a New Zealand flatworm in Fig. 2.2. Jones (1978*a*) has shown that in this type of locomotion, propulsion is still in the main provided by the action of cilia at the points of contact with the ground, but that the stationary wave increases speed to some extent. Flatworms moving in this way, like those showing continuous creeping, leave behind a mucous trail, but in this case the trail consists of a series of 'footprints' (Pantin, 1950). Besides increasing speed, the loss of mucus is therefore decreased, and Jones (1978*b*) has suggested that this type of locomotion may be adopted in low humidities as a method of water

conservation. In other terrestrial triclads the mucous trail left behind is, if anything, increased in volume, and takes the form of a thread so that the animal can lower itself through the air in much the same way as spiders (Hyman, 1951*a*).

The Bipaliidae are easily recognisable because of their expanded semicircular-shaped head bordered with numerous eyes. *Bipalium kewense*, although naturally restricted to tropical rain forest in Indo-China, has now been distributed to various parts of the world through the action of man. This species feeds solely on earthworms, and has become sufficiently abundant in the southern parts of the United States to become a minor pest of earthworm farms (Winsor, 1981). A species of *Bipalium* has been examined by Kawaguti (1932) with regard to water loss. It could lose up to 50% of its body weight and still recover. Beyond this limit some change occurred, as the rate of water loss with time was no longer linear. Since the body water was about 80% of the total weight, this suggests that about 60% of the body water was relatively 'free' compared to the rest. The animals regained water rapidly when placed on a damp substratum, apparently through the epidermis.

The general picture of terrestrial planarians presented above is one of strict confinement to damp habitats; but observations by Froehlich (1955) have made it quite clear that they may occur in such dry situations as under stones in burnt fields. At the same time, they are never found in very wet places, and are driven out of their refuges by heavy rains because these become waterlogged. The smaller forms such as species of *Microplana* retreat deep into the soil, using earthworm burrows when the upper layers of the soil dry up (Gislén, 1944). They are, in summary, very particular as to habitat, and are typical members of the cryptozoic fauna.

2.2   The nemertines
Because they possess such systems as an eversible proboscis, a circulatory system and a gut with a posterior anus as well as a mouth, the nemertines are regarded as being of a somewhat 'higher' grade of organisation than the flatworms. The terrestrial forms, however, occupy

Fig. 2.2. Terrestrial flatworm from New Zealand, showing locomotion by a combination of muscular waves and the action of cilia. The arrows show the points at which the animal is in contact with the ground. At these points, cilia beating in the layer of mucus propel the animal. Length *c*. 2 cm.

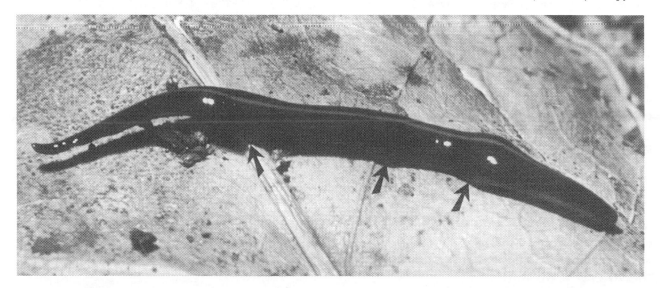

niches which are similar to, but often slightly damper than, those occupied by terrestrial triclads.

The majority of nemertines are marine, and the phylum is thought to be marine in origin; but some species, such as the common *Lineus ruber*, can live in brackish water, and some genera, e.g. *Potamonemertes* and *Prostoma*, inhabit fresh water. Many of the marine littoral forms, e.g. species of *Prosorhochmus*, live at quite high tidal levels, while a few of the terrestrial species (all of which were until recently placed in the genus *Geonemertes*) live in humid habitats near the top of the shore. This distribution has suggested to Pantin (1961, 1969) that the terrestrial forms have evolved from marine littoral ancestors, and this idea is supported by the occurrence of terrestrial species on isolated islands where there are no freshwater forms. Indeed, the terrestrial species are restricted to oceanic islands, apart from those found in New Zealand and Australia. Exactly why this should be so remains puzzling, except for the argument that there may be less competition from other animals in island situations. Such a distribution, together with the fact that many of the characters of the genus *Geonemertes* are those held in common by all generalised metanemertines, suggests that the genus may be polyphyletic in origin. Moore & Gibson (1981) have, as will be discussed later, confirmed that *Geonemertes* should be regarded as several genera, and this may go some way towards explaining the physiological diversity found within the terrestrial species. Nevertheless, it appears to be generally agreed that terrestrial nemertines have been derived from marine littoral forms, even if this evolution has occurred in parallel several times. We shall return to this point, but it is important to realise that there are some aspects which do not fit in well with the idea of a littoral origin for terrestrial species. In particular, it is important to consider the development of a system that has its own intrinsic interest, that of the protonephridia and excretory ducts.

Almost all nemertines possess flame cells or cyrtocytes, and an 'excretory' system. In marine forms this system is not very elaborate, and is restricted to the oesophageal region, and the flame cells are intimately connected with the blood system: in some species the flame cells actually project into the blood vessels. In the freshwater and terrestrial forms the number of flame cells is much increased – Schröder (1918) estimated that there were 35 000 in *Geonemertes palaensis* – and they are found either immediately underneath the dermal muscle coat, or scattered throughout the parenchyma over the whole body. There is also a tendency for groups of cells to open to the outside through individual ducts rather than through one common collecting duct, as in the marine forms. These tendencies suggest that the flame cell systems in freshwater and terrestrial forms are concerned with the elimination of water passing in through the body surface. Since the evidence for the function of these protonephridial systems comes from observations on various phyla, this discussion must include the freshwater rotifers as well as the freshwater flatworms and nemertines.

The first comprehensive summary of work on flame cells was that of Goodrich (1945). Flame cells were seen to be cells with a blind lumen in which beat a flame-like bunch of cilia. The lumen is continuous with a duct made up of further cells, leading to a nephridiopore. It was assumed by Goodrich that while some excretory granules pass into the duct from these latter cells, "the chief function of the 'flame' is to drive the fluid passing by osmosis through the thin wall of the chamber down the canal towards the nephridiopore". In 1947, Pantin studied the flame cells and nephridial ducts of *Argonemertes dendyi*, an Australian terrestrial nemertine now found in southern England. He concluded that the 'flame' of cilia beats inside the lumen of the cell in such a way that fluid could actually be forced down the nephridial tubule: when this was blocked, the cell and the duct became swollen because of accumulating fluid. In this species there is a complex duct leading to the exterior (Fig. 2.3*a*), in part of which is localised an alkaline

Fig. 2.3. Protonephridia in terrestrial nemertines. (*a*) *Argonemertes dendyi*, a Group 2 species (see Table 2.1) in which the flame cells are simple but the ducts are glandular and are probably concerned with the reabsorption of salts. This species can tolerate submersion in fresh water. (*b*) *Pantinonemertes agricola*, a Group 1 species in which the flame cells are binucleate and are strengthened by prominent ribs, while the ducts are not greatly differentiated. This species can tolerate submersion in sea water. (*a*) Redrawn from Pantin (1947); (*b*) redrawn from Coe (1930).

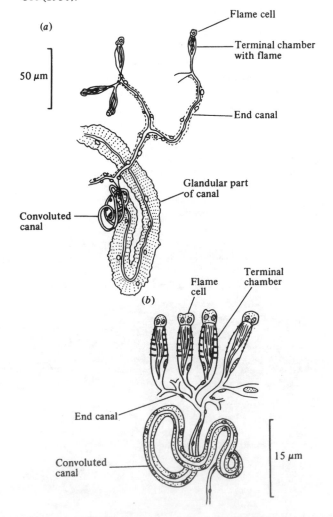

phosphatase, suggesting some functional similarity to
the reabsorptive kidney tubules of vertebrates. Following
this line of evidence that protonephridia might be osmoregulatory in function, further circumstantial evidence
has come from observing the rates of fluid output from
these systems when animals are placed in salt solutions
of various concentrations. In two species of the rotifer
genus *Asplanchna*, where fluid is excreted by a contractile
bladder, fluid output decreased as the salt concentration
of the medium was increased (Pontin, 1966). Proof of
the osmoregulatory function in *Asplanchna* came from
the micropuncture studies of Braun, Kümmel & Mangos
(1966), which showed that the fluid produced was
hypo-osmotic to the body fluid.

The mechanism by which flame cells operate is
not yet properly understood, but the current state of
knowledge has been reviewed by Wilson & Webster
(1974), Brandenburg (1975) and Kümmel (1975).
According to Wilson & Webster, each flame cell consists
in reality of two cells: the terminal cell which provides
the flame, and the first tubule cell which forms the
'barrel' inside which the flame beats. These two cells
interdigitate, and at their junctions are narrow fenestrations, which might act as an ultrafilter. In the planarian
Dugesia tigrina, for instance, the fenestrations are
formed by parallel slits 35 nm wide (McKanna, 1968).
Also in the genus *Dugesia* the cells of the nephridial
tubules show many complex infoldings associated with
mitochondria in much the same way as in structures
known in other animals to be reabsorptive. It therefore
appears that protonephridial systems could function by
producing an ultrafiltrate and reabsorbing salts from
this filtrate in much the same way as do other osmoregulatory systems.

The examples in which protonephridial systems
have been examined are those where they have been
much expanded in comparison with marine forms, and
this expansion appears to be an adaptation to life in
dilute aquatic environments. It is difficult to imagine
the selection pressure that would produce such adaptations in marine littoral nemertines, since few other
marine littoral invertebrates can osmoregulate well,
and the need to *expel* water in such habitats is usually
minimal. It is, therefore, necessary to consider the
possibility that some terrestrial nemertines may have
been derived from freshwater ancestors rather than
marine littoral ones, despite the lack of obvious present-
day freshwater relatives. On this basis, the modern
terrestrial nemertines that live near the shore might
represent the recolonisation of semi-marine habitats
by animals derived from fresh water.

The argument is complicated by the polyphyletic
origin of the terrestrial nemertines, and it is essential
at this point to anticipate later discussion by pointing
out that at least *Pantinonemertes agricola*, found in
Bermuda, has evolved independently from such forms as
Argonemertes dendyi, found in Australia. In this context it is interesting to note that while *P. agricola* cannot
withstand submersion in fresh water, it can survive long
immersion in sea water. In contrast, *A. dendyi* can withstand immersion in fresh water but not in sea water

(Hyman, 1951*a*). Some brief personal observations on
the osmotic pressure of the body fluids of two nemertines
are therefore relevant. Rhynchocoelic fluid from
A. dendyi had an osmotic pressure of 144 mOsm. Fluid
from the rhynchocoel of the freshwater nemertine
Prostoma jenningsi showed a similar value of 139 mOsm.
Both values are fairly high for soft-bodied invertebrates,
and their similarity could be taken as supporting evidence
for a freshwater origin of *A. dendyi*. However, recent
studies of the distribution of terrestrial and semi-terrestrial
nemertines suggest that a direct marine origin is more
likely, and this work must now be discussed.

Moore & Gibson (1973, 1981) and Gibson & Moore
(1976) have studied the anatomy and distribution of
freshwater and terrestrial nemertines. In so doing, they
have clarified the rather complicated picture in two
particular respects. First, they have established an
acceptable taxonomy for freshwater nemertines, and
have described two new genera. On the basis of this work
they have postulated two evolutionary lines within the
Hoplonemertini. As can be seen from Fig. 2.4, the
tetrastemmid line colonised brackish water, and produced
the genus *Prostoma* in fresh water. The prosorhochmid
line, in contrast, colonised first the marine supra-littoral
and then the land, and in this line freshwater genera
evolved from terrestrial ancestors. It is therefore apparent
that the freshwater genus *Prostoma* and the terrestrial
Argonemertes dendyi were derived from different
stocks, so that direct comparison of such factors as
the osmotic pressure of their body fluids is unlikely to
be particularly revealing. However, the evolution of
osmoregulatory mechanisms and of body fluid composition in the marine littoral zone has many complex
aspects, and further consideration is given to these in
Chapter 10.

The second respect in which the work of Moore
& Gibson has helped to clarify the situation concerns
the polyphyletic origin of terrestrial nemertines. From
consideration of a large number of characters, they
pointed out that '*Geonemertes*' consists of two groups
of genera. The distribution of these is given in Table
2.1. In Group 1 the genus *Pantinonemertes* has species
in the marine littoral zone, in mangrove swamps and in
fully terrestrial situations. One of the characteristics
of this group is the structure of the flame cells. These
are binucleate and are strengthened with bars, as shown

Fig. 2.4. Possible sequences of the evolution of freshwater and
terrestrial nemertines, after Moore & Gibson (1973). It is
suggested that fresh water may have been colonised by two
entirely different routes.

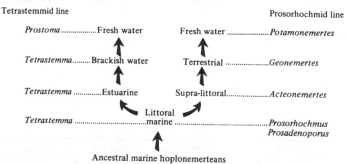

in Fig. 2.3*b*. The existence of species in this range of habitats must be taken as good evidence for a direct marine origin of the terrestrial species. In Group 2 one of the characteristics is the uni-nucleate flame cell without strengthening bars (Fig. 2.3*a*). *Acteonemertes bathamae* (Fig. 2.5) is found in the upper marine littoral of New Zealand shores (Pantin, 1961), and has now also been recorded on land at an altitude of about 30 m (Moore, 1973). This distribution also supports a direct origin of terrestrial forms from marine ancestors, independently of the Group 1 genera.

The arguments about the origins of terrestrial nemertines have been dealt with in some detail to show how many factors have to be considered in such complex cases. We can now describe briefly some of the ecology and physiology of terrestrial nemertines. A comprehensive account of nemertine biology is given by Gibson (1972). Terrestrial species occupy a variety of habitats, including damp situations under logs in forests, the leaf bases of the screw pine *Pandanus* in the Seychelles, and regions at the top of the marine littoral zone. All these situations have a high humidity, and the animals appear to venture out to capture prey at night only. They react negatively to light and are highly thigmotactic, in both respects resembling terrestrial planarians. All nemertines produce mucus which is used as a basis for movement so that they are unable to withstand desiccation. Pantin (1950) investigated movement in *Argonemertes dendyi*, and found that normal propulsion consisted entirely of ciliary gliding: the worm secretes mucus, especially at the head end, and this is carried back over the rest of the body. The worm therefore effectively swims in a tube of mucus. When strongly stimulated, however, the proboscis is everted. The tip of this adheres to the substratum and by muscular contraction of the proboscis the body is drawn·rapidly forward. A similar phenomenon is seen in the genus *Acteonemertes* (Fig. 2.5*b*).

Terrestrial species also use mucus during reproduction, when both worms are enclosed in one mucous sheath. The eggs and sperm are liberated into this and when the adults leave, the eggs develop inside it (Gibson, 1972). An exception is provided by *Pantinonemertes agricola*, which shows internal fertilisation and is ovovivi-

parous. The habit of viviparity is not widespread in nemertines, but has also been authenticated in some marine hoplonemerteans of the genera *Prosorhochmus* and *Poikilonemertes*, as well as in the heteronemertean *Lineus viviparus* (Gibson & Moore, 1976). Its absence in most terrestrial species is therefore perhaps surprising.

2.3 The nematodes

Animals living in the soil pose a problem for those who wish to define closely the terms 'terrestrial' and 'aquatic', as pointed out in Chapter 1. The smaller inhabitants of the soil, which are the most controversial in this sense, consist of protozoans, turbellarians ('alloeocoels' and rhabdocoels), nematodes, ostracods, copepods, rotifers, tardigrades and mites. Of these, the tardigrades and mites are considered later; the remainder are active only when covered by a thin film of water, and from many points of view can be regarded as aquatic. Some of them, however, occur in vast numbers and many are of great economic importance. In this section, therefore, the nematodes are considered as examples of essentially aquatic animals living on land. An attempt is made to see how they are adapted to what is, in fact, a very specialised environment.

A short summary of the conditions imposed upon soil-dwelling animals is appropriate at this point. More details can be found·in Kevan (1962), Russell (1957) and Wallwork (1970). Soils and their structure vary enormously, but essentially they consist of mixtures of mineral substrate, usually derived from the underlying rock, and dead organic material or humus. These two components are mixed to varying extents, forming 'crumbs' which have small spaces inside them, and which also have spaces between them. The size of the small spaces is often less than 20 μm and indeed half of the pore space of the soil may be in this fraction. This is a most important factor to bear in mind when considering the distribution of nematodes, most of which have diameters of 20–50 μm, and which are therefore excluded from the very fine spaces. The effect of pore size on movement of nematodes has been discussed by Nicholas (1975) and Lee & Atkinson (1976). A good example is provided by the work of Wallace (1958) on larvae of the eelworm *Heterodera schachtii*, which has a diameter

Table 2.1. *The distribution of terrestrial nemertines and some of their relatives*

	Lower marine litforal	Upper marine littoral	Terrestrial
Group 1	*Pantinonemertes enalios* (in silt under boulders, Australia)	*Pantinonemertes winsori* (mangrove swamps, Australia) *Pantinonemertes californiensis* (saltmarshes, North America)	*Pantinonemertes agricola* (Bermuda) *Geonemertes* (Indo-Pacific Islands and Caribbean)
Group 2		*Acteonemertes bathamae* (shingle shores, New Zealand)	*Acteonemertes bathamae* (Auckland Island) *Antiponemertes* (New Zealand) *Argonemertes* (Australia) *Leptonemertes* (Tristan da Cunha) *Katechonemertes* (Azores)

Information from Moore & Gibson (1981) and Gibson, Moore & Crandall (1982).

of 15–18 µm. The optimum particle size for dispersal of these larvae was 250–500 µm. Dispersal was restricted in soils where the pore size was less than 30 µm, and stopped in pore diameters less than 12 µm.

Much of the soil water may be held within the fine pores by surface tension so that it is not available to animals of nematode size. In fact water does not normally drain from spaces of less than 30 µm in diameter (Russell, 1957), and soil water can exist in three phases: water that drains gravitationally; water held in capillary spaces; and hygroscopic water (Vannier,

1978, 1983). The supply of water and the supply of air in the soil naturally vary inversely, since water entering the soil displaces air which is not trapped in blind cavities. The degree of saturation of the soil is highly important to the animals because with complete saturation comes complete deoxygenation: the oxygen in the percolating water is soon used up, and the rate of diffusion from the air is too slow to replace it. In such cases of flooding, large proportions of soil animals die. The opposite extreme, that of desiccation, may also produce high mortalities, although many animals

Fig. 2.5. *Acteonemertes bathamae*, a high-shore nemertine from New Zealand. (*a*) Crawling undisturbed (length *c*. 3 cm). (*b*) After sudden stimulation, the proboscis has been everted, and the tip can be seen adhering to the rock. Muscular contraction of the proboscis can pull the body rapidly forward. At the point of stimulation, the body wall has ruptured slightly.

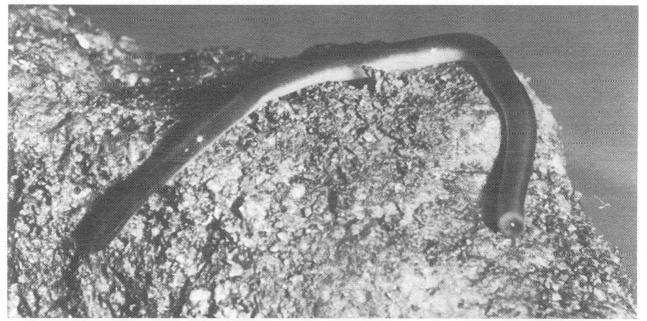

can burrow to deeper levels, and some can withstand being dried.

Diffusion normally ensures replacement of oxygen used up in the soil and removal of carbon dioxide produced. Locally, however, and especially at lower levels and during flooding, carbon dioxide may build up. In different soils a rise in pCO_2 will have vastly different effects: if the soil is acid, carbon dioxide concentration will rise because of a low capacitance for it, but in calcareous clay carbon dioxide levels stay fairly constant (Vannier, 1983). Besides being a direct problem in respiratory terms, this carbon dioxide build-up may cause local pH changes, and many invertebrates are very sensitive to these. The pH of different types of soil varies from about 3 to over 8, and each type of soil possesses a characteristic fauna. In general the more acid soils have a more sparse fauna than neutral or alkaline ones. The pH is partly dependent upon the nature of the underlying rock, but sometimes the soil above a particular rock type can be leached out, and it may then have an apparently anomalous pH value – as shown, for instance, in the acid heaths sometimes developed over limestones. The pH of the soil in turn affects the fate of any available humus: in acid soils decomposition is slow, the extreme being shown in peat bogs where organic remains can be preserved for thousands of years. In alkaline soils decomposition is rapid so that the humus is mixed with the mineral substances.

All the factors just considered show great variation, depending on the external conditions, but the soil does in some ways act as a buffer: temperature variations, for example, are much reduced, and at a depth of 20 cm the diurnal change is negligible. When soils are very wet, they have a higher specific heat, and temperature variations are smaller still. Light penetration into soils is usually limited, and most soil animals may be considered to be living in the dark. This in turn means that coloration is often reduced or absent in animals that never come to the surface.

As emphasised above, nematodes are essentially aquatic animals, and they are in fact never active unless covered by a film of water. Nevertheless, they are extremely abundant in terrestrial soils as well as in marine and freshwater sediments. Densities of the order of 20 million/m^2 have been recorded in woodland soils and in estuarine sediments (Wallwork, 1970, 1976; Platt & Warwick, 1980). Numbers of species are also estimated to be high: Croll & Matthews (1977) suggested that there may be 6500 marine species, 8000 species in the soil and fresh water, and 15 000 parasitic species, and other estimates are higher (see Crofton, 1966). Although many species fall into the general categories of bacterial feeders and plant parasites, the number of species is probably high because of resource partitioning, particularly in terms of feeding (Platt & Warwick, 1980). The importance of nematodes in every aspect of marine littoral biology has been emphasised by Platt & Warwick, and the group is probably marine in origin (Hyman, 1951*b*; Crofton, 1966), although the view that freshwater and terrestrial nematodes are primitive has also

been expressed (Riemann, 1977). Crofton (1966) has discussed the origin of soil forms, and concluded that some have been derived from marine ancestors while others have a freshwater origin. The situation is further complicated because of the probable recolonisation of freshwater and marine environments by terrestrial lines. As a result, the present-day distribution of nematodes in various habitats bears little relation to systematic grouping. Many individual species are also remarkably tolerant of environmental conditions and are found in diverse habitats. For example, a common species of soil and moss, *Aphelenchoides parietinus*, has been found in Denmark, the Pamir mountains of central Asia, the Baltic Sea, thermal springs in New Zealand and inland saline waters in Germany (Nicholas, 1975). The successful colonisation of soil habitats by nematodes is at least partly owing to two specific properties: the ability to remove excess water from the body in hypo-osmotic environments, and the ability to withstand some degree of desiccation as the soil dries up. Neither of these two properties is well understood, but we can consider each briefly.

Osmoregulation in nematodes has been reviewed by Lee & Atkinson (1976) and by Wright & Newall (1976, 1980). Estimations of changes in body volume in solutions of various osmotic pressures suggest that marine nematodes are iso-osmotic with sea water. Judging from changes in total sodium content – and *not* from sodium concentrations in any specific fluid – the marine *Enoplus communis* does not regulate internal ionic composition in changing external concentrations of sea water, whereas the brackish water *Enoplus brevis* maintains internal sodium at somewhat higher levels than those in the environment (Wright & Newall, 1976). Of the soil and freshwater forms, all are thought to be hyperosmotic to the environment, and some at least can regulate their volume well. Little is known of how this occurs, but both the gut and the tubular excretory organs are implicated. In *Rhabditis terrestris*, a species which spends part of its life cycle in the earthworm, free-living stages were able to regulate their volume to some extent after initial swelling, when transferred from their normal habitat to distilled water (Stephenson, 1942). This may have been partly due to removal of water through the anus by contractions of the posterior body wall, but volume reduction began *before* any water was seen to be expelled in this way. The tubular excretory system may therefore also be involved, and in three other species there is an inverse correlation between the rate of pulsation of the ampulla in this system and the external osmotic pressure (Wright & Newall, 1976). In the genus *Nippostrongylus* the excretory system may be involved in both osmoregulation and the removal of nitrogenous waste (Lee, 1970), but in general the gut is thought to be the major site of fluid excretion.

The ability of some nematodes to survive in a desiccated state, usually referred to as cryptobiosis or anhydrobiosis, is shared with some rotifers, tardigrades and insect larvae. The record length of time for a nematode remaining alive while dry appears to be held by *Tylenchus polyhypnus*, which survived on a dried herbarium specimen for 39 years (Evans & Perry, 1976). However, survival for up to 10 years is not uncommon for other phytoparasitic forms. The ability to tolerate lack of moisture is associated with the ability to tolerate very high and low temperatures: some species can withstand the temperature of liquid air, and others temperatures as high as 80–90 °C. Doubtless these extremes have no biological significance as such, but they do show that some nematodes can be extremely resistant to changes in external conditions. Such resistance permits them not only to withstand seasonal changes, but also provides an effective means of dispersal, since the dried stages are extremely small and light.

Two mechanisms appear to allow nematodes to undergo cryptobiosis. The first is provided by egg shells and cysts in larval stages, and by the cuticle in adults: all these structures help to slow down the rate of water loss, and this is essential because very rapid dehydration is lethal (Evans & Perry, 1976). The second mechanism which allows survival in the desiccated state is not well understood, but requires a certain time to come into effect. Quite probably this mechanism involves the production of substances which can protect the cell membranes during dehydration. Substances such as inositol and glycerol are synthesised during the onset of cryptobiosis (Crowe, Madin & Loomis, 1977; Evans & Womersley, 1980), and concentrations of these correlate well with survival rates. In *Aphelenchus avenae*, Crowe, O'Dell & Armstrong (1979) found that during rehydration, cryptobiotic specimens showed much lower rates of leakage of primary amines and of sodium than did quick-dried specimens. They suggested that the synthesis of glycerol during the induction of crypto-biosis in some way stabilised the cell membranes, since these remained intact in cryptobiotic specimens, but were disrupted when animals were dried quickly. Regard-less of the mechanisms involved, the phenomenon of cryptobiosis allows nematodes to survive in a very much wider set of habitats than would otherwise be possible. Their abundance, in contrast to the limited distribution of terrestrial flatworms and nemertines, emphasises the success in the soil environment of the nematode organisation.

2.4   Discussion

Although the organisation and life styles of the three phyla considered in this chapter are very different, all have in common the need for a copious supply of water during activity. Terrestrial triclads and nemertines move primarily by the beating of cilia in a layer of mucus, and the production of mucus depends upon an adequate external source of water. Nematodes are essentially swimmers, and require a water film for move-ment. It is therefore probable that whether evolutionary lines have moved directly from the sea to land, or indirectly through brackish water and fresh water, they have always maintained very close contact with the groundwater *en route*. While the larger forms may have stayed close to the surface of the substratum, the relative constancy of the soil environment has probably meant that many of the smaller forms have remained

within the interstitial environment. This environment in terrestrial soils shows many similarities to the condi-tions found within freshwater and marine sediments, and because these are directly connected with the terrestrial groundwater, the possibility of the colonisation of land through this interstitial route must be discussed.

In both marine and freshwater sands there is a well-developed interstitial fauna, or psammon (Pennak, 1951; Swedmark, 1964). As early as 1936, before much was known of these interstitial animals, Pearse con-sidered the possible origin of soil invertebrates from burrowing forms. More recently the fundamental unity of the subterranean habitat has been emphasised by Delamare Deboutteville (1960). In particular, the idea that animal groups may have migrated in what is essentially a continuous film of water linking marine, freshwater and terrestrial sediments may apply to small forms such as protozoans, rotifers, nematodes, copepods and ostracods; but it is possible that the route could have been utilised by slightly larger animals such as flatworms and nemertines. Although Kühnelt (1976) emphasised the unique character of soil organisms, Ghilarov (1956) raised the possibility that the soil acted as an intermediate environment for animal groups which have moved from an aquatic to a truly terrestrial life. Although he was referring mainly to insects, he also suggested that many groups, including flatworms and nemertines, may have moved *into* soils before moving to a more terrestrial life *above* them.

The junction between marine and terrestrial groundwater therefore has great interest. Remane (1951) found that in this region of coastal subsoil water, or '*Küstengrundwasser*', there were many species of fresh-water and terrestrial origin, as well as strictly marine species. The gradients from marine to terrestrial condi-tions in sedimentary environments are often rather sudden in tideless seas (e.g. Fenchel, Jansson & von Thun, 1967), but in tidal areas they may be extremely gradual (McIntyre, 1969). In this case the possibility of the penetration into terrestrial soils and the evolution of appropriate physiological mechanisms to deal with changing conditions seems more likely. The purely marine meiofauna shows some modifications, especially in reproductive habits, which could have preadapted them for existence in terrestrial soils: direct develop-ment is common, and pelagic larvae are scarce; copula-tion and hermaphroditism are common; and brood protection is widespread (Swedmark, 1964; McIntyre, 1969). Further adaptations necessary would therefore have been mostly concerned with mechanisms allow-ing tolerance of changes in salinity and temperature.

The meiofauna of present-day marine sands contains enormous numbers of nematodes, many of which are cosmopolitan (Swedmark, 1964). In coastal subsoil water strictly marine species of nematodes are mixed with those belonging to terrestrial genera (Delamare Deboutteville, Gerlach & Siewing, 1954), and it is reasonable to assume that this coastal ground-water forms a transition zone for the movement of nematodes between sea and land, and possibly in the reverse direction. Interstitial turbellarians are also abundant in sands of the marine littoral zone. Most of these, however, are not triclads but smaller forms such as the Otoplanidae, which are predators living fairly high on the beach (Swedmark, 1964). It seems quite possible that terrestrial rhabdocoels and 'alloeocoels' could have been derived from some similar marine interstitial ancestor. The terrestrial triclads, however, are more commonly associated with the soil surface or cryptic habitats such as leaf litter, and marine and freshwater triclads are also associated with the surface of the substratum rather than with the interstices of mud or sand. This suggests that the interstitial habitat has never been important in their

evolution. Within the nemertines, 27 species have been recorded from the marine interstitial zone, most of them hoplonemertines of the genus *Ototyphlonemertes* (Kirsteuer, 1971). As discussed earlier, terrestrial nemertines probably evolved from ancestral marine hoplonemerteans, and the presence of a number of hoplonemerteans in marine interstitial habitats suggests that this environment could have been an important one in providing a route on to land. The possibility that other animal groups may have used the same route is considered later in the book, and the ways in which physiological mechanisms might have evolved in the interstitial environment are discussed in Chapter 10.

3

Annelids

'Vermis the worm is an animal which is mostly germinated, without sexual intercourse, out of meat or wood or any earthly thing. People agree that, like the scorpion, they are never born from eggs. There are earth worms and water worms . . .'

From *The book of beasts* by T. H. White, 1954, Jonathan Cape, London.

Although the annelids are soft-bodied worms, they have occasionally been preserved as fossils, and undoubted polychaete remains have been recovered from Cambrian rocks (see Clark, 1969), so that the origin of the phylum Annelida, and of related phyla such as Echiura and Sipuncula, must date from Precambrian times. The long subsequent period of evolution and adaptive radiation has produced a great variety of types adapted to various modes of life, and it is therefore difficult to come to any unequivocal conclusions about the origins of the phylum. The conventional picture used to consider the archiannelids to be the most primitive of modern annelids, and these were therefore supposed to represent the stock from which arose in turn the polychaetes and then the oligochaetes and leeches. Hermans (1969), however, has suggested that the archiannelids are primarily adapted for an interstitial life, and that their structure reflects this mode of life rather than a primitive simplicity. The suggestion is supported by the idea that the coelom evolved as a hydrostatic organ which aided movement in fairly large animals (Clark, 1964): since the archiannelids are all small, and small size usually leads to structural simplification, it is difficult to account for the existence of their coelom except as a remnant derived from the functional space in larger annelids (Clark, 1969). Many of the archiannelids are members of the interstitial fauna, and their characters are in general similar to those of other animal groups found in the same specialised environment.

If it is accepted that the archiannelids are not particularly primitive members of the phylum, we must look elsewhere for annelid origins. The conventional derivation of the oligochaetes from polychaete stock has been reappraised by Clark (1969, 1978) who has considered the structural organisation of annelids from a functional viewpoint. Most annelids are segmented, with a well-developed coelom, so that this seems likely to have been the basis of the primitive body plan. In most of the animal kingdom, such a plan is usually utilised for peristaltic burrowing, and Clark suggested that this would have been the primitive state of the phylum – a state very much akin to that of present-day oligochaetes. The development of parapodia and consequent wave-like movement, as found in many polychaetes, would have been a later modification. Such an argument leads naturally to the suggestion that the oligochaetes are more primitive than the polychaetes; but the situation is complicated by the long period of adaptive radiation undergone by the annelids. During this period the polychaetes have in the main exploited marine habitats, while the oligochaetes have radiated into freshwater and terrestrial environments, with many consequent modifications, especially in their reproductive and excretory systems. Apart from their method of movement, the polychaetes have therefore retained many primitive features, while the oligochaetes – and the leeches which have been derived from them – have been modified in many respects. There appear to be no modern oligochaetes which might be regarded as forming the primitive stock of the phylum Annelida; but after at least 500 million years of evolution, this is perhaps hardly surprising.

The leeches are closely related to the oligochaetes, but appear to have been greatly modified owing to the change from a burrowing to a swimming and crawling mode of life, and the adoption of carnivorous and parasitic habits. The method of movement, using suckers and a looping motion, has allowed the reduction of the coelom which is represented by a series of canals only. Similarities to the oligochaetes include the development of hermaphroditism and the presence of a clitellum, and various groups of oligochaetes foreshadow in many ways the situation in the leeches.

A fourth group of annelids, the myzostomarians, consists of parasites on echinoderms, and we shall not be concerned with them. They are thought to be possibly related to the polychaetes, but are placed in a separate class, or sometimes even in a separate phylum.

With such difficulties in deciding about the relationships of the various groups of annelids, there must obviously be some argument concerning the formal classification of the phylum, and any scheme must be somewhat tentative. In the present account the system of Clark (1969) is adopted for the polychaetes and leeches, and that of Brinkhurst & Jamieson (1971) for the oligochaetes.

Having thus rapidly scanned some of the complexities of annelid classification, it is appropriate to mention briefly the distribution of the members of the various classes in terms of environment, before considering each class in detail. In general the polychaetes are marine, but about 12 species are found in fresh water. Only a few species are semi-terrestrial. The oligochaetes, in contrast, are widely distributed in fresh water and in soil, and a number are found in the sea – many of these in the littoral zone. All these marine members are thought to have moved back secondarily to the sea from fresh water or from land. The leeches are, in the main, found in fresh water, but one group has moved back into the sea, where its members are all parasites of fish; several have taken up a terrestrial life.

With this introduction to the phylum, we can now move on to consider in turn the three main annelid groups.

3.1 The polychaetes

Very few polychaetes are found on land, even in the dampest of environments. Only two families, Nereidae and Stygocapitellidae, have terrestrial members although several other families contain species found in fresh water (Table 3.1). It is most convenient to consider the various groups in three sections.

3.1.1 The Nereidae

The family Nereidae contains relatively large worms, most of which are marine including such common species as *Nereis virens* and *Perinereis cultrifera*. Many species are also found in brackish water, and some, e.g. *Nereis diversicolor*, are usually restricted to estuarine conditions. Whereas the marine forms cannot osmoregulate, *N. diversicolor* can maintain the osmotic pressure of its coelomic fluid above that of the external medium in concentrations lower than about 36% sea water (see, e.g. Oglesby, 1969). The adult of *N. diversicolor* can in fact live in fresh water, but it cannot reproduce there. Another species, *Nereis limnicola*, is found in the freshwater Lake Merced in California as well as in brackish water. Lake Merced is probably a marine relict lake and contains several relict species. It also has the rather high chloride content of about 2.5 mM/litre, probably due to airborne salt (Hutchinson, 1967). *N. limnicola* appears to be able to maintain itself in fresh water because the larvae are retained inside the coelom until a much later stage than in *N. diversicolor*, and when released they can osmoregulate. *N. limnicola* is also exceptional among polychaetes in being hermaphrodite. Several other species also occur in fresh water; these include *Namanereis quadraticeps* and *Namalycastis abiuma*, found mainly in brackish water, but living in fresh water in such places as Sumatra.

The situation with regard to the semi-terrestrial species is somewhat confused. Hartman (1959) has suggested that they can all be assigned to one of the two freshwater species *Namanereis quadraticeps* or *Namalycastis abiuma*, but from the original accounts of Lieber (1931) and Pflugfelder (1933) this may be questioned. *Lycastis terrestris* lives in burrows in dry soil in Sumatra (Pflugfelder, 1933). The habitat is covered by the sea at high spring tides. *Lycastis ranauensis*, also from Sumatra, was found in fresh water and produces a freshwater nectochaeta larva (see Lieber, 1931). Hartman placed both these in one species, *Namalycastis abiuma*. She also included in this species *Lycastis indica*, which has been studied by Krishnan (1952). *L. indica* has larger nephridia

Table 3.1. *Orders of the class Polychaeta and distribution of the families*

Order	Sea water	Fresh water	Land
Amphinomorpha	3 families	—	—
Eunicimorpha	Histriobdellidae (parasitic) + 7 families	Histriobdellidae (parasitic)	—
Phyllodocemorpha	Nereidae	Nereidae	Nereidae
	Nephtyidae + 22 families	Nephtyidae	
Spiomorpha	Stygocapitellidae + 12 families	—	Stygocapitellidae*
Drilomorpha	5 families	—	—
Terebellemorpha	4 families	—	—
Oweniimorpha	1 family	—	—
Sternaspimorpha	1 family	—	—
Flabelligerimorpha	2 families	—	—
Serpulimorpha	Sabellidae	Sabellidae	—
	Serpulidae	Serpulidae	
Psammodrilomorpha	1 family	—	—
Archiannelida	Protodrilidae	? Protodrilidae	—
	Nerillidae + 1 family	Nerillidae	

* The family Stygocapitellidae includes the genus *Parergodrilus*, which has often been considered as belonging to the oligochaetes.

than purely marine species, and can tolerate both brackish and fresh waters. Unfortunately the larva of these various 'species' is known only for *L. ranauensis*.

The ultrastructure of some of the air-breathing species from Sumatra has been studied by Storch & Welsch (1972). The integument of *Lycastis terrestris* is in general terms very similar to that of marine forms such as *Dendronereides heteropoda*, but in some areas of the body it has more epicuticular projections, with microvilli terminating in the furrows between projections. The general level of morphological change seems, however, to be slight.

A species of *Lycastopsis* from Amboina is described by Lieber (1931) as having separate sexes. Hartman (1959) grouped this with various other species, which are hermaphrodite, into the one species *Namanereis quadraticeps*. *Lycastopsis* sp. is perhaps the most interesting of these semi-terrestrial species, since it is found at altitudes of nearly 500 m, and at distances of more than 2 km from the sea. According to Lieber, it lives in the moist spaces between the leaves and the trunk of coconut palms, and is not normally found in water. While it will survive in fresh water, it shrivels and dies rapidly in sea water. Its terrestrial nature is fully indicated by the fauna associated with it: *Bipalium*, *Geonemertes*, *Peripatus*, *Periplaneta*, scorpions, myriapods, etc. Its mode of reproduction is not known in detail, but it produces very few, large, yolky eggs, so that the free-swimming larval stage has presumably been eliminated. Lieber suggested that such 'arboricolous' forms as *Lycastopsis* have evolved from nereids inhabiting mangrove swamps, while the freshwater forms such as *Lycastis* have their origin in estuarine ancestors which have remained mud dwellers. While both types have developed powers of osmoregulation, only *Lycastopsis* has eliminated the larval stage. Lieber also pointed out that the development of these semi-terrestrial forms is probably linked to the constancy of the terrestrial environment in equatorial regions, in contrast to that in the temperate zone. It is unfortunate that so little is known about this fascinating group.

3.1.2 The Stygocapitellidae

This family, created by Karling (1958), contains only two species, *Stygocapitella subterranea* and *Parergodrilus heideri*. *S. subterranea* is a member of the interstitial fauna of the marine littoral zone, and is found in the Bay of Kiel and on the southern Swedish coast. *P. heideri* lives in the leaf litter of forests in the Austrian Alps, and when it was first discovered was described as an archiannelid (Reisinger, 1925). Subsequently it was transferred to the enchytraeid oligochaetes, finally to be placed in the sedentary polychaetes by Karling (1958). More recently, Reisinger (1960) has confirmed this placing.

S. subterranea (Fig. 3.1*a*) reaches a length of about 3 mm, and bears relatively few long spatulate chaetae. It has separate sexes, each having only one pair of gonads. *P. heideri* (Fig. 3.1*b*) is very similar in organisation, but is only 1 mm long and is transparent. The animal is covered by a thick cuticle through which protrude short, simple chaetae. Some hermaphrodite specimens have been described, but separate males and females are also found, and there are doubts about the possibility of parthenogenesis. *S. subterranea* is much simplified in comparison with other polychaetes, as can be easily recognised by such factors as the lack of tentacles, the absence of parapodia, the small number of chaetae and the reduction in the number of gonads. These simplifications are similar to those found in other interstitial forms, and are presumably linked with both the habitat *per se* and the small size of the animals. It seems clear from Karling's (1958) study that *P. heideri*, although living in terrestrial leaf litter, has been derived from interstitial ancestors similar to *S. subterranea* by further reductions, and possibly by the adoption of hermaphroditism. The route by which *P. heideri* has colonised land has involved continental interstitial water and finally soil waters, a route suggested for other small forms in Chapter 2. Perhaps it is important in this context to emphasise that *P. heideri* is intolerant of both desiccation and submersion (Beauchamp, 1959), and therefore differs from many other small animals which were placed in the category of 'aquatic animals' (p. 3).

3.1.3 Other families containing freshwater species

Several other families contain species which live in fresh water (Hutchinson, 1967; Oglesby, 1978), although none has become even semi-terrestrial. Nevertheless, it is worthwhile considering these species briefly because it is important to attempt, finally, some analysis of the differences between polychaetes and oligochaetes which have led the latter to be so successful, and the former to be so rare, on land.

In the family Nephtyidae, several species have been found in fresh water, but these are probably better regarded as euryhaline brackish water species able to tolerate fresh water. In the sedentary Sabellidae the same probably applies, although two species may possibly be restricted to fresh water. The family Serpulidae also contains several euryhaline species, of which the most remarkable is *Mercierella enigmatica*. This is normally a brackish water species, but is reported from a freshwater lake in Tunis. It can withstand the enormous range from 150% sea water to fresh water, and in all these concentrations it maintains its blood hyperosmotic to the medium (Skaer, 1974*a*). In all except fresh water the osmotic pressure of the blood changes linearly in relation to external osmotic pressure, so that it is essentially an osmoconformer. Skaer (1974*b*) has suggested that in contrast to most animals, cellular electrical activity in *M. enigmatica* is dependent upon the ratios of ionic concentrations in the fluids bathing the cells, and not upon absolute concentrations. It may be this property which allows the species to survive in such a wide range of concentrations. Unfortunately nothing is known of ionic regulation in other freshwater polychaetes, although another species of serpulid, *Marifugia cavatica*, is found in underground waters in the karst region of Yugoslavia. In the genus *Mercierella* the nephridia may be involved in volume regulation (Skaer, 1974*a*), but no details of nephridial physiology are known. Not even in marine polychaetes is very much known about the ways in which nephridia function (Oglesby, 1978). It has been shown that in the marine genus *Sabella* the nephridia cannot transport sodium against large concentration gradients (Koechlin, 1979), but this is probably true of the osmoregulatory organs in most marine invertebrates.

Among the archiannelids, one species of *Protodrilus* may live in fresh water. In another archiannelid family, the Nerillidae, there is an interesting pair of species very comparable to *Stygocapitella* and *Parergodrilus* described above. *Thalassochaetus palpifoliaceous* is an interstitial form found in sublittoral marine sand, and *Troglochaetus beranecki* is found in hypogean waters in the alps (Hutchinson, 1967). It is supposed that *Troglochaetus beranecki* has reached its present habitat through interstitial waters.

Lastly, we must note that within the Histriobdellidae, represented by a marine species which is ectoparasitic on lobsters, there are also four species which are ectoparasites of freshwater malacostracan Crustacea.

To sum up the situation in the class Polychaeta, we may say first that there are very few freshwater species, and that of those that do exist a large proportion appears to be derived from the interstitial fauna. The number of terrestrial species is even smaller. Since this is in direct contrast to the situation found in the

oligochaetes, some comparison of the two groups is made at the end of this chapter, in an attempt to explain the differences in distribution.

3.2 The oligochaetes

The oligochaetes form a much more uniform group than the polychaetes discussed above. One of the features in which a fair amount of variation occurs is the position of the gonad-bearing segments, and this has often been used in schemes of classification. In most accounts the positions of testes, ovaries and spermathecae have been used, but Brinkhurst & Jamieson (1971) in their monograph showed that the position and number of the spermathecae are very variable. They also suggested that it is the relative position of the gonads rather than their exact location which is phylogenetically important. Using these features, as well as others such as the form of the blood system and the position of the clitellum, these authors tentatively suggested the classification shown in Table 3.2. They also suggested that the lumbriculids – predominantly freshwater forms usually a few centimetres long – retain some very primitive characters, but have evolved somewhat separately from other oligochaetes. The ancestors of the haplotaxids (worms of a similar size to the lumbriculids, but found in fresh water,

Fig. 3.1. Comparison of the organisation of (*a*) *Stygocapitella subterranea* (male) and (*b*) *Parergodrilus heideri* (male). *S. subterranea* is a member of the marine interstitial fauna and *P. heideri* lives in terrestrial leaf litter. It is probable that *P. heideri* has been derived from interstitial ancestors similar to *S. subterranea* by further simplification. The diagrams are not to scale, but *S. subterranea* is approximately 3 mm long, and *P. heideri* approximately 1 mm long. Redrawn from Reisinger (1960).

(*a*)

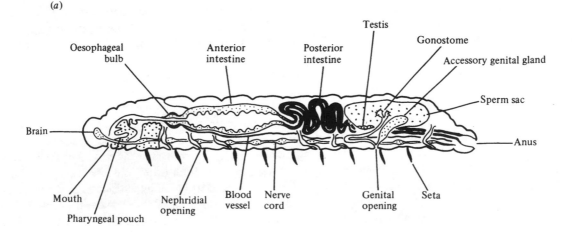

(*b*)

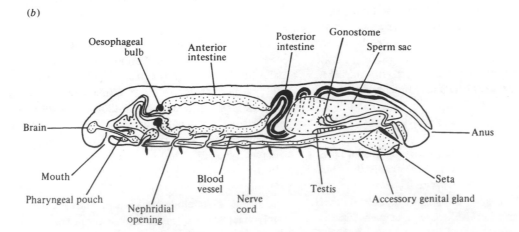

on land and occasionally in the littoral marine zone) provided the basal stock from which all other oligochaetes, apart from the Moniligastridae, were derived. More recently, Brinkhurst (1982) (Fig. 3.2) has derived *all* present-day orders of oligochaetes from haplotaxoid ancestors. One family, the Aeolosomatidae, was regarded by Brinkhurst & Jamieson as an enigma. In this family, the gametes escape through the nephridia, as in many polychaetes. It seems probable that there is some relationship with the Stygocapitellidae (see p. 19), in which case the aeolosomatids might be moved into the class Polychaeta. However this may be, Brinkhurst & Jamieson excluded them from their class Clitellata, and Brinkhurst (1982) placed them in a separate class, the Aphanoneura. We may, therefore, briefly consider them here before discussing the remaining groups of oligochaetes in more detail.

3.2.1 The Aeolosomatidae

Aeolosomatids are very small worms, all of them less than 1 cm long (see Fig. 3.3). They have no external grooves marking the segments, and usually no internal septa, but they bear groups of long hair setae (Stephenson, 1930). Reproduction is mostly asexual, as in the wholly aquatic Naididae: individuals are budded off from a fission zone roughly in the middle of the animal, giving rise to a chain of individuals. They live mostly in fresh or brackish water, but a few have been described from soil and leaf litter together with a species of naidid (Stout, 1952, 1956). It seems likely that in all cases they lead a primarily aquatic life, and their epidermis bears only a thin cuticle penetrated by microvilli as in purely aquatic forms (Potswald, 1971). They are, however, capable of encysting, and they overwinter in this form. The cyst is formed of mucus which hardens to form a membrane. Encystment occurs when temperatures are low, but sufficient food must be available to allow the worms to accumulate food reserves (Herlant-Meewis, 1950). Unlike the situation in some aquatic oligochaetes, encystment does not confer any resistance to desiccation.

3.2.2 Families transitional between aquatic and terrestrial habitats

The oligochaetes are sometimes considered in two groups, the Microdrili or limicoles, which are usually small and aquatic; and the Megadrili or terricoles, which are larger and are usually referred to as earthworms. Somewhat intermediate in form and habit are the Haplotaxidae and the Moniligastridae. The Enchytraeidae too, although normally placed in the Microdrili, contains some species which live more or less as earthworms. It is, then, convenient to consider these three families together before dealing with the earthworms proper.

The family Moniligastridae is found mainly in southern India. Some of its members are aquatic, and the terrestrial species are restricted to areas of great rainfall. They lack dorsal pores, which are usually found in terrestrial oligochaetes, and about which something is said later (see p. 24). One species, *Drawida grandis*, reaches lengths of over 100 cm.

The family Haplotaxidae is more widespread, but like the above family its members live in fresh water and very moist earth. Most species are small, but *Haplotaxis gordioides* may reach a length of 30 cm.

The small forms which constitute the family Enchytraeidae may be extraordinarily abundant, and up to 80 000 have been recorded from 1 m^2 of soil. Over 300 species have been described, so that in all senses this family is much more important ecologically than the two families previously mentioned. Enchytraeids are found in fresh water, in soil and in the marine littoral zone where they are usually common in rotting seaweed and under stones. Some of these marine species can withstand salt concentrations well above that of sea water, as well as fresh water. Measurements of weight changes after transfer between different salt concentrations show that they are not good volume regulators (Tynen, 1969), but no figures are available as regards osmoregulation. In this connection it may be noted that the nephridia are peculiar: the various coils of the nephridial tube are fused into one mass instead of lying free in several lobes. The

Table 3.2. *Classification of the oligochaetes and distribution of oligochaete families (class Clitellata, subclass Oligochaeta)*

Taxon	Marine	Fresh water	Terrestrial
Order Lumbriculida	Lumbriculidae	Lumbriculidae	—
Order Moniligastrida	—	Moniligastridae	Moniligastridae
Order Haplotaxida			
Suborder Haplotaxina	Haplotaxidae	Haplotaxidae	Haplotaxidae
Suborder Tubificina	Enchytraeidae	Enchytraeidae	Enchytraeidae
	Tubificidae	Tubificidae	
	Phreodrilidae	Phreodrilidae	
		Opistocistidae	
		Naididae	
		Dorydrilidae	
Suborder Lumbricina	Glossoscolecidae	Glossoscolecidae	Glossoscolecidae
	Megascolecidae	Megascolecidae	Megascolecidae
		Lumbricidae	Lumbricidae
			Eudrilidae

Fig. 3.2. The evolution of the oligochaetes, and their relationships with other annelids. The haplotaxoid stem form of the oligochaetes is envisaged as arising in estuaries or in groundwater. Four separate invasions of saturated soils and fresh water are indicated by the four arrows leading to numbers 1, 2, 3, 4, where: 1 is a proto-lumbricid (brinkhurstoid); 2 is

a haplotaxoid; 3 is a proto-tubificid (tiguassoid); 4 is a proto-lumbriculid. The width of each column indicates approximately the relative numbers of species within each family. Modified from Brinkhurst & Jamieson (1971) after R. O. Brinkhurst (personal communication).

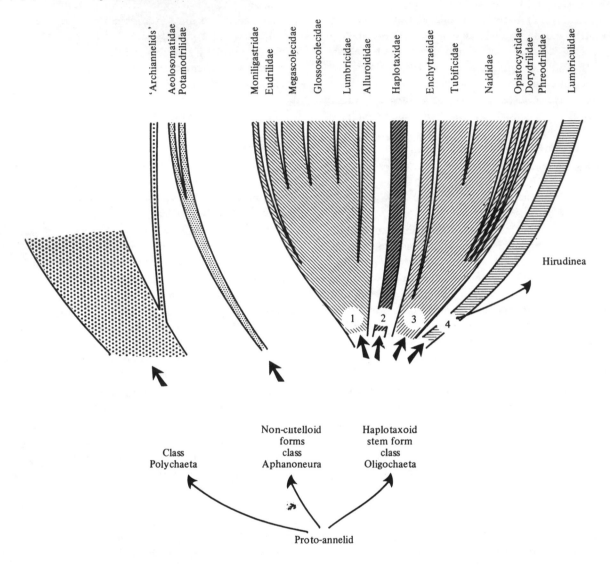

Fig. 3.3. The organisation of *Aeolosoma kashyapi*. This small species was collected from damp soil and from water retained by the leaves of epiphytes in rain forest. As with the terrestrial

Stygocapitellidae, an interstitial origin is likely. Note that there are no internal septa. After Marcus (1944).

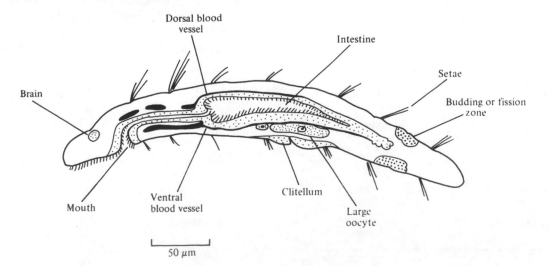

tolerance of enchytraeids to desiccation does not appear to have been investigated, but since littoral forms live in a fairly rigorous environment, it can be assumed that they can survive a good deal of dehydration. The resistance of earthworms to desiccation is discussed below.

3.2.3 The earthworms

Four families must be considered under this heading. The largest of these is the Megascolecidae, which is widespread in distribution but absent from northern Europe and North America. It contains such genera as *Pheretima,* common in India, and *Megascolides* in Australia. *Megascolides australis* can reach lengths of about 2 m. Related to the Megascolecidae is the family Eudrilidae, whose members are restricted to tropical and sub-tropical Africa. Little is known of their physiology and they will not be discussed further.

The members of the Glossoscolecidae are widely distributed in warm regions, being commonest in South America. Several species are aquatic, as in other families, but in this family some, such as *Alma emini*, are adapted for living in swamps. Others are terrestrial and may grow to giant size: *Rhinodrilus fafner* reaches 2 m and *Glossoscolex giganteus* grows to over 1 m in length. The aquatic forms are thought to have been derived from terrestrial ancestors, and not to be primitively aquatic (Stephenson, 1930).

The Lumbricidae, mostly found in temperate regions, contains such well-known species as *Lumbricus terrestris, Allolobophora caliginosa* and *Eisenia foetida*. Several species have been transported to tropical regions by man, and have successfully established themselves there in cultivated soil.

Because of the lack of detailed comparative knowledge of earthworms, it is convenient at this point to consider the group as a whole rather than to take each family in turn. Most earthworms live in burrows in the soil, feeding on the organic content within the soil as well as dragging down the remains of higher plants from the soil surface. Stephenson (1930) summarised previous views about the date of origin of terrestrial earthworms. He concluded that because most earthworms base their diet on dicotyledonous plants, they were not likely to have been widespread before Cretaceous times, when higher plants evolved. At the present time they are a most important constituent of the terrestrial fauna, as it is in large part owing to them that organic humus is mixed with the inorganic weathered rock to produce fertile soil. They do this by dragging organic material into their burrows, eating this and soil in the lower layers, and then defaecating at the surface. In this way a layer amounting to as much as 5 cm may be deposited on the surface in 10 years (Darwin, 1881). Darwin's account of the activities of earthworms remains one of the best sources of information on this subject.

Several characters concerned with the feeding of earthworms distinguish them from most of the freshwater oligochaetes (see, e.g. Michel & DeVillez, 1978). First, they possess a gizzard for trituration of food. This reflects their tendency to feed on higher plants, in contrast to the aquatic forms which feed on soft detritus.

The forms which are thought to have secondarily colonised swamps and fresh water (e.g. *Alma*) have lost the gizzard. Another organ which is normally present in the Megadrili (the earthworms) and not in the Microdrili (the freshwater families) is the dorsal longitudinal fold of the intestine, the typhlosole. Strangely, though, this is also lacking in the giant *Megascolides australis*. The function of the typhlosole is to increase the surface area for absorption, and its abundant vascularisation aids in this. In general this becomes necessary when the size of the worms increases, but the anomaly of *M. australis* remains to be explained. A third structure possessed by the terrestrial forms is absent in the freshwater ones, and is lost in the swamp-dwelling genera such as *Alma*: the calciferous glands, which are lateral swellings of the oesophagus secreting small crystals of calcium carbonate into the gut. The structure of the glands is shown in Fig. 3.4 from which it is apparent that they are secretory in nature, and well supplied with blood. Their function has been much discussed and disputed, but it is agreed that they produce numerous concretions containing almost entirely calcium carbonate, and that these concretions pass through the gut and are found in the casts. The carbonate contained in these concretions represents about 5% of the total carbon dioxide lost by the worm (Robertson, 1936, 1941). Although they are not, therefore, the main route by which carbon dioxide is lost from the body, they could be important in regulating the internal level of both calcium and carbon dioxide, thereby maintaining a constant internal pH. Recent work on the structure of the calciferous glands (van Gansen, 1959; Nakahara & Bevelander, 1969) has supported the idea that they are mainly concerned with acid–base balance, but does not explain why they are absent in

Fig. 3.4. The calciferous glands of the earthworm *Allolobophora caliginosa*. The figure shows a transverse section through the oesophagus with the glands surrounding it. The glands produce concretions rich in calcium carbonate which are released into the gut, and these may be important in acid–base balance. Redrawn from Stephenson & Prashad (1919).

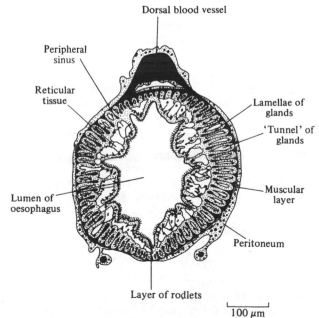

Dorsal blood vessel

Peripheral sinus

Reticular tissue

Lamellae of glands

'Tunnel' of glands

Muscular layer

Lumen of oesophagus

Peritoneum

Layer of rodlets

100 μm

aquatic forms. It is possible that because of their diet of higher plants, the earthworms take in more calcium than detritus feeders (Microdrili), and that this has to be excreted. Evidence summarised by Laverack (1963) suggests that the glands may clear excess carbon dioxide when this builds up inside the earthworm burrows (and consequently in the tissues of the earthworm), and this would be a problem not usually faced by the small aquatic forms. It cannot be said, however, that this difference between aquatic and terrestrial forms is well understood.

The tolerance of the different groups of earthworms to different degrees of soil hydration has already been briefly mentioned. Some, such as the Glossoscolecidae, require high rainfall. Others, such as some of the Lumbricidae, can remain active in soils which during the summer become very dehydrated. The situation has been reviewed by Edwards & Lofty (1972). Studies both on vertical distribution and activity of earthworms in Sweden (Nordström, 1975; Rundgren, 1975) have demonstrated that some lumbricids remain near the soil surface throughout the year, while others burrow quite deeply in summer. Species of *Dendrobaena* which did not migrate vertically were active at temperatures as low as 0–2 °C, and survived, although in an inactive state, when the ground was frozen.

Even in moist soils earthworms are said to be partially dehydrated. Adolph (1927) found that when earthworms were taken from soil and placed in water they usually gained weight. Careful controls showed that the uptake of water was temperature-dependent, and that when the transfers were made without change in temperature, no weight gain occurred; but later authors have found average increases in weight from 5–15% (Carley, 1978; Oglesby, 1978). However, earthworms. are very tolerant of changes in water content. Both *Lumbricus terrestris* and *Allolobophora chlorotica* can lose up to 60% of their body weight and survive, and this represents a loss of 70 and 75% of their body water, respectively (Roots, 1956). At the other extreme, most earthworms can withstand submersion if the water is well oxygenated.

Under natural conditions, when the soil becomes particularly dry, some species can aestivate. In species of *Allolobophora* the onset of aestivation is caused by desiccation, and the worms roll up into an earth cell, having evacuated their alimentary canal (Michon, 1949). This state may last for at least two months, during which time the worms are insensitive to external stimuli. Aestivation is ended by rehydration of the soil. Amputation of the caudal end of the worm will also provoke aestivation (Saussey, 1963), but in this case rehydration does not end it. Some species, e.g. *Lumbricus terrestris*, do not aestivate but remain active in soils which have water contents as low as 10% or less (Zicsi, 1958; Gerard, 1967).

The routes by which earthworms gain and lose water have been investigated by a number of people. Worms may lose up to about 13% of their body weight on repeated handling, and Wolf (1940) has shown that most of this fluid is lost through the nephridiopores. A small amount is also lost through the mouth and the anus, and through the dorsal pores. Before proceeding to

a detailed consideration of the role of the nephridia, we may pause to consider briefly these dorsal pores.

The dorsal pores open from the coelom to the exterior, usually in the dorsal midline, and are closed by sphincter muscles. They are found in many, but not all, earthworms and in some enchytraeids. Only a few species of the Moniligastridae possess them. Coelomic fluid is ejected through the pores when the worm is irritated. It has been supposed that the function of this ejected coelomic fluid is mainly to keep the skin moist and suitable for respiratory exchange (Stephenson, 1930). This might well account for the absence of pores in aquatic forms, which do not need to moisten the skin.

The control of water balance is, as in most animals, bound up with osmotic and ionic regulation. Earthworms have generally been shown to have osmoregulatory mechanisms much like those of freshwater invertebrates. *Lumbricus terrestris*, for example, is able to maintain the osmotic pressure of its coelomic fluid well above that of dilute media, but as the external osmotic pressure rises, it becomes an osmoconformer (Ramsay, 1949a; Dietz & Alvarado, 1970). This hyperosmotic regulation in earthworms is achieved partly by active uptake of chloride across the skin (Dietz, 1974) and partly by the production of hypo-osmotic urine from the nephridia (Bahl, 1947; Ramsay, 1949b; Boroffka, 1965), just as in most freshwater invertebrates. However, *L. terrestris* also appears to have properties not found in typical freshwater animals. If it is kept in 100% relative humidity (RH) – a habitat much more akin to a natural one than that of fresh water – there is an initial drop in weight which soon stops (Carley, 1978). This presumably reflects the cessation of urine production. Carley also showed that *L. terrestris* could reduce its integumental permeability to water when the relative humidity was reduced. Added to this, it can withstand enormous changes in internal osmotic pressure – a range of 154 to 371 mOsm is quoted by Oglesby (1978). The possibility that such properties may reflect an origin other than one directly from fresh water is discussed in Chapter 10. For the moment it is necessary to concentrate upon the structure and function of the nephridia.

Annelid nephridia usually consist of much convoluted tubules, well equipped with blood vessels, and leading from the coelom to the exterior (Fig. 3.5). The mechanism by which they function is unclear. To understand the situation we have to consider what is the ultimate origin of the fluid which emerges as urine. Since the internal end of the nephridium – the nephridiostome – is, in the case of *Lumbricus terrestris* at least, open to the coelom, it is reasonable to suppose that some coelomic fluid passes down the tubule. In turn, then, we must seek the origin of the coelomic fluid and its constituents. In *L. terrestris* the osmotic pressure of the blood is less than that of the coelomic fluid by approximately 18 mOsm (Ramsay, 1949a). A slight difference in chloride was hardly significant. Bahl's (1947) results for the megascolecid *Pheretima posthuma* were strikingly different. The osmotic pressure of the blood was reported to be higher than that of the coelomic fluid by approximately 80 mOsm. This remarkably high difference has never been reinvestigated. Further analyses by Bahl showed that in many respects the coelomic fluid was similar to the blood, but it had a much lower protein content, and no detectable glucose, amino acids or lipids. Analyses for *L. terrestris* (Kamemoto, Spalding & Keister, 1962) also showed similarities between the ionic composition of blood and coelomic fluid, and both Riegel (1972) and Zerbst-Boroffka & Haupt (1975) have suggested that coelomic fluid may be derived by ultrafiltration from the blood. Bahl (1946) regarded the blood as being used for distribution of nutrients, and the coelom as a 'private pond' into which excretory material is moved prior to its elimination from the body by the nephridia.

The cilia of the nephridiostome probably prevent cells in the coelomic fluid from moving down the nephridia; but the coelomic fluid itself, poor in nutrients although containing similar quantities of ions and nitrogenous waste to the blood, passes down the nephridia – at least in *L. terrestris* where the nephridiostome is open. Salts are reabsorbed in the last half of the nephridial tube (Ramsay, 1949*b*), and the final urine is therefore hypoosmotic to both the coelomic fluid and the blood. The factor which has not been much studied is the importance of the blood supply to the nephridia. This blood may merely provide the mechanism for carrying away the substances reabsorbed from the urine; but equally it may itself contribute to the urine. Studies of the fine structure do not support this view (Graszynski, 1963), but no physiological work has involved this blood supply.

The nephridia must alter the rate of production of urine with changes in the level of hydration of the worms. Roots (1955) has shown that the cilia of the nephridiostome are more active in dilute than in concentrated media, but it is difficult to evaluate the significance of this result. During aestivation the production of urine presumably ceases. As discussed above, cessation of urine production seems to occur even when *L. terrestris* is maintained in 100% RH (Carley, 1978). There are unfortunately no observations on the nephridia of aquatic oligochaetes, so that no comparisons can be made.

Not all earthworm nephridia are like those of the genus *Lumbricus*. Bahl (1947) has described and attempted to classify a large number of different types. Of chief interest among these are the nephridia with no internal opening or nephridiostome – the 'closed' nephridia – and those which open not to the outside but into the gut – the 'enteronephric' type. The closed nephridia have never been investigated physiologically, but since they have no connection with the coelom, we may implicate the blood in the initial process of urine formation. The enteronephric system (see Fig. 3.6) is found mostly in the Megascolecidae, where the nephridia open into an excretory duct which runs longitudinally above the intestine and opens into it in each segment. In the lumbricid *Allolobophora antipae* the nephridia open into lateral canals which run posteriorly and themselves open into an ampulla above the intestine, near the rear end of the worm. The ampulla then opens into the intestine. Bahl (1947) has discussed the function of the enteronephric system, basing his argument on a comparison between the genus *Pheretima*, which has enteronephric nephridia, and the genus *Eutyphoeus*, which lives in the same area but has exonephric nephridia. *Pheretima* can live in much drier habitats, and is active during the dry season, whereas *Eutyphoeus* requires soft, wet soil, and burrows deeply when the soil dries up. The castings of *Pheretima* are

Fig. 3.5. Nephridium of the megascolecid earthworm *Hoplochaetella* sp. (*a*) Shows the general layout of the nephridium with the excretory canals folded back on themselves many times. (*b*) Shows the excellent blood supply to the nephridium.

Although the significance of this blood supply has not been investigated by physiologists, it may be involved in both production and modification of the urine. Redrawn from Bahl (1947).

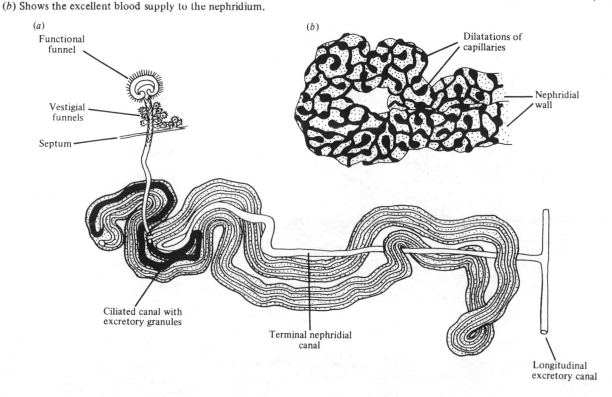

(*a*)

Functional funnel

Vestigial funnels

Septum

Ciliated canal with excretory granules

Terminal nephridial canal

(*b*)

Dilatations of capillaries

Nephridial wall

Longitudinal excretory canal

always drier than those of *Eutyphoeus*. Bahl considered, therefore, that the enteronephric system has evolved as a mechanism for water conservation in which the gut re-absorbs water that would otherwise be lost in the urine.

Although the foregoing discussion has been concerned solely with earthworms of terrestrial soils, it must be noted that several species of earthworms are found in the marine littoral zone. In particular, the megascolecid genus *Pontodrilus* is mainly littoral (Stephenson, 1930). *Pontodrilus matsushimensis* has been investigated by Takeuchi (1980). This species lives in sand at high tide mark, and survived in concentrations from distilled water to slightly concentrated sea water. At the top of this range it was a hyperosmotic conformer, but for more than half the range it showed marked hyperosmotic regulation. From this it is apparent that the oligochaete organisation is highly adaptable, and that further investigation of a wide variety of species is necessary before the generalisations derived from terrestrial earthworms are applied to the rest of the class.

Nitrogen appears to be lost from earthworms by three routes. In general, urea is excreted through the nephridia, ammonia through the gut (Tillinghast, 1967), and mucoproteins are lost in the mucus secreted by the

Fig. 3.6. Enteronephric nephridia in megascolecid earthworms. (*a*) Shows three different types of organisation. In the genus *Tonoscolex* the nephridia open into the intestine through a duct in each segment. In *Lampito* a longitudinal canal collects the urine from each segment. In *Pheretima* there are two longitudinal collecting canals. (*b*) Shows the organisation in *Megascolex*. Urine from the nephridial funnels is taken by the terminal canals to the septal excretory canal which opens dorsally into the intestine. Redrawn from Bahl (1947).

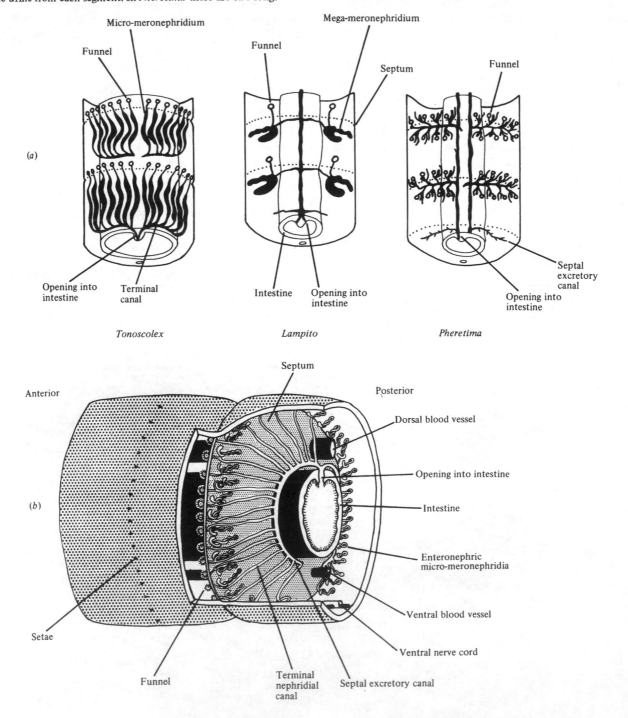

general body surface. The relative quantities of these substances excreted varies both with different species and in different conditions. Of the order of 50% of the nitrogen may be lost in the form of protein, and this is mainly found in the mucus (Needham, 1957). Large quantities of mucus are produced to keep the body surface moist, and to help in locomotion and in lining the walls of the burrow. The excretory output from the gut and nephridia of both *Eisenia foetida* and *Lumbricus terrestris* usually contains more ammonia than urea when the worms are feeding; but in starvation the urea output of *L. terrestris* becomes greater than that of ammonia (Cohen & Lewis, 1949; Needham, 1957; Waraska, Proos & Tillinghast, 1980). A small amount of uric acid and allantoin may also be excreted, but this does not represent an important fraction of the nitrogen lost.

In contrast to the situation in terrestrial and semi-terrestrial vertebrates, the use of urea in excretory metabolism is rare in invertebrates. Only the land planarians seem to have adopted urea as a major excretory product (see p. 8), although it is excreted in small amounts by some pulmonate gastropods (see p. 58). Needham (1957, 1970) has suggested that the switch to urea production in *L. terrestris* during starvation may be caused at least partly by problems of acid–base balance. The conversion of ammonia to urea lowers pH, while increase in ammonia production neutralises acids and raises the pH. If the urine of fasting *L. terrestris* is treated with urease so that urea is converted to ammonia, it becomes very alkaline indeed. The control of pH may therefore be, as in the vertebrates, a factor governing the relative production rates of urea and ammonia. As with many other topics, nitrogen metabolism has been investigated in the earthworms only, and there are no comparable studies involving the aquatic forms.

The control of acid–base balance must also be intimately linked with the production and removal of respiratory carbon dioxide. The calciferous glands may be involved here, as discussed above, but no detailed studies deal with internal carbon dioxide levels. Instead, respiratory studies have concentrated on morphological adaptations, and on the mechanisms involved in oxygen supply and consumption. The aquatic oligochaetes have few apparent structural modifications for respiration, but nevertheless show a variety of methods of obtaining oxygen. A few species, mostly in the Naididae, have external gills containing blood vessels. Some aquatic species of *Alma* (Glossoscolecidae) have dorsal gills. *Alma emini*, found in swamps in east Africa, can form a temporary 'lung' at the hind region of the body, and it protrudes this from its burrow into the air (Fig. 3.7). *A. emini* can in fact live without oxygen in the laboratory, but seems to use the lung to combat the conditions of anoxia and the high levels of toxic substances present in the papyrus swamps (Beadle, 1957). The lung is formed when water becomes stagnant, and its formation is not determined by falling pO_2 (Mangum, Lykkeboe & Johansen, 1975). Although the area of the lung is only 1.5% of the total body area, it can take up 50–60% of the oxygen normally taken up by the body surface because of its very vascularised surface and the high affinity for oxygen of the haemoglobin in its blood.

In many naidids and tubificids, and in the aquatic aeolosomatids, water is drawn into the posterior end of the intestine by antiperistaltic waves and by cilia. Stephenson (1930) has termed this 'intestinal respiration', and has pointed out that such phenomena are also found in the polychaetes. In the majority of oligochaetes, gas exchange takes place across the body wall. Aquatic species such as *Tubifex tubifex* possess some epidermal

Fig. 3.7. The posterior 'lung' formed by the swamp-dwelling oligochaete *Alma emini*. (*a*) Shows the extrusion of the hind end from the burrow. In (*b*) the hollowing of the dorsal surface can be seen. In (*c*) the final position has been reached, with the body

retracted into the burrow again, but with the tubular lung open to the air. (*d*), (*e*) and (*f*) show surface (dorsal) views of the hind end. Redrawn from Beadle (1957).

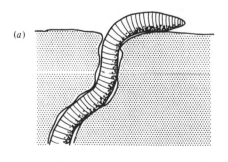

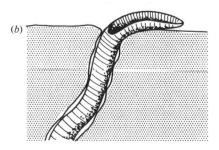

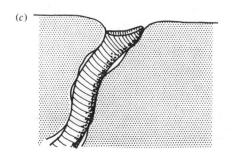

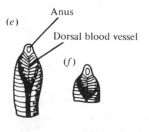

capillaries, but these are much less highly developed than in terrestrial forms such as *Lumbricus terrestris* (Palmer, 1966). Some aquatic forms have no capillaries in their body wall, and this probably reflects to a large extent their small size and hence their large surface : volume ratio. *T. tubifex* has been much studied because of its tolerance to waters of low oxygen content and high toxicity (Palmer, 1968). It possesses haemoglobin in the blood, with a high affinity for oxygen, and in this it is similar to the terrestrial oligochaetes. The oxygen tension at which the blood is 50% saturated is of the order of a few mmHg or less, both in *T. tubifex* and in earthworms of various species, and there has been much speculation about the function of the high-affinity pigment involved. This discussion has been hampered by lack of information about normal oxygen levels in earthworm blood. The work of Johansen & Martin (1966) on the giant earthworm *Glossoscolex giganteus* from Brazil is therefore of great interest. They have measured the oxygen content of the blood in the dorsal vessel, and found that this varies enormously. The average value is 41% oxygen saturation, but on contact with moist soil the oxygen saturation in one worm increased from 46% to 77%. Injections of adrenalin further showed that the cutaneous vessels are extremely sensitive, and that their diameter can change rapidly, thus effecting immediate changes in the rate of oxygen uptake. In the dorsal vessel the oxygenated blood from the epidermal capillaries is mixed with deoxygenated blood from the intestinal vessels, so that the blood distributed to the body has a diminished oxygen content however effective the respiratory exchange may be. In *G. giganteus* this diminished content is approximately 75% of saturation – a level at which oxygen will readily be given up to the tissues because it is on the steep part of the oxyhaemoglobin dissociation curve. Johansen & Martin suggested that the high-affinity pigments have arisen to counteract the dilution of oxygenated blood with deoxygenated blood caused by the structure of the blood system. Such a theory seems more likely than past suggestions that the haemoglobin acts as an oxygen store which is utilised only in times of severe oxygen shortage.

One further area in which one might expect the earthworms to show differences from the freshwater oligochaetes is that of reproductive mechanisms. Yet this does not seem to be so, as will be evident from a brief discussion. First of all, we must distinguish between the two processes of copulation and fertilisation, which are, in the earthworms, usually separated by about a day. All oligochaetes are hermaphrodite, and in the Microdrili (the freshwater forms) copulation proceeds as follows. The male apertures of each individual make contact with the spermathecal apertures of the other. In some cases penes are present, and these may be inserted into the spermathecae. Spermatozoa are then transferred between the two individuals, and are stored in the spermathecae. Fertilisation occurs later, during the formation of a cocoon, secreted by the clitellar region. As the cocoon passes forwards along the body, albumen is first secreted into it by the clitellum; then ova are introduced from ovisacs, and lastly sperm from the spermathecae. Ferti-

lisation, then, occurs in the cocoon. The same general mechanisms of copulation and fertilisation occur in most of the Megadrili (earthworms). In the genus *Eutyphoeus* (Megascolecidae) the process of copulation takes about 1 hour (Bahl, 1927), whereas in *Pheretima* (Megascolecidae) it takes from 4 to 5 hours (Dales, 1963). The story is different, however, in the Lumbricidae, where a more complex copulatory mechanism has evolved (see summary in Stephenson, 1930). Here each of the worms is enclosed in a mucous envelope. The individuals face in opposite directions, and have their ventral surfaces apposed (see Fig. 3.8). Spermatozoa are passed backwards along an external seminal groove, inside the mucous envelope, and then cross through the envelope of the other worm to enter the spermathecal opening. The duration of copulation in *Lumbricus terrestris* is 2 to 3 hours. The evolution of this mechanism of sperm transfer is difficult to explain, since it requires the use of large quantities of mucus and exposes the worms to the possibility of desiccation, and to predators, for a long period of time.

The cocoons of earthworms are deposited in soil, usually near the surface. When fresh they are white in colour, but later they harden and become yellower, probably by some tanning mechanism. Hatching usually occurs within a few weeks, but the cocoons themselves are very tough and may last in the soil for up to 3 years.

Fig. 3.8. Coition in the lumbricid earthworm *Eisenia foetida*. (*a*) Diagram of the two worms together during coition. The arrows show the course of the seminal fluid from the aperture of the vas deferens to the clitellum. (*b*) Transverse section through the two worms in the region of the clitellum of one worm. The coition slime tube can be seen surrounding the upper worm. Fine stipple shows the body of the upper worm, coarse stipple that of the lower worm. Redrawn from Grove & Cowley (1926).

(*a*)

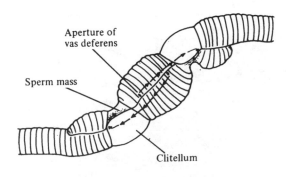

(*b*)

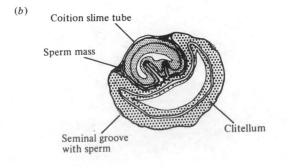

The eggs of earthworms have much less yolk than those
of freshwater oligochaetes, and the larvae emerge at an
early stage and feed on the albumen in the cocoon
(Dales, 1963).

3.3 The leeches

It is generally agreed that the leeches have evolved
from oligochaete stock by specialisation for a blood-
sucking and ectoparasitic habit (Clark, 1978). Several
features separate leeches from oligochaetes (Mann, 1962).
One is the constancy of the number of segments found
in leeches (there are always 34). Another is the reduction of the coelom, and its changes in function. It is
worthwhile outlining some of these changes briefly. In
oligochaetes the coelom is divided by septa and forms
a hydrostatic skeleton which allows changes in the shape
of individual segments. Movement is achieved by temporary attachment of some segments by chaetae, contraction of posterior segments and elongation of anterior
ones. Leeches, in contrast, have developed suckers for
attachment, and have lost all internal septa. The major
part of the body cavity is filled with mesenchyme cells,
and these form the hydrostatic skeleton. Movement is
carried out by stretching forward the anterior sucker and
after attachment drawing the posterior one up behind,
rather in the fashion of 'looper' caterpillars. Alternatively,
leeches can swim by dorso-ventral undulations of the
body. It can be seen that the coelomic fluid no longer
has the function that it possessed in oligochaetes and
polychaetes. Instead, it is reduced and has in most
leeches taken over from the blood system the functions
of a circulating body fluid. This has led to some confusion of terminology, since the coelomic fluid as well
as the fluid in the blood vessels (which some leeches
still possess) is called blood by some authors. Here we
shall distinguish the two fluids by their proper names of
blood and coelomic fluid.

The primitive leeches retain longitudinal blood
vessels which run within the coelomic sinuses. This blood
contains no pigment. More advanced forms, such as the
medicinal leech, *Hirudo medicinalis*, have no blood
vessels. Coelomic fluid circulates round the body as a
result of contractions of the lateral coelomic sinuses.
This fluid consists of a plasma with dissolved haemoglobin, and contains amoebocytes. It must therefore
be responsible for transporting dissolved gases and has
probably taken over some of the other functions of the
traditional blood system. In some species coelomic
'capillaries' penetrate between the epidermal cells (see
Harant & Grassé, 1959), where they act as a counterpart
of the epidermal blood capillaries found in oligochaetes.

In oligochaetes and polychaetes the nephridia are
closely associated with the coelom, opening from its
cavity to the outside. In most leeches the internal opening, the nephridiostome, has become closed off from the
nephridial canal, and its cilia beat into the coelomic
channels. There are often large numbers of nephridio-
stomes, found within the coelomic sinuses, and having
swellings or nephridial capsules at their bases. The whole
structure is known as a 'ciliated organ', and it is thought
that these organs are involved in manufacturing the coel-

omic amoebocytes. The remainder of the nephridium
still functions in excretion and in water balance, but the
mechanisms involved naturally differ somewhat from
those of oligochaetes. Urine appears to be formed by
a first stage of filtration from the coelomic sinus system
which closely invests the nephridium. From the filtrate
which reaches the connective tissue, a fluid is secreted
into the fine canaliculi of the nephridium, and these
eventually empty into the main nephridial tubule
(Boroffka, Altner & Haupt, 1970). The urine is retained
within a bladder before being excreted. Nephridial function is at least partly regulated by coelom volume, at least
in *Hirudo medicinalis* (Zerbst-Boroffka, 1978). Some
removal of nitrogenous waste is carried out by the
botryoidal tissue, which surrounds the gut, but the
final fate of this waste is not certain.

More contrasts between oligochaetes and leeches
are found in the methods of reproduction. Most leeches
produce their spermatozoa in spermatophores. While in
some species (including *H. medicinalis*) these are placed
directly within the female opening during copulation,
in others they are attached to the body wall, and the
spermatozoa have then to migrate through the epidermis
into the coelomic channels, and thence to the ovaries.

The specialisation of the leeches has been mostly
towards a blood-sucking habit, and nowhere is this more
apparent than in the digestive system. The members of
the order Rhynchobdellae have no jaws but a proboscis
with which they penetrate the tissues of the host animal.
Members of the order Gnathobdellae, however, have
three saw-like jaws with which to make incisions in the
host, and a powerful sucking pharynx to remove blood,
combined with a mechanism for preventing the blood
from clotting. The strangest characteristic of the digestive
tract is that it apparently possesses no digestive enzymes,
and it is thought that the digestion is performed by
symbiotic bacteria (Mann, 1962).

Despite the differences which have just been listed,
leeches are basically very similar to the oligochaetes.
This is emphasised by an animal which in some ways
forms a link between the two classes. *Acanthobdella
peledina* is a fish parasite found in Lake Baikal and in
Finland and Russia. Unlike other leeches it possesses
some chaetae, and its coelom is not much reduced in
size. It has no anterior sucker and the posterior sucker is
small. The remainder of the leeches form three orders:
the ectoparasitic Rhynchobdellae have a proboscis and
no jaws; the Gnathobdellae have jaws and are typified
by the genus *Hirudo*; and the Pharyngobdellae are
carnivores feeding on planarians, molluscs, etc.

Most leeches live in fresh water, but the range of
distribution is from sea water to land (see Table 3.3).
Many species which live part of the time in fresh water
can also be found in damp soil. Two examples are the
genera *Trocheta* and *Haemopis*. *Trocheta* (Erpobdellidae)
may be found in gardens, where it burrows and feeds on
earthworms and slugs. *Haemopis* (Hirudidae) may also
be found out of water, having a similar diet to *Trocheta*.
Haemopis can tolerate a loss of 80% of its body water
(Klekowski, 1961) and can also withstand anaerobic
conditions for some time.

The truly terrestrial leeches are almost all tropical. Most are placed in the family Haemadipsidae, which is closely related to the Hirudidae. Almost nothing is known about two other terrestrial families, Xerobdellidae (found in eastern Europe and in Central America) and Americobdellidae (a large form found in Chile). Haemadipsids are found in India and Indonesia, Australia, Madagascar, South America and some Pacific islands. They are found on shrubs and trees and in leaf litter on the ground, and feed by attaching themselves to passing animals, including man, and sucking their blood. The weakening effect of large numbers of these animals has been described by Decary (1950) who was attacked by species of *Haemadipsa* in Madagascar.

The accounts of species of *Haemadipsa*, found in India and Indonesia, differ widely from those of the genus *Chtonobdella*, found in Australia. *Haemadipsa* is said to live mainly on shrubs, where it sits erect on its posterior sucker. Sudden shadows activate the animal, and it will move towards the source of warm moist air currents such as human breath (Stammers, 1950). It will also turn towards a hand placed nearby (Fig. 3.9a) and Mathews (1954) has suggested that a chemical sense is involved. *Chtonobdella*, on the other hand, is found entirely in forest leaf litter, and does not respond to shadows or to moist air currents (Richardson, 1968). The feeding mechanisms in the two genera have been described as rather different, and altogether *Chtonobdella* seems to contain much less active animals than *Haemadipsa*.

Several modifications have been noted in land leeches in comparison with aquatic forms. One of these is the tolerance of long dry periods in a state called 'anhydrobiosis' by Richardson (1968). Species of *Chtonobdella* burrow in the soil and remain inactive and inert until brought into contact with water. Long periods of inactivity are usual for many leeches, including aquatic ones such as species of *Hirudo*, which may take up to 6 months to digest a blood meal; but tolerance of up to 5 months of desiccation is unusual in such soft-bodied animals. When species of *Chtonobdella* are not allowed to burrow, they quickly succumb to desiccation.

The nephridia are most important to the land leeches in several ways. In *Haemadipsa* the first pair open on to the anterior sucker and the last pair open above the posterior one, so that these important organs of attachment are always kept moist and functional (Mann, 1962). The nephridia also excrete a clear fluid during the taking of a blood meal (Fig. 3.9b; Worth, 1951), and this has been taken to indicate that excess water from the blood is rapidly removed to allow more blood to be taken up. No detailed study of the excretory products of land leeches has been made, but Bhatt (1963) has shown that both ammonia and urea are excreted when species of *Haemadipsa* are kept in water.

The reproductive adaptations of terrestrial leeches have not been properly described, but there is some suggestion that distance communication and courtship are involved (Leslie, 1951). Tapping by one Indian haemadipsid leech on a leaf apparently called a mate, and copulation was preceded by a courtship dance involving more leaf tapping and twining of the two bodies. A similar courtship procedure had been observed in Borneo (Harrison, 1953).

As yet the information about land leeches is so scanty, and the reports for different genera are so conflicting, that it is difficult to attempt any estimation of how successful these animals have been. They are restricted to damp environments, and are said to be patchy in distribution. The differences between the genera *Chtonobdella* and *Haemadipsa* are probably well reflected in the literature: while many writers have recorded the inconvenience caused by species of *Haemadipsa*, Richardson (1968) has found little reference to *Chtonobdella*.

3.4 Discussion

Some of the general characteristics and the distribution of members of the phylum Annelida can be summarised as follows. Most polychaetes are marine, have separate sexes and external fertilisation. They osmoregulate poorly if at all and their nephridia are often simple in structure. The oligochaetes are mainly found in fresh water and in the soil, and are hermaphrodite with fertilisation occurring more or less internally in a cocoon. They can osmoregulate in dilute solutions and have complex nephridia which are at least partly responsible for the retention of salts and the elimination of water. Leeches live mostly in fresh water but have some terrestrial representatives. They are closely related to oligochaetes and are hermaphrodite, but usually transfer their spermatozoa in spermatophores. Their nephridia are complex and function somewhat differently to those of oligochaetes.

After this summary one question in particular stands out: why are there so few freshwater and terrestrial polychaetes? Or, put another way, why has the

Table 3.3. *The distribution of the families of leeches (class Clitellata, subclass Hirudinea)*

Order	Marine	Fresh water	Amphibious	Terrestrial
Acanthobdellae	—	Acanthobdellidae	—	—
Rhynchobdellae	Piscicolidae	Piscicolidae		
		Glossiphoniidae	—	—
Gnathobdellae	—	Hirudidae	Hirudidae	Haemadipsidae
Pharyngobdellae	—	Erpobdellidae	Erpobdellidae	Xerobdellidae
			Semiscolecidae	Americobdellidae
		Trematobdellidae	Trematobdellidae	

Fig. 3.9. Terrestrial haemadipsid leech from Papua New Guinea. (*a*) With hind end attached to the substratum and anterior end stretching out towards a finger. Both temperature and chemical cues have been suggested for this response. Length of animal *c*. 13 mm. (*b*) Feeding on a human hand. The head is on the right, and in the loop formed by the body has appeared excretory fluid, shown by the intense reflections of the flash used for the photograph. Drops of this fluid are also outlined on the sides of the body by narrow white lines.

(*a*)

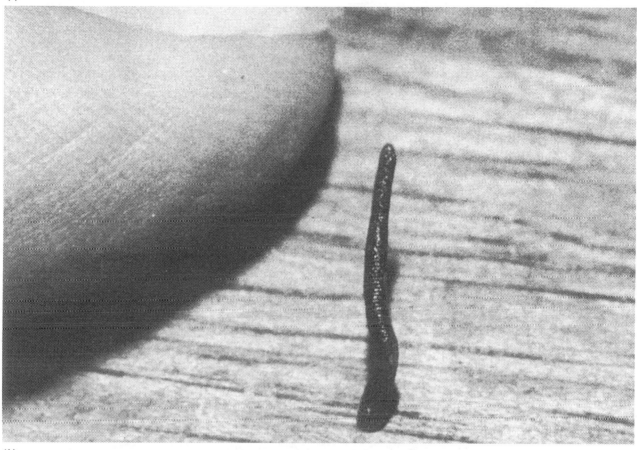

(*b*)

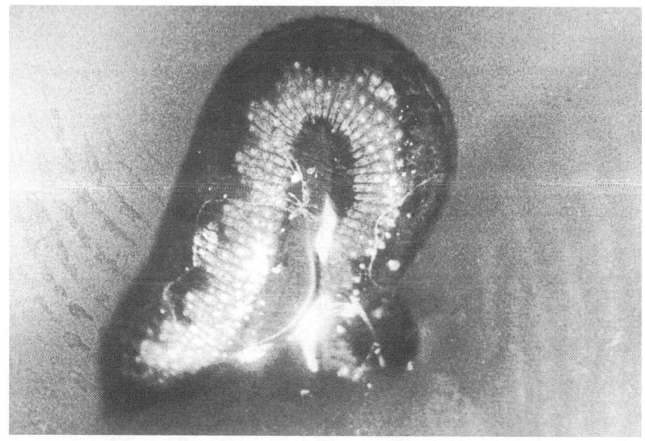

polychaete organisation not proved to be successful elsewhere than in the sea? Although this question cannot be simply answered, it merits some discussion.

First, it is appropriate to emphasise the basic differences between the polychaetes and oligochaetes. These differences lie in the methods of movement, excretion and reproduction. As regards movement, the oligochaete system of peristaltic contraction, using the coelom as a hydrostatic skeleton, is probably the most primitive found in the phylum (see Clark, 1969). Of the excretory mechanisms, however, those found in polychaetes are the simplest and, since they often contain protonephridia, they have usually been regarded as the most primitive (see, e.g. Goodrich, 1945). The presence of these simple nephridia also suggests a marine origin for the phylum. Similarly, it may be argued that external fertilisation is a primitive marine mechanism, and that internal fertilisation and hermaphroditism are specialisations for a life in such environments as fresh water. A hypothetical annelid ancestor would probably therefore have had such characteristics as oligochaete-type segmentation and movement, but polychaete-type nephridia and external fertilisation.

The evolution of the phylum can then be regarded as the exploitation of two different sets of habitats: one group exploited the marine habitat, meanwhile changing its method of movement to adapt itself to crawling over surfaces; the other invaded fresh water, and changed its methods of excretion and reproduction accordingly. It may then be argued that it would have been somewhat difficult for the polychaetes to invade fresh water because of the adaptive radiation undergone by the oligochaetes and the consequent lack of available niches. One may also suggest that since such an invasion necessitates the intermediate passage through estuarine conditions, where burrowing in the substrate is a great advantage in combating the violent salinity regimes, a group primarily adapted to burrowing would be more successful than one adapted to crawling.

An alternative approach to the problem is to suggest that the several distinct stems of the oligochaetes (Fig. 3.2) represent several distinct origins of the class from primitive annelid ancestors. On this basis, we may suggest that each time an annelid group invaded fresh water, it produced an 'oligochaete' form. In particular, such genera as *Stygocapitella* and *Aeolosoma* may have originated at a very early stage from primitive annelids, as they show both polychaete and oligochaete characters.

The only present-day polychaetes which are semi-terrestrial have reached their position through colonisation from brackish water or sea water, and not from fresh water. If the class had been able to invade fresh water, it would probably have produced many more terrestrial forms; polychaete distribution has been limited in the first instance by whatever factor(s) prevented the colonisation of fresh water. We may either regard this as due to competition with the forms which originally moved into fresh water, or as an inability of the majority to evolve appropriate excretory and reproductive mechanisms.

4

Molluscs

'. . . we noticed we were climbing a long gentle
slope. As we went upward it grew lighter. Finally
we saw that the snail had crawled right out of the
water altogether and had now come to a dead
stop on a long strip of grey sand.

Behind us we saw the surface of the sea
rippled by the wind. On our left was the mouth
of a river with the tide running out. While in
front, the low land stretched away into the mist . . .'

From *The voyages of Doctor Dolittle* by Hugh
Lofting, 1923, Jonathan Cape, London.

Present-day molluscs, most of which are characterised by
the possession of a mantle cavity and radula, and many
of which have a shell secreted by the mantle, are grouped
into seven classes. Of these only one, the Gastropoda, has
terrestrial members, and in fact gastropods are to be
found in practically every environment – salt water,
fresh water, streams, swamps, humid forests and even
deserts. The class Gastropoda also contains more members
than any other molluscan class, so that on at least two
counts it could claim to be the most successful molluscan
group.

The inability of the other classes (Monoplacophora,
Aplacophora, Polyplacophora, Scaphopoda, Bivalvia and
Cephalopoda) to colonise land is linked to various factors.
The most likely candidates would probably have been
the Polyplacophora or chitons, which are often found
intertidally and can withstand a great degree of desiccation (Wilding, 1968), and the Cephalopoda – the squids
and octopuses – which in many ways have become very
sophisticated animals. The Bivalvia have in the main
a limited power of movement, and are in general filter
feeders. A very few species have been able to adapt to
breathing air, and these include the common cockle
Cardium (Cerastoderma) edule. When this is exposed on
the shore it displays a pattern of valve movements which
expel water from the mantle cavity and draw in air
(Boyden, 1972). It then breathes by absorbing oxygen
through the mantle, and can survive for much longer
than species which do not air breathe. The Scaphopoda
possess a very specialised method of feeding which

probably limits them to the sublittoral. Of the Monoplacophora and Aplacophora we know very little concerning their mode of life, but both are thought to be
primitive, the Monoplacophora being considered by
some to represent forms close to the original molluscs.
The chitons also show many primitive characters (Morton
& Yonge, 1964), and in particular have retained external
fertilisation rather than any form of copulation. All three
of these classes are strictly marine. In the final chapter
some consideration is given to the reasons that may have
prevented the cephalopods from moving on to land. For
the present it seems more appropriate to consider the
forms that have colonised the land, the gastropods.

The class Gastropoda contains a wide variety of
types, and the first necessity is to outline their classification. Good general accounts of the class are available
(e.g. Hyman, 1967; Franc, 1968; Morton, 1979; Barnes,
1980), and the classification is outlined in Table 4.1. Of
a total of approximately 85 000 species of present-day
gastropods, there are about 55 000 prosobranchs,
10 000 opisthobranchs and 20 000 pulmonates. The
prosobranchs, typically with a coiled shell, include the
most primitive members of the class, and it is from
prosobranch stock that the other two subclasses have
evolved. The majority of prosobranchs are marine, but
several families are found in fresh water and on land,
all these being herbivores: none of the neogastropods,
which are carnivores, have left the sea.

The opisthobranchs in general have a reduced
shell or none at all, and are almost entirely marine.
Some have succeeded in establishing themselves in fresh
water, probably by the interstitial route (Hyman, 1967),
but these have not produced terrestrial groups. All are
specialised for life in the interstitial habitat, but their
existence in fresh water does add support to the idea,
suggested in Chapter 2, that movement from one
major habitat to another by the 'interstitial route' is
feasible.

The pulmonates show a striking contrast to the
other two subclasses in their distribution. They are
mainly terrestrial, and contain such well-known forms
as the terrestrial slugs and the majority of terrestrial
snails. However, they also have representatives in the

sea and in fresh water. The origin of these groups will be discussed briefly here, and in more detail in Chapter 10, where the recolonisation of aquatic habitats from the land is considered.

The gastropods are an old group, with the fossil record of their aquatic members extending back to the Palaeozoic, but the earliest records of terrestrial snails come from the Carboniferous period; there seems to have been no real expansion of terrestrial groups until the Jurassic and Cretaceous periods (see, e.g. von Zittel, 1927; Knight *et al.*, 1960), and they have only flourished within the last 40 million years. Although the oldest fossil records of terrestrial snails are those of pulmonates, it seems logical to begin this account with the prosobranchs, both because this group contains the most primitive members of the class, and because it shows a striking variety in the routes by which its members have colonised land. Following this account of the proso-

branchs will be a consideration of the pulmonates. In many ways these have become more successful on land, but much less is known about the origin of the terrestrial forms.

4.1 The prosobranchs

Only four superfamilies of prosobranchs are found on land, and the supposed relationships between these are shown in Table 4.2, based on Thiele (1931). One superfamily, the Neritacea, is placed either in the Archaeogastropoda, or in a separate order of its own (Morton & Yonge, 1964). The aquatic members of this group are found as fossils as far back as the Devonian period, and terrestrial members – the helicinids – are found in the Cretaceous (Knight *et al.*, 1960). The group has therefore been on land for over 100 million years, but is still in the main restricted to very damp habitats in the West Indies and in Pacific islands.

Table 4.1. *The distribution of gastropod families in different environments*

| Taxon | Number of families | | | |
	Marine	Fresh water	Amphibious	Terrestrial
Subclass Prosobranchia				
Order Archaeogastropoda	9	1	–	1
Order Mesogastropoda	32	5*	2	5
Order Neogastropoda	16	–	–	–
Subclass Opisthobranchia	45	1	–	–
Subclass Pulmonata				
Order Basommatophora	5	3	1	–
Order Systellommatophora	1	(1)	(1)	3
Order Stylommatophora	–	–	–	27

The use of brackets indicates doubtful records for the family Onchidiidae.
* There are also four families endemic to some of the ancient freshwater lakes (see, e.g. Boss, 1978).

Table 4.2. *The distribution of those prosobranch superfamilies containing terrestrial and amphibious species*

Taxon	Marine	Fresh water	Terrestrial
Order Archaeogastropoda			
Superfamily Neritacea	Neritidae	Neritidae	Helicinidae Hydrocenidae
Order Mesogastropoda			
Superfamily Architaenioglossa	–	Viviparidae Ampullariidae*	Cyclophoridae
Superfamily Littorinacea	Littorinidae Lacunidae	–	Pomatiasidae Aciculidae (= Acmidae)
Superfamily Rissoacea	Rissoidae Hydrobiidae Assimineidae	Hydrobiidae	Hydrobiidae Assimineidae
Superfamily Cerithiacea	Cerithiidae Potamididae* and others	Melaniidae	–

* denotes amphibious forms.

In the order Mesogastropoda the superfamily Rissoacea contains very few terrestrial species, belonging to the families Hydrobiidae (genus *Geomelania*) and Assimineidae (genus *Pseudocyclotus*). *Geomelania* spp. are confined to the West Indies, and very little is known about them. Since there are also very few species involved, they will not be considered further. The genus *Pseudocyclotus* is widespread in the Indo-Pacific region, and will be discussed later.

Two other superfamilies within the Mesogastropoda provide an interesting comparison. The superfamily Littorinacea is mostly marine with no freshwater forms, and has its origin in the Carboniferous. Its terrestrial forms – the Pomatiasidae – appeared first in the Cretaceous (von Zittel, 1927). In contrast, the superfamily Architaenioglossa contains mainly freshwater forms with no obvious marine ancestors, and no marine representatives. This group dates back to the Jurassic, but its terrestrial members – the Cyclophoridae – are all Tertiary-Recent forms. It will be seen later that the Littorinacea and Architaenioglossa show other interesting contrasts.

The only other superfamily to show any trend towards terrestrial life is the Cerithiacea. Characteristically these snails have long-spired shells and are marine, but in sub-tropical regions many live in estuaries. Members of the Potamididae are amphibious, and in some species the mantle cavity has been converted into a lung and the gill has been lost (Morton, 1979). *Cerithidea obtusa*, for instance, is essentially adapted to breathing air although it retains water in the mantle cavity when emersed (Houlihan, 1979), and the Mexican lagoonal species *Cerithidea mazatlanica* can aestivate through the dry season (Edwards, 1978). However, no species seem to have moved on to land from the semi-marine conditions of the delta muds and mangrove swamps of the Indo-West-Pacific, where the group flourishes (Macnae, 1968), and none can be considered truly terrestrial.

Because of the relatively large numbers of groups involved, it will be simplest to consider the four superfamilies in turn before going on to compare prosobranchs in general with the pulmonates.

4.1.1 The Neritacea

The marine neritids (family Neritidae) are characteristic of tropical shores from rocky habitats (e.g. Underwood, 1975; Fig. 4.1) to mangrove swamps (e.g. Berry, 1964). Many of them live high up in the intertidal zone, and are exposed to air for long periods. All of them retain a gill, but many also have a well-vascularised mantle floor (Fretter, 1965) and at low

Fig. 4.1. *Nerita melanotragus*, a marine intertidal neritid from New Zealand, actively crawling in air. Many marine neritids are able to respire out of water through the mantle epithelium. The eyes are on prominent tubercles at the base of the tentacles. Shell width approximately 15 mm.

tide they may draw air into the mantle cavity and utilise this for respiration. The mangrove species *Nerita articulata* is essentially an air breather (Houlihan, 1979). The neritids have also evolved a method of internal fertilisation, and the eggs are laid in jelly masses from which hatch free-swimming veligers. The evolution of internal fertilisation may be said to have in some measure preadapted the neritids for life on land. *N. articulata*, which lays its egg capsules on the stems of mangroves, appears to be more influenced in its reproductive cycles by such 'terrestrial' factors as rainfall than by marine factors such as the tides (Berry, Lim & Sasekumar, 1973).

The family Neritidae also contains freshwater genera, and these are usually known as neritinids, although apart from the difference in habitat they differ little from the marine neritids. The freshwater forms, too, are primarily tropical, but in Europe are represented by *Theodoxus* (*Neritina*) *fluviatilis*, a species which is found in calcium-rich streams and rivers.

The anatomy of such freshwater forms as *T. fluviatilis* is very similar to that of the marine genera (Bourne, 1908). All have two auricles to the heart, but only one kidney (Fig. 4.2). The upper internal whorls of the shell are dissolved away during development, providing one globular cavity instead of the normal spiral found inside most gastropod shells. As would be expected of freshwater animals, the neritinids osmoregulate to a certain degree (Neumann, 1960) and produce a hypo-osmotic urine from the kidney (Little, 1972). In this they differ markedly from the marine neritids, which do not osmoregulate, and regulate only very slightly the ionic content of their blood. The renal systems of marine and freshwater forms look very similar in terms of gross anatomy, but Delhaye (1974a) has shown that the 'glandular' part of the kidney in the freshwater genus *Theodoxus* has a reabsorptive epithelium, whereas in the marine *Nerita* the kidney is involved solely in excreting material in the form of concretions.

One form of neritinid, the genus *Neritodryas*, found in the Solomon Islands, spends most of its time out of water and thus provides a convincing link with the truly terrestrial forms (Cooke, 1895). These terrestrial forms (the Helicinidae; Fig. 4.3) show many similarities to the aquatic forms, but several striking differences. Thus, the basic plan of the helicinids is similar to that of the Neritidae, but the visceral mass has been rotated by about 90° (Fig. 4.2) so that the stomach comes to lie above the kidney instead of beneath it (Bourne, 1911). It is difficult to account for this rotation, but it seems possible that it is associated with the elongation of the mantle cavity, which is presumably in its turn caused by the need for increased respiratory surface area; for in these helicinids the single gill of the aquatic forms is lost and the mantle surface is well vascularised and used as a lung. Other differences involve the loss of the right auricle, and now the rectum no longer passes through the ventricle as it does in the Neritidae. The operculum also differs from that of the aquatic forms, and is horny instead of being calcareous. The kidney of helicinids

seems to function much like that of the freshwater neritinids, producing hypo-osmotic urine (Little, 1972). However, most reabsorption of salts apparently occurs not in the 'glandular' part of the kidney (Fig. 4.2) but in the bladder or 'ureter' (Delhaye, 1974a). This may indicate an independent evolution of the sites of ion uptake in freshwater and terrestrial neritaceans (Andrews, 1981). It is therefore interesting that Thompson (1980) has suggested that the helicinids should be regarded as a separate superfamily, to which the Neritacea is not ancestral. He has studied particularly the genus *Proserpina*, which has no operculum, and has concluded that the group is primitively non-operculate, and that opercula evolved secondarily.

Whether or not the helicinids belong in the Neritacea, their general mechanism of renal function is

Fig. 4.2. The organisation of the renal system in the Neritacea. (a) *Nerita fulgurans*, from the marine family Neritidae. Urine is produced by ultrafiltration through the wall of the heart, and the filtrate then passes through the glandular part of the kidney and the bladder, but ionic composition is not modified. (b) *Eutrochatella pulchella*, from the terrestrial family Helicinidae. Here the kidney occupies a position at right angles to that in the Neritidae, but the gross anatomy of the two families is similar. In the Helicinidae, however, the glandular part of the kidney reabsorbs ions, and a hypo-osmotic urine is produced. From Little (1972).

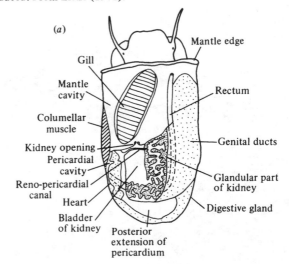

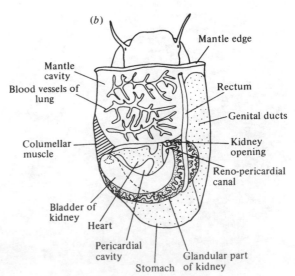

much more like that of freshwater forms than that of marine ones, suggesting a freshwater and not a marine ancestry. Further similarities can be found in the osmotic pressure and ionic composition of the blood (Table 4.3). In general, neritinids and helicinids resemble each other closely in these respects, and differ markedly from the neritids, which have a blood composition close to that of sea water. It will be seen later that in at least one line of land snails which are thought to have evolved directly from marine ancestors, the kidney cannot produce hypo-osmotic urine, and the osmotic pressure of the blood in this case is much higher than that of land snails that have evolved from freshwater ancestors.

Relatively little is known about the general natural history of the helicinids. In the main they are found on islands and on the coastal fringes of larger land masses in the tropics. They are commonest in the West Indies and Central America, and have another centre in the Philippines. They are very widely distributed on Pacific islands, and this island distribution is reminiscent of that of terrestrial nemertines. In the present case, however, although the matter has not been investigated in detail,

it seems likely that this distribution is due to dispersal from perhaps one centre, Central America, and not to independent evolution on individual islands (Bourne, 1911). This is understandable, since the nemertines are soft bodied and vulnerable, whereas the helicinids can isolate themselves from the environment by withdrawing into the shell and closing the aperture with the operculum, making passive dispersal through unfavourable conditions more feasible.

Most helicinids are tree dwellers, and most are active only when the surface is wet. I have seen them crawling on trunks streaming with water during a rain storm; but for the most part they are active at night when the dew falls. All species are herbivores, and presumably feed on the algae and fungae growing on tree barks. Some forms are found on limestone rocks in humid forests, where they must feed on the algal covering. Their habitat in this case closely resembles that of some of the marine intertidal neritids. Nothing is known about water loss by the helicinids, but presumably they are limited in their distribution by the conditions found within tropical rain forest.

Table 4.3. *Comparison of the composition of the blood of marine, freshwater and terrestrial members of the Neritacea (data from Little, 1972)*

Species	Osmotic pressure (mOsm)	Na^+	K^+	Ca^{2+}	Mg^{2+}	Cl^-	SO_4^{2-}	HCO_3^-
		(as % of cations)				(as % of anions)		
Nerita fulgurans (marine)	1060	83.2	2.2	2.0	9.4	95.0	5.1	—
Theodoxus fluviatilis (fresh water)	95	85.7	4.2	4.4	5.5	62.8	—	21.5
Eutrochatella tankervillei (terrestrial)	67	72.0	3.3	8.7	4.1	64.4	—	34.0

The figures for % cations and % anions were calculated using the numerical value for osmotic pressure as 100% in each case.

Fig. 4.3. A small terrestrial helicinid (shell diameter *c.* 8 mm) from Papua New Guinea. The tentacles in helicinids are particularly mobile, can bend as seen here, and in some species can even vibrate.

As a final note the genus *Smaragdia* should be mentioned. This is a neritid found on tropical shores that may have recolonised the marine environment from fresh water (Cooke, 1895).

4.1.2 The Architaenioglossa

The superfamily Architaenioglossa, or Cyclophoracea of some authorities, contains three large families (Table 4.2). One of these, the Viviparidae, contains freshwater snails. Another, the Ampullariidae, contains freshwater snails capable of aestivation. These may be termed amphibious since they usually climb out of the water to lay their eggs, and may also travel overland from one body of water to another. The third family, the Cyclophoridae, is made up entirely of land snails which are widespread in tropical forests. Some authorities have chosen to promote this taxon to the rank of superfamily (e.g. Taylor & Sohl, 1962), but others (e.g. Thompson, 1969) have pointed out that because the taxonomy of the group is at present in a confused state, it may be better to retain the family Cyclophoridae *sensu lato*. The latter view is adopted here, although it has been evident since the work of Tielecke (1940) that the family should be subdivided. Within the superfamily Architaenioglossa there are no marine species, and the group has no obvious close connections with any present-day marine mesogastropods.

The family Viviparidae contains medium-sized snails common in temperate regions, usually found in slow-running waters and swamps with muddy substrates. As the name suggests, the young develop in a uterus and leave the parent as miniature adults. At least one species, the European *Viviparus viviparus*, is partly a filter feeder, utilising the ciliary currents of its single ctenidium. There is one kidney, and this is used as an osmoregulatory organ (Little 1965b). It produces a hypo-osmotic urine which then passes along the ureter – a development from the mantle, and an unusual feature in prosobranchs (Fig. 4.9) – in which more salts are reabsorbed. Both kidney and ureter show cells typical of reabsorptive epithelia, with apical microvilli and basal infoldings associated with mitochondria (Delhaye, 1974b; Andrews, 1979). The kidney has been developed mainly from the nephridial gland of mesogastropod ancestors, and is the major site of ion reabsorption. Fluid is distributed throughout its sponge-like lumen by muscular contractions so that a rhythmic cycle of events draws fluid in from the pericardial cavity, reabsorbs ions, and expels hypo-osmotic urine to the pallial ureter (Little, 1965b). Excretion of organic molecules occurs in the form of concretions which are built up extracellularly, and contrary to earlier belief do not contain uric acid (Andrews, 1979). Uric acid previously thought to accumulate within the kidney (Needham, 1938) in fact is confined to connective tissue around the blood vessels, and its significance in nitrogenous excretion is unknown (Little, 1981). The digestive gland also accumulates uric acid (Spitzer, 1937), and this may be excreted through the alimentary canal.

The family Ampullariidae shows a wider distribution than that of the Viviparidae, various genera being found in tropical and sub-tropical swamps and slow-moving water. In particular there is *Pomacea* (=*Ampullaria*) of Central and South America, and *Pila* of Africa and Asia. Another genus, *Turbinicola*, is found in fast-flowing streams in India.

The ampullariids differ from the viviparids in several ways. They are not viviparous and most of them crawl out of the water to lay their eggs on the stems of reeds and grasses. These eggs are laid in batches and the shells harden after laying as the mucous coat dries. These shells must be very impermeable to water, as those of the genus *Pomacea* are a common sight in the Florida Everglades where temperatures are high, and the evaporation rapid. Ampullariids are well adapted for this amphibious life as they possess both a gill and a lung. The mantle cavity is divided, with the post-torsional left ctenidium occupying the right side; the left side, from which the ctenidium has been displaced, being occupied by much-vascularised pallial tissue which acts as a lung. This latter is mainly used when the snails are living in deoxygenated water, and in this case they come to the surface, form a siphon with the edge of the mantle, and draw air into the lung by a series of contractions of the body (McClary, 1964; Andrews, 1965). Even in species which seldom leave the water, such as *Marisa cornuarietis*, surface breathing is common and parallels that shown by many freshwater pulmonates (Akerlund, 1974). The lung is presumably also of major importance during aestivation.

The kidney of ampullariids is much more complex than that of viviparids, as it consists of two distinct parts (Fig. 4.9). The anterior part has been derived from the mantle, and may therefore be termed a pallial ureter (Andrews, 1981). However, it is much more highly specialised for ion reabsorption than is the ureter of species of *Viviparus*. Its lumen is tightly packed with lamellae covered by reabsorptive cells and provided with a good blood supply (Andrews, 1965, 1976; Delhaye, 1974b). This chamber is also muscular, and probably moves fluid through the renal system in a fashion analogous to that in *Viviparus* (Little, 1968). Hypo-osmotic urine emerges from the kidney pore, and it is also possible that the mantle epithelium is responsible for further ion reabsorption. The posterior chamber of the kidney, which represents the 'true' kidney, is not involved in ion reabsorption. Its functions appear to be correlated with the aestivating habit: it acts as a water store during aestivation (Little, 1968), and although it does not store uric acid, it provides a mechanism by which this is rapidly excreted when aestivation is terminated (Fig. 4.4). Uric acid is stored around the blood vessels in connective tissue, as in viviparids. Its excretion by the posterior chamber of the kidney is carried out by renal cells which form large concretions rich in purines (Andrews, 1981). Some uric acid is also excreted by a separate gland, the anal gland (Andrews, 1965). However, the snails may also excrete some of their nitrogen as ammonia gas: when keeping aestivating species of *Pomacea* in closed containers, I have noticed a very strong smell of ammonia after a short time, and this parallels observations on pulmonates which are discussed later. There is probably also some urea production, but the significance of this varies from species to species (Little, 1981).

The capacity for withstanding dry periods by burrowing into the mud and closing up is one of the most interesting abilities of the family. The snails burrow, close up the aperture with the tightly fitting operculum (which is horny in the genus *Pomacea* but calcified in *Pila*), and thereby seal themselves off from the outside world until the habitat becomes flooded again. This process of aestivation evidently has some variations, as *Pila virens* is said to aestivate anaerobically (Meenakshi, 1964), while *Pila ovata* aestivates aerobically (Coles, 1968). During aestivation the rate of water loss is very low, and the heartbeat is drastically reduced or stops. In both *Pomacea* and *Marisa* the metabolic rate drops to about 20% of the normal resting rate (Burky, Pacheco & Pereyra, 1972; Horne, 1979).

The mechanisms triggering the onset and termination of aestivation have been examined by various workers (Coles, 1968; Little, 1968). It has been suggested (Meenakshi, 1956) that a hormone is implicated in the onset, this being probably connected with the loss of a specific proportion of water. Termination can be brought about both by mechanical disturbance and by the presence of liquid water, and is also probably mediated by a hormone (Coles, 1969). The ability to aestivate probably arose early in the evolution of the Architaenioglossa, and has to some extent preadapted the aquatic forms for terrestrial life. However, it should be noted

Fig. 4.4. Changing composition of the body fluids during recovery from aestivation in *Pomacea lineata* (family Ampullariidae). (*a*) Osmotic pressure of blood and fluid in the posterior chamber of the kidney decrease at the same rate. Fluid in the anterior chamber of the kidney rapidly becomes hypo-osmotic and eliminates water entering osmotically. (*b*) Uric acid concentration in the fluid produced by the posterior chamber increases rapidly and some of the waste nitrogen stored during aestivation is excreted. Open circles, blood; triangles, fluid from posterior chamber; squares, fluid from anterior chamber of kidney. From Little (1968).

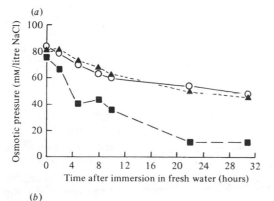

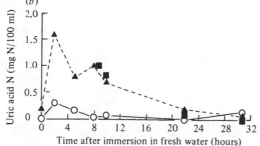

that aestivating ampullariids are often near their lethal physical limits, as is shown by two examples. Haniffa (1978), working with *Pila globosa* in India, found that mortality in natural populations could be as high as 30% in one dry season. Burky, Pacheco & Pereyra (1972), working with *Pomacea urceus* in Venezuela, showed that young snails were soon killed by temperatures tolerated by the adults. In this species the eggs are laid beneath the adult shell aperture at the start of the dry season. They hatch and then aestivate for about 4 months – but can only survive when protected by the adult shell.

The third family of the Architaenioglossa to be considered is the Cyclophoridae. These snails show some variety of shell form, illustrated in Figs. 4.5, 4.6 and 4.7, and on the basis of differences in reproductive systems the group should be divided into many families (see, e.g. Tielecke, 1940; Franc, 1968). Of these subdivisions, the cyclophorids *sensu stricto* (e.g. the genera *Cyclophorus* and *Leptopoma*) are an Old World taxon, ranging through Asia and Indonesia. The maizaniids are restricted to eastern Africa, and apart from them the group is absent from the African continent. The poteriids are found in Central and South America and Polynesia. The pupinids (e.g. *Pupina*) are small snails found in Asia and the Indo-Pacific. Lastly, the cochlostomatids are also small, and although mainly found in the Indo-Pacific, the European genus *Cochlostoma* is common in countries round the Mediterranean (von Prince, 1967; Kerney & Cameron, 1979). In general, the distribution can be seen to be tropical and sub-tropical.

None of the cyclophorids is conspicuous. Although some, such as the genus *Leptopoma*, are arboreal, and many are found in rocky habitats, most are restricted to forest leaf litter. They are therefore not for the most part exposed to desiccating conditions. Many probably aestivate in the dry season, and von Prince (1967) has shown that *Cochlostoma septemspirale* can survive at least 2 months of aestivation. Permanent changes such as the felling of forest trees, however, produce mass mortalities such as I have observed in Jamaica and in Papua New Guinea. Long droughts in particular years do the same, but even if the adults are not killed, the population will decline because most cyclophorids lay eggs in the leaf litter, and these cannot withstand desiccation. A few species are ovoviviparous (Tielecke, 1940). The young snails may grow very quickly, adding one visible growth ring to the shell each day under favourable conditions (Berry, 1962). The family is restricted to limestone areas, presumably because of the necessity to obtain calcium for shell growth. In Papua New Guinea, cyclophorids were also restricted to areas of high pH, and were found in regions of high, but not excessive, rainfall (Andrews & Little, 1982). Although cyclophorids were not present there at high altitudes, they have been found at over 1000 m in India (Kasinathan, 1975).

The cyclophorids have retained a fairly primitive structure in spite of having accomplished the transition from an aquatic to a terrestrial life (Morton, 1952). Nevertheless, they show specialisations in respiratory and renal structures and mechanisms, and these prompt discussion. The mantle cavity has become a lung, much as in other terrestrial prosobranchs. The anterior opening of the lung remains wide, and quite unlike that of pulmonates. No measurements of respiratory rate have been made, but most cyclophorids are slow-moving animals and probably have a low metabolic rate. The problem of respiration during aestivation is an intriguing one. Many species have evolved various types of slits, notches or pores at or near the aperture (Fig. 4.8), and these all appear to be modifications to allow diffusion of air to occur between the inside and the outside of the shell when the operculum is closed. This phenomenon is also commonly found in the tropical species of Pomatiasidae, the terrestrial relatives of the littorinids (see p. 44), so that the adaptation must be an important one. Rees (1964) has reviewed the occurrence of these breathing devices, and from their distribution he concluded that their presence is related to the presence of marked seasonal differences in the tropics. Their exact physiological significance in the dry and rainy reasons remains to be elucidated.

The renal system of cyclophorids produces a copious flow of hypo-osmotic urine (Andrews & Little, 1972). Most of the reabsorption of ions occurs within the lumen of the kidney itself, but the urine is ejected from the kidney into a posterior bay of the mantle cavity partly separated from the respiratory portion by a transverse septum (Andrews & Little, 1972, 1982).

Some further ion reabsorption probably occurs here, so that as in viviparids and ampullariids the mantle has become involved to some extent in renal processes. The urine produced is similar to that of other architaenioglossans (Table 4.4). Even its rate of production in the terrestrial forms is not much less than that of aquatic members of the superfamily. It therefore seems likely that in terms of overall ionic regulation, the cyclophorids behave just like the ampullariids and viviparids. In the initial process of urine formation, however, which occurs by ultrafiltration through the heart wall into the pericardial cavity, the cyclophorids do show one terrestrial adaptation (Andrews & Little, 1982): the filtration sites in the heart are reduced, thus presumably cutting down on the production of the primary ultrafiltrate. Also, the kidney entirely lacks mucoid cells, which in ampullariids and viviparids are responsible for the dual roles of secreting mucus and reabsorbing ions. It has been suggested by Andrews (1981) that the mucoid cells may be able to reabsorb ions without necessarily reabsorbing water. If this is so, their loss in terrestrial species and replacement by cells specialised for reabsorption of salts and water at the same time is consistent with terrestrial needs.

The kidney of cyclophorids is also involved in nitrogenous excretion. It normally contains quite high concentrations of uric acid (Rumsey, 1971), and both

Fig. 4.5. *Cyclophorus kubaryi* (family Cyclophoridae), a leaf litter species from Papua New Guinea. This species is typical of many large cyclophorids, being restricted to rain forest, consuming dead leaves, and seldom exposing much of its body outside the shell. Shell diameter 25 mm.

Fig. 4.6. *Leptopoma perlucidum* (family Cyclophoridae), an arboreal species from Papua New Guinea. During the daytime this species closes the operculum and hangs on the undersides of leaves. It is active after rain at night. Shell height 15 mm.

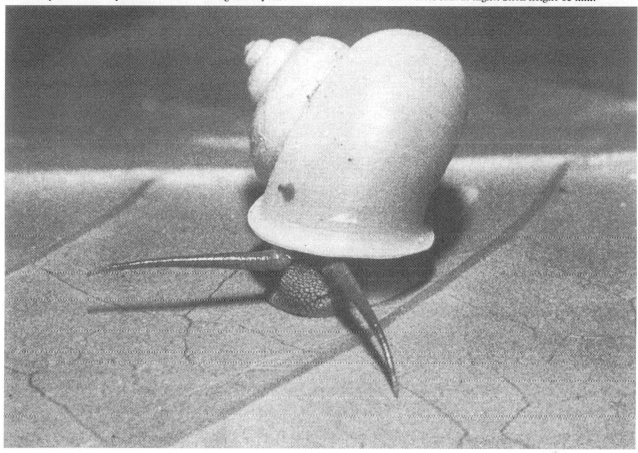

Fig. 4.7. A species of *Pupina* from Papua New Guinea (family Cyclophoridae). Like species of *Cyclophorus*, pupinids are leaf litter species, but they are often more active. Shell height 7 mm.

the concentration and amount of this rise during aestivation. While some of this uric acid probably accumulates in connective tissue around the blood vessels, the renal cells each form a large concretion rich in purines, just as in ampullariids (Andrews, 1981). An accessory excretory organ has also been developed from the hypobranchial gland in the mantle cavity (Andrews & Little, 1972). Normally, the molluscan hypobranchial gland secretes mucus, but in cyclophorids both purines and lipids are also produced. Unfortunately the relative importance of the hypobranchial gland and the kidney in nitrogenous excretion is not known, since it has become apparent that measurements of the concentration of purines in any particular organ do not necessarily reflect the organ's rate of excretion, but merely its capacity for storage, which may be a very different matter. However, it does seem that all through the Architaenioglossa there is a tendency to develop accessory organs which are concerned with both the storage and excretion of purines. The situation is complicated by the finding that gaseous ammonia, known to be an important excretory product in aestivating pulmonates (e.g. Speeg & Campbell, 1968a), may also be excreted in prosobranchs. Fig. 4.9 summarises the anatomical relationships of the various excretory organs known in the Architaenioglossa, but totally different organs may be involved in ammonia volatilisation.

Fig. 4.8. Breathing devices in some of the genera of the Cyclophoridae. (*a*) *Cyrtotoma*, showing dorsal and ventral views of a notch on the columellar side of the aperture. (*b*) *Spiraculum*, showing an imperfect tube and a more recently formed snout-like extension of the aperture. (*c*) *Rhiostoma*, with an imperfect breathing tube. (*d*) *Rhiostoma*, with a more completely formed breathing tube, the opening of the tube being protected by the wall of the preceding whorl. All these devices are thought to aid the transfer of air to the snail when the operculum is closed. Redrawn from Rees (1964).

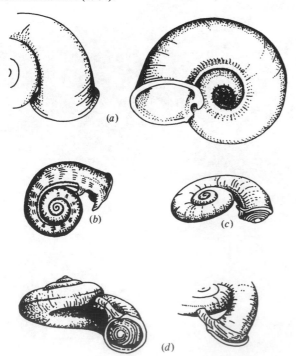

4.1.3 The Littorinacea

Of the four families making up the Littorinacea, two are marine and two terrestrial. The two marine families are the Lacunidae, which contains small littoral and sublittoral snails, and the well-known Littorinidae, widespread intertidally. The biology of both families has been summarised by Fretter & Graham (1980). The terrestrial littorinaceans are usually split into the Pomatiasidae (=Cyclostomidae) and the Aciculidae (=Acmidae), but the Pomatiasidae should probably be further divided to separate the chondropomids (e.g. Franc, 1968). So little is known of the biology of the chondropomids, however, that they will here be considered together with the pomatiasids. The family Pomatiasidae is widespread, but most species are found in the tropics. The members of the Aciculidae are all very small, and because their characteristic features are mostly related to this small size, their relationships are somewhat unclear. Some authorities place them in the Rissoacea.

The present-day members of the Lacunidae represent a family which has many more extinct members. They live mostly on seaweeds, are not resistant to desiccation, and are restricted to the lower range of the littoral zone and to the sublittoral. At low water of spring tides *Lacuna vincta* is often found on *Laminaria* fronds, crawling with a gait reminiscent of the bipedal motion of species of *Pomatias* (see p. 46). *Lacuna vincta* lays its egg masses on these fronds, and the eggs hatch into planktonic veliger larvae. In *Lacuna pallidula* the eggs hatch directly as young snails, and this abolition of the planktonic stage of larval development is repeated many times within the superfamily.

The family Littorinidae has been the subject of a multitude of studies, and the zonation of British littorinids on rocky shores has recently become the centre of renewed interest because of the recognition of several new species. The taxonomic status of these has been summarised by Fretter & Graham (1980). *Littorina littorea* is found from low water of spring tides up to high water of neap tides. It produces planktonic egg capsules from which hatch veliger larvae. *Littorina obtusata* and *Littorina mariae* are associated with the seaweeds *Ascophyllum* and *Fucus*, and probably are much protected from desiccation by this cover. In these species the eggs are laid on the algae and hatch as immature adults. *Littorina neritoides* is found on exposed shores and particularly at high levels. However, it produces planktonic egg capsules in a similar way to *L. littorea*, and so has to be covered with sea water during spawning. *L. neritoides* is extremely resistant to desiccation. The remainder of the species are members of the *Littorina saxatilis* complex, and the status of many of them is as yet uncertain. There seems to be no doubt that *Littorina nigrolineata* is a distinct species. It lives on the mid-shore and lays egg masses under stones. *Littorina neglecta* is at present mainly characterised by its small size. It lives in dead barnacle shells, and is ovoviviparous. *Littorina rudis* is by far the most widespread of the species in the complex, and is ovoviviparous. It apparently ranges from exposed rocky shores to saltmarshes, although

Table 4.4. *Urine production in some members of the Architaenioglossa*

Species	Osmotic pressure of haemolymph (mOsm)	Osmotic pressure of fluid from kidney (mOsm)	Osmotic pressure of final urine (mOsm)	Rate of urine production (μl g^{-1} min^{-1})	Reference
Viviparidae					
Viviparus viviparus	68.0	24.5	13.6	0.25–0.91	Little, 1965*b*
Viviparus malleatus	–	–	–	0.70	Monk & Stewart, 1966
Ampullariidae					
Pomacea lineata	117.4	47.2	27.4	0.96	Little, 1968
Cyclophoridae					
Poteria lineata	74.0	48.8	–	–	Andrews & Little, 1972
Poteria varians	84.0[1]	–	–	0.39[2]	[1]Rumsey, 1971; [2]Andrews & Little, 1972

Fig. 4.9. The organisation of excretory and respiratory systems in the Architaenioglossa. (*a*) A typical littoral monotocardian, showing the hypothetical ancestral condition. (*b*) A cyclophorid, adapted for terrestrial conditions. (*c*) A viviparid, adapted for fresh water. (*d*) An ampullariid, adapted for life in fresh water and also for aestivation. Of all these, only species of *Viviparus* possess a long ureter, but the anterior chamber of the kidney in ampullariids is also a development from the mantle. In the terrestrial cyclophorids the hypobranchial gland has been partly modified as an excretory organ. From Andrews & Little (1972).

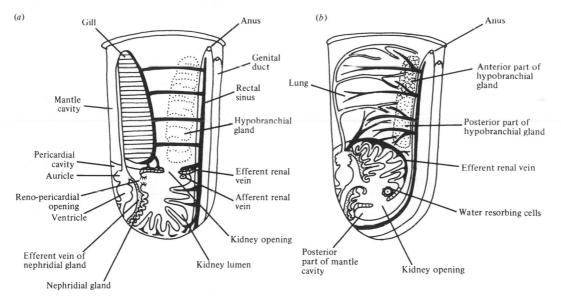

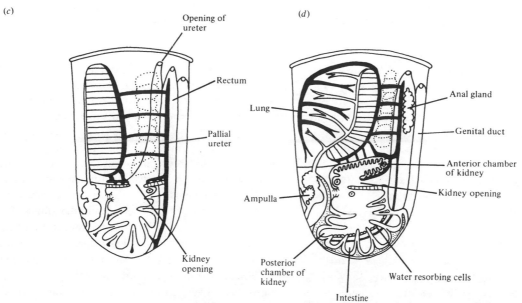

it is possible that the species should be divided, and that the inhabitants of rocky shores should be separated from those of shingle and boulder habitats. As mentioned above, *L. rudis* is ovoviviparous, but specimens with almost indistinguishable characteristics have been described laying egg masses at high levels on exposed shores. This oviparous species has been named *Littorina arcana*. There may also be further species within the complex, but even by considering the eight species listed above, it is apparent that no correlation between reproductive mode and level on the shore can be derived. As was emphasised by Remmert (1968) before the *L. saxatilis* complex was divided, the reproductive methods of littorinids do not therefore provide us with a model for the origin of land snails. However, the diversity of reproductive methods just described does express the plasticity of the genus, and it is worth considering the adaptations of the group as a whole to conditions of alternating submersion and emersion.

Littorina neritoides is the most resistant to desiccation of the British littorinids, and can withstand months without water. It has reduced gill leaflets and probably uses cutaneous respiration to a large degree when active out of water. The respiration of several littorinids has been investigated by McMahon & Russell-Hunter (1977). While sublittoral snails showed a lower uptake of oxygen in air than in water, intertidal littorinids maintained a fairly high uptake in air, presumably reflecting the use of cutaneous respiration. The oxygen consumption of intertidal forms also appeared to be more independent of temperature than that of subtidal species.

Although the uric acid content of the kidney does not necessarily reveal dependence upon uric acid excretion, it is interesting that *L. neritoides* has higher concentrations of uric acid in its kidney than the other species (Table 4.5). The idea that species living higher on the shore are more uricotelic than those lower down has received some support from Fischer & Brunel (1953) who found that while low-shore prosobranchs contained all the enzymes for the breakdown of uric acid to ammonia, high-shore species possessed only traces of these enzymes. However, Little (1981) has pointed out that the amount of uric acid found in the kidney at any one time is more likely to reflect the frequency of emptying than other factors, and such species as *L. neritoides* would empty their kidneys much less frequently than forms low on the shore: Daguzan (1971), for instance, showed that *L. littorea* emptied its kidney only when covered by the tide.

Tolerance of a wide range of salinities is also higher in upper-littoral species such as *L. neritoides* than in species lower on the shore (Avens & Sleigh, 1965). *L. neritoides* can withstand submersion in fresh water for over a week, and is active on rock surfaces after rain (Fretter & Graham, 1962). In contrast, *L. littorea* placed into 45% sea water loses magnesium from its tissues, and swells considerably while muscle tone is lost (Rumsey, 1973). None of the species of *Littorina* has been shown to be able to osmoregulate, so that they survive either by closing the operculum or by some mechanism of tolerance. *L. littorea*, at least, produces

urine iso-osmotic with the blood, and with a similar ionic composition (Rumsey, 1973). The external conditions in which *L. neritoides* and *L. littorea* are active are therefore very different in range, and there is no doubt that in terms of adaptation to the conditions of the terrestrial environment, upper-littoral forms are more advanced than those from the lower littoral. Some of the characteristics of British species are summarised in Table 4.5.

Many other species of *Littorina* are found in tropical regions. Some live high on the shore, but even so most of them produce planktonic egg capsules. One Caribbean species, *Littorina angulifera*, lives on mangrove trees. It retains its eggs until they hatch as veligers, and then it releases these directly into the water (Lebour, 1945). *Littorina scabra* (Fig. 4.10), which occupies a similar habitat in the Indo-Pacific, has exactly the same reproductive strategy. This species migrates vertically on the mangroves so that it avoids immersion (Nielsen, 1976), and although it releases its larvae directly into the sea, it leads an otherwise entirely terrestrial life.

Several other genera besides *Littorina* are found in the intertidal zone, and of these perhaps the most interesting is *Tectarius*. *Tectarius muricatus* is found high up on rocky shores in the Caribbean, usually above high tide mark, and often in the company of the species *Littorina ziczac*. *T. muricatus* is so resistant to desiccation that it can survive dry in the laboratory for up to 17 months; yet for all that time it is capable of responding to water within a few seconds and resuming activity (Mattox, 1949). It seems likely that death is due not to loss of water but to a final depletion of food reserves. As far as movement on to land is concerned, however, *T. muricatus* has probably proceeded to its limit, for like *Littorina neritoides* it produces planktonic egg capsules, from which hatch free-swimming veligers (Lebour, 1945).

Although the family Littorinidae is a marine group, with no common freshwater genera, there is one form which provides the exception to the rule: *Cremnoconchus syhadrensis* inhabits wet inland cliffs in India; but it is really amphibious rather than a freshwater or terrestrial form (Prashad, 1925), and never remains submerged. It possesses a ctenidium, with the leaflets much reduced as in *Littorina saxatilis* and *L. neritoides*, and not a lung. Indeed, in most points of its anatomy it resembles species of *Littorina* very strongly. Like *Littorina littoralis*, *C. syhadrensis* lays eggs, but these are much better provided with yolk than those of littorinids, probably as an adaptation to combat the periods of drought (Linke, 1935). Presumably *C. syhadrensis* has evolved directly from marine ancestors, since the family Littorinidae contains no freshwater members.

The terrestrial family Pomatiasidae, in which as explained above I here include the chondropomids, is found in the warmer parts of the Americas, in Africa and in the Mediterranean countries. A few have spread to India, but the group is not found in Indonesia or Australasia. The distribution extends as far south as

Table 4.5. *Some comparative data for British species of* Littorina

| Species | Normal level on shore | Production of young | Effect of desiccation (7 days at 18 °C) | | | Gill leaflets |
			Water loss (% original wt)	Mortality (%)	Uric acid in kidney (mg/g dry wt)	
Littorina littorea	Up to high water of neap tides	Planktonic egg capsules and veligers	37.5	70	1.5	Many
Littorina obtusata	Restricted to zone of *Fucus* and *Ascophyllum*	Fixed egg capsules. Young hatch as miniature adults	56.5	80	2.5	Many
Littorina mariae	As for *L. obtusata* but often lower	As for *L. obtusata*	–	–	–	Many
Littorina neritoides	Up to and above extreme high water of spring tides	Planktonic egg capsules and veligers	26.0	None	25	Considerably reduced
The *Littorina saxatilis* complex						
Littorina nigrolineata	Between mean high water neaps and mean low water neaps	Benthic egg masses. Young hatch as miniature adults	–	–	–	Reduced
Littorina neglecta	Restricted mainly to barnacle zones	Ovoviviparous	–	–	–	Considerably reduced
Littorina rudis	Widely distributed up to extreme high water spring tides	Ovoviviparous	39.7	8–17	5	Reduced
Littorina arcana	Usually between high water neaps and high water springs	Benthic egg masses	–	–	–	Reduced

Data from Fretter & Graham (1962, 1980); Needham (1938); Lewis (1964); Heller (1975).

Fig. 4.10. *Littorina scabra*, a high-shore littorinid common in mangroves of the Indo-Pacific. This species lives an almost terrestrial life, but releases its larvae into the sea. Shell length approximately 2 cm.

southern Africa (approximately 35° S), and as far north as southern Sweden (55° N). Pomatiasids are limited to calcareous soils, like most other terrestrial prosobranchs, and indeed they are often found on limestone cliffs and rocks. Some species are tree dwellers, and many live in the forest leaf litter, so that they are often found together with helicinids and cyclophorids.

Undoubtedly the best-known species is the European *Pomatias elegans* which lives in leaf litter and loose earth at the edges of woods and in hedge-banks in chalk and limestone districts (Kilian, 1951). As a guide to the biology of pomatiasids in general, it is therefore most convenient to summarise what is known about this one species. The anatomy is similar to that of the littorinids, but some differences are found. The foot is divided longitudinally, and the snail moves by using each side of the foot alternately in a stepping motion which can produce quite rapid movement (Lissmann, 1945*a*, *b*). The method of reproduction is not very different from that of some littorinids. After copulation the eggs are laid in soil, each in a capsule filled with albumen which supplies the embryo with food. The capsule is probably permeable to water and salts, since dyes will penetrate it, and the correct composition of external water is necessary for development (Creek, 1951).

The mantle cavity of *Pomatias elegans* has lost the ctenidium, and the mantle is vascularised, acting as a lung. Both the osphradium and the hypobranchial gland are retained. The kidney lies in the mantle skirt, but does not appear to be much specialised. *P. elegans* aestivates in the dry season and hibernates during the winter. During aestivation, uric acid was not found to accumulate in the kidney (Rumsey, 1971), and electron microscopy studies have shown that although the renal cells form excretory concretions, these contain lipids and phosphates, but no purines (Delhaye, 1974*b*; Martoja, 1975). Posterior to the kidney, however, lies a peculiar tissue known as the 'concretion gland', although it has an ill-defined outline and no separate lumen or opening. The function of this 'gland', present in all pomatiasids, is unclear. Its granules contain urates, glycogen and pigments, and are surrounded by bacteria (Meyer, 1925; Delhaye, 1974*b*). The size of the gland decreases during hibernation, but increases while the animal is active and feeding. It has therefore been suggested that it is some

kind of temporary storage organ which is also concerned with nitrogen excretion (Kilian, 1951). Martoja (1975) suggested that if urate from the gland were degraded as far as ammonia, this could be liberated in gaseous form from the lung, as it is in terrestrial pulmonates. Such a process could explain the lack of involvement of the kidney in nitrogen excretion, but there are some problems in applying such a model because of the low concentrations of the appropriate enzymes so far found in pomatiasids (see discussion by Little, 1981).

The rate of water loss from *P. elegans* during aestivation is very low, being 0.05–0.1 mg per animal per day (Rumsey, 1971). It can aestivate for several months, like many tropical pomatiasids, and this is to be expected because the group is typically found in fairly dry limestone areas. The barrier to water loss appears to be not so much the calcareous shell as the organic periostracum which covers it. Experiments by Rumsey (1971) showed that when the periostracum was worn away the rate of water loss rose dramatically. However, the physiological mechanisms by which the active snail maintains water and ion balance are at present poorly understood. The kidney does not act as an organ of osmoregulation, as it does in other terrestrial proso-branchs: the urine has the same osmotic pressure and ionic composition as the blood, although there is some slight reabsorption of ions in the mantle cavity (Rumsey, 1972). From the studies of Delhaye (1974*b*, *c*, *d*), the importance of another organ has emerged. Opening on to the sole of the foot of pomatiasids is a tubular gland, lined by cells specialised for ion reabsorption. At present it is not clear whether this gland produces a hyperosmotic secretion, or whether it is involved in absorbing salts and water from the soil. It seems likely, however, that it has replaced the kidney as an organ of osmoregulation.

The inability of *Pomatias elegans* and other pomatiasids to produce a hypo-osmotic urine from the kidney has probably been associated with the evolution of the group directly from marine forms, in which such mechanisms are not found. Presumably there was initially no strong selection pressure to modify the concentration of the urine, in contrast to the situation for snails colonising fresh water. The suggestion of direct colonisation from the littoral zone is supported by the relatively high osmotic pressure of the blood and its resemblance in composition to that of species of *Littorina* (Table 4.6), as well as the absence of

Table 4.6. *Comparison of the composition of the blood of marine and terrestrial members of the Littorinacea (data from Rumsey, 1972)*

Species	Osmotic pressure (mOsm)	Na⁺	K⁺	Ca²⁺	Mg²⁺	Cl⁻	SO₄²⁻	HCO₃⁻
		(as % of cations)				(as % of anions)		
Littorina littorea (marine)	1098	87.9	3.0	2.3	11.2	95.9	2.8	1.0
Pomatias elegans (terrestrial)	254	79.7	4.5	11.8	1.6	81.1	3.5	8.7
Tropidophora ligata (terrestrial)	267	81.0	3.3	6.9	2.1	78.0	2.0	–

The figures for % cations and % anions were calculated using the numerical value for osmotic pressure as 100% in each case.

modern freshwater littorinaceans. We may conclude that the pomatiasids are one of the groups to have colonised the land not via fresh water, but via the marine littoral zone.

The final family, Aciculidae, is represented in Britain by the tiny *Acicula* (=*Acme*) *fusca.* This species lives among leaf litter and logs in calcareous areas, and is exceedingly inconspicuous, as it grows to a shell height of only 2 mm. Indeed, the animal is so small that when active it probably leads an almost aquatic existence. Creek (1953) has suggested that although it has lost its gill and breathes air by using the mantle, the initial loss may have been associated with small size rather than with the terrestrial habit. A similar phenomenon occurs in the tiny marine littoral snails of the genera *Omalogyra* and *Rissoella.* In *Acicula fusca* the digestive system has also been simplified, and this again is characteristic of very small species. Together with the lack of physiological information, this effect of small size makes it difficult to speculate about the origins of the family.

4.1.4 The Rissoacea

Within the Rissoacea, the only family to receive much attention has been the Assimineidae. This family consists mainly of small snails (Assimineinae) which are common throughout the world at high tidal levels in estuaries and saltmarshes. It also contains a number of terrestrial Indo-Pacific genera placed in a different subfamily, the Omphalotropidinae (Thiele, 1927, 1931).

Assiminea grayana is the best-known member of the family. It lives from mean tide level to the level of high water of spring tides in European saltmarshes (Seelemann, 1968b). It therefore spends most of the time in air, and has lost the gill although it retains an osphradium (Fretter & Graham, 1962). The adults are tolerant of violent salinity extremes, but show only a slight ability to osmoregulate (Seelemann, 1968b; Little & Andrews, 1977). They remain hyperosmotic to the external medium over the range from fresh water to 200% sea water, and it has been suggested that this may be an adaptation to allow them to take up water osmotically from the substratum, since their coverage by sea water is minimal (Little & Andrews, 1977). The urine produced is slightly hypo-osmotic, and is therefore at least partly responsible for maintenance of a high osmotic pressure of the blood. The kidney is very large, ramifying between the organs of the body, and contains both reabsorptive cells and cells which produce excretory vacuoles.

Assiminea grayana lays benthic egg capsules in the mud high up on saltmarshes. These have a great

Fig. 4.11. *Pseudocyclotus* sp. (family Assimineidae) from Papua New Guinea. This terrestrial form is common on the undersides of leaves, with helicinids, in limestone areas. Note the very short tentacles in comparison to those of most other terrestrial prosobranchs. Shell height 7 mm.

tolerance to environmental salinity, but survive best
when they remain in the salinity in which they were
first laid. However, they produce normal veligers even
when laid on freshwater substrata (Seelemann,
1968*b*).

Very few terrestrial assimineids have been examined
in detail. As with *A. grayana*, they have no gill but have
retained the osphradium (Thiele, 1927). Their habitat
has not been described for many species, but *Pseudocyclotus* sp. (Fig. 4.11) was collected from the leaves
of bushes in limestone areas of Papua New Guinea,
where it was found with small helicinids (Little
& Andrews, 1977). This habitat is seldom utilised by
terrestrial prosobranchs, most of which are found in
forest leaf litter, presumably because it renders snails
liable to desiccation. The kidney of *Pseudocyclotus*
is large, as in *A. grayana*, and produces hypo-osmotic
urine. The osmotic pressure of the blood in active
snails was 103 mOsm, compared with 152 mOsm for
A. grayana living on a freshwater substratum (Little
& Andrews, 1977). The value for *Pseudocyclotus* is
higher than that found in terrestrial cyclophorids and
helicinids (Tables 4.3 and 4.4), which are thought to
have a freshwater origin, and together with the known
distribution of assimineids in estuaries suggests
a brackish water origin for the terrestrial forms.

4.1.5 The origins of terrestrial prosobranchs – summary

At this point it is appropriate to summarise the
conclusions reached about the origins of terrestrial
prosobranchs. The family Helicinidae has been derived
from the freshwater, and not from the marine representatives of the family Neritidae; or at least from some
similar freshwater ancestors. This derivation is supported
by similarities in the ionic composition of the blood of
Helicinidae and freshwater forms, and by the production
of a hypo-osmotic urine. The family Cyclophoridae is
probably related to the freshwater Viviparidae and the
amphibious Ampullariidae, and there are no related
marine groups. The Cyclophoridae have blood which
has a very low osmotic pressure, and produce a very
dilute urine in the same way as freshwater forms. The
family Pomatiasidae, in contrast, is related to marine
Littorinidae, and there are no freshwater members of
the superfamily. This fact, together with the evidence of
a high osmotic pressure of the blood, and their inability
to produce a hypo-osmotic urine, shows that the
Pomatiasidae must be derived from marine ancestors.
Finally, the family Assimineidae seems to have produced
terrestrial species from brackish water ancestors.

4.2 The pulmonates

In contrast to the prosobranchs, the pulmonates
form a very uniform group, being characterised by
a vascularised mantle forming a lung which is closed
except for the pneumostome, and which has the rectum
and ureter opening near it. Their mantle cavity therefore
constitutes a true lung much more than do the equivalent structures in the prosobranchs. In general, the
Basommatophora, i.e. those with eyes at the base of
the tentacles, are found in the marine intertidal zone

and in fresh water, while the Stylommatophora, with
eyes at the tip of the tentacles, are terrestrial forms
(Table 4.7). A third small group, the Systellommatophora,
contains slug-like forms that inhabit marine, freshwater
and terrestrial regions.

4.2.1 The Basommatophora
The Ellobiidae

The basommatophoran family Ellobiidae was
regarded by Morton (1955*b*) as containing the most
primitive of the present-day pulmonates. The family
has representatives not only in the littoral zone but in
estuaries and on land. Morton also suggested that the
ellobiids probably arose from the prosobranchs close
to the point at which the opisthobranchs arose, and
that besides this they may have provided the stock
from which the Stylommatophora have been derived.
The family is therefore of much interest and must be
considered in some detail.

Morton (1955*a*) recognised four ecologically
defined groups of the Ellobiidae, occupying supratidal
and estuarine habitats, intertidal and crevice habitats,
terrestrial habitats fringing the coast, and inland terrestrial
habitats. These ecological divisions correspond to some
extent with the arrangement of the various genera, but
when it comes to the overall evolutionary trends within
the family, it is found that each genus has a 'mosaic or
blend of basal and more advanced characters' (Morton,
1955*a*). Thus while one genus such as *Ovatella* (including
the common *Ovatella myosotis* of European saltmarshes)
has an 'advanced' character in the closed external seminal
groove, it also shows 'primitive' characters such as the lack
of internal reabsorption of the shell, and a lack of muscular
development of the stomach. As another example,
Carychium (including the minute *Carychium minimum*
found in European beechwoods) shows advanced features
in the development of special pedal glands and concentration of the nervous system, but primitive features in
the unspecialised penis and the lack of a transverse pedal
groove. It is therefore necessary to consider the general
biology of a number of different ellobiids before discussing their relevance to the movement of pulmonates on to
land.

The supratidal and estuarine category contains
most of the forms found in Europe and North America.
Typical of these is *Ovatella myosotis* (Fig. 4.12), found
high up on saltmarshes, often under thick vegetation. In
this position it is only infrequently covered by the tide,
and like other ellobiids it breathes mainly through a lung.
As in most Basommatophora this lung is not widely open
to the exterior, as is the mantle cavity of most prosobranchs. Instead, it opens through a small muscular
sphincter or pneumostome. *O. myosotis* can withstand
very large changes in the salt concentration of the mud
on which it lives – from fresh water to a salinity of
90$^0/_{00}$ (Seelemann, 1968*a*). Like most estuarine
molluscs, it does not osmoregulate except at low
salinities, but it does maintain its blood very hyperosmotic to all media. This unusual situation resembles
that in the prosobranch *Assiminea grayana*, found in
similar habitats, where increased osmotic pressure may

Table 4.7. *The distribution of the pulmonates (after Hubendick, 1978; Solem, 1978)*

Taxon	Marine	Fresh water	Terrestrial
Order Basommatophora			
Superfamily Ellobiacea	Ellobiidae Otinidae	–	Ellobiidae
Superfamily Amphibolacea	Amphibolidae	–	–
Superfamily Siphonariacea	Siphonariidae Trimusculidae	–	–
Superfamily Lymnaeacea	–	Chilinidae Latiidae Acroloxidae Physidae Lymnaeidae Ancyloplanorbidae	–
Order Systellommatophora			
Superfamily Onchidiacea	Onchidiidae	? Onchidiidae	Onchidiidae
Superfamily Soleolifera	–	–	Veronicellidae Rathousiidae
Order Stylommatophora			
Suborder Orthurethra	–	–	e.g. Vertiginidae Achatinellidae
Suborder Mesurethra	–	–	e.g. Clausiliidae Strophocheilidae
Suborder Sigmurethra	–	–	e.g. Succineidae Athoracophoridae Achatinidae Endodontidae Arionidae Zonitidae Limacidae Helicidae

Fig. 4.12. *Ovatella myosotis* (family Ellobiidae), crawling on saltmarsh mud. The kidney is visible through the shell as a white, wedge-shaped area. Shell height 7 mm.

be related to ability to take up water from the sub-stratum (see p. 47). It is not known how *O. myosotis* maintains an osmotic difference from external media. The kidney is a simple sac lined by cells which contain excretory vesicles, and which may also be able to reabsorb salts and water, but it has no ureter (Delhaye & Bouillon, 1972*a*). Certainly there is no part of the kidney especially adapted for salt and water reabsorption. In contrast to the prosobranchs, but like all other pulmonates, *O. myosotis* is an hermaphrodite, each individual being able to produce both eggs and sperm. The reproductive system is relatively simple, although copulation occurs so that fertilisation is internal as in all other pulmonates. The snails lay eggs, and no ellobiids have developed viviparity, although this is found in some of the Stylommatophora.

In other parts of the world different genera occupy equivalent positions in saltmarshes. New Zealand has *Ophicardelus costellaris* (Fig. 4.13) which lives on the upper parts of saltmarshes, especially on *Juncus*. In North America the commonest species is *Melampus bidentatus*. This species is well adapted to breathing air, and when desiccated can survive a loss of nearly 80% of its body water (Price, 1980). Although it can also tolerate submersion, its rate of oxygen consumption is reduced under water, in contrast to the situation in littoral proso-branchs such as species of *Littorina* (McMahon & Russell-Hunter, 1981). It can also withstand a wide temperature range, including freezing, and generally appears in terms

of physical tolerance to be well adapted to a terrestrial life. However, its reproductive habits link the species firmly to a marine littoral existence. On spring tides it lays gelatinous egg masses on the surface of the marsh, from which hatch veliger larvae (Russell-Hunter, Apley & Hunter, 1972). Hatching takes place about 14 days after laying so that it coincides with the next series of spring tides. The period in the plankton also lasts about 14 days, so that the larvae settle on to the marsh on the third consecutive series of spring tides. The life history is therefore perfectly matched to a marine, high-shore life.

The intertidal and crevice-dwelling category is well exemplified by *Leucophytia bidentata* (Fig. 4.14*a*), found typically in crevices on rocky shores. Normally this species breathes through the lung, but it can also use a lobe of the mantle skirt to breathe under water, and in this it resembles several other types of Basommato-phora. Morton (1955*b*) considered this species to have more advanced characters than *Ovatella myosotis*, and these are presumably related to its adoption of a some-what more specialised habitat. Most of the internal whorls of the shell are dissolved away to produce a single internal cavity rather similar to that in the Neritacea (see p. 36). In the digestive system the stomach shows more muscular development than *O. myosotis*, and the reproductive system has become more specialised.

The two terrestrial categories show little relation-ship to one another, and have arisen independently. Species of *Pythia*, relatively large coastal terrestrial forms (up to 2 cm shell length), are found in tropical forest bordering the coastline in the Indo-Pacific. Almost nothing is known about their way of life or physiology.

Fig. 4.13. *Ophicardelus costellaris* (family Ellobiidae) from high up in the *Juncus* region of a New Zealand saltmarsh. The faecal string can be seen trailing behind the snail. Shell height 10 mm.

Species of *Carychium*, in contrast, have a shell length less than 2 mm, and are found inland in Europe in leaf litter and other moist microhabitats on calcium-rich soils. Their structure (Fig. 4.14*b*) shows little apparent adaptation to a terrestrial life, but this is probably due to the tendency to live where the relative humidity is always high, and even small size does not lead to desiccation. However, the genus does have one major reproductive modification: the eggs are surrounded by a leathery egg capsule, from which the young snails hatch directly (Morton, 1954).

From this brief consideration of some of the ellobiids, it can be seen that the change from an intertidal life to a terrestrial one can be carried out with very little structural change. What is more necessary is sufficient physiological adaptation to allow the animals to tolerate the harsh conditions of the upper shore, and in this case particularly the very high and very low salinities of estuarine saltmarshes. This must be coupled with behavioural responses allowing them to select the most appropriate microhabitats. Such adaptations have allowed many ellobiids to become relatively independent of the sea, except for their mode of reproduction. However, the development of an egg protected from external conditions by a tough capsule and provided with sufficient food reserves to allow the embryo to develop into a young snail is a prerequisite for total independence from the sea, and very few present-day ellobiids have this type of egg. Nevertheless, if the ellobiids or related forms gave rise to the ancestors of the Stylommatophora, it is the extensive mechanisms of physiological tolerance developed in the upper part of saltmarshes which has preadapted them for terrestrial life. The ancestry of the Stylommatophora thus contrasts markedly with that of

most terrestrial prosobranchs, and the consequences of this different ancestry are discussed at the end of this chapter.

The freshwater Basommatophora

The three most primitive freshwater basommatophoran families are the Chilinidae (South America), Latiidae (New Zealand) and Acroloxidae (North America) (Hubendick, 1978). These all have specific adaptations to fresh water in the development of a separate region of the kidney specialised for the reabsorption of salts and water (Delhaye & Bouillon, 1972*a*). The kidney of *Chilina* is a simple sac with internal lamellae that are lined with reabsorptive cells near the kidney opening. In *Latia* the excretory and reabsorptive parts of the kidney are more distinctly separated, and in *Acroloxus* the reabsorptive region forms a long duct. A ctenidium is not present in any of the three families, but while species of *Chilina* have no gill at all, those of *Acroloxus* have a secondarily developed external gill or pseudobranch. This secondary development of gills suggests that the groups have re-invaded water from an air-breathing existence, although it has also been postulated that gills could have been lost in relation to small size (Fretter, 1975), in which case an intermediate air-breathing stage does not have to be postulated. If the ellobiids really do represent a primitive state within the Basommatophora, however, air breathing has been a fundamental characteristic of the group, and it will be assumed here that the invasion of fresh water was carried out by snails that were fundamentally air breathing.

Many of the more advanced freshwater Basommatophora such as the genera *Planorbis* and *Ancylus*

Fig. 4.14. The general structure of two ellobiids. (*a*) *Leucophytia bidentata*, a high-shore marine species usually found in crevices. The stomach and genital ducts are to some extent specialised.

(*b*) *Carychium tridentatum*, a very small species found on land. Here the genital system is primitive, but the stomach is simplified, possibly because of the animal's small size. From Morton (1955*a*).

(a)

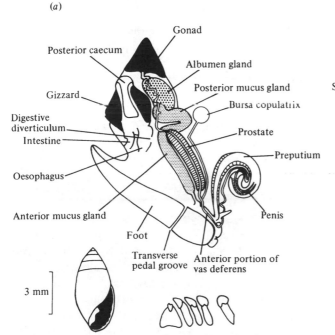

(b)

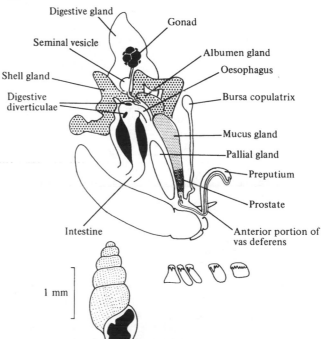

have also developed secondary gills, but others such as
Lymnaea rely on the skin and the lung for oxygen
uptake. The origins of secondary adaptations to an
aquatic life are discussed in more detail in Chapter 10.
Here it is appropriate to note that the relative importance of lung, gill and cutaneous respiration varies
markedly from one species to another, and that these
variations can be related to the animal's life style.
A comparison in this respect was made for *Lymnaea
stagnalis* and *Planorbis corneus* by Jones (1961). In
both species the importance of pulmonary respiration
increases as the external oxygen tension falls (see Fig.
4.15). However, at all oxygen tensions the lung is more
important in *L. stagnalis* than in *P. corneus* reflecting
the development in the latter, but not in the former,
of an accessory gill.

Like the three primitive freshwater families, the
advanced freshwater Basommatophora – Lymnaeidae,
Physidae and Ancyloplanorbidae – have kidneys which
are functionally divided into an excretory portion and
a part specialised for reabsorption (Delhaye & Bouillon,
1972*a*). Kidney function and osmoregulation have been
investigated in detail only in the Lymnaeidae. In
Lymnaea stagnalis the osmotic pressure of the blood
is 127 mOsm (Greenaway, 1970), while that for the
amphibious *Lymnaea truncatula* is 148 mOsm (Pullin,
1971). This compares with a range of 68–117 mOsm
in freshwater and amphibious prosobranchs (Tables 4.3,
4.4). The renal system of *L. stagnalis* appears to function
in a manner similar to that of the freshwater prosobranchs
(van Aardt, 1968). Urine is produced by ultrafiltration
through the heart into the pericardial cavity, and this
filtrate is then modified in the kidney by the reabsorption of ions. The rate of urine production is similar to
that in prosobranchs, being 0.93–1.84 μl g^{-1} min^{-1}. This
similarity would certainly be expected, since the two
types have much in common in terms of osmotic
requirements.

Fig. 4.15. Oxygen consumption in the aquatic basommatophoran
Lymnaea stagnalis. Open circles show total oxygen uptake;
closed circles show oxygen taken up by the lung; triangles show
cutaneous uptake. Unbroken lines are regression lines and
dotted lines show confidence limits. At low oxygen tensions the
lung is most important in oxygen uptake, but as oxygen tension
rises the relative importance of the lung declines while the
importance of cutaneous uptake increases. From Jones (1961).

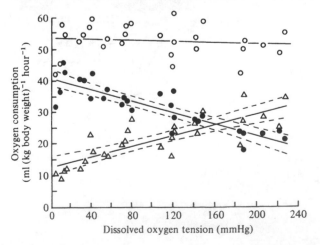

The tolerance to desiccation shown by freshwater
pulmonates has not been examined in great detail, except
where the snails are of economic importance, but some
effort has been concentrated on the vectors of schistosomiasis. For example, *Tropicorbis centimetralis*, one of
the vectors of *Schistosoma mansoni* in Brazil, has been
found to live for up to a year when placed on soil which
was gradually dried out (Olivier & Barbosa, 1956).
Besides this capacity of tropical forms to withstand
long periods of desiccation, temperate forms can also
tolerate a certain degree of dehydration. *Planorbis
corneus*, for example, can live for over 40 days in about
20% RH, and during this time the osmotic pressure of
the blood doubles (Klekowski, 1963). Further examples
are given by Machin (1975). Even more extreme tolerance
of desiccation is exhibited by the Stylommatophora, as
will be seen later.

The marine Basommatophora

Finally, the Basommatophora also contains
a number of marine families apart from the high-shore
ellobiids considered earlier. These include the small
Otina otis found in crevices high up on the shore. The
family shows many primitive characters, and may be
grouped with the Ellobiidae (Hubendick, 1978).
Another family, the Amphibolidae, contains snails
that live on estuarine sand and mud flats. The members
of the Siphonariidae, in contrast, are limpet-like in form,
and live on rocky shores. Since in some sense all these
species can be said to have recolonised the marine intertidal zone from a semi-terrestrial existence, discussion
of their adaptations is deferred until Chapter 10.

4.2.2 The Systellommatophora

Although this group is in no way intermediate
between the Basommatophora and the Stylommatophora, it is convenient to consider it here. It contains
three families: Onchidiidae, Veronicellidae and
Rathousiidae. All have a slug-like form, and although
at one time *Onchidella celtica* was placed in the
opisthobranchs (Fretter, 1943), there is now little
doubt that all three families have a pulmonate grade
of structure (Solem, 1978). Members of the Onchidiidae
mostly live in the marine intertidal zone, and more is
known about them than about the other families. In
Europe, *O. celtica* is found widely distributed in restricted
areas of southwestern England and the western coast of
France. It lives in crevices above mid-tide level, and
makes feeding excursions to browse on encrusting algae
when uncovered by the tide. Individuals of this and
other species (Arey & Crozier, 1921) home to specific
sites. During activity air is taken in through the lung,
which opens posteriorly (Fretter, 1943), but cutaneous
respiration has also been shown to be important (Arey,
1937). The kidney also opens posteriorly, but its involvement in nitrogen excretion has yet to be assessed.
Onchidella verruculatum excretes most of its nitrogen
as ammonia, but also produces urea (Deshpande,
Nagabhushanam & Hanumante, 1980). In *O. celtica*,
egg masses are laid on the shore and the veliger stage is
passed inside the egg capsule, but in other species such

as the New Zealand *Onchidella nigricans*, the adult of
which is shown in Fig. 4.16, a free-swimming veliger is
produced (Morton & Miller, 1973). Reproductive
processes are probably under hormonal control
(Nagabhushanam, Deshpande & Hanumante, 1981).

Although most onchidians are marine, some
species have been described from fresh water and the
land. Plate (1894) described *Onchidium montana* from
specimens found living on rocks and tree trunks high
up on an island in the Philippines. Another species,
Onchidium typhae, was recorded from between the
leaves of *Typha elephantina* in fresh water, and others
have been recorded as semi-terrestrial. Unfortunately
all these records are very old, and new descriptions of
the conditions in these habitats are needed. However, it
is certain from Plate's description that the onchidians
have produced at least one terrestrial representative.
Although there are no details of physiological adaptations, it seems likely that a parallel may be drawn with
the ellobiids, which have produced terrestrial forms
without any gross structural or physiological modifications of their organisation.

The other two families contained within the
Systellommatophora are entirely terrestrial. Little is
known of their biology or their relationships (Solem,
1978). The Veronicellidae are herbivores found in the
tropical parts of America, Africa and Asia (Runham
& Hunter, 1970). They have no mantle cavity and

presumably respire through the body surface. The
Rathousiidae are found in Indonesia – which is also
the main centre of the Onchidiidae (Hoffmann, 1929) –
and are carnivorous (Solem, 1974). Many of the
characteristics of these terrestrial forms are related to
their adoption of the slug habit. Their lack of a shell,
and consequent susceptibility to desiccation, has limited
them to humid habitats such as tropical and southern
temperate forests.

4.2.3 The Stylommatophora

The basic organisation of the stylommatophoran
body is fairly uniform. Typical external appearance is
illustrated in Fig. 4.17. Stylommatophorans are all
terrestrial, almost all breathe through a lung, and they
are all hermaphrodites. They are characterised by the
presence of two pairs of tentacles, the posterior of
which bears eyes and can be withdrawn into the body.
Despite this basic uniformity, however, the stylommatophorans present a variety of external form. There
are shelled species, with the shell ranging from high
spired to discoidal, and there are several lines of terrestrial slugs in which the shell has been reduced. Some of
the shells are thick and calcareous, while others are
thin and horny. Most land pulmonates do not close the
aperture of the shell except during aestivation, but one
group, the Clausiliidae, has a kind of 'sliding door' or
clausilium (Fig. 4.18) which closes the aperture. Snails

Fig. 4.16. *Onchidella nigricans* (family Onchidiidae), a marine
intertidal systellommatophoran slug from New Zealand. Here it
is crawling on the mud surface, but it is commoner on purely rocky shores, and retreats to a 'home' in rocky crevices as the shore begins to dry up later in the tidal cycle. Length 15 mm.

of the genus *Thyrophorella* have a flap of shell which serves the same function, and many species have ridges which partially block the aperture (Fig. 4.18). Besides this variety of shell form, there is a wide variety of habit from the common herbivorous type to several specialised carnivores. Members of the Stylommatophora are also found in a great range of habitats. They occur predominantly in humid tropical forests and in humid niches in the temperate zones, but are also found as far north as Greenland, and in hot deserts such as the Sahara. Thus although they are not as well equipped for life on land as some of the arthropods, they have representatives in almost every terrestrial environment.

Their classification is normally based upon the organisation of the renal system. Although some authorities consider that there should be five major subdivisions, Solem (1978) has reduced this to three, and his scheme will be followed here, although not all his rankings are adopted (Table 4.7). A comparison of four classification schemes is given by Fretter & Peake (1978). The Orthurethra possess an elongate kidney, of which the distal portion, or ureter, is specialised for reabsorption and opens near the pneumostome (Delhaye & Bouillon, 1972*b*). Orthurethrans are mostly small snails which live in a variety of habitats. Many are only a few millimetres high and have high-spired shells. Some of the largest are the Hawaiian tree snails of the family

Achatinellidae (Solem, 1974). Although abundant, almost nothing is known about their physiology.

In the Mesurethra the kidney opens directly into the mantle cavity with either no ureter or a very short one. The group contains two superfamilies of very different aspect. In one of these is the family Clausiliidae, which contains small snails represented in Europe by many species. *Clausilium bidentata*, for example, is common in moist leaf litter, especially on chalklands. In the other superfamily is the family Strophocheilidae, which consists of large snails inhabiting tropical rain forest in South America.

In the Sigmurethra the anatomy of the reabsorptive ureter is very variable, but normally it is considerably elongated and is reflected to form a primary and a secondary arm (Fig. 4.21). Delhaye & Bouillon (1972*b*, *c*) viewed the Sigmurethra as continuing the trends of kidney evolution seen in the Mesurethra, with no clear division between the two orders. In contrast, the members of the Orthurethra constitute a separate evolutionary line. The suborder Sigmurethra contains the bulk of stylommatophorans. It therefore includes all the commonest land snails and most of the slugs, and no detailed subdivision will be attempted here. It must be stressed, however, that the members of the group are very varied and that many have undergone striking changes from the hypothetical basic form. According

Fig. 4.17. *Xesta* sp., a typical shelled stylommatophoran pulmonate from Papua New Guinea. The large tentacles bear eyes at the tip. The pneumostome is wide open, allowing air to be drawn into the lung. Shell diameter 2.5 cm.

to Solem (1978), the group contains the family
Succineidae, which some would place in a separate
order, the Heterurethra, because the kidney and ureter
lie obliquely across the mantle cavity. The succineids
are adapted to life in very moist habitats, and although
some authorities have considered their organisation
a primitive one (e.g. Rigby, 1965; Minichev & Slavoshevskaia, 1971), Solem's (1978) arguments that the
adaptations are secondary are convincing. Also included
in the Sigmurethra by Solem is the family Athoracophoridae, containing slugs with a very long and much
folded ureter. Besides these, the sigmurethrans have
produced many more slug types which have radiated
in moist environments, as well as snails with an amazing
tolerance of desiccation which can live in deserts. In
comparison with the sigmurethrans, very little is known
about the other two orders, and discussion will of
necessity concentrate upon the most advanced group.
For these, as for all terrestrial pulmonates, the most
important problem is undoubtedly that of prospective
water loss, and this will be considered first.

Many snails are protected to some degree by
a relatively impermeable shell covering much of the
body; but water loss from the exposed parts can be
very rapid, and it is not surprising to find that stylommatophorans are active only in high humidities.
Behavioural patterns ensure that they emerge from their
hiding places after rain or dew has fallen, or when, at
night, the relative humidity rises (Hunter, 1964). This
response, which has often been considered a direct
response to increasing relative humidity, often appears
to be provoked in reality by falling temperature. Slugs,
for instance, become active in falling temperatures,

while their activity cannot be correlated with any
particular degree of humidity (Dainton, 1954). Other
reactions such as movement away from strong air
currents tend to keep pulmonates in areas of high
humidity (e.g. Runham & Hunter, 1970).

Nevertheless, terrestrial pulmonates often experience desiccating conditions, and lose water both by
evaporation and in the mucus secreted by the pedal
glands. Indeed, in the slugs this water loss can be
immense: rates of loss from inactive slugs at 45% RH
were between 3 and 5% of the body weight per hour
(Dainton, 1954). Active slugs lost up to 17% of their
body weight through mucus in only 40 min, even in
a saturated atmosphere, and such large losses then led
to reduced activity.

During activity it is thought that water evaporates
from molluscan skin much as it does from a free water
surface, and that mucus does not reduce the rate of
water loss (Machin, 1964). Eavporative water losses are
in fact positively related to the rate of mucus secretion,
because the mucus keeps the skin moist, and therefore
allows evaporation to continue. In *Helix aspersa* (Machin,
1964) the mucus collects mainly in grooves in the skin,
and is thinly spread over the raised areas; but incipient
drying out of the skin induces further mucus production,
and local muscular undulations then spread the mucus
out from the grooves.

High rates of water loss mean that land pulmonates
must be able to tolerate large changes in the water content
of their bodies. Some of these fluctuations have been
examined in detail by Howes & Wells (1934), and are reviewed by Burton (1983). In *Helix pomatia*, for example,
the body weight was found to fluctuate periodically by
as much as 50% (Fig. 4.19). These fluctuations occurred
in constant humidity and temperature, and even in
100% RH, and did not appear to be correlated with
changes in external conditions. The snails did tend to
aestivate when their weight was low, however, and to be
active when weight was high. Similar types of fluctuations

Fig. 4.18. Some adaptations of stylommatophoran pulmonates
for protecting the tissues by closing the aperture. (*a*) *Thyrophorella*, showing the flap of shell which closes the aperture.
(*b*) *Clausilium* sp. with the shell cut open to show the
clausilium, a flap which is hinged to the columella and can slide
across the aperture. (*c*) *Endodonta*, showing well-developed
ridges which reduce the area of the aperture. (*a*) and (*c*) after
Thiele (1931).

Fig. 4.19. Weight records of two young specimens of *Helix
pomatia*. Open circles show active snails; closed circles show
aestivating ones; circles enclosing dots show snails withdrawn
into their shells. Despite the fact that the snails had continuous
access to food and water, they showed wide and fairly regular
fluctuations in weight. In general the snails were heaviest when
they were active, and had a low weight during aestivation. From
Howes & Wells (1934).

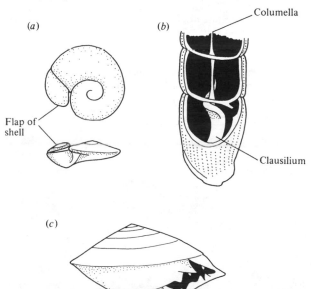

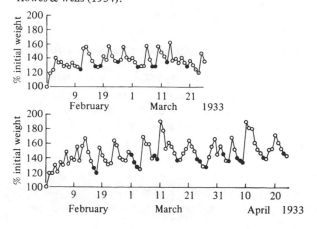

were found in other species by Blinn (1964). Presumably these weight changes could be accounted for by a number of variables which make up a 'weight budget'. Evaporation and loss of mucus have already been emphasised. There will also be loss in the faeces and through the kidney, and the total loss must be balanced by uptake in the diet and by any absorption of water that occurs through the skin. Uptake from a saturated atmosphere has been demonstrated for slugs by Dainton (1954), and there is evidence that at least some snails normally drink (Blinn, 1964). To complicate this budget, some species carry round a water store within the mantle cavity (Blinn, 1964; Smith, 1981). In the American family Polygyridae this water accumulates when the snails are hydrated, and decreases in volume during dehydration. Its osmotic pressure was found always to remain below that of the blood (Smith, 1981), suggesting that it may originate as urine. The pallial water is evidently important as a reserve in times of desiccation, and it is interesting that it has also been observed in other snails such as *Otala lactea* and *Helix pomatia* (Blinn, 1964). If it should be shown to be of importance in a wide variety of stylommatophorans, it must count as a very significant factor in allowing them to cope with terrestrial conditions (Solem, 1978).

Terrestrial snails may interrupt their normal activities when conditions become unfavourable by withdrawing into their shells and sealing the aperture with a secretion from the mantle. In winter this process is referred to as hibernation, and the aperture is then usually closed with a calcareous plate, the epiphragm. In summer the withdrawal is often of shorter duration, the aperture is closed by a thin film of dried mucus, and the process is called aestivation. The adaptations for aestivation are extreme in some cases, as shown for a number of helicid snails. In *Helix aspersa*, Machin (1966) showed that the rate of water loss during aestivation was very low. Part of this reduction from the normal rate was due to the film of mucus, but the major part was related to active retention of water by the mantle epithelium. In *Otala lactea* still lower rates of water loss were found, with a mean of $16 \mu g \, cm^{-2} hour^{-1} (mmHg)^{-1}$ (Machin, 1972). This rate is comparable with that of many insects (see Table 8.4). The barrier to water movement is not understood, but appears to be near the apical surface of the cells of the mantle epithelium (Appleton, Newell & Machin, 1979). In some way this barrier must prevent water from being mobilised from underlying parts of the cells.

Similarly reduced rates of water loss have now been reported from other snails (e.g. Riddle, 1975), and the whole problem of water balance in pulmonates has been reviewed by Machin (1975). So far the ability to reduce water loss to extremely low levels has been demonstrated only in the family Helicidae, and one further example shows the amazing desiccation tolerance of the family. *Sphincterochila boisseri* is found in the Negev desert where there may be a year between rains. The rate of water loss when the snails are completely undisturbed is so low that if a 4-g snail could tolerate losing half its body water, and could also with-

stand the high temperature, it could survive for at least 4 years (Schmidt-Nielsen, Taylor & Shkolnik, 1971). The rate of water loss across the mantle in this species is so low that even the loss across the shell itself is a significant fraction of the total loss (see Machin, 1975). The limit to aestivation is therefore probably set by the amount of food reserves stored in the snail's body rather than its water content. This species and another, *Sphincterochila prophetarum*, are active for a maximum of less than 30 days in the year, during the winter rains (Shachak & Steinberger, 1980; Steinberger, Grossman & Dubinsky, 1981, 1982). In this time they feed on algae growing on the desert soil, thus increasing their carbohydrate and protein content. They also mate during this period and lay eggs about a month later.

During the dry months the surface temperature of the soil in the Negev may rise to 65 °C (Schmidt-Nielsen, Taylor & Shkolnik, 1971). When aestivating, *S. boisseri* withdraws its tissues from the lower, body whorl, which rests on the ground, so that they are effectively insulated from the ground by a large air space (see Fig. 4.20). Of equal importance is another factor, dependent upon the very thick, white, calcareous shell. This produces a very high reflectivity so that about 95% of the incident radiant energy is reflected. The net result is that although the exposed soil may be at 65 °C, the snail's tissues never reach a temperature higher than 50 °C, which is within the tolerance limits for the species.

The termination of aestivation in helicids seems to be controlled by factors at least partly different from those acting on amphibious prosobranchs such as species of *Pomacea* (see p. 39), which require the presence of liquid water, or severe mechanical disturbance. In *Otala lactea* a rise in humidity of the air to 85% RH will trigger arousal (Herreid, 1978). Although physical stimulation enhances arousal and increases the rate of water loss (Machin, 1975), neither this nor the presence of liquid water seems to be essential.

Fig. 4.20. Temperature distribution and heat flow in and around the desert pulmonate *Sphincterochila boisseri*. The direction of heat flow is indicated by long dashes. Solar radiation is shown by short dashes. The heavily stippled area shows that part occupied by the tissues, which remains much cooler than the desert surface. Redrawn from Schmidt-Nielsen, Taylor & Shkolnik (1971).

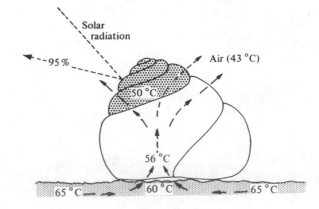

The observations on aestivation so far discussed all refer to sigmurethrans. The situation in other orders is probably very different, but unfortunately we have very little information on the subject. The tropical Strophocheilidae (Mesurethra) also aestivate, but they produce no epiphragm or film of mucus, and during aestivation the pneumostome remains open (de Jorge, Petersen & Ditadi, 1970). This suggests that the whole process of aestivation may bear little relationship to that in higher forms.

It is to be noted that during aestivation species of *Sphincterochila* at least do not use any type of evaporative cooling, and indeed any such device would cause too rapid a loss of water. During activity, however, both

slugs and snails may be cooled quite considerably by evaporation. The slug *Arion ater*, for example, may be some 12–13 °C below the external temperature at 24 % RH (Hogben & Kirk, 1944). Such a low humidity would certainly be unusual in the slug's natural environment, but significant cooling occurs in more normal conditions, and similar effects have been found with active snails. Evaporative cooling must, of course, be limited in extent even in species from damp environments because of the simultaneous need for water conservation.

Closely involved in water loss are the characteristics of kidney function and nitrogen excretion. Unfortunately, renal mechanisms have been examined only in the sigmurethrans, and this account must therefore be limited to them. The anatomy and function of the kidneys of the genera *Helix* and *Achatina* have been investigated by Vorwohl (1961), Martin, Stewart & Harrison (1965), Skelding (1973a, b) and Newell & Skelding (1973). Kidney anatomy is shown in Fig. 4.21a. The lumen of the kidney is connected to the pericardial cavity by a reno-pericardial canal. Unlike the situation in basommatophoran pulmonates and in prosobranchs, however, urine is not formed by ultrafiltration through the heart wall, although fluid can at times pass through the reno-pericardial canal (Vorwohl, 1961). Instead, urine appears to be produced by filtration directly into the kidney lumen, possibly via the intercellular spaces in the renal epithelium (Skelding, 1973a). The kidney sac also contributes concretions to the urine which are concerned in nitrogen excretion, as discussed later. In the primary and secondary ureters, salts and water are reabsorbed and pH is modified (Fig. 4.21b). A copious flow of hypo-osmotic urine is produced when the snail is in wet surroundings, but all the initial filtrate can be reabsorbed in dry conditions (Vorwohl, 1961; Skelding, 1973b). As discussed later, there are important implications in the differences between this stylommatophoran system and that in terrestrial prosobranchs, the pulmonate system being

Fig. 4.21. The renal system in the genus *Helix*. (a) The general anatomy, showing the reflexed ureter. (b) The physiological processes involved. Iso-osmotic primary urine is formed by some type of ultrafiltration into the kidney sac, and exchange of salts and water in the ureter converts this to hypo-osmotic urine with an adjusted pH. Uric acid is produced in the kidney sac.
(a) Partly after Skelding (1973a); (b) redrawn from Vorwohl (1961).

(a)

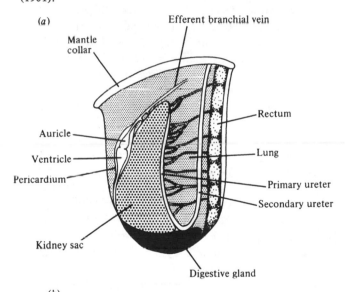

(b)

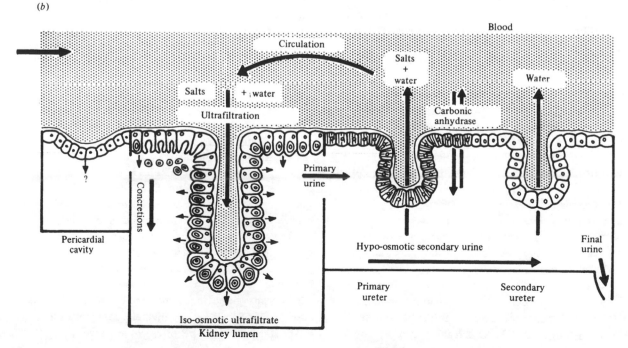

much better adapted for terrestrial life because of the spatial separation of the sites of uric acid production and fluid reabsorption. However, even the kidney of stylommatophorans does not regulate the volume and composition of body fluids particularly efficiently. In *Helix pomatia* the blood volume may vary by a factor of four, and the associated variation in sodium concentration is from 46 to 129 mM/litre (Burton, 1965, 1983). The concentrations of other ions also vary tremendously, and although some changes can be correlated with feeding and aestivation, it is difficult to predict the ionic composition of the blood of any individual snail. Some of the average values for a wide variety of pulmonates are given by Burton (1968, 1983).

Many stylommatophorans are thought to be predominantly uricotelic, or rather purinotelic, with most of their waste nitrogen being either stored or excreted in the form of uric acid, xanthine and guanine (e.g. Florkin, 1966). The renal cells of *H. pomatia*, for example, form and excrete large concretions containing calcium urate, uric acid and lipid (Bouillon, 1960). Very little is really known about the relative importance of storage and excretion, but some short-lived pulmonates do not appear to produce excreta, and presumably store all their nitrogenous waste (Badman, 1971). Early work suggested that excretion of urea was very important, and although this work has for the most part been shown to be in error because of the analytical techniques involved, some recent authors have shown that urea is synthesised by terrestrial pulmonates (Horne & Barnes, 1970), and that it may accumulate to high levels during aestivation (Horne, 1971). In the slug *Limax flavus* 59% of the excreted nitrogen was in the form of urea, 41% as purines, and no ammonia was formed (Horne, 1977). There still seems to be disagreement about the amount of urea in the urine of species of the mesurethran pulmonate genus *Strophocheilus* (de Jorge, Petersen & Ditadi, 1969; Tramell & Campbell, 1970). The production of urea in large quantities would certainly be an interesting parallel to the vertebrate situation.

One further aspect of nitrogen excretion concerns the liberation of ammonia. It has generally been thought that since ammonia is highly toxic it is not a likely end product of nitrogen metabolism in terrestrial forms. This idea is supported by the fact that during activity ammonia accounts for only about 5% or less of the total nitrogen excreted. During aestivation, however, such snails as *Otala lactea* and *Helix aspersa* have been shown to produce large quantities of gaseous ammonia, which diffuses out through the shell and which may account for 30% of the total excreted nitrogen (Speeg & Campbell, 1968a, b). Loest (1979a, b) has shown that very many species of shelled pulmonates produce ammonia gas during aestivation, whereas all the slugs investigated actually absorbed ammonia from the air if it was available. He supported the hypothesis of Speeg & Campbell (1968a) that ammonia is derived from urea in the mantle and then, at least on some occasions, is involved in enhancing the rate of deposition of calcium carbonate in the shell. This occurs because the ammonia accepts H$^+$ ions from carbonic acid (H$_2$CO$_3$), leaving the CO$_3^{2-}$ to be

precipitated as calcium carbonate. At other times the ammonia is presumably released directly as a gas without influencing shell deposition. Here we have a most interesting example of the utilisation of what is primarily an aquatic end product as a specialised adaptation to life on land, and one which, as will be seen later, also exists in the isopod Crustacea. The diffusion of gaseous ammonia through the shell raises the interesting question of the permeability of the latter to various gases. The molecular diameter of ammonia is not as large as that of carbon dioxide, but it is larger than that of water. On this basis, water should readily diffuse out of the shell if ammonia can do so, but it is apparent that this does not happen. Consideration of this type of problem will recur in the discussion about the physiology of the various types of eggs which are resistant to desiccation, but which are able to respire aerobically (see, e.g. Chapter 9).

One of the major adaptations to terrestrial life seen in the pulmonates is, as we have seen in the Basommatophora, the change to aerial respiration, as indeed the name 'pulmonate' implies. Respiratory metabolism has been studied in a number of terrestrial pulmonate species and varies over a tenfold range (Ghiretti & Ghiretti-Magaldi, 1975). According to Riddle (1978) some part of this variation may be related to habitat: he found that oxygen consumption in desert species of *Rabdotus* was less than that of species of *Helix* both during activity and in dormancy, and suggested that this might be an adaptation to the desert environment. In all terrestrial stylommatophorans oxygen is taken up from a vascularised mantle as it is in the prosobranchs, but in pulmonates the opening of the 'lung' so formed is restricted to a small aperture, the pneumostome. This has allowed a fundamental difference to develop between the function of terrestrial prosobranch and terrestrial pulmonate lungs. While the former are 'diffusion' lungs, the latter are 'ventilation' lungs (Ghiretti & Ghiretti-Magaldi, 1975). That is to say, air is actively drawn into pulmonate lungs by the depression of the floor of the mantle cavity, and gas exchange does not depend solely upon diffusion.

An interesting development of the pulmonate lung has occurred in the slugs of the family Athoracophoridae (Fig. 4.22). These slugs have fine tubules extending into the mantle tissue from the cavity of the lung (Fig. 4.23), forming a superficial resemblance to insect tracheae (Hyman, 1967). For this reason the family used to be named the Tracheopulmonata. Unfortunately there is absolutely nothing known about the physiology of this family.

To consider the development of the lung is, of course, to take into account only one part of the respiratory process. Oxygen obtained from the lung is then carried round the body in the blood, and this carriage is often aided by the pigment haemocyanin. Very little is known of any differences between the haemocyanins of aquatic and terrestrial gastropods (Ghiretti, 1966), but there has been some work on the importance of haemocyanin in buffering the blood in *Helix pomatia* (Burton, 1969, 1983). The pigment contributes little to the stability of the pH of blood in active snails, and over 90% of the buffering is due to calcium and bicarbonate ions. During aestivation, however, when the pH of the blood falls and the bicarbonate ion concentration is also lowered, haemocyanin accounts for about 10% of the buffering capacity. It is also important in increasing the amount of carbon dioxide carried in the blood.

The change from aquatic to aerial respiration necessitates many changes in blood biochemistry, but for molluscs the investigation of acid–base balance is still in the early stages. The topic is reviewed by Burton (1983). One aspect of particular importance is the origin of the carbon dioxide utilised in shell formation (see Wilbur, 1964). This is thought to be derived from

Fig. 4.22. A New Zealand slug from the family Athoracophoridae, with its egg mass. The leaflike pattern on the body is typical. The black ring one quarter of the way down the body is the pneumostome which opens into the 'pseudotracheate' lung. Length approximately 5 cm.

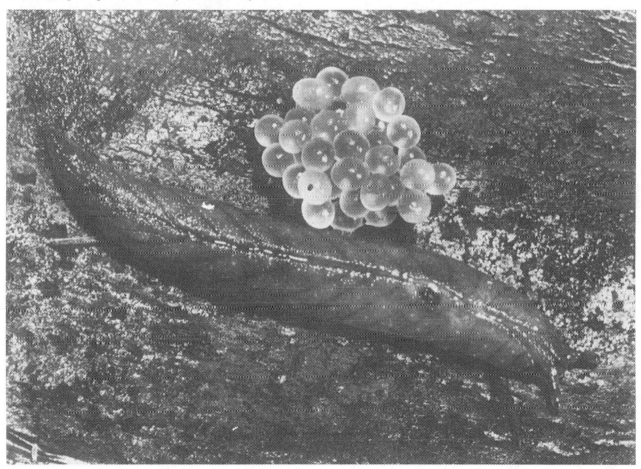

Fig. 4.23. The lung in the stylommatophoran slug family Athoracophoridae. (a) Longitudinal section through the lung of *Janella schauinslandi*, showing the fine diverticula which radiate out from the main cavity of the lung into the tissues. (b) Three hypothetical stages in the derivation of such a lung from the normal pulmonate lung. Redrawn after Plate (1898).

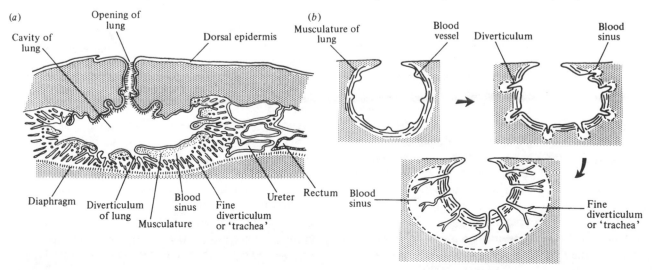

a carbon dioxide–bicarbonate pool within the mantle.
The shell is then formed in a thin layer of fluid between
the mantle and the inner shell surface: calcium carbonate
is precipitated in this extrapallial fluid, probably aided
by the diffusion of ammonia from the blood, as discussed
above. Apart from this effect of ammonia, which is
probably limited to terrestrial forms, no other differences between aquatic and terrestrial snails in the mechanism of shell formation have been described. However,
many terrestrial stylommatophorans have little calcium
in their shells, presumably due to the relative shortage of
calcium on land. Although the picture of land snail
distribution in relation to calcium supply and pH is not
a clear one, the numbers of species and individuals are
usually lower in areas low in calcium and with a low pH
(Peake, 1978). Calcium supply is therefore one of the
major factors affecting land snail diversity and abundance,
and Heller & Magaritz (1983) have shown that the
carbonate, and probably the calcium, used to form the
shell are derived from soil water.

In contrast to the Basommatophora the terrestrial
Stylommatophora have evolved complex behaviour
patterns in relation to reproduction, and in particular
show well-developed courtship (see, e.g. Hyman, 1967;
Duncan, 1975). For example, when two individuals
of *Helix pomatia* meet, and if they are ready for copulation, they raise their bodies up so that the foot soles are
pressed together at the anterior end. Then they exchange
'caresses' with the tentacles and oral lappets, and after
a time each may evert the 'dart sac', and eject a dart
into the partner. This dart is a short, pointed calcareous
shaft, and when it is received in the snail's epidermis
it produces a great deal of stimulation. Eventually, the
penis of each snail is protruded and cross-copulation
occurs: during this process the individuals exchange
spermatophores. In the slugs, copulatory behaviour is
even more complex, and *Limax maximus* is famous for
its copulatory position in mid-air, where two individuals
twine round each other on the end of a mucous strand.
It is somewhat difficult to explain the development of
such complex behaviour patterns in terrestrial forms
rather than aquatic ones. Presumably they prevent
interspecific copulation, but it is difficult to see why
the need for this should be greater on land than in water.
Possibly the explanation is much more indirect: because
of the greater variety of habitats on land than in fresh
water, terrestrial pulmonates have become much more
active, 'advanced' animals than their freshwater relatives.
Linked with this, their behaviour patterns in general
have become more complex, and the complicated
reproductive behaviour may be just one facet of this
'advance'. Whatever the reason for this complexity of
courtship, the reasons for the complexity of the reproductive anatomy are more easily explicable, and are
concerned with ensuring the safe transfer of sperm or
spermatophores, and with the parental care of the young.
Many pulmonates are oviparous, laying eggs in the soil,
under rotten logs, or in similar damp places. Others are
ovoviviparous, i.e. they retain the eggs inside the body
until they hatch. Others again are reported to be viviparous, i.e. they provide nourishment for the young,

which develop inside a uterus. These developments in
parental care have possibly arisen because the pulmonates
have not produced a 'cleidoic' egg as have the vertebrates.
The eggs of even the large *Achatina*, although calcified,
and therefore to some extent protected against predators,
are permeable to water, and the embryo must survive by
tolerance of desiccation rather than protection from it
(Duncan, 1975). The structure of egg capsules has been
studied in a number of terrestrial species by Tompa
(1976). All except the Succineidae, which are found
in very humid, semi-aquatic habitats, had a calcified
layer. This may reduce water loss. Many also had an
outer hygroscopic layer of jelly which took up water
from the environment and may also help to prevent
desiccation. However, it is in the evolution of life
histories as a whole, rather than in the development of
specific egg types, that the stylommatophorans are seen
to be so well adapted reproductively to life on land.
Peake (1978) has emphasised the mixture of *r*-strategists
and *K*-strategists found within the Stylommatophora.
Broadly the group has developed reproductive strategies
which have adapted some species to unstable environments and others to very stable ones such as tropical rain
forests. It is this broad spectrum which at least partly
accounts for their wide distribution.

One set of adaptations obviously needed in the
complex rituals of courtship and reproduction concerns
the sense organs. The main battery of sense organs is
located in the two pairs of tentacles, although there are
also tactile receptors in other parts of the body, and the
statocysts lie inside the body on the pedal ganglia. The
most obvious attribute of the posterior tentacles is the
pair of eyes. These are larger and more complex than in
the primitive ellobiids and other basommatophorans (see
Charles, 1966). The eye of *Helix pomatia*, for instance,
contains up to 4000 light-sensitive cells. However, it is
probable that these eyes cannot perceive much more
than light intensity. The eye of species of *Agriolimax*
has fewer light-sensitive cells and is also probably unsuitable for form vision (Kerkut & Walker, 1975). Therefore,
although *H. pomatia* can detect an object in its path
10 cm away, it probably does so by detecting changes
in light intensity caused by the object. This suggests
that complex behaviour patterns must to a great extent
depend upon tactile and chemoreceptors. The latter
appear to be well developed, since various slugs can detect
the presence of fungi from distances up to 120 cm (see
Runham & Hunter, 1970). The 'smell' receptors are
located on the optic tentacles, but there are also 'taste'
receptors on the anterior tentacles, operating as contact
receptors. Various other factors such as humidity and
temperature can be detected, and these may both be
mediated by the same receptors, since decreasing
humidity will produce increased evaporation and hence
a lowering of skin temperature. Because of the evaporation involved, temperature is probably more important
to terrestrial forms than to aquatic ones. Important for
the same reason are air currents, and slugs react to these
by moving down wind. The total effect of these various
responses, together with the commonly found negative
response to light, is to ensure that the active animals

stay in cool, humid microhabitats, where evaporation rates are low (Machin, 1975). This activity is also usually regulated by a diurnal rhythm which triggers the start of foraging at night.

It seems appropriate to end this section by considering briefly two groups of pulmonates that have lost many of the structural and physiological provisions against water loss. One of these is the Succineidae, many species of which live in vegetation overhanging water and have a very nearly aquatic existence. Succineids cannot even retract fully into their shells, so that their resistance to desiccation must be almost negligible. As discussed above, the possibility that these forms are primitive, and not secondarily readapted to semi-aquatic habitats, has been discounted by Solem (1978). The second group to be considered here comprises the terrestrial slugs. The five families of present-day stylommatophoran slugs have arisen independently from shelled families within the Sigmurethra. All are characterised by great reduction of the shell, but although this has become small and internal in many forms, it has only been completely lost in one family, the Philomycidae (Runham & Hunter, 1970). Most slugs have a mantle cavity that opens near the anterior end on the right side, and this includes the peculiar Athoracophoridae (Fig. 4.22). Others, however, such as the burrowing Testacellidae, have the mantle cavity at the rear, protected by the small external shell. The testacellids are streamlined for burrowing, and spend much of their time underground where they prey on earthworms, other slugs and various small invertebrates. The two best-known families are the Arionidae and the Limacidae. Both are herbivorous as is normal for stylommatophorans. The success of the slug form, which might be regarded as surprising in view of the use to which the shell is put in many other terrestrial pulmonates, is well testified by the enormous damage that slugs do to commercial crops (Hunter, 1978). Many slugs now show a distribution which relates to man's activity; however, the greatest diversity of slugs is found on the windward side of large mountain chains where the moisture supply is constant and calcium availability low because of the volcanic origin of the rocks (Solem, 1978). It can therefore be deduced that the slug form evolved in response to conditions where the shell was on the one hand not needed as a protection against desiccation, and on the other hand difficult to maintain because of the scarcity of calcium.

4.3   Discussion: prosobranchs and pulmonates

We may conclude this chapter by reviewing briefly the differences that have been found between terrestrial prosobranchs and terrestrial pulmonates, and some of the possible causes of these differences. First of all, it is almost certain that the pulmonates have been a terrestrial group for longer than the prosobranchs: their earliest fossils date back to the Carboniferous in contrast to the earliest land prosobranchs of the Cretaceous. It must be stressed, however, that the majority of land pulmonates did not appear until much later, and at about the same time as the early land prosobranchs; so that in all probability the main radiation of the stylommatophorans and of most land prosobranchs has occurred over roughly the same period. Yet it is apparent that the pulmonates are in many ways a more successful group on land than the prosobranchs – they are in general larger and more active, they are usually more common in terms of numbers of species, and they inhabit some very 'difficult' environments, such as deserts, where prosobranchs are not found. The obvious question, then, is why should the pulmonates have flourished on land more than the prosobranchs?

In many respects the two groups are strictly comparable: both have representatives that can aestivate for long periods in dry conditions; both have produced some form of lung; both have some tendencies towards the development of parental care. But there are also differences, even if it is difficult to determine the ones that are really significant. The lung, for instance, is a more complex organ in pulmonates, and since it is a ventilation lung, it probably allows better facilities for gas exchange and its control, and hence higher rates of metabolism and a more active life. All pulmonates are hermaphrodite, in contrast to the prosobranchs, which have separate sexes, and this may well have helped the pulmonates to spread, especially as in many species self-fertilisation occurs (Hyman, 1967) so that only one individual is needed to found a new colony.

In terms of water balance, the ability of the pulmonate mantle to lower the evaporation rate during aestivation counteracts the inability of either the epiphragm or the mucous film to prevent water loss. Although the prosobranchs can effectively isolate their tissues by closing the operculum, the development of many different types of 'breathing tubes' has shown that this also creates its own respiratory problems. The carriage of a water store by pulmonates may also be a large factor, in spite of the fact that it must be energetically an expensive mechanism.

The pulmonate kidney is also better adapted to terrestrial life than the various types found in the prosobranchs, and this is due in the main to the routes by which the various groups have moved on to land. The helicinid and cyclophorid prosobranchs, invading via fresh water, have kidneys which can remove excess water from the body, but because the processes of uric acid excretion and water reabsorption occur together in the kidney it is not possible for them to excrete uric acid without losing a great deal of water. The pomatiasids, invading via the littoral zone, cannot produce hyper-osmotic or hypo-osmotic urine from the kidney, but can probably utilise the pedal glands as osmoregulatory organs. It is not known how they excrete nitrogen, but it is interesting to note that the kidney does not excrete uric acid concretions perhaps because there is insufficient urine flow to flush the organ out. One of the conditions for a kidney that can successfully eliminate uric acid *and* control water balance on land is thus the spatial separation of the two processes. This separation is found in the Stylommatophora, in which the kidney excretes uric acid and the ureter reabsorbs water. In dry conditions the renal system can therefore excrete 'dry' urine, and obviously this is essential in any climate without constant moisture. It is unfortunate that no detailed comparisons can be made with the Pomatiasidae, which also have a marine origin. However, it seems likely that in this group also the functions of nitrogen excretion and water balance have been spatially separated. This would have been a response to selection pressures during early evolution when there would have been a high priority for the ability to be able to excrete nitrogen even in a dry environment. In the groups with a freshwater origin, in contrast, the highest priority was for the ability to produce a high flow rate of hypo-osmotic urine. This flushed uric acid out of the kidney and there was no selection pressure for the spatial separation of nitrogen excretion and water balance.

Also coupled with a marine or estuarine origin of the pulmonates is the greater tolerance of variations in the concentration and composition of the blood shown by them in comparison with the terrestrial prosobranchs. Although very few measurements are available to support this suggestion directly, the constancy of the osmotic pressure of the blood in terrestrial helicinids and cyclophorids kept in leaf litter is almost startling (Little, 1972; Andrews & Little, 1972). In contrast, Burton (1968)

when studying terrestrial pulmonates was obliged to make his snails crawl in a shallow layer of water to bring about reasonable consistency between individuals. The same point is emphasised by the large inherent variation in body weight of pulmonate slugs and snails found by Howes & Wells (1934), which presumably is accompanied by changes in the concentration of the body fluids. Also to the point is the fact that the prosobranchs have produced no slug forms – presumably they could not tolerate the variations in body water inherent in such types. In summary, the prosobranchs maintain a very constant blood concentration while external conditions are relatively constant, but they cannot tolerate large changes in the external environment since these are beyond the limits of their control systems. In contrast, the pulmonates have a very variable blood concentration, with wider limits to their control systems, so that they can tolerate greater external changes. This development of control systems with wider limits of operation has been at least partly responsible for the great success of terrestrial pulmonates; and the initial development of these systems can be traced back to the primitive estuarine origin of the group from ellobiid stock.

5

Crustaceans and the evolution of the arthropods

'And the Man said "If you choose, I will make a Magic, so that both the deep water and the dry ground will be a home for you and your children – so that you shall be able to hide both on the land and in the sea".'

From 'The crab that played with the sea' (*Just so stories*) by Rudyard Kipling, 1902, Macmillan, London.

While the phylum Mollusca contains something of the order of 80 000 present-day species, the phylum Arthropoda has a recorded number of about 800 000. A large proportion of these is terrestrial, so that the origins and adaptations of the various groups which constitute the phylum are of much interest in the context of the colonisation of land. The relationships between the classes of arthropods have been discussed and disputed by generations of zoologists, but as yet no general consensus of opinion has been reached. Nevertheless, it is necessary here to outline some of the possible links within the phylum, in order to be able to discuss the ways in which the several terrestrial arthropod groups have been derived. When this has been done, we can proceed with each group in turn.

5.1 The evolution of the arthropods

The five present-day classes which are usually included in the phylum Arthropoda are the xiphosurans or horseshoe crabs, the arachnids, the crustaceans, the insects and the myriapods. The onychophorans may be included, or may be given the rank of phylum. Many of these classes are sometimes subdivided into further classes: the pycnogonids may be separated from the arachnids, the myriapods may be considered as four classes containing the centipedes, millipedes, symphylans and pauropods, while the insects may be split into the proturans, collembolans, diplurans and 'insects proper' (including thysanurans). To this great variety must be added the fossil representatives: the eurypterids, sometimes placed with the horseshoe crabs, sometimes with the arachnids; the trilobites; and the pseudocrustaceans. These groups, with some of the links that have been

proposed by different zoologists, are listed in Table 5.1 together with the small groups which may have arthropod affinities, such as the tardigrades and pentastomids.

When considering the great variety of different groups, various authorities have, understandably, come to vastly different conclusions concerning evolutionary trends. While some believe the phylum to be monophyletic, others consider that it is equivalent to a grade of organisation, and that it is diphyletic or polyphyletic in origin. These viewpoints have often been reached after rather different approaches. Thus, of the recent venturers in this field, Sharov (1966) – a supporter of the monophyletic scheme – leaned heavily on the evidence from the fossil record, while Manton (1970, 1977) concentrated on the study of functional morphology. Anderson (1973), in contrast, attacked the problem from an embryological viewpoint. Patterson (1978), having discussed various definitions of polyphyly, concluded that for the arthropods a polyphyletic origin has yet to be convincingly shown.

To the outsider the array of different types and the varying conclusions reached by arthropod specialists may well appear alarming; but there are some points of relationship which, if not agreed by all zoologists, can at least command the support of the majority, and it is worth outlining these points briefly before going on to the differences between the monophyletic and the polyphyletic schemes. Most authors agree that the various arthropod groups were derived, initially, from some kind of 'lobopod' annelid ancestor. This showed metameric segmentation, and was very probably marine. Another point of agreement is the validity of the group Chelicerata, containing xiphosurans, eurypterids, arachnids and pycnogonids, although the exact relationship of these component classes or subclasses is disputed. In contrast, the validity of the group Mandibulata, which contains most of the other major arthropod classes, is far from agreed. While some would argue that the myriapods, insects and crustaceans should be united under this heading, others argue that the crustaceans show little similarity to the insects and myriapods, and should be placed in a separate group. The close relationship between the insects and myriapods, or rather between the smaller groups into which these have now been split, is, however, agreed by the majority of zoologists. Many would include the onychophorans in this group, the Uniramia, while others would exclude them from the phylum Arthropoda.

We are left, then, with the trilobites, the chelicerates, the crustaceans and the insect–myriapods as the largest universally accepted groups within the arthropods, and three much smaller groups, the onychophorans, tardigrades and pentastomids, which are often given the rank of separate phyla. These assemblages are viewed in rather different ways by different authorities. The advocates of a polyphyletic arthropod origin envisage three main lines of evolution (see, e.g. Tiegs & Manton, 1958; Manton, 1970, 1979; Anderson, 1973, 1979; Bergström, 1979; Ranzi, 1980). The form of the jaws and limbs, and the tagmata of the body, suggest that the fossil trilobites have some affinities with the chelicerates; the crustaceans stand alone; and the onychophorans, myriapods and insects constitute a natural group. Such a scheme runs into the difficulty that it requires many organs and structures to have evolved several times, and to be similar in different groups due to convergence. While this may be acceptable for such relatively simple organs as tracheae, Malpighian tubules and mandibles, there is much argument over the possibility of the convergent evolution of the complex compound eye in three different groups.

The contrasting monophyletic theories do not have the difficulty of explaining large degrees of convergence, but they run into other troubles. The theory of Snodgrass (1938), for example, begins with lobopod annelids from which primitive sclerotised 'Protarthropoda' are supposed to have been derived. From these arose on the one hand the trilobites and chelicerates, and on the other the Protomandibulata, which gave rise to the mandibulates. The latter consisted of the aquatic crustaceans, and a terrestrial group, the Proto-

myriapoda, which in turn gave rise to the myriapods and insects. This theory has much in its favour, but unfortunately, although fossil myriapods date from the Silurian, no traces of either the Protarthropoda or Protomandibulata have been found as fossils in the Palaeozoic. Since the theory also links the crustaceans with the insect–myriapods, it brings disagreement from those who consider that the mandibles were acquired independently in the two groups.

Another, more recent, monophyletic theory is that of Sharov (1966). Sharov derived the arthropods from lobopod annelids via the Proboscifera, a fossil group which he linked with the pycnogonids. From the Proboscifera he derived the trilobites, and from these in turn the chelicerates and crustaceans. The insect–myriapods were descended from the crustaceans, and the onychophorans were independently derived from lobopod annelids, in parallel with the arthropods. One of the weaknesses of this scheme is the concentration upon fossil forms, of which we know little in detail, and instead have to rely upon dubious reconstructions. Another is the disregard for the differences between the insect–myriapods and the crustaceans, and for the similarities of the insect–myriapods and the onychophorans. Nevertheless, a number of authors favour a monophyletic origin of the arthropods. This has been argued from eye structure (Paulus, 1979), segmental structures (Boudreaux, 1979*a*, *b*; Weygoldt, 1979), visceral anatomy (Clarke, 1979) and haemocyte types (Gupta, 1979).

In conclusion, it can be seen that there is no current consensus of opinion on the subject of arthropod evolution. The impossibility of relating features in the crustaceans, chelicerates and insect–myriapods balances the difficulty of postulating the convergent evolution of many structures. In this account the crustaceans, chelicerates and insect–myriapods will be treated as relatively separate groups. The position of the onycho-

Table 5.1. *The major subdivisions of the phylum Arthropoda*

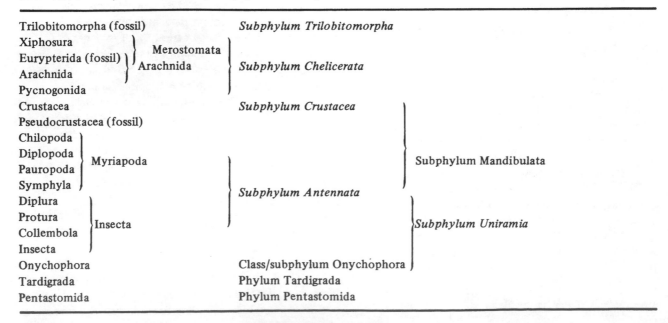

Major groups over which there is some general agreement are in italics.

phorans remains uncertain, although some relationship with the insect–myriapods seems likely. Since present-day onychophorans did not give rise to the insect-myriapods, however, it is appropriate to include them as another, but related, group within the phylum Arthropoda.

The various groups of the arthropods are now considered in turn, beginning, in this chapter, with the crustaceans, which are mostly aquatic; then going on to the chelicerates, which have very old terrestrial representatives; and ending with the myriapods, onychophorans and insects.

5.2   Introduction to the Crustacea

It is clear from Table 5.2 that the crustaceans form predominantly an aquatic group, and of the 26 000 described species, only about 5% are terrestrial or semi-terrestrial. The vast majority of terrestrial species are found in the order Isopoda (about 1000 terrestrial species), and almost all the others are found either in the order Decapoda or the order Amphipoda. Most of the discussion will therefore centre on these three groups, for which much information was sum-marised at a symposium of the American Society of Zoologists (see, e.g. Edney, 1968). Many of the other crustacean groups have representatives which are semi-terrestrial or cryptobiotic, however, and these may be considered as a preliminary to the more truly terrestrial forms.

For convenience, the members of the subclasses other than the Malacostraca are termed the 'entomo-stracans', although this name does not indicate any close evolutionary ties between the subclasses. Most entomo-stracans are purely aquatic. The barnacles (Cirripedia),

for instance, are all marine or estuarine, although one species, *Balanus improvisus*, has been reported to live in some areas which have predominantly fresh water (Newman, 1967). No barnacles are found above the splash zone, since they breed by liberating gametes directly into the sea, and obtain food by filtering sea water. Yet many of the sessile species are found in company with such gastropods as *Littorina neritoides* (see Chapter 4), in situations where they are subject to a high degree of desiccation (see, e.g. Lewis, 1964). Those sessile barnacles which do live high up on the shore, e.g. *Balanus balanoides* and *Chthamalus stellatus*, can tightly close their opercular plates, while species such as *Balanus crenatus*, which live lower down, are unable to do this (Barnes, Finlayson & Piatigorsky, 1963). As might be expected, *B. balanoides* and *C. stellatus* are able to survive much longer in air than *B. crenatus*, and this is at least partly due to the lower rates of water loss shown by them (Foster, 1971). During their initial exposure to air, *B. balanoides* and *C. stellatus* expel the water from their mantle cavity and obtain oxygen from the air through a small 'micropylar' opening between the opercular valves. This is probably the normal procedure between periods of high tide (Grainger & Newell, 1965). When exposure is prolonged over a period of days, however, the micro-pylar opening closes further, and the barnacles respire anaerobically, producing lactic acid. When aerobic condi-tions are restored, very little of the accumulated oxygen debt is repaid, and most of the lactic acid is excreted. Despite this apparent lack of adaptation to long-term isolation from the external environment, however, 50% of a population of *C. stellatus* survived for over 5 days in dry air (Barnes *et al.*, 1963). Indeed, Monterosso

Table 5.2. *Crustacean groups and their habitats*

	Marine	Fresh water	Terrestrial
Subclass Cephalocarida (1 order)	x		
Subclass Branchiopoda (4 orders)	(x)	x	(Cryptobiotic)
Subclass Ostracoda (4 orders)	x	x	(x)
Subclass Mystacocarida (1 order)	x		
Subclass Copepoda (7 orders)	x	x	(Cryptobiotic)
Subclass Branchiura (1 order)	x (Ectoparasitic)	x	
Subclass Cirripedia (4 orders)	x		
Subclass Malacostraca:			
Superorder Phyllocarida (1 order)	x		
Superorder Syncarida (1 order)	(x) (Fossils)	x	
Superorder Hoplocarida (1 order)	x		
Superorder Peracarida:			
Order Isopoda	x	x	x
Order Amphipoda	x	x	x
Other orders (5)	x	x	
Superorder Eucarida:			
Order Euphausiacea	x		
Order Decapoda	x	x	x

'Entomostracans' brackets the first seven subclasses (Cephalocarida through Cirripedia).

Parentheses show that the distributions in the habitat shown is to be qualified in some way, e.g. branchiopods may have reached the sea secondarily; ostracods are found on land, but are active only in water.

(1930) was able to keep some specimens of *C. stellatus* dry in his laboratory for as long as 119 days, although this may well have been at a high relative humidity.

The stalked barnacles show many similar properties to those of the sessile forms, but also some differences. *Pollicipes polymerus* is common on the upper two thirds of the shore on the west coast of North America. It is tolerant of a wide range of temperatures, and evaporation from the peduncle cools the animals. *P. polymerus* therefore shows increases in the osmotic pressure of its haemolymph during desiccation, and as well as tolerating these it can withstand some dilution by rain (Fyhn, Petersen & Johansen, 1972). Unlike the sessile barnacles, it shows *greater* uptake of oxygen in air than in water, indicating a gaseous exchange mechanism which is efficient in air. At least part of the extra uptake in air occurs through the peduncle (Petersen, Fyhn & Johansen, 1974). As in many terrestrial and semi-terrestrial animals, pCO_2 rises when the barnacles are out of water. At the same time, pH of the haemolymph falls, but this is not due to accumulation of lactic acid, and may be related purely to the increased pCO_2. The differences from the sessile forms are therefore in the most part to be related to the properties of the peduncle, coupled perhaps with the generally larger size of stalked forms, which means that they can retain an area of fairly permeable integument without losing too large a proportion of their body water during desiccation.

Most copepods live either in the sea or in fresh water, but several small harpacticoid species are found in the soil and in leaf litter. They may be abundant in some areas, and are found both in the tropics and in such temperate regions as England. Their distribution is further discussed in Chapter 10. These animals are of the order of 1 mm in length, and they crawl and swim in the water films surrounding leaves and soil grains. In dry conditions they can form cysts in which they survive until moisture returns, as can some true fresh-water species (Elgmork, 1967). They can therefore be placed in either the category of 'aquatic' animals or that of 'hygrophilic' animals suggested in Table 1.3.

The ostracods, similarly, form an almost totally aquatic group, but several species have now been described from forest humus in South Africa (Harding, 1953) and Madagascar (Danielopol & Betsch, 1980). The genus *Mesocypris* has many fine hairs on parts of the carapace which attract water from the soil by capillarity. Its relatives are all found in fresh water, and Harding (1953) suggested that this genus has become terrestrial from fresh water by colonising mosses. *Terrestricandona* has also been derived from a freshwater group, but probably in this case from interstitial forms (Danielopol & Betsch, 1980).

The branchiopods are in the main freshwater animals, the best known being the water fleas, such as *Daphnia*. Other members of the subclass live in temporary freshwater and saline inland pools, and some, such as the brine shrimp *Artemia salina*, are famous both for their tolerance of a wide range of external salinities and for their osmoregulatory powers. Species of *Artemia* can survive in concentrations from 10% sea water to crystal-

lising brine (about 600% sea water), and the osmotic pressure of the haemolymph remains remarkably constant over this range (Croghan, 1958). Relatives of *Artemia*, such as species of the fairy shrimp *Branchinecta*, can tolerate concentrations over 50% sea water, but cannot maintain the osmotic pressure of their body fluids below that of the external medium, as can *A. salina* (Broch, 1969). Species of the tadpole shrimp *Triops* can withstand concentrations of up to 16% sea water, but even in this medium 50% of the experimental population died after 2 days (Horne, 1968a).

Branchiopods are interesting not only for the salt tolerance of the adults, but also for the resistance of their eggs to extreme environmental conditions, in the state of cryptobiosis. When the eggs of *Artemia salina* have been desiccated, they can survive in a dry state for years. Even after 15 years, a small fraction will still hatch out when given the appropriate conditions (Clegg, 1967). The eggs contain a high proportion of trehalose, which is the embryo's only energy source; yet the amount of trehalose does not appear to decline over a storage period as long as 28 years – by which time none of the eggs is still viable. From these facts it would seem that the eggs – or encysted embryos – show no sign of life when desiccated; but a recent study (Sundnes & Valen, 1969) has shown that over a period of 780 days, the rate of respiration is measurable, even at $0\,^{\circ}C$. Eggs of other branchiopods are equally resistant: those of *Triops*, for instance, withstand temperatures up to, but not including, $100\,^{\circ}C$ (Carlisle, 1968). Since they are killed when subjected to boiling under reduced pressure – i.e. at a reduced temperature – it is presumably the boiling process itself which is lethal. The ultrastructure of the eggs of *A. salina* has been investigated by Morris & Afzelius (1967). The shell of the cysts consists of an outer chorion and an inner embryonic cuticle, the whole being initially covered by an outer cellular membrane. The dehydrated cysts produced by *A. salina* float on the water surface, and do not become rehydrated until this outer cellular membrane is damaged – usually this occurs when the cysts are blown to the shore. Subsequent hydration occurs rapidly when the cysts again come into contact with water. It has been postulated by Clegg (1981) that during this rehydration water is first taken into the 'bound' state, and that following this a 'vicinal water phase' is established in close proximity to the various intracellular membranes. It is in this phase that the cellular enzymes are thought to operate. Only when the vicinal water phase is complete is 'bulk water' absorbed. These conclusions are important in the consideration of all cryptobiotic animal stages. If the three phases are dehydrated in the reverse order during the onset of the cryptobiotic state, they may also suggest some insight into how the enzyme systems remain viable. If these are restricted to membrane surfaces, their biochemical organisation, and that of the cell, will remain intact as long as the membranes themselves are not damaged.

This is perhaps an appropriate point at which to mention the cryptobiotic properties of tardigrades, although they are not crustaceans, and may not even belong in the phylum Arthropoda. Many of the animals are found in mosses and lichens, often in the gutters of houses, together with the rotifers made so famous by the observations of Leeuwenhoek. Some tardigrades form cysts when external conditions become unfavourable, and although these have a lower metabolic rate than active animals, the rate is far from negligible (see review by Keilin, 1959). Others, however, do not form cysts but can become desiccated without harm. In this case the metabolic rate drops enormously. The rate depends on the ambient humidity, and therefore presumably on the water content of the tissues. The tolerance to desiccation also depends upon the external relative humidity, and this is because tardigrades must be dried slowly if they are to survive (Crowe, 1972). Apparently they alter their body shape as drying proceeds, withdrawing the more permeable parts of the cuticle so that these do not contact the external air. The cuticle also becomes less permeable as desiccation proceeds. When the cryptobiotic state is reached, the structure of the tissues bears little resemblance to that in active life, since all the cell organelles become crowded together into an almost unrecognisable mass (May, 1949), but this structure can return to normal in a very short time. Similarly, the tissues of cryptobiotic nematodes have been shown to return to normal in the extraordinarily short time of 2 min after rehydration.

5.3 The Malacostraca

Terrestrial malacostracans, in contrast to the terrestrial entomostracans, show no tendency to be cryptobiotic, and most are large enough to be independent of water films. Many of them do live in leaf litter and humus, however, and in other humid environments which are to some degree isolated from the general macrohabitat. They are then referred to as cryptozoic, a term discussed in detail by Lawrence (1953) with respect to forest faunas. Within the malacostracans, three orders, the amphipods, the isopods and the decapods, have terrestrial members.

5.3.1 The amphipods

There are over 60 families of amphipods, but of these all except one are exclusively aquatic. The exception is the family Talitridae, in which some members are found inhabiting the marine littoral zone and the shore above the surf line, and others form part of the cryptozoic fauna of forest leaf litter. Those which live high on the beach, such as species of *Talitrus*, *Orchestia*, *Orchestoidea* and *Talorchestia*, usually form some kind of burrow in sandy substrates, or live under piles of seaweed on the strand line (Fig. 5.1). *Orchestoidea californiana*, for instance, forms burrows approximately 30 cm deep, with the surface opening plugged during the daytime (Bowers, 1964). Most species of hoppers usually remain hidden during the day, when external relative humidities are low, but many emerge at night and range over the beach scavenging. This type of behaviour has the double advantage of protecting the

hoppers both from low relative humidities and from daytime predators such as birds. Other behavioural characteristics distinguish the semi-terrestrial forms from their aquatic relatives, but in general there are very few structural or physiological modifications.

The intolerance of desiccation is a good example of this lack of physiological adaptation. Williamson (1953) showed that the survival of three species of high-shore talitrids was considerably reduced at 95% RH compared with that at 100%. At RHs lower than 50%, survival was for less than 2 hours. Some species are tolerant of wide ranges of salinity, however: *Orchestia platensis* can survive on substrata containing salinities of $5-40^0/_{00}$, but dies if fresh water is used (Bock, 1967). This species could therefore probably not survive in purely terrestrial habitats. Others, such as *Orchestia bottae* and *Orchestia cavimana*, have been found at the edge of freshwater habitats, and are more nearly terrestrial (Reid, 1947; Arendse & Barendregt, 1981). Most species will survive immersion in sea water. Another physiological problem has been discussed by Wieser (1967). While marine forms can obtain the copper necessary for haemocyanin by uptake from sea water, semi-terrestrial and terrestrial forms must rely more on uptake from food. More work in this field has been done with isopods, however, and the situation is discussed in that section.

Behavioural adaptations include changes in the mating procedure (Williamson, 1951), an ability to select high humidities (Williamson, 1953) and the development of complex orientation responses (see summary in Newell, 1979). It is appropriate to discuss these in turn.

In aquatic gammarids and in the primitive aquatic talitrids such as species of *Hyale*, the male carries the female below him, dorsal surface up, and in the same plane. This carrying may continue for a long period before copulation occurs. In the semi-terrestrial species the carrying period is very short. The males are attracted to newly moulted females, and the female is then carried lateral surface up, and transversely (Williamson, 1951). In this position the male can walk and jump quite effectively, whereas this would be impossible in the position used by aquatic forms. At copulation, sperm are placed on the ventral surface of the female's abdomen, and the eggs are fertilised in the brood pouch. Egg laying may occur from a few hours to 4 days after copulation, since it is governed by the time of moult. This is quite different from most aquatic forms, in which egg laying usually takes place almost immediately after copulation.

When talitrids are placed in chambers where there is a humidity gradient, they select the area with the highest humidity (Williamson, 1953). This selection is achieved by two forms of behaviour: an increase in speed in drier air (hygro-orthokinesis) and an increase in the frequency of turning in dry air (klinokinesis). No humidity receptors have been located, and it has been suggested that the animals respond to the rate of water loss from the whole body.

Many beach talitrids have been shown to be capable of visual orientation (Newell, 1979). There is some disagreement about the exact mechanisms involved, but detection of the sun, of polarised light, and of the silhouette of the shore have all been implicated, and a magnetic sense may also be involved (Arendse & Barendregt, 1981). Certainly there is a considerable difference between the behavioural complexity of the semi-terrestrial forms and that of the purely aquatic ones. Although this may be coupled with improved sense organs, no such improvement has as yet been shown. It may also be coupled with the rate and mechanism of movement. Aquatic amphipods are primarily swimmers, although they crawl at times, and may·jump when exposed or disturbed. The semi-terrestrial talitrids can both jump and walk, and when walking they maintain themselves dorsal surface upwards, in contrast to the aquatic forms which lie on their side. Even under water, talitrids normally walk. Despite this, Edney (1960) was of the opinion that, in comparison with isopods, the amphipod type of locomotion is in general unsuitable for life on land because of its inherent instability. This is caused by the lateral compression of the body and the movement of the legs in a vertical plane, in contrast to the dorso-ventral flattening of isopods and the flexure of the isopod legs under the body.

Concerning the terrestrial talitrids (Fig. 5.2), most of which live in forest leaf litter, surprisingly little is known (for a review see Hurley, 1968). The leaf litter habitat provides a very constant environment with a high humidity, and a large number of groups of animals are well represented in its cryptozoic fauna (see Lawrence, 1953). Among these groups the amphipods may be

found in high densities in tropical forests and in some cold to temperate forests, where there is high rainfall. In Australian rain forest, *Talitrus sylvaticus* has been found in densities of up to $4000/m^2$ (Birch & Clark, 1953). Most species are found around the Indo-Pacific region, but there are also records for the Atlantic and the Caribbean. Hurley (1968) believed that the evolution of terrestrial forms has been relatively recent, and that the overall distribution of species can best be explained by dispersal along recently broken land bridges, by local origins from littoral species, and by accidental human distribution. An alternative view places the origin of the group further back in time, and invokes the aid of the theory of continental drift to explain present-day distributions (Bousfield, 1968).

The distribution of a Japanese species, *Orchestia platensis*, strongly supports a recent littoral marine origin (Tamura & Koseki, 1974). This species contains two subspecies. One, *Orchestia platensis platensis*, is found on marine beaches among damp seaweed. The other, *Orchestia platensis japonicus*, is morphologically indistinguishable from the first, but is found in the leaf litter of damp evergreen and deciduous forests throughout Japan, reaching a maximum population density of $400/m^2$. It therefore seems probable that in this case the terrestrial form has a recent direct marine littoral origin.

Talitrids inhabiting the leaf litter have few apparent adaptations to life on land. Their lateral flattening helps them to utilise the spaces between leaves in much the same way as in crevice dwellers. In their general morphology, the major difference from supra-littoral talitrids is the change from thick, heavily spined appendages to

Fig. 5.1. The marine amphipod *Talitrus saltator* living on the strand line. During the day *T. saltator* burrows in sand above high water, but at night it emerges to feed on the beach, and particularly on the strand line. Body length 15 mm.

slender, finely spined ones; many of these fine spines may be concerned with tactile responses. In comparison with terrestrial isopods, talitrids have a rather thin cuticle and are somewhat fragile.

A few species have been found to colonise grassland as well as forest leaf litter. In New Zealand, *Orchestia hurleyi* lives in grassland (Duncan, 1969) in company with *Orchestia patersoni*, which is mainly a forest dweller. *O. patersoni* lives under the canopies of the grass plants, while *O. hurleyi* is found for the most part outside the canopies. These distributions are related to the amount of rainfall intercepted by the grass leaves: underneath the canopies, rainfall is low but fairly constant, while outside the canopies it is higher but more variable. *O. hurleyi* has the ability to climb, and hence it can avoid waterlogging, but death due to osmotic stress does occur. Other causes of mortality include predation during ecdysis, occasionally desiccation, and starvation in the autumn when the population eats out the supply of leaf litter.

Little is known about talitrid physiology. Terrestrial forms have enlarged gill plates compared to those of the supra-littoral species, but in many the pleopods have been reduced in number and size. In aquatic forms these are used for drawing water over the gills and for swimming. In terrestrial forms, where they are still present, they may be used for agitating and oxygenating the water which is retained inside the brood pouch (exosomatic water). Breeding behaviour appears to be similar to that in beach talitrids. The number of eggs produced is usually small (1–10), and when the young emerge they are relatively large and well developed. No detailed work has been done on osmoregulation, but personal observations on some Australian terrestrial amphipods showed the osmotic pressure of the haemo-

lymph to be about 400 mOsm in animals kept in damp leaf litter. This value compares with 274 mOsm in the freshwater gammarid amphipod *Gammarus pulex* (Lockwood, 1961). It is therefore relatively high, and may be taken to support the idea of a marine rather than a freshwater origin. The suggestion by Dresel & Moyle (1950) that nitrogen metabolism has been suppressed in terrestrial isopods and amphipods has been proved wrong in isopods (Wieser, Schweizer & Hartenstein, 1969), and it is unlikely to apply to amphipods. The apparent reduction in excreted nitrogen probably reflects the use of gaseous ammonia as an end product, although there are no measurements to confirm this.

To sum up, it can be said that talitrids have colonised the leaf litter habitat mainly by behavioural adaptations. Such osmoregulatory adaptations as are necessary have not been investigated. In many ways the terrestrial forms are identical to the supra-littoral species, and this has suggested that colonisation of leaf litter occurred directly from the shore. Particularly in the tropics, the forests often reach right down to the sandy zones occupied by beach talitrids, so that there would have been little barrier to such a movement. Any colonisation via fresh water is somewhat unlikely, since no present-day members of the family Talitridae are found in fresh water.

5.3.2 The isopods

Most of the nine suborders of the order Isopoda are marine, and all the terrestrial forms are contained within one suborder, the Oniscoidea. It is rather difficult to define this suborder, but its members all differ from their aquatic relatives in having a much reduced antennule. The difficulty in defining the suborder arises

Fig. 5.2. A terrestrial litter-dwelling amphipod from New Zealand. In tropical and some southern temperate forests

amphipods are probably very important in the breakdown of leaf litter. Body length 10 mm.

because it is probably polyphyletic in origin. Vandel (1943, 1965) believed that it had three separate origins from marine ancestors, represented by three different 'series': Tylian, Trichoniscian and Ligian. Each series has modern aquatic and terrestrial forms, with the most primitive ones in general living on the sea shore or in caves. It now seems to be generally accepted that all terrestrial isopods have been derived directly from marine ancestors.

There are over 20 present-day families of terrestrial isopods, containing over 1000 species, with representatives living in an astonishing range of habitats. The Tylian series has species in sandy marine beaches (*Tylos*), and in the humus of inland oakwoods (*Helleria*). None of these is very resistant to desiccation, and although abundant round the Mediterranean, none is found in Britain. Species of *Tylos* show many similarities in their way of life to the amphipod *Talitrus saltator* (see Newell, 1979). They burrow above high water mark, sometimes to a depth of 30 cm, and emerge at night to scavenge on the strand line. They have orientation mechanisms which can utilise the sun and the moon, and a lunar day rhythm which controls their emergence, activity patterns and respiratory rate (Marsh & Branch, 1979). Although well adapted to the marine environment, Verhoeff (1949) believed them to have recolonised the intertidal

zone from a more terrestrial situation. They possess pseudotracheae in their pleopods (Fig. 5.11), which are discussed later. Verhoeff pointed out that these are essentially a terrestrial adaptation.

The Trichoniscian series is also restricted to humid habitats, but species are found in such widely different places as the littoral zone and terrestrial soil (*Trichoniscus*), caves (*Speleonethes*), and even in the snow zone (*Oritoniscus*). The population dynamics of *Trichoniscus pusillus* in limestone grassland were studied by Sutton (1968). This species was much less tolerant of freezing and desiccation than *Philoscia muscorum*, a member of the Ligian series. However, it was numerous, occasionally reaching population densities as high as $8000/m^2$. In response to cold or dry conditions it burrowed into the soil, while *P. muscorum* remained on the surface where it appeared to be able to tolerate freezing as well as high summer temperatures.

The Ligian series is the largest group, and contains all the desiccation-resistant forms. *Ligia oceanica* (Fig. 5.3) is found in the rocky marine intertidal at high levels, often in crevices or in shingle. Some species of *Armadillidium* are found in littoral marine sands, while *Armadillidium vulgare* (Fig. 5.4) is common in grassland on calcareous soils. Species of *Ligidium* occur in very damp habitats such as marshes and river banks, while more terrestrial species live in garden soil and woodlands (*Porcellio*), in caves (*Cylisticus*), in ants' nests (*Platyarthrus*), sand dunes (*Porcellio*) and even in deserts (*Armadillo, Venezillo, Hemilepistus*). The ecology of

Fig. 5.3. *Ligia oceanica*, a marine isopod living high on the shore, usually in crevices, or as here under shingle. Body length 25 mm.

the desert species *Hemilepistus reaumuri* has been investigated by Shachak, Steinberger & Orr (1979) and Shachak (1980). This species lives in burrows, is monogamous and enlarges the burrow as the young hatch. It has a high fecundity and shows parental care, bringing food for the newly hatched young. These features contrast markedly with most mesic isopods. They adapt

H. reaumuri to the desert environment and show that it is essentially an *r*-selected species.

A brief classification of the Oniscoidea is given in Table 5.3. The division of the suborder into Synochaeta, Diplochaeta and Crinochaeta is derived from the disposition of the genital apertures. The further division of the Crinochaeta into Atracheata and Pseudotracheata

Table 5.3. *Classification of the suborder Oniscoidea, showing the position of species mentioned in the text (adapted from Vandel, 1960, 1965)*

Series	Tribe		Family	Genus
Tylian			Tylidae	*Tylos, Helleria*
Trichoniscian	Synochaeta	Styloniscoidea	Styloniscidae	
			Titaniidae	
			Schöbliidae	
		Trichoniscoidea	Trichoniscidae	*Trichoniscus, Speleonethes, Oritoniscus*
			Buddelundiellidae	
Ligian	Diplochaeta		Ligiidae	*Ligia, Ligidium*
			Mesoniscidae	
	Crinochaeta	Atracheata	Stenoniscidae	
			Tendosphaeridae	
			Spelaeoniscidae	
			Rhyscotidae	
			Squamiferidae	*Platyarthrus*
			Oniscidae	*Oniscus, Philoscia*
		Pseudotracheata	Cylisticidae	*Cylisticus*
			Porcellionidae	*Porcellio, Tracheoniscus, Hemilepistus*
			Atlantiidae	
			Armadillidiidae	*Armadillidium*
			Eubelidae	
			Actoeciidae	
			Armadillidae	*Armadillo, Venezillo*

Fig. 5.4. *Armadillidium vulgare*, a common terrestrial isopod found mostly on calcareous soils. Body length 15 mm.

depends on the presence or absence of pseudotracheae in the pleopods. This is discussed later, and is in great measure related to the degree of 'terrestrialness' shown.

The origins of the Oniscoidea are very much in doubt. According to Vandel (1943, 1965) each of the three series has evolved independently, and often in parallel, but the time at which they originated from marine ancestors can only be guessed. Although the first fossil oniscoids are known from only as far back as the Eocene, fossil isopods are known from the Trias, and even by that time all the major aquatic subdivisions of the order had appeared. It has therefore been argued that terrestrial forms probably date from around this time, at the end of the Palaeozoic, and that woodlice are among the oldest terrestrial invertebrates (Vandel, 1965). The present distribution of the Trichoniscian series, in which two primitive members are found separated by the Atlantic, could be explained by their origin in one area followed by separation due to continental drift at the end of the Palaeozoic. This again would suggest an ancient origin.

The diversity of niches occupied by woodlice has already been emphasised. The exact relation of particular species to their habitats has not been defined in many cases, but some interesting studies relate both to animals living in damp habitats and to those found in drier areas. In the Netherlands *Porcellio scaber* has been extensively studied by Den Boer (1961). *P. scaber* was more active at night than during the day, as are most woodlice. Den Boer suggested that this activity, which resulted in more animals climbing up trees, was not due to a search for food, but was best interpreted as a means of finding a less humid environment, and therefore of losing excess water gained in the more humid leaf litter during the day. Mayes & Holdich (1975), however, have shown that *P. scaber* is in equilibrium with saturated air, and is unlikely to accumulate excess water in leaf litter unless this is waterlogged. The basis for vertical migration therefore requires further study.

The activity of *Armadillidium vulgare* has been examined in a drier grassland habitat in California (Paris, 1963, 1965). *A. vulgare* also was more active at night than in the day, but this was because low humidity and high temperatures produced desiccating conditions in the daytime, and feeding could be better carried out at night. Activity was definitely correlated with low vapour pressure deficit (Fig. 5.5), although from the studies of Warburg (1965c) it is evident that *A. vulgare* can withstand a wide range of physical conditions (Fig. 5.6).

Desiccating effects become more extreme in desert environments, and observations on two desert forms are therefore most interesting. *Hemilepistus reaumuri* lives in vertical burrows in north African deserts, where conditions above ground may reach extremes of 23 to 42.5 °C and 29 to 46% RH (Cloudsley-Thompson, 1956). The conditions inside the burrows are less extreme, and measurements by Edney (1968) showed that 5 cm below the surface the relative humidity was always at least 80%. In Arizona, USA, Warburg (1965c) has shown that extremes which were tolerated by *Venezillo arizonicus* included 45 °C and 15% RH.

These differences between different species, living in a variety of habitats, indicate that isopods have a variety of physiological and behavioural mechanisms which are used to combat or evade terrestrial conditions. In order to understand further how isopods manage to live on land, we must now consider what is known of their physiological and behavioural adaptations.

The osmoregulatory mechanisms of isopods have not been investigated in great detail, but information is available for marine, brackish water and freshwater forms, as well as for terrestrial species. The marine isopod *Idotea granulosa* is slightly hyperosmotic to normal sea water and can osmoregulate to some degree in dilute sea water (Todd, 1963). Specimens of the glacial relict species *Mesidotea entomon* taken from the Baltic osmoregulate well, but cannot live in fresh

Fig. 5.5. The activity of *Armadillidium vulgare* on dry (left) and moist (right) nights. Triangles and circles show two separate surveys. Open symbols show vapour pressure deficit.

Closed symbols show isopod density. Activity is much higher on the moist nights, i.e. when saturation deficit is low. Redrawn after Paris (1963).

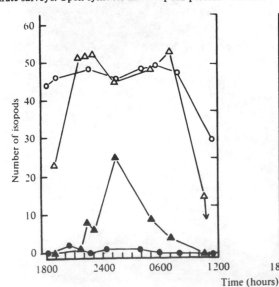
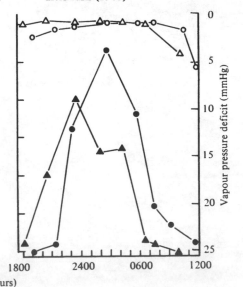

water; while specimens of the same species from fresh-water lakes are also able to live in full strength sea water, indicating a wide tolerance to a range of external and internal osmotic pressures. As *M. entomon* is believed to have colonised fresh water since the last ice age (Lockwood & Croghan, 1957), the development of the necessary osmoregulatory mechanisms has presumably been a relatively recent step. *M. entomon* has maintained a high osmotic pressure even in fresh water and its adaptation has probably involved increased efficiency in the mechanisms of salt uptake, as well as reduced permeability of the external surfaces to salts (Croghan & Lockwood, 1968). The situation in the freshwater *Asellus aquaticus* may be somewhat different, since the osmotic pressure of the haemolymph is much lower (Lockwood, 1959*a*). In *A. aquaticus* a large store of sodium is maintained in Zenker's organ (Lockwood, 1959*b*), and this may be available in times of need; *M. entomon* has no such extra-haemolymph store. *A. aquaticus*, however, is unable to survive in concentrations higher than that of its normal haemolymph, and in this it is more representative of freshwater animals than is *M. entomon*.

This general background of wide osmotic tolerance coupled with some osmoregulatory ability is also found in the various species of *Ligia* that have been examined (Parry, 1953; Todd, 1963; Wilson, W. J., 1970). At least some species of *Ligia* can also maintain their haemolymph hypo-osmotic to that of the external medium at high concentrations, and this has not been shown for other isopods – although it is common in many crustaceans. The ability to hyporegulate is most important for animals that are exposed to desiccation, including the truly terrestrial forms, and maintenance of a constant osmotic pressure in the face of desiccation has been shown for *Porcellio* and *Armadillidium*. Horowitz (1970) showed that when *P. scaber* was exposed to desiccation, the weight diminished in a linear fashion; but the osmotic pressure, after rising a small amount, remained constant for many hours before rising steeply (Fig. 5.7). Similar results were obtained for *A. vulgare* by Price & Holdich (1980). Since Holdich & Mayes (1976) showed that most water lost during desiccation was lost from the haemolymph, and not from the cells, the regulation of osmotic pressure must imply either the sequestration of osmotically active substances from the haemolymph, or the addition of water from some source. Lindqvist & Fitzgerald (1976) have demonstrated that in *P. scaber* the osmotic pressure of fluid in the gut rises during desiccation while that of the haemolymph hardly changes. Some degree of regulation could therefore occur by pumping ions into the gut, or possibly by absorbing water from the gut into the haemolymph. Whatever the process, it could be closely related to that involved in the hypo-osmotic regulation by marine forms. It is here important to note also that in *Oniscus asellus*, which inhabits damper environments than *P. scaber* or *A. vulgare*, no regulation has been detected (Bursell, 1955; Price & Holdich, 1980).

Fig. 5.6. Thermohygrograms showing conditions in various microhabitats in the Chiricahua Mountains, Arizona, where *Armadillidium vulgare* was common. Fine vertical hatching, air at a height of 1 m. Coarse vertical hatching, ground shaded by leaves. Fine stipple, under a log. Coarse stipple, under a large stone. The numbers show times in Mountain Standard time. *A. vulgare* was occasionally active even in the most extreme conditions, possibly when it was changing habitats. Most individuals, however, emerged from beneath rocks and logs only when humidity was high. Redrawn from Warburg (1965*c*).

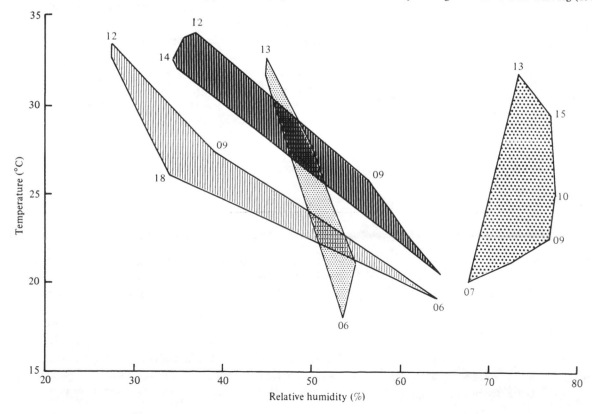

Bearing this varying capacity for osmoregulation in mind, it is informative to compare the composition of the haemolymph of terrestrial forms with that of aquatic isopods. In Table 5.4 some figures are given for two marine forms, *Ligia oceanica* and *Tylos latreilli*, the freshwater form *Asellus aquaticus*, and two terrestrial woodlice, *Porcellio scaber* and *Oniscus asellus*. From these figures it is evident that besides being able to maintain the osmotic pressure of its haemolymph considerably above that of normal sea water, *L. oceanica* is able to regulate some of the component ions: the concentrations of magnesium and sulphate ions are low, and those of calcium and potassium high, in comparison with sea water. *T. latreilli* forms a striking contrast in being unable to maintain magnesium at a low internal concentration. From the situation in *L. oceanica* it is straightforward to derive that in *O. asellus* and *P. scaber*, since all the measured ions, as well as the osmotic pressure, are approximately halved. Simple dilution could therefore explain the composition. This is obviously not so in *A. aquaticus*, however, where the osmotic pressure is only a quarter of that in *L. oceanica*, while the potassium concentration in the haemolymph is no less than in the terrestrial forms. *A. aquaticus* has, therefore, developed further ion regulatory mechanisms in response to its dilute environment, while the terrestrial isopods have remained very similar to the ancestral *L. oceanica* type. A direct origin of the terrestrial forms from marine ancestors is the simplest way of explaining the situation.

The problems involved in the maintenance of a constant haemolymph composition are rather different for the terrestrial forms and for *L. oceanica*. For the latter, salts and water are available at least periodically from the sea. Water is probably taken up passively, because of the high osmotic pressure of *L. oceanica*

Fig. 5.7. Changes in the weight and in the osmotic pressure of the haemolymph in *Porcellio scaber* during desiccation at 87% RH. While the weight diminished regularly, the osmotic pressure reached a plateau, suggesting some regulatory process. Open circles show no significant difference from the 1-hour group, while closed circles show a significance ($P<5\%$). Vertical lines show S.D. Redrawn from Horowitz (1970).

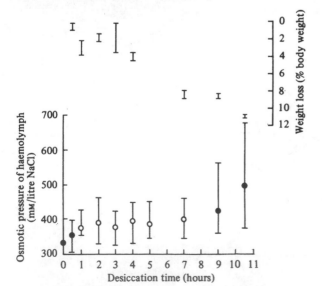

haemolymph. For terrestrial forms, salts are scarcer, but are often available in the substratum and in the food. Liquid water, when present, often contains salts in solution, and Kümmel (1981) has shown that most terrestrial woodlice have transporting epithelia on their pleopods so that they are able to take up ions actively. The water supply on land may be scarce, resulting in desiccation, or at times may be in excess, resulting in osmotic stress or drowning. In both terrestrial and littoral forms, water balance must be achieved by equalising loss rate and uptake rate. It is therefore of interest to examine the various routes by which water is gained and lost by a variety of oniscoids.

Water is lost through the integument, through the respiratory membranes, with the faeces, and presumably from the excretory organs. Transpiration through the general integument has been measured in representatives from six families, as shown in Table 5.5. The rates of water loss vary by more than an order of magnitude, the lowest being from *Venezillo arizonicus*, in which the rate is similar to that in insects, and the highest from *Ligia oceanica*. Edney (1968) has concluded that there is a general correlation between moistness of habitat and permeability of the integument, but has pointed out two important qualifications to this observation. One is that in attempting to define the general conditions in which an animal lives, a local macroclimate is usually described. This, however, consists in reality of a series of microclimates, and as far as woodlice are concerned most species normally occupy a saturated atmosphere, venturing out into unsaturated air from time to time. The length of time that can be tolerated away from a saturated atmosphere is probably the most important factor. The second point concerns size. In an unsaturated atmosphere large animals are at an advantage because of their small surface : volume ratio. This means that even if they lose water at the same rate per unit area as a small animal, they will lose less per unit weight, and the loss will have less effect on their water content. In fact, although *L. oceanica* loses more water per unit area than *Porcellio scaber*, it survives longer in unsaturated air simply because it is larger.

In any one species, several other factors affect the rate of water loss by transpiration, and the overall rates have been investigated by a number of workers. In general it appears that at any one temperature, the rate of loss is proportional to the saturation deficit – the amount by which the water vapour present in a sample of air falls short of the saturation value (see Edney, 1957). Temperature itself generally has a relatively small effect, although there are exceptions (Warburg, 1965a). The situation is complicated by the fact that the rate of water loss decreases with time in most isopods (Bursell, 1955). This apparent decrease in permeability is independent of the degree of hydration of the cuticle. Indeed, the water content of the cuticle does not alter during desiccation (Lindqvist, Salminen & Winston, 1972). The osmotic pressure of the haemolymph, however, increases as desiccation proceeds, as discussed above. Bursell showed that the increasing osmotic pressure caused shrinkage of the cuticle, and suggested that this in turn caused a decrease in permeability because it brought lipid molecules in the cuticle closer together.

The importance of lipids in the cuticle of isopods is not, however, entirely clear. Bursell (1955) produced good evidence that in *Oniscus asellus* there is a lipoid layer within the endocuticle which acts as a barrier to water loss. As with work on insect cuticles, this was initially shown when investigating the effect of temperature on transpiration rate. When *O. asellus* was exposed to different temperatures at a constant relative humidity (not at a constant saturation deficit), the transpiration rate increased with increasing temperature, but two breaks occurred in the curve (see Fig. 5.8), at approximately 25 and 35 °C. These temperatures agreed well with the melting points of two types of lipoid extracted from the cuticle (23 and 37 °C). In *O. asellus* there is no evidence that any epicuticular or exocuticular lipoid layer is present, as found in insects.

Not all isopods, however, are similar to *O. asellus* in this respect. Warburg (1965*b*) studied the desert species *Venezillo arizonicus*, which has a very low rate of transpiration, and a high temperature tolerance. Its rate of water loss increased suddenly at 38–40 °C, even when the saturation deficit was kept constant (see Fig. 5.9). This contrasts with other isopods, even such desert

forms as *Hemilepistus reaumuri*, where temperature has little effect at constant saturation deficit (e.g. Coenen-Stass, 1981), and resembles the situation in insects. Since treatment of the exocuticle of *V. arizonicus* with chloroform increased the transpiration rate, there may well be an external lipid layer in this species, either as well as, or instead of, the endocuticular layer.

The temperatures measured by Warburg (1965*b*) were actually those of the animal, not of the surrounding air, and this raises another important point. When he compared the body temperatures of four species with the external temperature, he found that in *V. arizonicus* and *Armadillidium vulgare* there was no depression of body temperature. In *Oniscus asellus*, however, body temperature was depressed by about 1 °C at 38 °C, and that of *Porcellio scaber* was depressed by as much as 12 °C. Although the ability to reduce the body temperature by high rates of transpiration of water is not directly related to survival at high temperatures, the reduction in temperature may be important for short periods. The whole situation has been examined in more detail by Edney (1953) who measured temperatures of woodlice and of *Ligia oceanica* in the sun. Dead, dry animals remained at air temperature, but living ones had lower temperatures than the air. In shade the depression of temperature was small, but in sun *L. oceanica*

Table 5.4. *The inorganic composition of isopod haemolymph*

Species	Osmotic pressure (mOsm)	Na$^+$	K$^+$	Ca^{2+}	Mg^{2+}	Cl$^-$	HCO$_3^-$	SO$_4^{2-}$
		(mM/litre)						
Sea water (1)	*c.* 1070	497	11	11	56.5	581	–	30
Ligia oceanica (1)	*c.* 1160	586	14	36	21	596	–	4.5
Ligia italica (2)	–	613	16.5	34.6	19	704	–	–
Tylos latreilli (2)	–	577	23.9	25.5	55	636	–	–
Asellus aquaticus (3)	273	137	7.4	–	–	125	–	–
Oniscus asellus (4)	577	230	8.2	16.7	9.1	236	11.5	–
Porcellio scaber (4)	701	227	7.7	14.7	10.9	279	–	–

Data from: (1) Parry (1953) – *L. oceanica* kept in sea water.
(2) Lagarrigue (1969) – *T. latreilli* kept on sand containing 10–20% sodium chloride.
(3) Lockwood (1959*a*) – *A. aquaticus* kept in fresh water.
(4) R. T. Barrett (unpublished) – *O. asellus* and *P. scaber* kept on damp leaf litter.

Table 5.5. *Comparative rates of transpiration in woodlice (from Edney, 1968; data taken from various authors)*

Species	Family	Temperature (°C)	Rate of transpiration (μg cm^{-2} hour^{-1} (mm Hg)$^{-1}$)
Ligia oceanica	Ligiidae	30	220
Philoscia muscorum	Oniscidae	30	180
Oniscus asellus	Oniscidae	23	165
Cylisticus convexus	Cylisticidae	30	125
Porcellio scaber	Porcellionidae	30	110
Porcellio dilatatus	Porcellionidae	28.7	104
Armadillidium vulgare	Armadillidiidae	30	85, 78
Hemilepistus reaumuri	Porcellionidae	19	23
Venezillo arizonicus	Armadillidae	25	15

was as much as 10 °C cooler than the air. Under similar conditions, the order in which temperatures were depressed was: *L. oceanica* > *O. asellus* > *A. vulgare* > *P. scaber*. It might be expected that this order would follow that in Table 5.5 showing relative rates of transpiration, and this is so except for the position of

A. vulgare. This may be due to other factors than evaporation. It must be remembered that when an animal is in air hotter than itself, the heat balance equation will be:

$$\text{Radiation} + \text{Conduction} + \text{Convection}$$
$$= \text{Evaporation.}$$

In *A. vulgare* the cuticle is very shiny and less radiant heat will be absorbed than in the species with dull cuticles. Other factors such as shape and size are also important, but as a general summary it is fair to say that those species which have a high transpiration rate can depress their body temperatures more than those with a low transpiration rate. The graphic diagram by Edney (1953) (Fig. 5.10) showing some appropriate temperatures for *L. oceanica* and its habitats shows the advantages of this capacity for temperature depression.

　　Work by Lindqvist (1971, 1972) has highlighted yet another complication in loss of water by evaporation. It has long been known that the outer surface of the pleopods is kept moist, and that water is conducted to them by a series of capillary channels. Verhoeff (1917) thought that this water was collected from the substratum by the uropods, but the channels are now known to conduct water from the anterior of the animal over the under surface of the body. While it was previously thought that fluid from the mouth was expelled into the capillary system (Lindqvist, 1971), it is now thought that it is urine from the maxillary glands which spreads throughout the system (Hoese, 1981). This keeps the pleopods moist and, as will be discussed later, allows the excretion of nitrogen in the form of ammonia gas. In *L. oceanica* the water entering the capillary system cannot be regained by the animal, but in *P. scaber* and other terrestrial woodlice it is reabsorbed by being taken into the rectum. Water can therefore be circulated outside the animal between the excretory organs and the anus. This means that many of the published figures for permeability of the cuticle

Fig. 5.8. The effect of temperature on the rate of water loss from woodlice. Closed circles show *Oniscus asellus*, exposed to different temperatures at a constant RH of 41%. Triangles show *O. asellus* exposed to different RHs at a constant temperature of 23 °C. Open circles show *Porcellio scaber* exposed to different temperatures at a constant RH of 41%. Breaks in the curves suggest that transpiration is reduced by the presence of lipoids within the cuticle. From Bursell (1955).

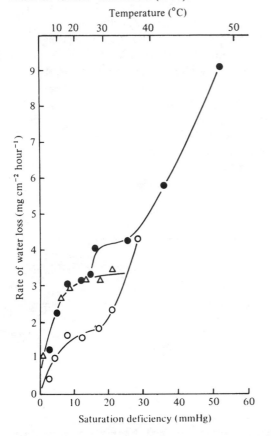

Fig. 5.9. The rate of evaporation from three isopods. Solid lines show the rate in dry air. Broken lines show the rate at constant saturation deficit. The sharp transition shown by *Venezillo arizonicus* at constant saturation deficit suggests the presence of an epicuticular lipid which limits transpiration. From Warburg (1965b).

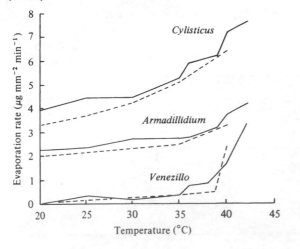

Fig. 5.10. A diagrammatic section to show microclimatic conditions in regions inhabited by the high-shore isopod *Ligia oceanica*. Although temperatures outside the cryptozoic niches are high, evaporative cooling reduces body temperature considerably. From Edney (1953).

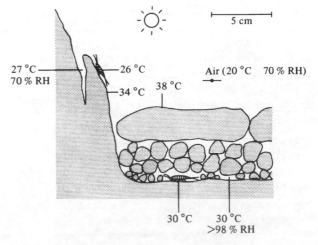

may be too high, since conversion from measured transpiration rates to calculated permeability does not take into account water which has not passed through the cuticle but has been added from the outside.

Since most of the measurements of water loss discussed above relate to loss by whole woodlice, they do not differentiate between water lost through different areas of cuticle. Edney (1951), however, measured the loss from the dorsal surface, the ventral surface and the ventral region which bears the pleopods or 'gills'. Although approximately the same total amount of water was lost from each of these areas, the amount lost per unit area was much greater from the gill area than from the other areas. Partly this is a reflection of decreased permeability of the dorsal integument in association with the development of such features as tubercles (Schmalfuss, 1975), and partly it must relate to the respiratory function of the gills. The latter is considered later in the context of respiratory adaptations.

Water loss in the faeces must be important to isopods, but has been looked at in detail in only two species of *Hemilepistus* (Coenen-Stass, 1981). Kuenen (1959) observed that the faeces may be drier than the food eaten, and Berridge (1970) suggested that the rectal epithelium may be involved in reabsorbing water. The structure of this epithelium has now been shown to be specialised for water and ion transport, the cells having extensive infoldings associated with mitochondria at both apical and basal surfaces (Witkus, Grillo & Smith, 1969; Vernon, Herold & Witkus, 1974). In *Hemilepistus reaumuri* water absorption by the hindgut plays an important part in water balance, and it will be considered under the heading of water gain.

The last route of possible water loss, that of the excretory organs, has not been investigated. Because urine may be 'recycled' in the manner described above, the rate of final loss from the animal may not, in any case, bear a direct relationship to the rate at which fluid is excreted from the maxillary glands.

Since it has been shown that woodlice lose water whenever they are not in saturated air, the mechanisms by which they gain water must be of great importance to them in their natural habitats. The possible sources of water appear to be: as a free liquid, in soil or sand, on damp surfaces and as water vapour. As mentioned above, Verhoeff (1917, 1920) thought that water could be picked up from the substratum by the uropods, but now that water is known to be supplied to the capillary system from the maxillary glands, there must be some doubt about his conclusions. In *Ligia oceanica*, however, the sixth and seventh walking legs have a capillary system of small hairs which allows them to absorb water from the substratum (Hoese, 1981). *L. oceanica* and the terrestrial forms can also take up water from damp porous surfaces and from liquid films through both the mouth and the anus (Spencer & Edney, 1954), but no uptake through the integument has been shown. Unlike a few well-documented species of insects and ticks, woodlice cannot absorb water vapour below 98% RH (Spencer & Edney, 1954). However, water vapour can be taken up in saturated air, and in the

burrowing desert species *Hemilepistus reaumuri* and *Hemilepistus aphganicus* this and the water obtained by eating damp sand form the bulk of the water intake (Coenen-Stass, 1981). The animals retreat to their burrows and when the water content of the sand is 10% or greater, they can absorb water through the cuticle as well as gaining water by reabsorption from ingested sand in the rectum. The mechanism of cuticular water uptake is not known, but it is possible that the transporting epithelia of cells on the pleopods described by Kümmel (1981) are involved.

The respiratory function of the pleopods or gills has already been mentioned. In many aquatic isopods these appendages beat regularly causing a current of water to flow over them, so that they and the ventral surface of the abdomen are well supplied with well-oxygenated water. In littoral isopods such as *Ligia oceanica* the pleopods are kept damp and oxygen is absorbed through them and through the ventral abdominal wall (Edney & Spencer, 1955). When the pleopods were blocked with emulsion paint, the rate of oxygen uptake by *L. oceanica* in moist air fell to about 50% of the normal rate, and the same applied to the terrestrial form *Oniscus asellus*. The remaining uptake was presumably through the ventral cuticle. In dry air the rate of oxygen uptake is much lower in normal *L. oceanica* and *O. asellus*, and when the pleopods were blocked in this case, it fell to about 25% of the already reduced value. These results may be contrasted with those from equivalent experiments with *Porcellio scaber* and *Armadillidium vulgare*. In these, the rate of oxygen uptake was hardly diminished when the animals were moved from moist to dry air, perhaps reflecting the efficiency of the water capillary system in these two species. However, blocking the pleopods in this case reduced the oxygen uptake rate to about 30% of the normal rate in moist air, and to about 6% in dry air, showing that cutaneous respiration is less important than in *L. oceanica* and *O. asellus*. Some detailed figures are given in Table 5.6. It is apparent that the pleopods of *P. scaber* and *A. vulgare* are much more effective at extracting oxygen from dry air than are those of *L. oceanica* and *O. asellus*. This property has been linked with the development in the two former species of tubular invaginations of the pleopod cuticle known as pseudotracheae (Fig. 5.11). An examination of the various types of these structures and their distribution is therefore appropriate at this point. They occur in two groups of oniscoids. The members of the Tylian series possess pseudotracheal pockets on pleopods 2 to 5, opening through a number of individual stigmata on each pleopod (Fig. 5.11). In this group the pseudotracheae are found in the forms which live high up in the marine littoral zone, such as *Tylos latreilli*, as well as in the forest dwellers. Pseudotracheae are also found in the Ligian series, but only in the terrestrial forms. Among these there are two distinct types of organisation of the respiratory organs: in one type (e.g. *Hemilepistus, Trachaeoniscus, Venezillo*) all five pleopods bear pseudotracheae, but the branching of the tubules is only poorly developed. In the other type (e.g. *Porcellio, Armadillidium*) only pleopods 1 and 2 bear pseudotracheae, but the tubules are well branched. No correlations with the environment have yet been properly determined.

Nevertheless, it appears that some forms of pseudotracheae are much more efficient than others – although how they aid respiration in dry air remains unknown. It is possible that they allow oxygen to diffuse inwards for only limited times, so that outward diffusion of water vapour is minimised, but this would depend upon some mechanism for closing the apertures, as is present in insects. No such closing mechanism has been described. The pseudotracheae do not seem to be particularly important in the outward diffusion of carbon dioxide, which presumably occurs through the general integument.

The pleopods are well vascularised, and oxygen diffuses from the pseudotracheae into the haemolymph, or directly across the external cuticle. The oxygen carrying capacity of the haemolymph is increased by

Table 5.6. *Oxygen uptake by woodlice in moist and dry air, and with normal and blocked pleopods (data from Edney & Spencer, 1955)*

Species	Moist air		Dry air	
	O_2 uptake ($mm^3\ mm^{-2}\ hour^{-1}$)	O_2 uptake with pleopods blocked (% normal rate)	O_2 uptake ($mm^3\ mm^{-2}\ hour^{-1}$)	O_2 uptake with pleopods blocked (% normal rate)
Ligia oceanica				
Normal	0.155	52	0.020	<25
Pleopods blocked	0.081		<0.005	
Oniscus asellus				
Normal	0.075	52	0.023	22
Pleopods blocked	0.039		0.005	
Porcellio scaber				
Normal	0.084	35	0.075	<7
Pleopods blocked	0.029		<0.005	
Armadillidium vulgare				
Normal	0.086	26	0.081	<6
Pleopods blocked	0.022		<0.005	

Fig. 5.11. Some isopod pseudotracheae. (*a*) Exopodite of first pleopod of a terrestrial isopod of the Ligian series, *Porcellio scaber*. One large internal cavity has many short pseudotracheal tubules. (*b*) Second pleopod of a high-shore marine isopod of the Tylian series, *Tylos latreilli*. The exopodite has many separate pseudotracheal openings. (*c*) Exopodite of third pleopod of a terrestrial isopod of the Tylian series, *Helleria brevicornis*. The exopodite has two separate openings and well-developed pseudotracheal tubules. (*a*) After Verhoeff (1920); (*b*) and (*c*) after Vandel (1960).

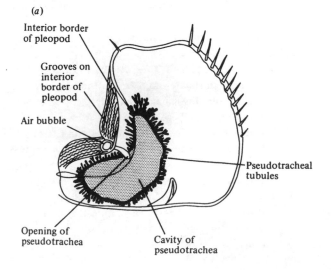

(*a*)

Interior border of pleopod

Grooves on interior border of pleopod

Air bubble

Opening of pseudotrachea

Cavity of pseudotrachea

Pseudotracheal tubules

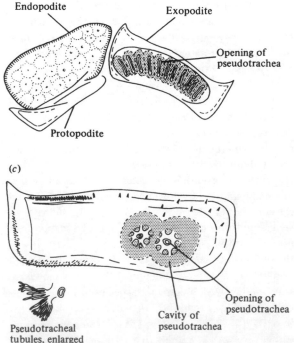

(*b*)

Endopodite

Exopodite

Opening of pseudotrachea

Protopodite

(*c*)

Pseudotracheal tubules, enlarged

Cavity of pseudotrachea

Opening of pseudotrachea

the presence of haemocyanin, as in most crustaceans. In the Ligian series the oxygen affinity of haemocyanin decreases with increasing 'terrestriality', i.e. from *Ligia* to *Porcellio* to *Armadillidium* to *Armadillo* (Sevilla & Lagarrigue, 1979). However, because the oxygen affinity of haemocyanin from the marine *Tylos* is even lower than that of *Armadillo*, it is difficult to evaluate the significance of this.

   Wieser (1967) has drawn attention to the problems of obtaining enough copper to ensure an adequate supply of haemocyanin. In marine isopods, copper is probably taken up directly from sea water, where it is in plentiful supply. Relatively little is absorbed from the food, because in living plant cells most copper may be present in very stable organic complexes. In terrestrial forms, copper lost in the faeces cannot be replaced by uptake from solution, and Wieser has shown that at least under some conditions uptake from leaf litter cannot compensate for the loss. However, isopods are generally able to remain in copper balance when feeding on decayed leaves (Coughtrey *et al.*, 1980). It is when they feed on dried but undecayed leaves that they lose copper, and in this situation they can regain it only by re-ingesting their faeces (Wieser 1968, 1979; Dallinger & Wieser, 1977). The availability of copper on land may therefore be somewhat sporadic, and terrestrial isopods have large stores of copper to tide them over periods of low availability. These stores are mainly in the hepatopancreas, and they are more strictly regulated than those in marine forms. The copper is held in membrane-bound vesicles, and some of the hepatopancreas cells appear to be almost filled with these under normal conditions, although the copper can be mobilised during the moulting process.

   In many terrestrial animals, the necessity of conserving water has led to the development of insoluble, non-toxic nitrogenous end products (Needham, 1938). All isopods, however, have been shown to be primarily ammonotelic (Dresel & Moyle, 1950). It was thought initially that the adaptation of woodlice to a terrestrial habit had involved a general reduction of protein metabolism, rather than a change of nitrogenous end product, but it has now been shown that this is not so (Wieser & Schweizer, 1970). Woodlice release the major portion of their excretory nitrogen as ammonia gas, so that no great water loss need be associated with removal of nitrogen. The total amount of nitrogen lost is similar to that lost by aquatic isopods. It is probable that gaseous ammonia is released from the water film over the pleopods by a rise in pH (Hoese, 1981), in a manner similar to that found in terrestrial snails (see p. 58). It is not known whether ammonia is initially excreted in the urine, or whether it diffuses across the integument of the pleopods from the haemolymph. In either case the ion transporting structures on the pleopods described by Kümmel (1981) may be involved in the alkalisation process. The release of ammonia gas occurs in bursts and at particular times of day (Wieser, Schweizer & Hartenstein, 1969), and the timing is related to the possibility of water loss (Fig. 5.12). Most ammonia is released during the day, when woodlice are usually in humid

microclimates and there is little danger of desiccation. Little release occurs at night when they are normally foraging for food. Although Wieser (1972) found that the high-shore marine species *Ligia beaudiana* excreted ammonia only when sufficient sea water was present to keep the animal's ventral surface moist, Kirby & Harbaugh (1974) found diurnal patterns of ammonia release in a number of littoral isopods. This may suggest that the release of ammonia during the day by isopods was evolved in the littoral zone, and may therefore be regarded as a preadaptation to life on land.

   Other nitrogenous substances excreted by isopods include amino acids, urea and uric acid. None of these appear to be of great importance compared with ammonia, but the distribution of uric acid is intriguing. Some figures for concentrations in whole animals are given in Table 5.7. Although these concentrations do not necessarily indicate importance as excretory products, the

Table 5.7. *Uric acid content of whole isopods and amphipods (from Dresel & Moyle, 1950)*

Species	Habitat	Uric acid content (mg/g wet weight)
Marinogammarus spp.	Marine littoral	0.07–0.10
Orchestia sp.	Supra-littoral	0.08
Ligia oceanica	Supra-littoral	0.06–0.08
Oniscus asellus	Terrestrial	0.08–0.09
Porcellio laevis	Terrestrial	0.06–0.22
Armadillidium vulgare	Terrestrial	0.4–1.2
Gammarus pulex	Fresh water	0.06
Asellus aquaticus	Fresh water	6.1–8.1

Fig. 5.12. The release of gaseous ammonia by terrestrial isopods, in relation to time of day. Open circles, *Porcellio scaber* in dry conditions. Filled circles, *Oniscus asellus* in dry conditions. Half-filled circles, *O. asellus* in moist conditions. Vertical bars show S.E. Most ammonia is released during the day, although this is not so clear in moist conditions. Under natural circumstances the woodlice would then be in humid microclimates, so that evaporation of ammonia would not cause excessive water loss. Similar patterns are now known for intertidal isopods. From Wieser, Schweizer & Hartenstein (1969).

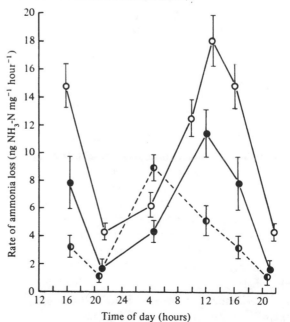

levels do increase from marine forms and terrestrial forms found in moist habitats, such as *Oniscus asellus*, up to *Armadillidium vulgare*, which is tolerant of much drier conditions. *Asellus aquaticus*, however, provides a striking exception to this sequence. In this species, most of the uric acid is found in Zenker's organs (Lockwood, 1959*b*), where its presence may be linked with the retention of sodium in a sodium store.

Problems of water loss are important in many other processes, especially in reproduction. Most crustaceans have separate sexes, and copulation is the general rule even in aquatic forms. The brooding of young in some way is also widespread, so that in these two respects crustaceans are preadapted to life on land. In woodlice the sperm are transferred to the genital opening of the female by the modified pleopods, or stylets, of the male (see Sutton, 1972). The fertilised eggs are passed to the brood pouch, which is a cavity developed below the thorax in ripe females. The pouch is formed from broad, leaf-like plates extending horizontally inwards from the base of the limbs, and initially it is filled with fluid. As the young develop, however, this fluid disappears. Finally the young are released to the outside through gaps between the plates, and at the next moult the female sheds the modified cuticle and reverts to the normal body form. Clearly the whole process provides good protection from desiccation at all stages. The stage most susceptible to desiccation, disease and predation is probably that at which the young are liberated. These become more resistant after the second moult (Cloudsley-Thompson, 1968).

From the above examination of some of the physiological characteristics of terrestrial isopods, it is apparent that while woodlice are common, and can in many ways be called successful terrestrial animals, their physiological adaptations to terrestrial life are not greatly developed. It is now widely agreed that although some basic physiological adaptations are evident, the success of woodlice is largely dependent upon behavioural, rather than physiological, adaptations. Linked with behavioural responses are modifications of the sense organs, although little is known about these.

There are several behavioural responses of woodlice which are concerned with external temperature and humidity. As physiological work has shown, these two factors are important in controlling the distribution of woodlice. One of the commonest responses, not limited to woodlice, is that of conglobation, or curling up to form a ball. Many species of the aquatic genus *Sphaeroma* do this, as also do the intertidal isopod genera *Campecopea* and *Tylos*. Conglobation cuts down the rate of water loss by protecting the thin, permeable pleopods. It is also important in giving protection against predators, and in this many woodlice are aided by the development of various types of tubercles on the dorsal surface (Schmalfuss, 1975), and by the secretions from the tegumental glands, which make them distasteful to a number of predator species, especially spiders (Gorvett, 1956). Conglobation is found in the Armadillidiidae and in *Cylisticus*, but in these cases the antennae are not enclosed within the ball. In the Armadillidae

(*Venezillo, Armadillo*) the ball closes firmly, with the antennae inside. Other reactions, such as thigmokinesis and a negative photoreaction, bring the animals into places where in general they find higher humidities and lower temperatures. Specific reactions to humidity and temperature have also been much investigated, and these reactions are often interdependent, so that animals react to a complex of circumstances rather than to a single factor (Cloudsley-Thompson, 1977). *Armadillidium vulgare*, for instance, is normally positively hygrokinetic, but at high temperatures it becomes negatively hygrokinetic. *Venezillo arizonicus*, which lives in arid environments, is normally negatively hygrokinetic, and stays in medium humidities. It becomes positively hygrokinetic only at low temperatures (Warburg, 1968). The responses to temperature usually take the form of increases in speed and in changes of direction, with increasing temperature. An exception to this is *V. arizonicus*, which shows fewer shifts in direction at higher temperatures. This is presumably an adaptation to a hot arid climate. Another feature which may be of help at high temperatures is the stance of the animal: the desert form *Hemilepistus reaumuri*, for instance, walks with the body well raised when the ground is hot (Edney, 1968).

The sense organs necessary for these behavioural responses have not, in most cases, been identified. For instance, the reaction to light may normally be mediated through the eyes, but blind species such as *Platyarthrus hoffmannseggi* are very responsive to light (Cloudsley-Thompson, 1968). The eyes of most terrestrial isopods are well developed, however, and Thiele (1971) has shown that even in *Ligia oceanica* the eyes are bigger, and have more ommatidia, than in the low-shore genus *Idotea*. Responses to humidity are mediated by hygroreceptors (Jans & Ross, 1963), but there is no information about how these function. Chemoreceptors probably play a large part in directing isopod behaviour, and a contact chemoreceptor has been described in the woodlouse genus *Metoponorthus* (Mead, Gabouriaut & Corbière-Tichané, 1976). This consists of a tuft of hairs projecting from the antenna. In the homologous organ in *Orchestia* and *Talorchestia* the hairs are much longer, and the reduction in *Metoponorthus* may be an adaptation to life in terrestrial leaf litter.

As a general summary of behavioural mechanisms, it can be said that woodlice become less active the more the external conditions become more favourable, and therefore they tend to accumulate in places which are dark, humid and cool. Species from different environments, however, have very different preferred limits to these conditions. Fig. 5.13 summarises some of the relevant behavioural reactions of a number of species, together with some of their physical adaptations to different environments.

5.3.3 The decapods

Within the order Decapoda there are approximately 70 families, and of these about 55 are exclusively marine. It is, then, only a minority of groups within the order which has penetrated into fresh water and on to

Fig. 5.13. A summary of some of the morphological and behavioural characteristics of isopods in relation to habitat. Redrawn from Warburg (1968).

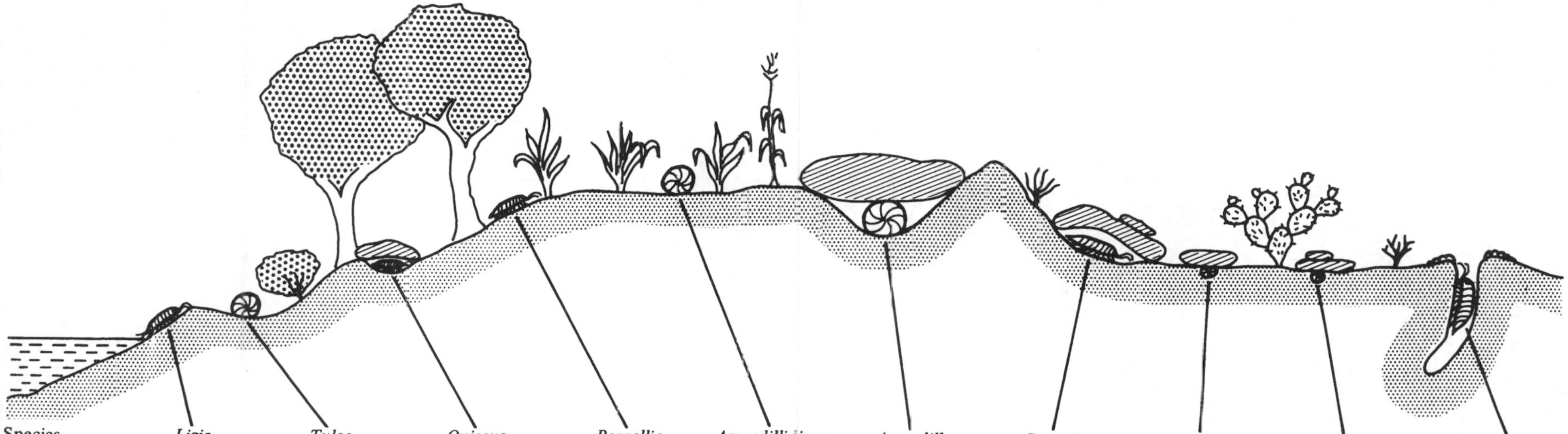

Species	Ligia italica	Tylos latreilli	Oniscus asellus	Porcellio scaber	Armadillidium vulgare	Armadillo officinalis	Porcellio olivieri	Venezillo arizonicus	Armadillo albomarginatus	Hemilepistus reaumuri
Family	Ligiidae	Tylidae	Oniscidae	Porcellionidae	Armadillididae	Armadillidae	Porcellionidae	Armadillidae	Armadillidae	Porcellionidae
Conglobating	No	Yes	No	No	Yes, but excluding antennae	Yes, including antennae	No	Yes, including antennae	Yes, including antennae	No
Respiration	Pleopods and tegument	Pseudotracheae on pleopods 2–4, and with several microstomata	Pleopods	2 pseudotracheae	2 pseudotracheae	5 pseudotracheae	5 pseudotracheae	5 pseudotracheae	5 pseudotracheae	5 pseudotracheae
Capillary water system	Yes	—	Yes	Yes	Yes, but reduced	No	Yes, but reduced	No	No	No
Water loss rate	High	—	High	High	Limited	Limited	Limited	Very limited	Very limited	Very limited
Body temperature	Evaporative cooling	—	Evaporative cooling	Evaporative cooling	Limited evaporative cooling	No evaporative cooling	Limited evaporative cooling	No evaporative cooling	No evaporative cooling	No evaporative cooling
Hygroreaction	Positive	—	Positive	Positive, increased when desiccated	Positive except at high temperatures	—	—	Negative except at high temperatures	—	—
Photoreaction	Negative except at high temperatures	—	Negative	Negative except in dry air and high temperatures	Positive except at low temperatures	Negative	Negative except at high temperatures	Negative	Negative	Positive except at high temperatures
Thermoreaction	—	—	Positive	Positive	Positive	—	—	Negative	—	—

land. Only nine families have terrestrial members, and
it is quite apparent from the distributions given in Table
5.8 that some of these are of direct marine origin, while
others have migrated to land from freshwater habitats.
The former include the terrestrial hermit crabs, the
Ocypodidae, Gecarcinidae and Grapsidae, while the
latter take in the Potamoidea, Parathelphusoidea,
Pseudothelphusoidea and the terrestrial crayfishes.
All these are included in the suborder Reptantia (forms
which are basically adapted for walking) and there are,
reasonably enough perhaps, no terrestrial members
of the suborder Natantia (forms basically adapted for
swimming).

Since one would expect the major problems faced
by decapods colonising the terrestrial environment to be
concerned with salt and water balance and with reproduction, it should be instructive to compare the
approaches taken to these problems by colonists from
fresh water and from the sea. Within the Astacura all
the terrestrial forms have freshwater ancestors, and

many are themselves little different from freshwater
forms. In the Anomura, some of whose members are
highly adapted to a terrestrial life, there are almost no
freshwater members. It is in the Brachyura that we find
the greatest diversity and the widest variety of habitat.
Here there are families with members living in fresh
water and on land, families with members in the sea,
brackish water and on land, and one family, the
Grapsidae, with representatives in the sea, brackish
water, fresh water and on land. A consideration of the
Brachyura is therefore perhaps the most instructive in
discussing routes of colonisation. But first we must
examine the Astacura and Anomura: as mentioned
above, these provide us with examples of the least, and
perhaps the greatest, adaptations found within the
order.

Astacura

Little attention has been paid to the semi-
terrestrial crayfishes and such differences as may exist

Table 5.8. *Distribution of the Decapoda Reptantia*

Taxon	Habitats				Genera
	Marine	Brackish water	Fresh water	Terrestrial	
Palinura					
3 families	x				*Palinurus*
Astacura					
2 families	x	x			*Homarus*
Astacidae		x	x	(x)	*Cambarus*
Parastacidae			x	x	*Cherax*
Austroastacidae			x	x	*Austroastacus*
Thalassinidae	x			(x)	*Thalassina*
Anomura					
10 families	x				*Eupagurus*
Coenobitidae	(x)			x	{ *Coenobita* *Birgus*
Aeglidae			x		*Aegla*
Brachyura					
17 families	x	x			*Carcinus*
Hymenosomidae	x		x		
Potamoidea			x	x	*Potamon*
Pseudothelphusoidea			x	x	*Potamocarcinus*
Parathelphusoidea			x	x	{ *Holthuisana* *Paratelphusa*
Ocypodidae	x	x		x	{ *Macrophthalmus* *Ocypode* *Uca*
Mictyridae	x	x			
Grapsidae	x	x	(x)	x	{ *Sesarma* *Aratus* *Pachygrapsus* *Metopaulias*
Gecarcinidae	x	x		x	{ *Cardisoma* *Gecarcinus*

Parentheses indicate semi-terrestrial forms.

between them and their truly freshwater counterparts. Yet all three families of crayfishes have species which crawl about on land, especially in wet weather; and all have species which make burrows either in dry soil or in places where the water supply dries up at certain times of year.

In the Astacidae the common British species is *Austrapotamobius pallipes* (formerly *Astacus fluviatilis*). This lives in freshwater streams and rivers, but can also live on land in damp conditions for several days. In North America, species of *Cambarus* and *Procambarus* burrow in wet pastures and swamps. Although *A. pallipes* can walk on land, its speed is much reduced (Pond, 1975), probably because its apparent weight in air is 400–600% greater than that in water. Movement on land is also jerky for the same reason. The time spent on land is, however, probably limited by respiratory rather than mechanical problems. When placed in air *A. pallipes* shows increased pCO_2 and decreased pH in the haemolymph (Taylor & Wheatly, 1981), a phenomenon found in many semi-terrestrial invertebrates and vertebrates. The acidosis is partly respiratory but partly metabolic, reflecting build-up of lactic acid. The gills of this species apparently do not allow sufficient gas exchange to permit permanent life on land. The renal system is also adapted to a freshwater existence, and the antennal glands can produce hypo-osmotic urine (Riegel, 1963), unlike those of most crabs, which are involved in ionic, but not osmotic, regulation.

Two other families of freshwater crayfishes, the Parastacidae and Austroastacidae, are found on mainland Australia and in Tasmania (Clark, 1936). All the species, both aquatic and terrestrial, are burrowers. Some terrestrial species are solitary, and the burrows may reach depths of 2 m. Others form community burrows which converge on a central pool of water. As far as is known, all species burrow down to water level, or to the level of very damp soil. In Australia, when they form dense communities their burrows may riddle the ground so that it collapses under the weight of cattle and horses. Breeding occurs in the spring, after moulting. The eggs are attached to the female, where they hatch and then attach themselves to her swimmerets, as in freshwater crayfishes of other families (see, e.g. Huxley, 1881). In general it must be assumed that terrestrial crayfishes maintain themselves in extremely moist conditions, and therefore show few physiological or structural adaptations to a land existence; but the lack of investigation on this score must leave this an area of doubt.

One further interesting animal within the Astacura is *Thalassina anomala* (Thalassinidae). This lives in burrows at or above high water mark in the Indo-Pacific (Pearse, 1911; Macnae, 1968). The burrows are marked by mounds nearly 1 m in height, and are plugged when the entrance is not in use. All of them reach down to standing water, although some are dug in almost terrestrial grassy meadows. Since it is likely that much of this water is poorly aerated, it is interesting that *T. anomala* is able to ventilate its gill chamber by alternately compressing and relaxing the lateral branchiostegites. This method will probably be more effective than the normal ventilatory beating of the scaphognathites. *T. anomala* is a mud feeder, rather than a vegetarian, although Pearse (1911) suggested that it may also feed on higher plants. It can make sounds, possibly by stridulating in a fashion analogous to the ghost crabs, but nothing is known of their significance.

Anomura

The anomurans are probably not a natural group, but include a number of separately derived forms (Barnes, 1980). In all of these the abdomen is somewhat reduced in comparison with that of the Astacura, but not usually to such a degree as in the true crabs, the Brachyura. Most anomurans are marine, or are tied to the sea by their methods of reproduction, but there is one genus of galatheids (squat lobsters) found in fresh water. This genus, *Aegla*, is found in South American rivers. *Aegla laevis*, for instance, is found in Chile, in rivers with gravel or shingle bottoms (Bahamonde & Lopez, 1961). In the spawning period the females migrate to the river's edge, where they remain under stones, out of water, for long periods. During this time their associated fauna consists of arachnids, isopods and insects, and presumably they must be resistant to desiccation to some degree.

The best known anomurans are the hermit crabs, and a number of these have been investigated. Two separate anomuran lines have evolved the hermit crab habit (MacDonald, Pike & Williamson, 1957): the Paguroidea and the Coenobitoidea. The former are all marine, while many of the latter are terrestrial (Fig. 5.14).

Members of the Paguroidea are common in the marine littoral of the temperate zone. *Pagurus bernhardus*, for instance, inhabits gastropod shells and has some tolerance of desiccation, but shows little adaptation to terrestrial life. It has a complete set of 26 gills, broods a large number of eggs (12 000–15 000) before releasing them into the sea where they develop into zoea larvae, and is never found far away from the sea (Kaestner, 1970). Of the paguroids examined, none can osmoregulate (Young, 1979), but the shell is extremely important in delaying osmotic equilibrium with the external water, so that species such as *P. bernhardus* can tolerate very hypo-osmotic environments for short periods (Shumway, 1978). An analysis of the haemolymph of this species is interesting because although it shows a very low magnesium concentration (29 mM/litre, or half that in sea water), it has a very high sulphate level (about 38 mM/litre, or 138% of that in sea water) (Robertson, 1960). The low magnesium is a common feature in Crustacea, and it has been suggested that this is associated with increased activity; but most crustaceans also have a low level of sulphate, and the high level in *P. bernhardus* has not been explained.

Paguroids show no obvious structural adaptations for air breathing. In the Dry Tortugas, islands to the southwest of Florida, Pearse (1929a, 1936) studied a large subtidal species, *Petrochirus bahamensis*, and a smaller intertidal species, *Calcinus tibicen*. Both had

Fig. 5.14. A terrestrial coenobitoid hermit crab from Papua
New Guinea. (*a*) Emerging from the protective gastropod shell.
(*b*) Aperture of the shell closed by the chelae. Shell length 5 cm.

a full complement of 26 gills. However, although the subtidal genus *Pagurites* studied by Burggren & McMahon (1981*a*) showed a fall in rate of oxygen consumption when transferred from water to air, the intertidal species *Pagurus bernhardus* showed an increase in rate, suggesting that some species at least are physiologically adapted for aerial respiration.

The Coenobitoidea form a distinct contrast to the Paguroidea, and show several adaptations to a semi-terrestrial and terrestrial life. The genus *Clibanarius* is mainly intertidal, but *Clibanarius vittatus* migrates offshore in the winter (Fotheringham, 1975), presumably to avoid the harsh physical conditions on the shore. Unlike paguroids, species of *Clibanarius* can maintain hyperosmotic haemolymph when placed in diluted sea water (Young, 1979; Sharp & Neff, 1980). *C. vittatus* is also more tolerant of desiccation than paguroids (Young, 1978), and *Clibanarius tricolor* is adapted to air breathing by a reduction in the number of gills from 26 to 18 (Pearse, 1929*a*; Fig. 5.15). The genus is therefore to some extent adapted to the alternating conditions of submersion and emersion in the intertidal zone.

The genus *Coenobita* is abundant on tropical coasts and most species live an essentially terrestrial life except for their reproduction. *Coenobita perlatus*, for example, is common on the shores of such Pacific atolls as Eniwetok (Gross, 1964*a*). It is found on exposed parts of the shore and also in forested areas, and may be active during the day as well as during the night. The

osmotic pressure of its haemolymph usually remains above that of sea water. When offered a choice between sea water and fresh water, *C. perlatus* chooses sea water under normal conditions, and it usually moves down to the sea at night to fill its shell with sea water. When its haemolymph composition is altered from normal, however, it is able to take up appropriate proportions of fresh and salt water to bring this composition back to normal (Gross & Holland, 1960). *C. perlatus* cannot regulate its haemolymph hypo-osmotically, as can some brachyurans, but it is very tolerant of a range of internal concentrations. Like *Pagurus bernhardus*, it has a low concentration of magnesium in the haemolymph, but there are no figures for sulphate concentrations.

Coenobita brevimanus is usually found in wooded areas, or in piles of rotten coconuts. It is not active in such a wide range of conditions as *C. perlatus*. The osmotic pressure of its haemolymph is about 80% of that of sea water (Gross, 1964*a*). It has not been observed in the sea, and when offered a choice it prefers fresh water to sea water. It carries water in its shell, often in large quantities, and replenishes this from time to time. This has to be picked up from pools, and cannot be absorbed from damp substrates, in the way found among some isopods (see p. 77). As with *C. perlatus*, the osmotic pressure of the haemolymph is regulated by behavioural selection of uptake from salt or fresh water, but in this case the regulation is more effective, and the osmotic pressure is maintained within narrow limits.

Fig. 5.15. The distribution of hermit crabs in relation to tidal height on the Dry Tortugas Islands. Diagrams on the right show the number and distribution of the gills. *Petrochirus bahamensis* and *Calcinus tibicen* (Paguroidea) are sublittoral and have the full complement of gills. *Clibanarius tricolor* and *Coenobita*

clypeatus (Coenobitoidea) are found at the top of the shore and have fewer gills. This presumably reduces the surface area from which water loss can take place, and allows further development of the lung surface across which aerial gas exchange occurs. Redrawn after Pearse (1936).

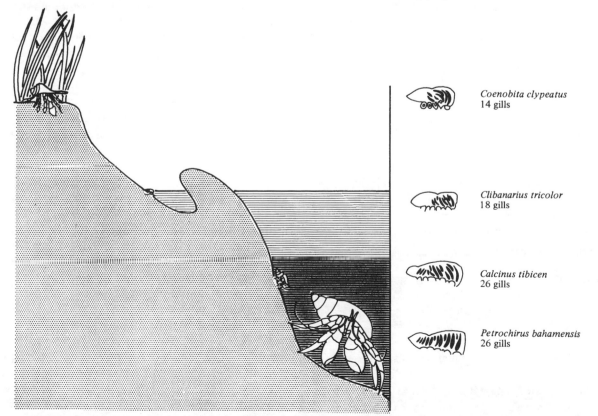

Coenobita clypeatus
14 gills

Clibanarius tricolor
18 gills

Calcinus tibicen
26 gills

Petrochirus bahamensis
26 gills

Other species of *Coenobita* have been investigated in different regions. In the Dry Tortugas, Pearse (1929b) found that *Coenobita diogenes* lives a purely terrestrial life, and its number of gills has been reduced to 14. The physiological adaptations of *Coenobita clypeatus* to terrestrial respiration have been examined by McMahon & Burggren (1979). The scaphognathites pump air through the gill chamber, probably aided by branchiostegal pumping as mentioned for *Thalassina*. Even so, pCO_2 in the haemolymph is high, and pO_2 is low. The reduced gill area and thickened respiratory surface reduce water loss, but also prevent rapid diffusion of oxygen. This is only partially compensated for by the presence of a very high affinity haemocyanin.

In the Red Sea, Niggemann (1968) has observed *Coenobita scaevola*, and has shown that its form of locomotion is different from that of aquatic hermits such as *Pagurus*. In the latter the leg joints are loose and the legs can be moved in a variety of directions; when the animal walks on a dry surface, the body sags and touches the ground. In *Coenobita* the leg joints are firmer, and the legs can only move in a sagittal plane. They are also held nearer the body, and the animal stays off the ground. Its sequence of leg movements is rather like that of an insect, but it also uses its large claws for balance. *C. scaevola* is also able to detect moving objects at a considerable distance in air, and this is a further difference from aquatic forms. Another hermit found on the shores of the Red Sea is *Coenobita jousseaumei* (Magnus, 1960). This species either hides in holes or rocks, or burrows into the sand during the day. Since it burrows to a depth of about 7–13 cm, it avoids the extreme temperatures of up to 33 °C found in the surface sand. At a depth of 10 cm temperatures do not rise above 28 °C.

Birgus latro, the coconut or robber crab, has been more fully studied than other terrestrial hermits (see Storch, Cases & Rosito, 1979). It does not inhabit a gastropod shell except as a juvenile (Harms, 1938), and the abdomen, instead of being spirally coiled to fit such shells, has become secondarily symmetrical again (Smith, 1909). Large calcified plates protect the abdomen and take the place of the shell. Another adaptation is found in the respiratory system: unlike *Coenobita*, which retains gills, *B. latro* has a modified branchial chamber, with two parts. The dorsal part has well-vascularised folds and ridges, and functions as a lung, while the ventral part contains rudimentary gills. These have a thick cuticular covering, and appear to be involved in ion and water transport, whereas the epithelium of the lung is thin and is kept moist by glandular secretions (Storch, Cases & Rosito, 1979). The haemolymph pressure is high, and may help to keep the lung vessels inflated. The scaphognathites pump air through the respiratory chamber and do not reverse their beat as in aquatic crabs (Cameron & Mecklenburg, 1973). In spite of this, however, extraction efficiency is low (2–7%) compared with aquatic forms, where it is 20–70%.

The respiratory adaptations of *Birgus latro* were investigated on a cruise of the *Alpha Helix* to the Palau

Islands in the Pacific (see Cameron, 1981a). The gills occupy only 12 mm^2 g^{-1} weight of crab, compared with over 1000 mm^2 g^{-1} in aquatic crabs. The animal is therefore essentially an air breather, and it uses ventilatory compensation to regulate its acid–base status, rather than exchanging H^+ and HCO_3^- across the gills, which is the normal method of regulation in aquatic animals (McMahon & Burggren, 1981). In this it resembles terrestrial vertebrates (see Chapter 9), but in contrast to them the rate of ventilation is controlled not by pCO_2 but by either pH or HCO_3^- concentration (Smatresk & Cameron, 1981). Besides using these ventilatory mechanisms, it can also redistribute bases between intracellular and extracellular compartments, but it does not use renal mechanisms to regulate acid–base balance (Cameron, 1981b). During dehydration, oxygenation of the haemolymph was not greatly affected because although metabolic acidosis occurred, tending to shift the oxygen equilibrium curve to the right (i.e. to decrease oxygen affinity of haemocyanin), the increased concentration of ions in the haemolymph caused by dehydration tended to shift the equilibrium to the left (i.e. to increase oxygen affinity) (Burggren & McMahon, 1981b). *B. latro* is therefore well adapted to air breathing, and shows parallels in respiratory mechanisms with those found in terrestrial vertebrates.

B. latro lives in similar habitats to those of *Coenobita brevimanus*, but also in some regions climbs coconut trees, from which it gains its name. It feeds on the juice of the nuts, and uses the husks to line its burrow. Moulting occurs in these burrows, and exuviae are never seen because they are eaten (Held, 1963). The habits of *B. latro* probably differ in different regions, however, and it is common in regions lacking coconut palms. There it eats carrion, acts as a general scavenger, and consumes the pith of the sago palm (*Arenga*) and the screw pine (*Pandanus*) (Gibson-Hill, 1947). *B. latro* is much larger than other terrestrial hermits, often reaching more than 30 cm in length, and no doubt this is one reason why more is known about it than about *Coenobita*. The osmotic pressure of the haemolymph is maintained at a lower level than that of sea water (averaging 74.7% sea water), and given a choice of fresh and salt water *B. latro* can regulate this osmotic pressure in the same way as *C. brevimanus* (Gross, 1964a). *B. latro*, however, has no shell to dip into the sea and in which it stores water. Instead, liquid is picked up by the chelipeds, transferred to the maxillipeds, and then in turn to other mouthparts. Also, the last pair of thoracic appendages, usually held inside the branchial cavity, can be pushed forward to take water from the maxillipeds. They then pass it into the branchial cavity (Gross, 1955), where presumably it is used to keep the respiratory membranes moist. It is not known whether *B. latro* has a high rate of urine production (Kormanik & Harris, 1981), although it has a high clearance rate for inulin. It has been suggested that it may re-ingest urine, thereby conserving sodium. It can tolerate a 22% loss of its body water, and during dehydration urinary filtration ceases (Harris & Kormanik, 1981). The importance of urine in nitrogenous excretion is unknown, but the concentrations of both ammonia and uric acid in the haemolymph are high. This has led Henry & Cameron (1981) to suggest that the mechanism of nitrogenous excretion could involve the evolution of gaseous ammonia as shown in terrestrial isopods and terrestrial pulmonate molluscs.

These brief descriptions of several species show that while *Coenobita perlatus* is little changed from 'normal' marine animals, except for its behaviour, *C. brevimanus* has become somewhat adapted for a terrestrial life, and *B. latro* has become even more so. Indeed, *B. latro* and at least some species of *Coenobita* will drown if kept submerged for any length of time (Gross, 1955). According to Calman (1911), some species of *Coenobita* have not only very vascularised abdomens, but a pair of contractile vesicles at the base of the abdomen to act as accessory hearts and provide an increased circulation through these abdominal vessels. In this way the soft abdomen acts as a kind of extra lung. All terrestrial hermits, however, are linked to the sea by their breeding behaviour. Ovigerous females migrate to the sea and deposit the hatchlings in sea water. The young hatch as zoeas, and then pass through successive moults to become glaucothoe larvae. It is this stage which is amphibious and crawls out on to land to enter small gastropod shells (Reese, 1968).

Brachyura – the 'freshwater crabs'
The true crabs will be considered under four headings. First come the 'freshwater crabs', some of which show extreme adaptations to life on land. Next are members of the Grapsidae, which have become adapted to a wide variety of habitats. Third are the members of the Ocypodidae, which have not moved inland from the strand line, but which are such a common feature of tropical and semi-tropical shores. The fourth section considers the Gecarcinidae, the family containing many species highly adapted to land. Lastly, having considered the groups in turn, it will be possible to make some general comparisons between them.

Many genera which are usually known as 'freshwater crabs' have members which spend a good deal of time on land as well as those which are aquatic. These freshwater crabs were previously all referred to the family Potamonidae, or to three families, Potamidae, Pseudothelphusidae and Trichodactylidae (Kaestner, 1970). Bott (1970), however, considered that they should be grouped into three superfamilies, characteristic of different geographical regions: Potamoidea (five families) found in Europe, Africa, the Middle East and Far East; Parathelphusoidea (three families) in India, the East Indies and Australia; and Pseudothelphusoidea (three families) in Central and South America. This grouping is used here.

Potamon niloticus is an aquatic species of the Potamoidea. It is common in lakes and rivers in eastern Africa, and its osmoregulatory physiology has been investigated by Shaw (1959). *P. niloticus* differs from many other freshwater animals in that it has a low permeability of the body surface to salts and to water, whereas other animals have a low permeability to salts

only (Potts, 1954b). *P. niloticus* is therefore able to survive while maintaining haemolymph with a high osmotic pressure (492 mOsm), and producing very small quantities of iso-osmotic urine, although there may also be some extra-renal excretion of water. Salt lost by diffusion is balanced by active uptake. Three amphibious species of *Potamon* were investigated by Dandy & Ewer (1961). These showed relatively little adaptation to terrestrial life. Although they did not have high rates of evaporative water loss, they could withstand total losses of only 15–19% of their body weight. Similarly in *Sudanonautes*, lethal water loss was only 20% of the body weight (Lutz, 1969). Much of the water loss in species of *Potamon* occurred through the integument as well as through the gills, showing that the crabs had no external waterproofing layer. Although the potamonoids are therefore well adapted to a freshwater existence, and can survive as amphibious forms, those species investigated have only a limited ability to live on land.

The respiratory adaptations of South American pseudothelphusoids have been studied by Díaz & Rodríguez (1977). Some of these species are able to stand long periods out of water when rivers and streams dry up. In the family Trichodactylidae the area of the branchial chamber above the gills has a smooth lining, but in the Pseudothelphusidae this has a unique perforated region which appears to provide increased surface area for oxygen uptake. Little general information is available about the pseudothelphusoids, but it is interesting that *Potamocarcinus richmondi* is able to stridulate, and this presupposes a certain degree of communication between individuals. The sounds are produced by the movement of the second and third maxillipeds, and this appears to be unique among the decapods: other crabs almost always employ the cheliped (Abele, Robinson & Robinson, 1973).

Of the Parathelphusoidea, the most studied example is *Holthuisana transversa*, a crab found in arid zones of Australia, where annual rainfall is less than 200 mm. This species lives in burrows which do not reach down to water, except during the rainy season when it emerges and feeds (Greenaway & Taylor, 1976). Like the potamonoids, *H. transversa* has haemolymph with a fairly high osmotic pressure (525 mOsm), and produces iso-osmotic urine in small quantities, and it probably also has an extra-renal mechanism of eliminating water (Greenaway, 1980, 1981). It has a mechanism for sodium uptake in the gills which has a very high affinity for sodium, and a low rate of sodium loss. When in air it has a low rate of evaporative water loss (MacMillen & Greenaway, 1978). It can tolerate a water loss of 31% of its body weight (45% of its body water), which is much greater than the potamonoids. It can drink water to recover from dehydration, but more importantly for its normal existence, it can absorb moisture which condenses in the burrows as the temperature falls at night. It can therefore stay in water balance in regions where conditions in the macroclimate are exceedingly severe, by exploiting the use of its burrow (Greenaway & MacMillen, 1978).

The adaptations of *H. transversa* to respiration in air are also more extreme than those of other freshwater crabs (Greenaway & Taylor, 1976; Taylor & Greenaway, 1979). The scaphognathites do not produce an air flow through the branchial chambers as they do in semi-terrestrial anomurans such as *Birgus latro*. Instead, a tidal flow is produced by lateral movements of the thoracic wall so that the visceral mass oscillates slowly from one side to the other, expanding and contracting the branchial chambers alternately on each side. The cuticle in the lung is very thin, and the distance across which oxygen has to diffuse from air to haemolymph is only 0.2 μm, which is similar to that found in vertebrates. In comparison, the air to haemolymph distance in most semi-terrestrial crabs is 1–3 μm, and the water to haemolymph distance in the gills of *H. transversa* is 5–8 μm.

Other semi-terrestrial species of parathelphusoid also have haemolymph with a high osmotic pressure (Shaw, 1959; Padmanabhanaidu & Ramamurthy, 1961), but little else is known about their osmoregulatory abilities. *Paratelphusa hydrodromous* has been shown to excrete primarily ammonia in fresh water, but to switch to urea production as salinity rises (Krishnamoorthy & Srihari, 1973). Although this phenomenon is also seen in some brackish water vertebrates (see Chapter 9), it is usually accompanied by the retention of urea in the blood so that although the osmotic pressure rises, most of the increase is made up by urea and not by salts. In *P. hydrodromous*, however, the urea is not retained and the significance of the switch is unknown. Species of *Paratelphusa* are common in tropical Asia where they burrow into the banks of streams and the 'bunds' separating paddy fields (McCann, 1937; Fernando, 1960). Some species, e.g. *Paratelphusa guerini*, are active on land both by day and night, where they catch and eat a variety of animals: frogs, lizards, beetles, moths, grasshoppers, etc., as well as scavenging for carrion. The burrows vary in shape and dimensions with the species, but all reach down to the water table. The crabs probably copulate inside the burrows, and the females emerge with the young crabs already hatched and held beneath the abdomen by the pleopods. The young are therefore produced and hatched without return to a river or stream, but they are soon liberated into water by their mother. They then spend some time leading an aquatic life before venturing back on to land.

The three superfamilies of freshwater crabs show a wide range in their degree of emancipation from aquatic requirements. Their movement on to land has in each case occurred from fresh water, and not directly from the sea. However, in contrast to most other freshwater invertebrates colonising the land, they have retained haemolymph with a high osmotic pressure and iso-osmotic urine. This is related to their development of an integument with an extremely low permeability, not found in most freshwater invertebrates. The low permeability has therefore in some sense preadapted them for a terrestrial life. Their respiratory adaptations are in some cases better fitted for aerial breathing than those of other terrestrial crabs. In their

production of young, some species at least are almost independent of a water supply, and this contrasts strongly with all other terrestrial decapods. More information on reproductive strategies within the group would be of the greatest interest.

Brachyura – the grapsids

The grapsids are more widely distributed in terms of habitat than any other family of crabs, being found in the sea, the littoral zone, brackish and fresh water and on land. Some of the species found in the littoral and sublittoral zone in Tasmania have been investigated by Griffin (1971), who detailed the adaptations and preferences of species found from exposed rock platforms to sheltered marshes. As in other groups of crabs, there is some tendency for species living higher on the shore to have lower gill volumes relative to body volume than those species of the sublittoral. Griffin pointed out, however, that because some species seek shelter on the shore while others do not, desiccation stress is not closely related to position on the shore, and consequently the relation of gill volume to tidal height is not a close one. This emphasises the need to look at many aspects of biology in relation to movement out of the sea.

The best known and most studied grapsid is *Pachygrapsus crassipes*, the lined shore crab of the west coast of North America. *P. crassipes* is found from low water up to extreme high water, usually where the substratum is rocky. It is most active at night, but also hunts its prey in the daytime, using visual rather than chemical clues: its eyestalks can move forward and laterally, and can raise the eyes above the level of the carapace (Hiatt, 1948). It also uses side to side movements of the eyes to scan the field and to improve the visual acuity, a phenomenon investigated in more detail in another rocky shore grapsid, *Leptograpsus variegatus* (Sandeman, 1978). It is interesting to compare *P. crassipes* with the portunid *Carcinus maenas*, which is common intertidally on the east of North America, and in Europe. *C. maenas* has receptors responsive to chemicals such as amino acids (Case & Gwilliam, 1961) and, rather than using its eyes, probably finds its food mainly through chemical stimuli. While *P. crassipes* normally moves sideways like most crabs, it can also move directly forwards, unlike *C. maenas*, which is unable to move other than sideways. *P. crassipes* is a fairly fast runner, as might be expected in a predator hunting by eyesight, and reaches speeds of about 1 m/s. This is slow, however, when compared with more terrestrial grapsids such as those in the genus *Sesarma*, which have been reported to run at more than twice this speed (Hiatt, 1948). *P. crassipes* differs from grapsids of the lower littoral in not being gregarious, although it has only a poorly developed degree of the territoriality often found in solitary species. Many other grapsids are common intertidally, especially in the mid to upper littoral, on both soft and hard substrata. In New Zealand, *Cyclograpsus lavauxi* (Fig. 5.16) is found at or above high water mark under shingle, often together with the semi-terrestrial nemertine *Acteonemertes*. *Helice crassa* (Fig. 5.17) is

Fig. 5.16. *Cyclograpsus lavauxi*, an upper-shore grapsid from shingle shores in New Zealand. Regions of geniculate hairs, which aid circulation of water films over the body, can be seen on the branchiostegites, below the eyestalks. Carapace width 2.5 cm.

Fig. 5.17. *Helice crassa*, an upper-shore grapsid from sandy shores in New Zealand. The eyestalks can be raised nearly vertically, improving all-round vision. Carapace width 15 mm.

also found high on the beach, but in mud and sand where it forms a burrow. Two low-shore American species invite comparison with *P. crassipes*: *Hemigrapsus nudus* occurs in sandy areas, and *Hemigrapsus oregonensis* in mud. Of these, *H. nudus* has in comparison to *P. crassipes* a rather coarse set of hairs at the entrance to the respiratory channel, while *H. oregonensis* has a well-developed sieve of very fine setae in the same position (Hiatt, 1948). These are undoubtedly adaptations for preventing either sand, or mud, respectively, from entering the respiratory chamber, and *P. crassipes*, which lives on clean rocky shores, has no well-developed filter.

When given a choice of salinities, *P. crassipes* prefers 100% sea water to more dilute or more concentrated media (Gross, 1957). If allowed to visit both saline and fresh water, it can maintain its haemolymph with normal levels of sodium and potassium, in the same way as discussed for some hermits. It is capable of hypo-osmoregulating in concentrations of up to 190% sea water, however, which hermits cannot do, and it hyperosmoregulates in dilute media (Jones, 1941; Gross, 1957). Sodium regulation is particularly similar to that shown by *Carcinus maenas*, since below an external concentration of 250 mM/litre sodium, the internal concentration parallels that of the external medium. In very high concentrations of sodium a similar pattern is seen, and here it appears that extra-renal excretion must account for sodium loss. It is possible that the crabs may drink to maintain their body water, and excrete excess salt extra-renally (Rudy, 1966). One difference from *C. maenas* is that in dilute sea water, *P. crassipes* loses more than twice as much sodium in its urine, suggesting that the urine flow rate is greater

Fig. 5.18. Osmoregulation in a variety of crabs. The thick line is the iso-osmotic line and also represents the curve for *Cancer antennarius*. Below, the crabs are placed in estimated order of 'terrestrialness', and this can be seen to correlate well with osmoregulatory ability. *Cancer antennarius* (Cancridae): subtidal; *Hemigrapsus oregonensis* (Grapsidae): usually submerged, found in brackish and hypersaline water; *Pachygrapsus crassipes* (Grapsidae): semi-terrestrial, mainly nocturnal; *Grapsus grapsus* (Grapsidae): semi-terrestrial, more diurnal than *P. crassipes*; *Ocypode ceratophthalma* (Ocypodidae): high on the beach, nocturnal; *Uca crenulata* (Ocypodidae): high on the beach, diurnal; *Gecarcinus lateralis* (Gecarcinidae): terrestrial, rarely enters water. Redrawn from Gross (1964b).

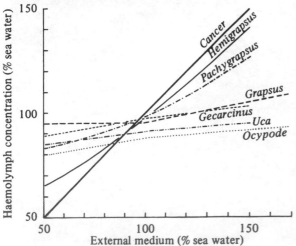

in this species. This in turn suggests a more permeable body surface. Osmoregulation is similar in *Hemigrapsus oregonensis*, but the latter cannot regulate as well in either dilute or concentrated media (Gross, 1964b). *Grapsus grapsus*, on the other hand, which is found at the same tidal level as *P. crassipes*, but is more diurnal, can regulate to a greater degree than *P. crassipes* (see Fig. 5.18). Such findings have suggested to Gross (1964b) that there is a strong correlation between 'terrestrialness' and ability to hypo-osmoregulate. This hypo-osmoregulation is not carried out by the antennal glands, which produce iso-osmotic urine, but these glands are responsible for differential excretion of ions, especially of magnesium and sulphate. In *Goniopsis cruentata* the urine has high concentrations of magnesium and sulphate, which account for the low concentrations of both ions in the haemolymph (Zanders, 1978). In *H. oregonensis*, the normal magnesium concentration of the haemolymph is 23.5 mM/litre, and of the urine 130.5 mM/litre (Gross, 1961). In *P. crassipes*, haemolymph magnesium is only 10 mM/litre, while urine magnesium is 118 mM/litre.

Fig. 5.19. Concentrations of (a) sodium and (b) magnesium in the haemolymph and urine of five crabs. All specimens were immersed in sea water. In the sequence *Cancer antennarius*, *Hemigrapsus oregonensis*, *Pachygrapsus crassipes* and *Uca crenulata*, there is a trend of decreasing sodium and increasing magnesium in the urine. In *Gecarcinus lateralis* the sodium in the urine is high and the magnesium relatively low. While this has suggested that *G. lateralis* may have a different origin from other terrestrial crabs, it may be that it has merely adapted further to the terrestrial habitat. From Gross (1964b).

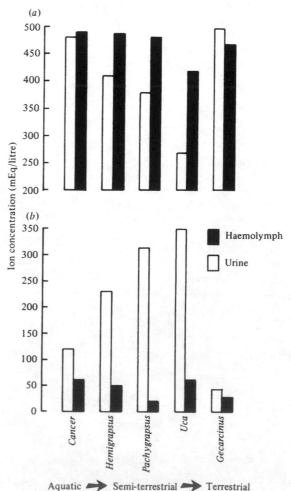

H. oregonensis therefore does not keep its haemolymph magnesium as low as that of *P. crassipes* (Fig. 5.19), and this difference ie even more evident in concentrated sea water. Gross (1961) has suggested that the considerable osmotic and ionic capacities of many of the semi-terrestrial decapods reflect evolution in areas of widely fluctuating salinity, such as coastal lagoons, which would have produced selective pressures favouring hypo- and hyperosmotic regulation. In contrast to this, *Carcinus maenas* has evolved in areas in the temperate zone such as estuaries, where there is little chance of finding hypersaline waters. Perhaps as a result, it can hyper-osmoregulate, but cannot hyporegulate. Support for this suggestion of the differences in origin of *C. maenas* and *P. crassipes* can be derived from a comparison of magnesium regulation in the two forms (Lockwood & Riegel, 1969). *C. maenas* regulates magnesium levels when living in dilute sea water rather better than does *P. crassipes*, but in high concentrations *P. crassipes* is the better regulator. It is unfortunate that little information is available for the even better osmoregulator, *Grapsus grapsus*.

Since the antennal glands are not concerned with osmoregulation, extra-renal mechanisms must be well developed. Salt uptake in animals showing hyperosmo-regulation is probably through the gills (Robertson, 1960), but little is known of the mechanism of hypo-osmotic regulation. An investigation of two grapsids, *Cyclograpsus punctatus* and *Plagusia chabrus* (Heeg & Cannone, 1966), suggested that the diverticula of the gut may be involved in osmoregulation, since ligaturing them upset osmoregulatory ability. The size of the diverticula is apparently correlated with ability to regulate hyperosmotically. Epithelial structure is typical of those involved in transport (Mykles, 1977).

Besides exploiting habitats at the top of the rocky intertidal zone, grapsids are frequent in mangrove areas. In Jamaican mangroves, for instance, five species of grapsid are commonly found (Warner, 1969). The distribution of these species is governed by tidal levels, just as on open shores, even though the tidal rise and fall may be slight: factors such as salinity range, moisture content of the soil and desiccation appear to control this distribution. It is important, however, to be able to distinguish the micro- and macrohabitats in which the various species live. Under the surface of the swamp there exists a maze of 'crab runs' excavated by a few species but occupied by many. There appear to be two reasons for the occupation of this maze. Some species, notably the xanthids, are active only at high tide, i.e. they are not tolerant of desiccation and have to hide during the period when they would otherwise be exposed. Other species are nocturnal, and hide in the crab runs during the day – e.g. the grapsids *Pachygrapsus gracilis* and *Sesarma* spp. Some grapsid species, however, are active in the day, even at low tide, e.g. *Goniopsis cruentata* and *Aratus pisoni*. *G. cruentata* is found mainly in lower, wetter parts of the swamp, while *A. pisoni* extends to the higher, drier parts. *A. pisoni*, called the tree crab, avoids predation and competition from other crabs by climbing mangrove trees, a habit which is uncommon amongst crabs. It is small (up to 26 mm across the carapace) and lightly built, like many grapsids, and is a key member of the arboreal community, eating mangrove leaves and also preying on crickets, beetles and caterpillars (Beever, Simberloff & King, 1979). It breeds continuously throughout the year (Warner, 1967), in synchrony with a lunar rhythm. Females carrying eggs which are about to hatch migrate to the seaward edge of the swamp and deposit the newly hatched prezoeas into the water. Other nearly terrestrial grapsids are common in the Indo-Pacific (see, e.g. Fig. 5.20). Some of these show similar reproductive rhythms to those of *A. pisoni*, reminiscent of

Fig. 5.20. A terrestrial grapsid from the coastal fringe in Papua New Guinea, found with terrestrial insects and snails in tropical forest. Carapace width 1.5 cm.

saltmarsh pulmonates discussed in Chapter 4. Even species of *Sesarma* which release their larvae into rivers instead of into the sea, do so at the time of high water on the local coastline so that the larvae spend as short a time as possible in fresh water (Saigusa & Hidaka, 1978; Saigusa, 1981). All semi-terrestrial marine grapsids are therefore firmly tied to the sea by their reproductive processes.

Some of the respiratory adaptations of *Goniopsis cruentata* and *A. pisoni* have been studied by Young (1972). In considering a series of (mostly unrelated) crabs, he suggested that the characteristics of oxygen affinity of the various haemocyanins could be related to the degree of 'terrestrialness'. In particular, he suggested that the oxygen tension for 50% saturation of haemocyanin (P_{50}) tended to increase in more terrestrial species (see Fig. 5.21). Since, however, P_{50} is similar in the transitional *Goniopsis cruentata* and in the distinctly terrestrial *Gecarcinus lateralis*, this idea can hardly be accepted without qualification. When comparison is restricted to the two grapsids examined, it is seen that the more terrestrial *A. pisoni* has a P_{50} much *lower* than that of *Goniopsis cruentata* which is found lower down the shore. Presumably, as Young has suggested, the real significance of increasing P_{50} values is that they are associated with better systems for gaseous exchange. In some circumstances this may mean the development of lungs instead of gills, but in

many cases it may be adaptive for various other environmental stresses (Warner, 1977). In the comparison of *Goniopsis cruentata* and *A. pisoni*, in fact, neither species appears to have well-developed lung structures. An investigation of the oxygen exchange mechanisms in these two crabs would be of interest. All that is known of *Goniopsis cruentata* at present is that either air or water can be passed over the gills. This may well parallel the system in *Carcinus maenas*, where breathing occurs in air as well as in water, although with a slightly reduced rate of oxygen uptake (Newell, Ahsanullah & Pye, 1972). The normal mechanism of ventilation by *C. maenas* utilises a countercurrent flow system, but even so the percentage utilisation of oxygen is low, probably because of the impermeable gill epithelium (Hughes, Knights & Scammell, 1969). The relatively thick cuticle over the gills does, however, give them support, and because the lamellae are widely spaced the gills are structurally suited for respiration in air as well as in water (Taylor & Butler, 1978). *C. maenas* ventilates the gill chambers in both forward and reverse directions in air, unlike the hermit crabs examined. In *Aratus pisoni* ventilation in air is effected by recirculation of water from the gills over channels on the antero-ventral surface of the body, where it can be reoxygenated. This method of providing an oxygen supply has been considered in detail by Verwey (1930) and by Alexander & Ewer (1969). The latter in particular have considered the grapsids *Sesarma* spp. and *Cyclograpsus punctatus*. When these species are in air, the water in the branchial chambers is expelled through the exhalant openings by the scaphognathites. Initially some of this water may pass over the dorsal surface of the carapace, but most passes over areas of the branchiostegites which are covered by geniculate hairs (see Fig. 5.22). These hairs act as an 'inverted plastron', retaining the water as a thin film and ensuring efficient gas exchange with the air. The water passes back into the gill chamber through openings at the bases of the legs. The circulation time is very small – of the order of seconds – and the route differs from that used under water, where water returns only through the openings at the base of the chelipeds. If the gaseous exchange is as efficient as in the freshwater grapsid *Eriocheir sinensis*, one circuit of the water approximately doubles its oxygen content, and increases the pH by up to 0.3 units, so that the carbon dioxide content must be considerably reduced (Olthof, 1936).

As the water used in the process described above evaporates, water loss occurs from the gill membranes. Indeed, the gills tend to dry up relatively faster than other tissues, at least in *Pachygrapsus crassipes* (Gross, 1955). At the same time, oxygen consumption falls during desiccation because of the increasing impermeability to oxygen of the drying membranes. The tissues which line the gill chamber, however, do not dry, and these assist in gaseous exchange. The drying of the respiratory membranes may explain why *P. crassipes* spends only relatively short periods on land, and frequently seeks re-immersion in the sea. However, most of the water loss from crabs in fact occurs through the

Fig. 5.21. The P_{50} values of haemocyanin from a variety of crabs. Values are given at their physiological pH and at 28°C. Vertical bars show S.D. Numbers show mean physiological pH. (a) *Callinectes sapidus*; (b) *Goniopsis cruentata*; (c) *Aratus pisoni*; (d) *Cardisoma guanhumi*; (e) *Gecarcinus ruricola*; (f) *Gecarcinus lateralis*. The species are arranged in increasing order of 'terrestrialness' from left to right, and within the Gecarcinidae, P_{50} may increase in this order. Overall, however, P_{50} is probably related to a number of complex factors, including the efficiency of gaseous exchange with the external environment, and therefore internal pO_2. From Young (1972).

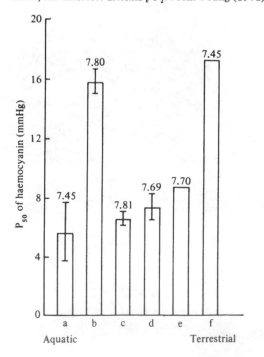

integument (Herreid, 1969*a*, *b*). Thus if the external surface of the gecarcinid *Cardisoma guanhumi* is covered with petroleum jelly, the initial rate of water loss is halved; after 3 to 4 hours the loss is less than a third of the normal rate (Fig. 5.23). Similar results were obtained for other semi-terrestrial crabs. Measurements on the isolated integument showed that the littoral grapsids *Sesarma reticulatum* and *Sesarma cinereum* had permeabilities greater than those of the terrestrial *Gecarcinus lateralis*, but lower than those of sublittoral crabs.

In many situations continued water loss from a crab will lead to desiccation and death – values of 12–25% water loss have been found to be lethal for a variety of (non-grapsid) crabs (Herreid, 1969*a*). Many grapsids, however, are able to take up water from very small sources. *Sesarma meinerti*, for instance, is able to take up water from moist sand (Alexander & Ewer, 1969). It squats down on the substratum and water is picked up by a capillary system of hairs on the ventral surface of the body. These conduct the water to the leg bases, from where it is drawn into the branchial cavity by the pumping of the scaphognathites. When pumping is well established the carapace is raised and water runs out of slits at the posterior end of the branchial chambers. This water runs down and joins that on the animal's ventral surface, ensuring that the circulation of water is not broken. The posterior slits are not involved in water intake as suggested by Verwey (1930).

The grapsid crab best known to northern Europeans is the Chinese mitten crab, *Eriocheir sinensis* (see Kaestner, 1970). Introduced from China early in the present century, *E. sinensis* has penetrated many European rivers; but it can only breed in the sea, and when mature the adults migrate down-river again. The larvae are planktonic, and the young remain near river mouths until they are about two years old. At this stage they migrate up-river again to mature in freshwater ponds and ditches. Most of their life is spent in water, but they can, and do, migrate over land to bypass obstacles. The osmoregulatory adaptation of *E. sinensis* to fresh water has been examined by Shaw (1961), who showed that while the permeability to sodium is decreased in comparison with *Carcinus maenas*, it is not as low as in *Potamon niloticus* (see p. 87). Nevertheless, the sodium loss is adequately compensated for by an increase in the capacity for active uptake of sodium, and without the production of hypo-osmotic urine. Another species, *Eriocheir japonicus*, is found in fresh water in Japan.

Truly freshwater grapsids are uncommon, and nothing is known of their physiology. Four species are known from Jamaica (Hartnoll, 1963), where their presence has been linked with the absence of potamonid crabs. One species of *Sesarma* is cavernicolous, while two more live in rivers and streams, and on wet rock surfaces, where they can complete their life cycle (Abele & Means, 1977). The most interesting species is *Metopaulias depressus*, which passes its entire life

Fig. 5.22. Anterior view of the grapsid *Sesarma catenata*, showing details of grooves and distribution of geniculate hairs on the antero-lateral surface of the left side of the branchiostegite. From the deep groove which originates at the exhalant opening of the branchial chamber run three grooves: one leads to the orbit, the second runs across the pterygostome, and the third runs almost vertically down and terminates near the opening of Milne Edwards, at the base of the cheliped. These grooves and the fields of geniculate hairs help to direct and maintain a flow over the body surface which aerates branchial water. Redrawn after Alexander & Ewer (1969).

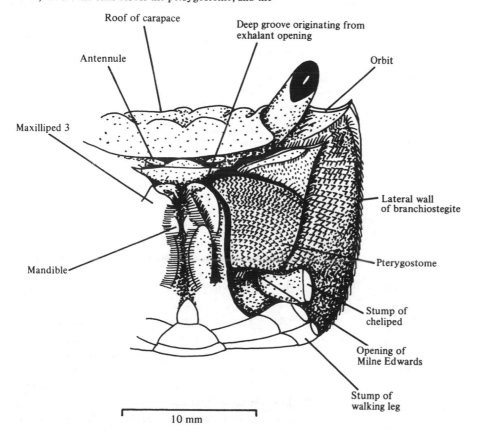

cycle in the large epiphytic bromeliads that retain water
in their leaf axils all the year round. *M. depressus* is
a small crab, up to 2 cm across. It is restricted to altitudes
over 300 m and to areas of limestone, presumably
because of its need for skeletal calcium. Its genital
system is somewhat different from that of *Sesarma*: the
female genital aperture is protected by a hinged operculum arranged in such a way that both copulation and
egg laying can occur at any phase in the moulting cycle,
and are not restricted to newly moulted specimens
(Hartnoll, 1963). In species of *Sesarma* copulation can
only occur when the female is soft because at other times
the operculum is fused in a closed position. However,
according to Bliss (1968), several other species of
grapsids can copulate when the shell is hard. In
M. depressus the eggs are bigger than those of marine
grapsids, and while the young hatch as zoea larvae, the
development is much abbreviated so that these zoeas
moult to form megalopa larvae directly, instead of
passing through several zoea stages. This reduction in
the number of larval stages is a phenomenon carried
further by some potamonids, as discussed above (p. 88),
and is characteristic of many freshwater invertebrates.
It does not occur in terrestrial crabs which return to
the sea to breed, and so can be regarded as a true freshwater adaptation.

In considering the four species of freshwater
grapsids in Jamaica, Hartnoll (1963) has suggested that
there has only been one invasion of fresh water, and
that the ancestral stock was probably in the subgenus
Sesarma which has representatives both in the marine
littoral zone and in fresh water. The fact that two of
these 'freshwater' species can live on land (Abele
& Means, 1977) provides a most interesting contrast
with species of *Sesarma* which have evolved to a semi-

Fig. 5.23. Water loss from *Cardisoma guanhumi*. Filled circles
show normal crabs and open circles show crabs with petroleum
jelly rubbed over the body surface. All experiments were at 25%
RH and 35 °C. Vertical bars show S.E. After the initial 3 hours
treated crabs lost water at approximately one third of the rate of
normal crabs. This demonstrates that most water is normally
lost by evaporation from the general body surface rather than
from the branchial cavity. From Herreid (1969*b*).

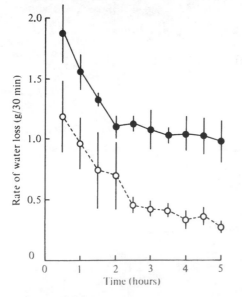

terrestrial habit directly from the marine littoral zone.
Other parallel series may exist elsewhere, as Fimpel
(1975) has recorded two species of *Sesarma* from
bromeliad reservoirs in Brazil.

Brachyura – the ocypodids

Most of the ocypodids are found in the Indo-West-
Pacific region, with comparatively few representatives in
the Atlantic. Much more is known about the biology of
Atlantic shores, however, the best known genera there
being *Uca*, the fiddler crabs, and *Ocypode*, the ghost
crabs (Fig. 5.24). Both of these live relatively high on
the shore, and *Uca* in particular is abroad during daylight;
but in the Indo-West-Pacific many other genera are found
in the lower intertidal regions. Of these, sand-dwelling
forms include *Scopimera* and *Dotilla*, while mud dwellers
include several species of *Macrophthalmus* and *Australoplax*.

The majority of ocypodids are substratum feeders;
that is to say, they feed by picking up either sand grains
or mud, consuming or cleaning some of this substratum,
and rejecting the rest in the form of pellets. The whole
process is somewhat reminiscent of the formation of
pseudofaeces by bivalve molluscs. The pellets are common
on tropical beaches, laid out in patterns characteristic of
each species. The water currents from the exhalant
openings of the branchial cavity aid in the sorting
process which leads to their formation. Many species
are very specialised with respect to the type of substratum on which they feed (see, e.g. Barnes, 1974;
Icely & Jones, 1978), and their distribution is therefore
closely governed by the type of substratum available.
Since ocypodids are also common in estuaries, however,
the external salinity is also an important factor, and
both these points have been examined in an Australian
estuary (Barnes, 1967). Four species of ocypodid were
examined, together with one mictyrid, *Mictyris longicarpus*. Three of the ocypodids – *Macrophthalmus
setosus*, *Australoplax tridentata* and *Paracleistostoma
mcneilli* – live in mud or sandy mud, and since this is
abundant in the Brisbane River and its estuary, substratum type is not limiting for them. The limits of
their distribution appear to be set by their differing
powers of hyperosmotic regulation, and to a lesser
extent their powers of hyporegulation. *Macrophthalmus
crassipes* and *Mictyris longicarpus* both live in muddy
sand, but *Macrophthalmus crassipes* does not reach into
the Brisbane River itself, while *Mictyris longicarpus*
reaches only 4 km upstream. *Macrophthalmus crassipes*
has very poor powers of osmoregulation and narrow
limits of salinity tolerance so that its distribution is as
expected; but *Mictyris longicarpus* has wide tolerance
limits and a greater power of hyperosmotic regulation.
Its lack of penetration up the Brisbane River must
therefore be attributed to lack of suitable sandy
substrata, and in adjacent primarily sandy estuaries
it penetrates well upstream (Barnes, R. S. K., personal
communication).

The ocypodids just discussed have fair powers of
hypo-osmotic regulation, but these are very much less
than those of some of the grapsids. The more terrestrial

forms, however, e.g. *Uca crenulata* and *Ocypode cera-tophthalma*, have excellent regulatory capacities (Gross, 1964b). *O. ceratophthalma*, for instance, shows a change in osmotic pressure of the haemolymph equivalent to that of only 10% sea water when the external salinity changes from 50% to 170% sea water (Fig. 5.18). As in grapsids, the antennal glands do not control the osmotic pressure of the haemolymph, but they may regulate its ionic composition. Sodium is reabsorbed from the urine under some circumstances (Schmidt-Nielsen, Gertz & Davis, 1968). The magnesium concentration of urine is high, whereas in the haemolymph it is low. In different species, though, higher magnesium concentrations in urine are not necessarily correlated with lower haemolymph magnesium, and this has led Gross (1964b) to suggest that urine composition may not be the major factor controlling magnesium in the haemolymph. It seems likely, however, that the rates of urine production will be different in different species, and these different rates could account for the apparent lack of a constant haemolymph : urine ratio. There is no information about rates of urine production by species of *Uca*, but in *Ocypode albicans* urine production stops when the crabs are placed on dry sand (Flemister, 1958), and this is due to the cessation of the initial filtration process. This species is usually only active at night and at this time it immerses itself briefly in sea water. It is this sea water which provides the crab with the water necessary for urine production.

Many species of *Uca*, in contrast to most species of *Ocypode*, are active during the day and are subjected to hot, dry conditions. Edney (1961, 1962) has studied the water and heat relationships of a number of species of *Uca* at Inhaca Island in Mozambique. Here there are five species of fiddler crabs living in and around the mangrove swamps. In saturated air the body temperature of the crabs is the same as the ambient temperature. In unsaturated air (about 80% RH) their body temperature is depressed by as much as 2 °C. In low relative humidities of 25%, temperature depressions of as much as 6 °C were observed. These temperature depressions were greater in species with higher transpiration rates, as would be expected. In experiments in the natural environment, the temperatures of living and dead crabs exposed to direct sunlight showed differences of a similar order, and it was assumed that such differences (6–8 °C) were due to transpiration. The temperature of crabs near their burrows in the shade was sometimes as high as the ground temperature (42 °C), which for *Uca annulipes* is lethal for an exposure time of 15 min. In sunlight, body temperatures were lower than that of the ground by as much as 10 °C, due to a combination of convection and transpiration. Temperatures inside the burrows were always much lower than either the air or the surface temperatures, so that the crabs could escape from the macro-climate at any time (see Fig. 5.25). Similar temperature relationships have been found for *Uca panacea* and *Uca virens* in North America (Powers & Cole, 1976).

Fig. 5.24. A juvenile of *Ocypode* sp. standing at the entrance to its burrow. The combination of light pigmentation, giving cryptic colouration, and the vertically raised eyestalks with 360° vision make this genus exceptionally well adapted to a carnivorous and scavenging life on tropical sandy shores. Carapace width 10 mm.

During exposure to unsaturated atmospheres, species of *Uca* lose considerable quantities of their body weight as water. Thus *Uca annulipes* exposed to less than 25% RH lost 10% of its body weight in 7.5 hours, and *Uca marionis* lost 24% in the same period. These figures can be compared with that for *Uca minax* which died after a loss of 18% of its body weight (Herreid, 1969*a*). Because of the possibility of retreat into the burrow, death by desiccation does not seem likely in nature; but the crabs are living very close to their lethal temperatures and death from hyperthermia must be at least a possibility. The site of water loss was not investigated by Edney (1961, 1962), but Herreid (1969*b*) has shown that over 90% is lost through the general integument, as for other semi-terrestrial crabs (see p. 92). Much may also be lost simply from the external surface of the integument, since individuals of *Uca pugilator* in their natural habitat are often moist (Smith & Miller, 1973). These moist crabs may be as much as 4 °C cooler than dry ones. This species returns to burrows approximately every 20 min (Wilkens & Fingerman, 1965), and it may be that this is to remoisten the body surface. If this is so, the long-term laboratory experiments on water loss are somewhat irrelevant to the natural situation. *U. pugilator* also uses other mechanisms to regulate its temperature. Its orientation to the sun determines how much heat is absorbed, and its carapace colour is also involved: above 15 °C the black pigment concentrates, and above 20 °C the white pigment disperses, so that the crabs become 'blanched'. This blanching depresses body temperature in sunlight by about 2 °C (Wilkens & Fingerman, 1965; Smith & Miller, 1973). Orientation to the wind and the position of the chelipeds also determine how much heat is lost by convection. In all, *U. pugilator* has a sequence of complicated behavioural patterns which serves to maintain its temperature below the lethal one. *Uca rapax*, which lives

rather more in mangrove swamps than on open sandy beaches, has similar patterns but depends more upon retreat into the shade or into the burrow (Smith & Miller, 1973).

Even greater extremes of temperature can be tolerated by *Ocypode gaudichaudii* (Koepcke & Koepcke, 1953). With complete immersion in water, this species died in 40 min at 41–42 °C, but in air it survived indefinitely at 50 °C, and for 15 min it tolerated 60 °C without harm. Such tolerance is undoubtedly owing to the evaporation of water and consequent cooling, but unfortunately there are no figures for body temperatures of these experimental crabs.

When aquatic ocypodids such as *Macrophthalmus* emerge from water, they pump water out of the branchial cavity and so aerate it in much the same way as some grapsids (Verwey, 1930). The more terrestrial forms such as members of the genera *Uca* and *Ocypode* do not pump water. Instead, water is retained in the branchial cavity, and air is pumped through this by the scaphognathite. Both the course of this circulation of air and that of water when the crabs are submersed differ from that in the more aquatic forms because of a special opening developed between the bases of legs 3 and 4. This opening (Müller's opening) often acts as the main source of inhalant water, while water may be expelled through the normal exhalant opening, through the opening at the base of the chelipeds (Milne Edwards' opening) or through both. When in air, species of *Ocypode* normally take in air through Müller's opening, but air may pass either way through this opening and through Milne Edward's opening. From descriptions of Müller's opening (Stebbing, 1893) it appears that its design allows the efficient entry of air while preventing water from draining out of the branchial chamber. The nearness of this opening to the ground is advantageous in two ways: first, air drawn in through it is

Fig. 5.25. Temperatures in the natural habitats of some species of *Uca*. (*a*) *Uca annulipes* near *Avicennia*. (*b*) *U. annulipes* in the open centre of the mangrove swamp. (*c*) *U. annulipes* near a footpath. (*d*) *Uca chlorophthalmus* near a stream in the *Rhizophora* zone. (*e*) *Uca marionis* near a stream in the

Rhizophora zone. Note that the vertical scale refers to the burrows only. The crabs' temperatures are usually below those of the ground, but even so may be near their lethal level. Burrows and pools form retreats if body temperatures become too high or the crabs lose too much water. Redrawn after Edney (1961).

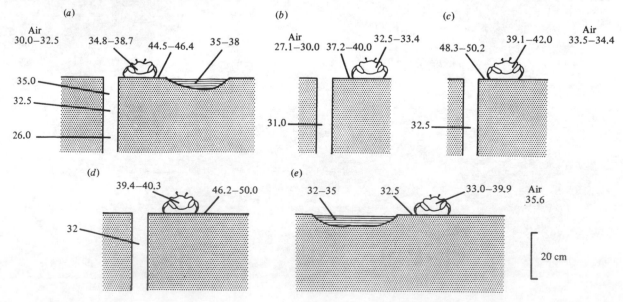

likely to come from near the mud surface, and may be
damper than surrounding air; and second, water can
be taken up through it. To facilitate this the margins of
legs 3 and 4 are fringed with long hairs so that water can
pass up between the legs and into the branchial cavity
(Verwey, 1930; Fig. 5.26). Water can also be taken up
from the soil capillaries by an extraordinary suction
mechanism (Wolcott, 1976): pressure in the branchial
chamber is reduced by as much as 40 mmHg, sucking
moisture from the sand. It has also been reported (Rao,
1968) that water taken up in this way can be stored
in the pericardial sacs of *Ocypode cordimana* during
proecdysis, but this is presumably to be interpreted as
meaning that the water is first absorbed into the haemolymph, and that increased haemolymph volumes can be
accommodated in these pericardial sacs. The surface area
of the sacs in *O. cordimana* (a semi-terrestrial species) is
greater than that of the sacs of two intertidal ocypodids,
thus suggesting that their development is related to
O. cordimana's relative independence from the sea.

The branchial cavities of species of *Uca* and
Ocypode differ from those of aquatic ocypodids in
being at least partially separated into two chambers: an
upper 'lung chamber', and a lower gill chamber (Verwey,
1930; see Fig. 5.27). The upper chamber is invested with
numerous capillaries, and presumably aids respiration in
air: the capillaries form a spongy mass which is quite
unique in semi-terrestrial crabs (von Raben, 1934). These
lung capillaries have a thin epidermis covered by a thin

cuticle, and resemble the respiratory epithelia of many
other animals (Fig. 5.28; Storch & Welsch, 1975). The
gills, in the lower chamber, show no gross histological
differences from those of aquatic forms (von Raben,
1934), but they contain podocytes, which are normally
associated with epithelia involved in filtration (Fig. 5.28;
Storch & Welsch, 1975). These are presumably involved
in the production of a filtrate of the haemolymph, and
in this situation they could be responsible for keeping
the surface of the gills moist in air. Since the gills also
show some of the structures typical of reabsorptive
epithelia, they probably are the site of uptake of salts
and may be concerned with the osmoregulation discussed
earlier. Species of *Ocypode* have the normal complement of six pairs of gills, but the last two are squeezed
together and reduced (Koepcke & Koepcke, 1953). This
reduction facilitates the absorption of water through
Müller's opening, since the last gills lie directly above
it. Most species of *Uca* follow *Ocypode* in this, but
Uca consobrinus has proceeded one stage further, and
has only four pairs (Verwey, 1930). In general, the
weight of the gills relative to body weight is reduced in
species living higher up the beach (Pearse, 1929b, 1936),
suggesting that the lung becomes more important for

Fig. 5.27. Internal anatomy of (a) a male *Gecarcinus lateralis*
and (b) a male *Ocypode quadrata*. In both cases the hepatopancreas has been removed. *G. lateralis* has much larger pericardial sacs than *O. quadrata*, with a posterior extension bearing
setae. It also has a more expanded epibranchial area. Both these
adaptations reflect the more terrestrial existence led by
G. lateralis. Redrawn from Bliss (1968).

(a)

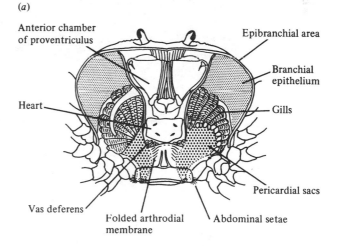

(b)

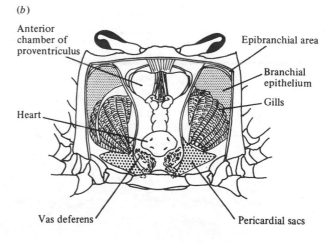

Fig. 5.26. Abdominal setae of crabs, used in water uptake.
(a) *Gecarcinus lateralis*, ventral aspect, with parts of the
abdomen and carapace cut away to show the posterior extension
of the left pericardial sac and the setae at its tip. External setae
and setae on the posterior margin of the carapace also facilitate
water uptake. (b) *Ocypode quadrata*, ventral aspect. The setae
at the bases of the second and third walking legs are indicated.
On the left the legs have been separated; on the right they are in
the normal resting position. Redrawn from Bliss (1968):

(a)

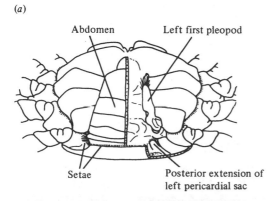

(b)

respiration in these species. However, this correlation is at least partly determined by factors of size, since it is the larger crabs which have relatively smaller gills.

Some of the behavioural mechanisms by which species of *Uca* regulate their body temperatures have already been described. Further behaviour patterns are concerned with intraspecific reactions and with the daily cycles of activity which are geared to the tides. These differ markedly in various species of ocypodid, and may best be described by beginning with the relatively aquatic forms and then moving on to *Uca* and *Ocypode*. *Dotilla fenestrata*, for example, lives on sandy east African shores between the mean low water of neap tides and mean tidal level (Hartnoll, 1973). One to several hours

after being uncovered by the tide, *D. fenestrata* emerges, repairs its burrow, and feeds on the organic particles left by the last tide. In doing this it parcels the sand into small pellets after extracting the organic fraction, using water expelled from the branchial cavity. This water is replaced with the aid of capillary setae on the surface of the abdomen: when this is pressed against the sand, water is taken up and runs forward along grooves at the edge of the abdomen into the branchial chamber. *D. fenestrata* carries out this action on the sand surface, and indeed since the burrows do not reach down to the water table no help would be gained by entering them. When the tide is about to cover the burrows, the crabs enter and seal the opening. Some species form an 'igloo'

Fig. 5.28. Semi-diagrammatic view of the ultrastructure of lung and gill epithelia in *Ocypode ceratophthalma*. (*a*) Lung, showing the thin cuticle and thin layer of epidermal cells, producing a small air to haemolymph distance. (*b*) Gill, showing in contrast the thick cuticle and large water to haemolymph distance. The basal

infoldings of the epithelial cells, associated with mitochondria, are typical of structures concerned with the transport of ions and water. On the left is a podocyte, indicating that ultrafiltration may keep the gill membranes moist. Redrawn from Storch & Welsch (1975).

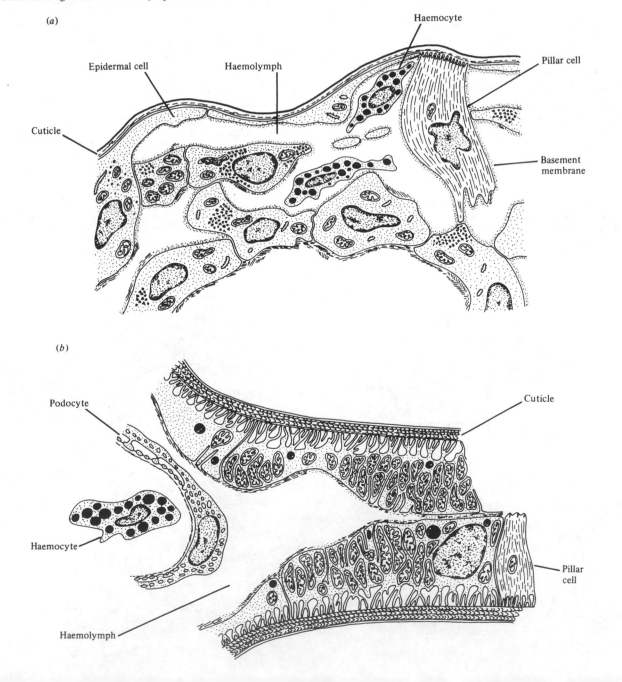

of sand pellets over the entrance to the burrow (Tweedie, 1950; Altevogt, 1957) when they enter it, but the significance of this is not clear.

The Australian *Scopimera inflata* leads a similar life, but lives between mean tidal level and the mean high water of neap tides (Fielder, 1971). A large volume of water is carried in the branchial cavity, and a film of water covers the carapace. In this species the branchial water is replaced in the burrow, which extends down to the water table. Tufts of hairs between the second and third pairs of legs aid this, and there are no modified setae on the abdomen. *S. inflata* has good vision, and can perceive movements over distances of at least 20 m.

Species of *Uca* normally live in dense populations, so that interactions between individuals are common, and social behaviour is said to be more complex in this genus than in any other crustacean. The general rhythm of behaviour is geared to that of the tides, as in *D. fenestrata* and *S. inflata*: the crabs emerge as the tide ebbs, undertake burrow repairs, make feeding excursions and are involved in courtship; as the tide returns, they go back into burrows. These rhythms differ somewhat in species living at different tidal levels: while those at lower levels have a regime strictly geared to the tides, those high up on the beach may spend several days in their burrows at periods of neap tides, or may climb up mangrove trees (Hagen, 1970a). During their periods of activity species of *Uca* are often seen moving over the sand flats in large numbers. *Uca pugilator* flocks normally move up and down the shore, and movements to the top of the beach are enhanced if predators appear. Their orientation in

these movements is partly achieved using the sun and the polarised light pattern of the sky, but they also move towards prominent landmarks (Herrnkind, 1968, 1972), in the way suggested for amphipods (see p. 68). The eyes are well developed in *Uca*, and are held upright on eye-stalks when the crabs are in air, as they are in most ocypodids. Both vision and a vibration sense are important in the complex courtship behaviour, and it is thought that this is in contrast to the more aquatic species, where chemical and tactile stimuli are involved. This apparent difference could, however, be due to lack of study of visually directed courtship patterns in aquatic species. Certainly the cues used for orientation by *Uca* under water are similar to those used in air (Young & Ambrose, 1978).

The courtship display of *Uca*, in which the male has one cheliped much enlarged, is well known (Fig. 5.29). In 'primitive' species, such as some with a narrow carapace found in the Indo-Malayan region, a simple waving of the cheliped leads after a relatively short sequence to copulation on the surface of the beach. In more 'advanced' species with wider carapaces, many of which are found in Central and South America, the male's display is more conspicuous and continues for much longer, and the female is usually enticed into the burrow before copulation occurs (Crane, 1941). Most males display in front of their burrows, and may face in any direction, but in *Uca terpsichores* some males build 'shelters' over the burrows and display only to the angle which is still in the field of view, i.e. about 180°. This has been construed as a mechanism to lower aggression

Fig. 5.29. Sexual display in ocypodids and in the grapsid *Goniopsis cruentata*. (*a*) Beckoning of *Uca rizophorae* male, showing a 'vertical wave'. This is the primitive form of display. (*b*) Beckoning of *Uca annulipes* male, showing 'lateral wave'.

This is a more advanced display. (*c*) Beckoning of *Uca pugilator* male, with laterally held chelipeds. (*d*) Waving of *Dotilla blanfordi*. (*e*) Waving of *G. cruentata* male: 1, from the front, and 2, from the side. Redrawn from Schöne (1968).

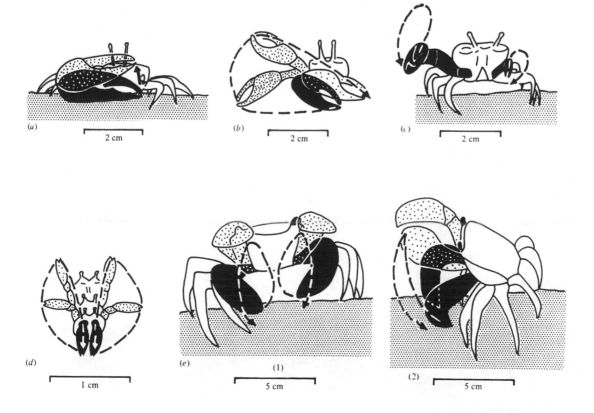

(*a*) 2 cm   (*b*) 2 cm   (*c*) 2 cm

(*d*) 1 cm   (*e*) (1) 5 cm   (2) 5 cm

in areas of high population density, while allowing the stimulating effect of one male upon another to continue (Zucker, 1974). The stimulating effect sometimes results in synchronous claw waving by large numbers of males (Gordon, 1958). Many species also produce sounds which are involved in courtship. In *Uca tangeri* some of these are substitutes for claw waving, either when in vegetation or at night. They are made either by the walking legs, which produce a 'honking' sound, or by the rapping of the cheliped against the ground (Salmon, 1971; Salmon & Horch, 1972). The sounds involved are not, however, 'heard' in the strict sense, as species of *Uca* are sensitive only to vibrational stimuli transmitted through the substratum (Horch & Salmon, 1969).

Much of the behaviour of *Ocypode* is somewhat different from that of *Uca*, and is associated with its faster and more predatory way of life. *Ocypode ceratophthalma*, for example, may travel 400–500 m from its burrow in search of food on the intertidal sand flats (Hughes, 1966). Although it is partly a scavenger and may also use the sand pellet feeding found in *Uca*, it preys mainly on *Macrophthalmus* and on the burrowing bivalve *Donax*. It is also aggressive towards members of its own species (see Fig. 5.30), and cannibalistic. Its visual perception is excellent, and it can detect large objects moving at a distance of over 100 m. Like *Uca*, species of *Ocypode* produce sounds, but these are in the main stridulations, made by rubbing tubercles on the propodus of the larger cheliped against a ridge on the ischium. Perhaps linked with this rather more refined method of sound production is the capacity for detecting airborne sounds (Horch & Salmon, 1969). The sound receptor is a myochordotonal organ in the merus of the leg (Horch, 1971), and similar receptors detect substrate-borne vibrations. The stridulations are produced only in the burrows and only by male crabs. They are made primarily at night and attract females to the burrows (Hughes, 1973). Other factors are also important in attracting females, however, especially in the regions where the crabs are

active during the daytime or early evening: in this case the males form the sand which is excavated from their specially constructed 'copulation burrows' into pyramids, at distances of 30–50 cm from the entrance. These pyramids serve both to stimulate other males to build, and to attract females (Linsenmair, 1967). It is thought that copulation normally occurs in the burrows constructed by the males, but it has also been observed on the beach surface (Hughes, 1973). As is typical of land crabs, copulation occurs when both male and the female have hard shells, in contrast to many aquatic crabs, where the female has usually just moulted before copulation.

Ocypode ceratophthalma is regarded as the fastest of the terrestrial decapods. Not only can it achieve speeds well over 2 m/s, but it can also accelerate rapidly, and reverse direction without losing speed (Koepcke & Koepcke, 1953). When moving at only moderate speeds, two pairs of legs alternate: the crab moves sideways, and while legs 2 and 4 on one side are put down together, they alternate with legs 3 and 5. The chelipeds are not involved, so that the crab is effectively eight-legged, and the sequence is similar to that of other crabs such as *Carcinus maenas*. At higher speeds, however, leg 5 on each side is raised so that it becomes effectively six-legged. At maximum speed only three legs touch the ground: legs 2 and 3 on the trailing side move alternately to provide thrust, while leg 3 on the leading side acts as a skid to provide balance (Burrows & Hoyle, 1973; Fig. 5.31). In terms of thrust, the crab is therefore acting as a two-legged runner. Its speed is increased by the fact that it effectively leaps through the air, so that its step length is greater than could be estimated by measuring the dimensions of the crab.

Such a rapid form of locomotion suggests the necessity for rapid acting sense organs. Relatively little is known about the details of the neurophysiology of *Ocypode*, but the excellence of its visual system has already been mentioned. As in many semi-terrestrial forms, the eyestalks can be raised vertically, and indeed

Fig. 5.30. Agonistic display in ocypodids. (*a*) to (*e*) *Ocypode quadrata*. (*a*) to (*c*) shows a sequence of threat postures taken from a film, the numbers indicating time from the start of filming. In (*a*) the chelae are in threat posture; in (*b*) the crabs are rising on their walking legs and lifting their first walking legs; in (*c*) one has sunk to the ground. (*d*) Shows a formalised fight with chelae pushing. (*e*) Shows a wild attack. (*f*) *Ocypode saratan*, showing threat posture. Redrawn from Schöne (1968).

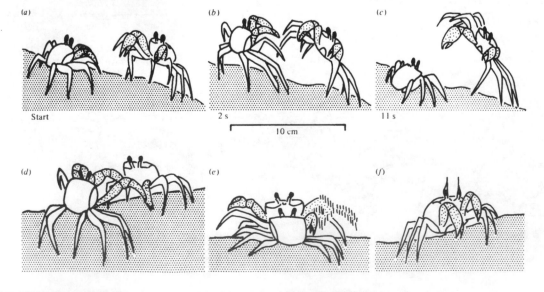

the family Ocypodidae is noted for the many species in which the eyestalks are greatly elongated. This elongation (macrocaulia) is also found in aquatic decapods, but in a few species of ocypodids it is associated with projections of the peduncle beyond the cornea (stylophoria). The raising of the eyes provides obvious benefits such as periscopic vision in shallow water and an extended field of view in relatively flat areas such as sand and mud flats (e.g. Hagen, 1970b). Since most decapods with elongated eyestalks are burrowers, another advantage is the possibility of receiving visual information from above the substrate – especially concerning predators – while remaining hidden, and this is probably the original selective pressure for the evolution of such modifications (Barnes, 1968a). The further extension of the peduncle is probably also concerned with a sensory function, but has not been satisfactorily explained. It bears hairs (Dembowski, 1913), and where investigated these resemble the aesthetascs of other decapods which are thought to be chemosensory (Ghiradella, Case & Cronshaw, 1968).

Visual perception has been examined only briefly in *Ocypode* and *Uca,* but some extrapolation may be made from more detailed observations on other crabs. The number of ommatidia in each eye varies from about 6000 in *Cancer* and 9000 in *Uca* to 16 500 in *Ocypode arenarius*. This large number in *O. arenarius* is exceeded by other terrestrial crabs such as *Grapsus grapsus* which has 17 000 (Nunnemacher, 1966). The ratio of nerve fibres in the optic nerve to number of ommatidia is variable, but is exceedingly low in *Uca*, suggesting perhaps that more integration occurs in the optic ganglion in this genus than in such forms as *Cancer*. The angle surveyed by each ommatidium is of the order of 2–3.6° in *Uca* (Altevogt, 1957), but the visual acuity is equivalent to much less than this (Clark, 1935; Altevogt, 1957; Korte, 1966). This difference is presumably linked with the behaviour of the eye, which in crabs such as *Carcinus maenas* has been shown to perform scanning movements

of 0.5–1°, and to exhibit a tremor of 0.05–0.2° in amplitude (Burrows & Horridge, 1968). The tremor sharpens the perception of contrasting edges. Both movements are driven by visual stimuli acting via the motoneurones which control the complicated musculature of the eye cup. Presumably such eye movements occur in *Ocypode* and *Uca*, but there appear to be no observations on the details of such phenomena in terrestrial as opposed to aquatic crabs.

Another property of the eyes of ocypodids which has been investigated is the flicker fusion frequency. For *Uca* the maximum rate of flicker which can be resolved by the eye is 90–100/s, and for *Ocypode* 170/s, which accords well with the differences in speed and reaction of the animals in these genera. Both these values lie between the extremes for 'slow' and 'fast' eyes in insects (Altevogt, 1957), and are higher than values for other crustaceans (see Waterman, 1961). The overall field of view of the eyes in *Uca* is nearly 360°, and in *Ocypode* the elongated eyestalks allow uninterrupted overlapping fields of vision (Barnes, 1968a), so that some degree of binocular vision occurs. The capacity for depth perception that this confers is, however, unknown. It is also likely that some crabs are capable of colour discrimination, since Wald (1968) has shown that even in *Carcinus maenas* there are red and blue photoreceptors. The perception of polarised light has been referred to above, and both *Uca* and *Ocypode* have been shown to respond to changes in the plane of polarisation of light in experimental situations (Korte, 1965; Daumer, Jander & Waterman, 1963). The mechanisms of detection of polarised light remain elusive as they do in other animals (Waterman, 1961).

Brachyura – the gecarcinids

Of the families of crabs discussed so far, the Grapsidae and Ocypodidae are probably closely related. The Gecarcinidae are also said to be closely related to the Ocypodidae, with genera such as *Ucides* being

Fig. 5.31. Fast running in *Ocypode ceratophthalma*. The movements made by legs L3 and R3 (i.e. the left third and the right third) are shown, (a) and (b) being two separate recordings. The tips of the dactylopodites of both legs were dipped in paints of different colours, and the crab was allowed to run to its right across a sheet of paper. The tip of the dactylopodite of trailing leg 3 was put down precisely at each step, leaving neat spot impressions (circled). This leg provides the motive force, alternatively with L2. Leading leg 3 leaves a long smear impression, indicating that it is used as a skid for balance without contributing sideways force to the movement. From Burrows & Hoyle (1973).

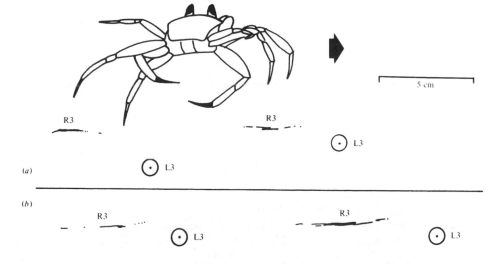

transitional between the two families (Tuerkay, 1970). All members of the Gecarcinidae are relatively terrestrial, from *Ucides* which lives in brackish water mangrove swamps, through *Cardisoma* which often lives some distance from the sea, to *Gecarcinus* and *Gecarcoidea* which are probably the most terrestrial of the true crabs. All these genera are tropical or sub-tropical in distribution. *Gecarcinus* is found in the Americas, *Gecarcoidea* in the Indo-Pacific and *Cardisoma* is worldwide.

Ucides cordatus burrows in the soft mud of mangrove swamps, and is not found either above high water mark or in regions where the mud is able to support insect larvae (de Oliveira, 1946). It normally tolerates salinities of 25–30$^0/_{00}$, but should the water in its burrow be diluted to 10$^0/_{00}$, it moves to another region. Apart from these movements it does not stray far from its burrow.

Cardisoma guanhumi is the commonest semi-terrestrial crab of the southern United States. It grows to a weight of 500 g, inhabits mangrove swamps, open fields and coconut plantations, and may be found as far as 8 km inland. Individuals have also been found in the sea, several metres off-shore (Gifford, 1962*b*). *Cardisoma lateralis* lives in colonies, burrowing into the soil down to water level in regions of both salt and fresh water. In general the colonies are most active at night but at certain times of year they migrate, and in this case the migrations continue in daylight. These migrations sometimes involve all the adults, but the spawning migrations which occur at the times of full moon involve only adult ovigerous females which travel to the sea to release their eggs. *Cardisoma carnifex* lives in similar environments in the Indo-Pacific (Cameron, 1981*a*).

Gecarcoidea humei natalis is the commonest crab on Christmas Island in the Pacific, where it is found in jungle (Gibson-Hill, 1947). It is a vegetarian but also takes carrion. Unlike other crabs, it will drink from standing water by using the chelae in a fashion similar to that described for *Birgus latro* (see p. 86). In *G. humei natalis* the spawning migrations involve males and females, and copulation occurs very near the sea. The spawn is deposited in shallow water where the crabs hatch after a month. The young spend some time in the littoral zone and then migrate inland. *Gecarcoidea lalandii* is also wholly terrestrial as an adult, and is found in the drier parts of islands such as the Palau Islands in the Pacific (Cameron, 1981*a*).

Gecarcinus lateralis is found in the Caribbean region, where it lives above high water mark in grassy areas, sometimes far inland (Bliss, 1968, 1979). Its burrows do not reach groundwater, in contrast to those of many other terrestrial crabs. However, the burrows do provide protection from temperature fluctuations. For example, when the external temperature at Bimini (Bahamas) varied from 13 to 29 °C, the burrow temperature varied only from 21 to 24 °C. The crabs remain in these burrows during the day, often cleaning them out, and emerge at night (Bliss, 1968). As in *Cardisoma*, the females carry round the developing young and these are deposited in the sea as prezoeas or zoeas (Klaassen, 1975).

Apart from its necessity to return to the sea for reproductive purposes, *G. lateralis* lives the life of a completely terrestrial animal. The physiological mechanisms which allow it to do this have been discussed by Bliss (1979) together with morphological and behavioural adaptations. The relations of these to adaptations in other gecarcinids are of great interest, and it is particularly disappointing that we know so little of the more aquatic genus *Ucides*. One of the prime considerations must be the regulation of salts and water. In this respect both *Cardisoma guanhumi*, and *Cardisoma armatum* from Africa, show similarities to other terrestrial crabs. They are good osmoregulators, being able to regulate both hyper- and hypo-osmotically (Herreid & Gifford, 1963; de Leersnyder & Hoestlandt, 1963; Gross *et al.*, 1966). They have low magnesium concentrations in the haemolymph, and as usual this is probably accounted for by excretion through the antennal glands. *G. lateralis* shows some striking differences from *Cardisoma* (Gross, 1963, 1964*b*; Gross *et al.*, 1966). It will drown if kept completely submerged. It osmoregulates well, but dies if kept for long periods with access only to sea water. If given fresh water, or sea water diluted to 30$^0/_{00}$, it survives well. It maintains potassium and calcium in the haemolymph at a more or less constant level, but while it has a low magnesium concentration, it does not have high concentrations of this ion in the urine. *Cardisoma carnifex* in sea water may have a urine : haemolymph ratio for magnesium of 5 : 1, but in *G. lateralis* it is never more than 2 : 1 and is often less than this. This difference has suggested that while *Cardisoma* may have evolved in hypersaline lagoons like many of the grapsids, *Gecarcinus* may have a different origin (Gross, 1963; see Fig. 5.19). It may equally well be argued, however, that in *G. lateralis* the ability to concentrate magnesium has been moved from the antennal gland to some other organ in order to conserve water. The necessity for excreting magnesium must, in any case, be a spasmodic one since *G. lateralis* rarely experiences full strength sea water. The similarities of *G. lateralis* to *Cardisoma carnifex* in other details of regulation such as the remarkably constant levels of potassium and calcium suggest that *Gecarcinus* has become further adapted to a terrestrial life, but that the route taken is similar to that of *Cardisoma*.

The rate of urine production in gecarcinids is surprisingly high, being 3.2% of the body weight per day in *Cardisoma carnifex*, 1.9% in *C. guanhumi*, 3.4% in *Gecarcoidea lalandii*, and 10.5% in *Gecarcinus lateralis*. These values compare with only 0.05–0.6% of the body weight per day in freshwater species of *Potamon* (Harris, 1977; Kormanik & Harris, 1981). Nevertheless, the rates can be reduced during desiccation (Flemister, 1958; Harris & Kormanik, 1981), probably by a reduction in primary filtration rate rather than by reabsorption in the antennal glands.

During desiccation *Gecarcinus lateralis* continues to regulate the composition of its haemolymph remarkably well. The potassium level, for instance, stays constant as it does under changing conditions of external salinity (Gross, 1963). Nevertheless, loss of water is obviously important, and gecarcinids can tolerate only a loss of 15-18% of their body water (Harris & Kormanik, 1981). As in other crabs, this loss occurs in the main from the general integument (Herreid, 1969a,b; Fig. 5.23), but the rate of loss is slightly lower in *G. lateralis* than in *C. carnifex*, and in both it is much lower than in crabs from other families. Such water loss as occurs is counteracted by a number of methods. Water is taken up from damp substrates by capillary action, as in other terrestrial crabs. In gecarcinids the pericardial sacs are enlarged and they have a posterior extension fringed with setae (Fig. 5.26). When the abdomen is placed on the ground, water is conducted by these setae into the branchial cavity, particularly to the posterior gills (Bliss, Wang & Martinez, 1966; Bliss, 1968). The walls of the pericardial sacs are convoluted, and this appears to aid the flow of water. A suction mechanism which acts like the one in *Ocypode* is also present (Bliss, 1979). The posterior gills of *G. lateralis* are different from the anterior pairs (Copeland, 1968). In the anterior pairs the platelets have small knobs and projections which keep them apart so that they present a large surface area even in air (von Raben, 1934); but these projections are lacking in the posterior gills so that the platelets are close together. The gill epithelium in this region is specialised for the transport of salts and water: it is 10 to 20 times as thick as the respiratory epithelium and consists of cells with complex foldings and interdigitations with many mitochondria. Such a structure is typical of transporting epithelia. Presumably these posterior gills take up salts and water. It may also be that they are the site of magnesium regulation in this species, although this possibility has not been investigated.

Water which is taken up is stored as haemolymph, and the blood volume of gecarcinids is increased by the pericardial sacs mentioned previously. These organs can therefore be thought of as water storage accessories as they are in some other crabs. Such storage is necessary particularly for *Gecarcinus* because water is required at the moult when the crab may be far from a water supply. Since calcium is also required for the new exoskeleton, it also is stored, and can be found before the moult as gastroliths between the epidermis and cuticle of the proventriculus (Bliss, 1968). At moulting, water is moved into the proventriculus, and gas accumulates there (Mantel, 1968). This combination is responsible for expanding the new cuticle and producing the much-inflated epibranchial areas so typical of gecarcinids (see Fig. 5.27), and many other terrestrial crabs.

The epibranchial area of the carapace covers the large branchial chambers. Neither *Cardisoma guanhumi* nor *Gecarcinus lateralis* carries water in these, although *C. guanhumi* may flush the chambers with water and then allow it to drain out (Gifford, 1962b). *C. carnifex*, however, does retain some water in spongy tissue above the leg bases (Wood & Randall, 1981a). While it is likely that the anterior gills of both genera are able to function in air because of the separation of the gill platelets mentioned above, it is probable that more oxygen is taken up across the wall of the branchial chamber, which is covered with a network of vessels. Indeed, the gill area per unit body weight in *G. lateralis* is reduced to only 60% of that in *C. guanhumi* illustrating the decline in importance of gill breathing (Bliss, 1968). Díaz & Rodríguez (1977) have emphasised that an enlarged branchial chamber is particularly characteristic of gecarcinids, which have no folding of the respiratory membrane that might increase its area. Air is passed through the branchial chamber by the action of the scaphognathites, whose beating is intermittent as in aquatic forms (Cameron & Mecklenburg, 1973).

The haemolymph of both *Cardisoma* and *Gecarcinus* has a high oxygen capacity, and most oxygen is carried as oxyhaemocyanin. As in grapsids, there are now known to be wide variations in the P_{50} values of the haemocyanins of both aquatic and terrestrial decapods (Taylor & Davies, 1981). Although there is therefore no conclusive evidence that terrestrial forms have, in general, higher values of P_{50} than aquatic forms, the P_{50} is closely related to *internal* oxygen tensions. It would, therefore, be interesting to attempt to relate P_{50} values to gill/lung structure and the efficiency of gaseous exchange, rather than directly to the environment as attempted in Fig. 5.21. In *G. lateralis* the haemocyanin is 64% saturated with oxygen when it reaches the lungs, and 95% saturated when it leaves them (Taylor & Davies, 1981), and the extraction efficiency is low – only 2%. In *C. guanhumi* the extraction efficiency is also normally low (3.8%), but in *C. carnifex* it is much higher at 11.4% (Herreid, O'Mahoney & Shah, 1979; Wood & Randall, 1981a). These extraction efficiencies probably reflect in the main an inefficient circulation of air within the branchial chambers, so that although the ventilation rate is high, little of the air contacts the respiratory membranes (see Taylor & Davies, 1981). In contrast, high ventilation rates in aquatic forms result in high extraction efficiencies.

The response of ventilation rate to hypoxia in *Cardisoma* and *Gecarcinus* is small, but an increase in pCO_2 greatly increases ventilation (Cameron, 1975), as does exercise (Herreid, Lee & Shah, 1979). The response to pCO_2 is typical of air breathers rather than water breathers, yet the resting pCO_2 in the haemolymph of *C. carnifex* is in fact low for an air-breathing species – 15 mmHg (Wood & Randall, 1981b). However, *C. guanhumi*, in contrast, has a pre-branchial pCO_2 of 30 mmHg, compared with the aquatic species *Callinectes sapidus*, where it is only 6.5 mmHg (Young, 1972). It is not known why the two species of *Cardisoma* are so different in this respect. In *C. carnifex*, exercise produces a build-up of lactic acid which causes mild metabolic acidosis, but regulation of haemolymph pH is in fact good. How this is carried out is not entirely clear, although bicarbonate may be mobilised from the shell, and there is some evidence for ventilatory regulation of pCO_2 (Cameron, 1981b; Henry et al., 1981; McMahon & Burggren, 1981; Randall & Wood, 1981). In summary, the mechanisms of acid–base balance within the gecarcinids show some variation, but essentially they are adjusted to deal with a situation in which the exchange of ions across the gills is no longer possible. Their use of ventilatory regulation and the mobilisation of calcium carbonate reserves therefore show some interesting parallels with mechanisms found in the semi-terrestrial vertebrates (see Chapter 9).

Relatively few investigations have been concerned with the excretion of nitrogen by gecarcinids. It is known that in aquatic marine crabs such as the portunid *Carcinus maenas* a large fraction of ammonia is excreted, together with a small amount of uric acid (see Parry, 1960). In the one gecarcinid to be investigated, *Cardisoma guanhumi*, there is a high concentration of ammonia in the haemolymph, but little in the urine (Gifford, 1968; Horne, 1968*b*). Nothing is known of the route by which ammonia is excreted, but the gut may be implicated because ammonia concentrations in the stomach are even higher than in the haemolymph. Alternatively, it may be that the high concentration of ammonia in the haemolymph allows ammonia to be released directly to the air as demonstrated in snails and woodlice. In this connection it may be important that *Cardisoma carnifex* has high concentrations of carbonic anhydrase in the gills of at least some individuals (Randall & Wood, 1981). *C. carnifex* also shows another possible parallel with the gastropods in the storage of uric acid: this is deposited throughout the haemocoel, often in large quantities. Since it is not found in recently moulted specimens, but does not appear to be actually cast with the old exoskeleton, its mode of elimination from the body remains unknown. A similar situation probably occurs in *Gecarcinus lateralis* because this also accumulates a white deposit in the haemocoel. The fact that *C. guanhumi* can also excrete urea after feeding on nitrogen-rich diets (Horne, 1968*b*) complicates the picture, and it is evident that much remains to be elucidated.

The details of adaptations in behaviour, movement and sensory perception are also little known in gecarcinids. This is rather surprising, since *Cardisoma guanhumi* has an impressive range of types and mechanisms of moving: it can move laterally, backwards, forwards or diagonally, and it can climb vertical surfaces. In these movements stepping is highly irregular and appears to be controlled on a 'cycle to cycle' basis rather than having a basic uniformity in each cycle as do insects such as the cockroaches. Also, unlike these insects, the coupling between contralateral pairs of legs in *Cardisoma guanhumi* is weak or even non-existent, allowing a great variety of movement sequences. During lateral walking, for instance, the leading legs can step at a higher frequency than the trailing ones (Evoy & Fourtner, 1973), a phenomenon also observed in the aquatic *Callinectes sapidus*.

Many of these movements can be observed in the ritualised agonistic and courtship behaviour of *Cardisoma*, although it is apparently the posture of the chelipeds that is of most importance here (Wright, 1968). These display patterns have overall similarities with those of other terrestrial crabs, and probably entail communications which are primarily visual. Nothing is known about the possible existence of true hearing in gecarcinids, but substrate-borne vibrations are undoubtedly important in *Gecarcinus*: sequences of pulses produced by stridulation are transmitted via the substratum and are specific for

threat, appeasement or sexual display (Klaassen, 1973, 1975).

Brachyura – general conclusions

All the terrestrial families of crabs avoid extreme values of temperature and humidity by digging burrows and emerging from these either at night or for relatively short periods during the day. This difference from the terrestrial amphipods and isopods is forced upon them because of the size difference: whereas members of these two groups weigh of the order of 0.05 to 0.5 g, crabs may reach weights of over 500 g. It is therefore impossible for them to find shelter by hiding under debris and so on. Despite the fact that their surface area : volume ratio is reduced in comparison with the woodlice, crabs lose a great deal of water through the carapace. In the strictly marine forms the rate of transpiration is more than an order of magnitude greater than that found in the terrestrial species (Table 5.9). The rates measured in *Gecarcinus lateralis* and *Cardisoma guanhumi* are not strictly comparable with those in other experiments because the saturation deficit was not given. However, using an approximate value for this, the rates were of the order of 60–80 μg cm^{-2} hour^{-1} (mmHg)$^{-1}$. This is comparable with rates in some insects (Table 8.4) and some terrestrial isopods (Table 5.5). The structural basis for the barrier to water loss which is present in gecarcinids is not known, but since lipid layers have been found in insects and isopods, it seems likely that at least *Gecarcinus* has developed some equivalent.

The problems of obtaining water have been solved in similar ways by all the groups of crabs. Setae on the underside of the crabs allow water to be drawn up into the branchial cavity by capillarity, and in this way water can be extracted from substrates which are merely damp. This parallels the mechanisms used by isopods which are also able to extract water from damp surfaces. The ability of some species of decapods to extract water from the soil capillaries, using a suction method, is surprising, and parallels that used by spiders (see Chapter 6). Nevertheless, only *Holthuisana* and *Gecarcinus* have been shown to make burrows that do not extend down to the

Table 5.9. *The rate of transpiration from the integument of decapods*

Species	Transpiration rate (mg cm^{-2} hour^{-1})
Gecarcinus lateralis (Gecarcinidae)	1.2
Cardisoma guanhumi (Gecarcinidae)	1.4
Ocypode quadrata (Ocypodidae)	2.6
Uca pugilator (Ocypodidae)	3.2
Sesarma reticulatum (Grapsidae)	6
Callinectes sapidus (Portunidae)	14
Panopeus herbstii (Xanthidae)	16
Menippe mercenaria (Xanthidae)	21

Data from Herreid (1969*b*). Measurements were made on excised pieces of integument.

water table, so that retreat to the burrow can in most cases provide a respite from desiccation.

Osmoregulation, both hyper- and hypo-osmotic, is well developed in the more terrestrial species, and this may suggest an origin for most groups in littoral saline lagoons. Since such environments are mainly restricted to the tropics and sub-tropics, the equatorial distribution of terrestrial crabs may be related to their origin, as well as to the more equable conditions found in these regions than in temperate zones. The mechanisms of osmoregulation are unknown but are extra-renal. Even the crabs which have moved into fresh water have haemolymph with a fairly high osmotic pressure, as can be seen from the figures given for various decapods in Table 5.10, and they do not produce hypo-osmotic urine. This is in direct contrast to most other terrestrial invertebrates of freshwater origin, and reflects the great reduction in permeability to both salts and water associated with the arthropod cuticle. The reduction in osmotic pressure that has occurred appears to be almost entirely due to reduction in the concentrations of sodium and chloride. Magnesium is also low, but this is so even in *Gecarcinus*.

Respiratory mechanisms show a variety of modifications in the different families. While the gills are sometimes reduced in volume, the more general change is the development of increased vascularisation of the walls of the branchial cavity, and the enlargement of the epibranchial region to give a greater surface area for gaseous exchange. In some species water is retained in the branchial cavity, while in others it is not, but in all cases the respiratory membranes must remain moist even in air. At least in *Ocypode*, it appears that the gills may be kept moist by ultrafiltration from the haemolymph, but the way in which the walls of the cavity retain a water film is not known. The methods by which acid–base balance is regulated vary between different decapod groups, but all show the development of mechanisms which are independent of exchanges across the gills. Parallels with the terrestrial vertebrates, such as the use of ventilation rate to control pCO_2, are found in some decapods, but none appears to use the renal system to counter acid–base problems arising from aerial respiration.

All the decapods except the 'freshwater crabs' and a very few species of grapsids are linked to the sea by reproductive processes. Even *Gecarcinus* has to return to the sea to liberate its larvae. In freshwater crabs it is strange that while the larvae develop in contact with the mother in her burrow, they are then liberated, *as young crabs*, into fresh water, where they remain for some time. There appears to be some block to the emancipation of the decapods from an aquatic development, but why this should be so when both amphipods and isopods exhibit forms independent of water is at present inexplicable. Some changes in the reproductive mechanisms, such as the possibility of mating when both adults have hard shells, have occurred, but at some stage all egg-bearing females must migrate back to water. The families Ocypodidae and Gecarcinidae, and most of the Grapsidae, must remain restricted to the coastal fringe, while the inland Grapsidae and the freshwater crabs require bodies of fresh water to complete their reproductive cycle.

Nevertheless the adaptations of behaviour and of sensory physiology make these crabs conspicuous members of the coastal zones, and invite comparisons with primarily terrestrial groups such as the insects. The visual systems, for instance, have some characteristics comparable with both insects and vertebrates (Table 5.11). For example, although the visual acuity has only been estimated in two crabs, in one of these the acuity is better than that of insects. The flicker fusion frequency is comparable with that found in isopods, and lies between that of insects such as cockroaches and blowflies. These few figures suggest that semi-terrestrial crabs are very much aware of their surrounding environment and the action taking place in it. In conjunction with this, the running speeds of *Ocypode* and some of the grapsids exceed the top speeds of insects such as cockroaches (see Hughes & Mill, 1974), and the complexity of social interactions within species parallels that of some of the insects. Communications involve

Table 5.10. *Haemolymph composition of some terrestrial and semi-terrestrial decapods*

Species	Osmotic pressure (mOsm)	Na$^+$	K$^+$	Ca^{2+}	Mg^{2+}	Cl$^-$	Reference
				(mM/litre)			
Anomura							
Coenobita perlatus	–	465	10.5	14.7	30.9	–	Gross, 1963
Brachyura							
Sudanonautes africanus (Potamoidea)	480	207	6.0	11.8	10.6	241	Lutz, 1969
Holthuisana transversa (Parathelphusoidea)	525	270	6.4	15.7	4.7	266	Greenaway & MacMillen, 1978
Ocypode albicans (Ocypodidae)	960	449	7.0	15.7	28.6	475	Gifford, 1962a
Gecarcinus lateralis (Gecarcinidae)	765	468	12	17.3	7.6	–	Mason, 1970

movement and vision in most crabs, but the importance of sound production and reception is now being realised, and here again there are striking parallels with the insects.

Semi-terrestrial crabs possess an ability to respond rapidly to visual stimuli and exhibit behaviour patterns which appear to be more complex than those of aquatic crabs. Similar suggestions could be made about the semi-terrestrial amphipods in comparisons with the purely aquatic ones. Indeed, it has often been postulated that since visual communication is more efficient over long distances in air than it is in water, visual displays and visually oriented behaviour in general might be expected to be more complex on land. In water it might be expected that chemical and tactile stimuli would be more appropriate. While such postulates may seem entirely reasonable, they have yet to be firmly established for the Crustacea because so little is known of the behaviour of the purely aquatic forms.

Table 5.11. *Characteristics of decapod eyes compared with those of other groups*

Species	Visual acuity	Minimum angle detectable	Flicker fusion frequency (flashes/s)
Decapods			
Goniopsis cruentata (Grapsidae)	0.400	2.5$'$	–
Uca sp. (Ocypodidae)	0.0042	4$°$	90–100
Ocypode sp. (Ocypodidae)	–	–	170
Isopods			
Ligia occidentalis	–	–	120
Insects			
Periplaneta americana	0.167	6$'$	50
Calliphora sp.	0.167–0.083	6–12$'$	300
Vertebrates			
Homo sapiens	2–2.5	0.5$'$	50

Note: because visual acuity has been measured in different ways, the figures quoted should be regarded as approximate. Visual acuity is defined as the reciprocal of the angle in minutes of arc subtended at the eye by the minimum detail. Data from Dethier (1963), Waterman (1961) and Mazokhin-Porshnyakov (1969).

6

Chelicerates

'When first seen, they come from the deeper water, the male, which is almost always the smaller, grasping the hinder half of the carapax of the female with the modified pincer of the second pair of feet. Thus fastened together the male rides to shallow water . . . I have never seen the couples come entirely out of the water, although they frequently come so close to the shore that portions of the carapax are uncovered.'

From 'The embryology of Limulus' by J. S. Kingsley, 1892, *Journal of Morphology*, 7, 35–68.

Unlike the situation in most of the groups so far considered, there is serious debate about the place of origin of the chelicerates. While some theories suggest an origin in the sea, others postulate terrestrial beginnings with subsequent colonisation of the freshwater and marine habitats. Much of this confusion and disagreement occurs because of the ancient origin of the group: the horseshoe crabs (Xiphosura) have been recorded from Cambrian rocks, the pycnogonids may have fossil forms dating from the Devonian, and all the major orders of the Arachnida are represented in the Carboniferous.

 Another problem with the chelicerates is the definition of the group: which animals are sufficiently closely related to be included, and which are related in some rather more vague way? The 'basic' chelicerates may be said to comprise the eurypterids (all fossil), the horseshoe crabs and the arachnids. Some authors would also include the pycnogonids (e.g. Barnes, 1980), and others would suggest a more distant relationship with the trilobites (e.g. Fage, 1949; Kraus, 1976). Other views again would separate off the trilobites, and would also place the pycnogonids as a separate group with some chelicerate affinities. These various groups are listed in Table 6.1.

 Even concerning the terrestrial members, i.e. the Arachnida, there is no agreement about evolutionary relationships. Here perhaps the most realistic approach is to regard each order as having a mosaic of characters, some primitive and some more specialised, in the same way that ellobiid molluscs were considered (see p. 48).

Thus the scorpions are said to show the greatest combination of primitive characters, but in terms of segmentation other orders provide more evidence of the primitive segmental pattern. At the anterior end, for instance, most orders have a uniform carapace covering the dorsal surface of the head, but the Schizomida, Solifugae and Palpigradi have three head tergites. On the ventral surface the Schizomida, Uropygi and Palpigradi also have persistent head sternites behind the mouth. Similar contrasts are found in the opisthosoma. Here the primitive number of segments is thought to be 11, posterior to the pregenital somite or pedicel (Savory, 1964). In some, e.g. the scorpions, all these segments are present. In other orders the last three somites tend to form a narrow 'pygidium', apparent in the Pedipalpi, Palpigradi, Ricinulei and, to a small extent, the Araneida. In most other orders this pygidium is lost.

 Primitive characters are thus seen to be common not only in scorpions, but also in Schizomida, Palpigradi, Uropygi, Amblypygi and Solifugae. All of these are rare in the fossil record, but this may be simply because some of them are small and little likely to be preserved, while others live in habitats where fossilisation is unlikely. The survival of the scorpions as fossils in earlier rocks than any other terrestrial arachnid may therefore, it could be argued, be misleading in suggesting scorpions as the first terrestrial arachnids. The great resemblance between scorpions and the fossil eurypterids has usually been accepted as evidence of the origin of the former

Table 6.1. *Chelicerates and related groups*

[Subphylum Trilobitomorpha]	All fossil	Marine
Subphylum Chelicerata		
Class Xiphosura	Present day/ fossil	Marine
Class Eurypterida	All fossil	Fresh water/ marine
Class Arachnida	Present day/ fossil	Mainly terrestrial
[Class Pycnogonida]	Present day/ ? fossil	Marine

Square brackets indicate doubtful affinity.

from the latter; the scorpions are then viewed as a basal group from which many of the terrestrial arachnids may have radiated. If, on the other hand, orders other than the scorpions show many primitive characters, it could be that the first terrestrial arachnids were not scorpions, and that this order represents a subsequent development from terrestrial forms not preserved in the fossil record. Such an argument has been carried to its extreme by Versluys & Demoll (1923), who suggested that the aquatic eurypterids have been derived from the terrestrial scorpions.

With such basic disagreements about the origin of the terrestrial arachnids, it is impossible at present to come to any final conclusion on the subject. Since, however, the present-day arachnid assemblage is predominantly a terrestrial one, it seems appropriate to consider the terrestrial adaptations of its members before returning to the subject of its origin.

6.1 The arachnids

The orders of the class Arachnida are listed in Table 6.2. From this table it is apparent that the vast majority of species (over 95%) come from only a few orders: spiders, mites, harvestmen and pseudoscorpions. In the main, research has concentrated upon the spiders and mites because of their medical and agricultural importance, but recently more information is being obtained about scorpions, pseudoscorpions and solifugids. Little is known yet about the small groups (see Savory, 1964; Kaestner, 1968).

Before discussing the terrestrial adaptations of arachnids, it may be useful to say something briefly about each of the orders to introduce them. Of these orders, the largest animals are found in the Scorpiones, in which one fossil species reached a length of 36 cm. The smallest is a present-day species with a maximum length of only 13 mm. Scorpions are easily recognised by the

Table 6.2. *The orders of Arachnida (from Savory, 1964)*

Order	Approximate number of species
Present-day orders	
Scorpiones (scorpions)	700
Pseudoscorpiones (pseudoscorpions)	1 700
Opiliones (harvestmen)	3 700
Acari (mites and ticks)	6 000
Palpigradi (micro-whipscorpions)	20
Uropygi (whipscorpions)	100
Schizomida	20
Amblypygi (whipscorpions)	60
Araneida (spiders)	40 000
Solifugae (sun spiders/wind scorpions)	800
Ricinulei	14
Fossil orders	
Architarbi	
Haptopoda	
Anthracomarti	
Trigonotarbi	
Kustarachnae	

(Uropygi, Schizomida, Amblypygi are grouped as 'Pedipalpi')

poison sting at the end of the tail, and the large chelae on the pedipalps. They are found mainly in warmer climates, but spread from a southern limit in Tasmania to a northern one on the north Mediterranean coast. They are present in the fossil record from the Silurian onwards, being most common in the Carboniferous. They resemble eurypterids in general appearance, and it seems likely that one evolved from the other although as already mentioned there is dispute about which is the more primitive.

Pseudoscorpions are, in contrast to most scorpions, very small, none being longer than about 7 mm. They are crevice dwellers and are common in leaf litter and under bark, although as in many other arachnid orders, they show a great overall diversity of habitats, and can be found from the intertidal zone to desert environments (see Weygoldt, 1969). They resemble scorpions in that they have large pedipalps bearing chelae; in pseudoscorpions, however, these chelae contain poison glands, and there is no poison sting at the animal's posterior end.

Harvestmen are large enough to be well known, and are easily recognised from their short round bodies and long thin legs. They require a humid environment and tend to be found particularly in grasslands and forests. They are carnivores like most other arachnids, eating snails, earthworms, millipedes and so on, but for predators are surprisingly delicate: the legs are easily lost, and loss of the second pair (which are the longest and normally lead the walking rhythm) can be fatal (Savory, 1964; Sankey & Savory, 1974).

Mites can rightly be termed ubiquitous, since they are found on land from freezing to tropical climates, in the marine littoral zone, in fresh water and even in the aerial plankton (Hughes, 1959). Their diversity of life styles is as remarkable as their variety of habitats: some are free-living vegetarians, carnivores or detritus feeders, while others are parasitic to a varying degree and for differing parts of the life cycle. Many are well known because they cause a horticultural nuisance (e.g. *Eriophyes*, which produces 'big bud' on blackcurrant bushes) or because they are ectoparasites on man or his domestic animals (e.g. *Ixodes*, which sucks the blood of sheep and cattle). In all Acari, segmentation is much reduced and the body assumes a sac-like shape with a great variety of sculpture and decoration.

The palpigrades are small, uncommon animals which live in the humid environment underneath stones in warm climates. They have a number of primitive characteristics, and possess a long jointed 'telson' attached to the last segment of the opisthosoma.

The three small groups Uropygi, Schizomida and Amblypygi were formerly placed in one order, the 'Pedipalpi'. While they show distinct differences from each other, they are all tropical or nearly so, and they all hold their first pair of legs out in front of them to act as sensory organs. The Uropygi have a long telson, and secrete formic acid or acetic acid from special glands at the posterior end of the opisthosoma. The Schizomida have only a short telson, and the Amblypygi have no telson, but extremely long first legs and large powerful pedipalps. The members of all three groups are found in humid places or under stones, and cannot withstand much desiccation.

The spiders need little introduction, since they are the most widespread and plentiful of arachnids. They have a worldwide distribution and are characterised above all by their use of silk for a great variety of purposes: while mites and pseudoscorpions also secrete silk, it is in the spiders that the development of silk to produce webs, cocoons and draglines has reached its peak. Spiders inhabit a vast range of environments, from deserts and tropical forests to the sea shore and fresh water. Much of our knowledge of arachnid biology has been obtained from work on spiders.

Solifugids are typically found in deserts. They are large predators (up to 7 cm in length) with very powerful biting chelicerae, but no poison glands. They can run very rapidly and when doing so they use only three pairs of legs: as in the Pedipalpi, the first pair are held out in front and act as sense organs. Solifugids are usually active at night and spend the daytime in burrows which they excavate themselves (see Junqua, 1966).

The final group is the order Ricinulei. This is a very small group, about which very little is known. The animals are heavily sclerotised, and none is longer than 1 cm. They have a very scattered distribution in the tropics and individuals are exceedingly rare. They are found in leaf litter and other humid places such as caves, and are intolerant of desiccation.

6.1.1 Water balance and temperature control

Following this brief introduction to the various orders of arachnids, we can consider some of the ways in which various types are adapted to a terrestrial life. We know far more about the factors which control water balance than we do about other adaptations, so that this is a sensible place at which to begin. Nevertheless, information on basic points such as normal rates of water loss is available only for representatives of five groups (see Table 6.3), and these results probably give a rather false impression of arachnids as a whole since most workers have concentrated upon desert species. It is certain that water is lost from members of the smaller groups such as pseudoscorpions and the Pedipalpi at a very much faster rate. The measurement of water loss from arthropods raises a number of complicated questions, as pointed out by Edney (1957). First of all, water loss should be measured in moving air at a known humidity and temperature. It may then be expressed in its simplest form, as % loss per unit time, as in Table 6.3. This does not, however, fully account for variation due to the animal's size because of the increased surface : volume ratio in smaller animals. It would be best, if figures were available, to compare loss rates per unit area of animal. Water loss is also governed by temperature effects, and measurements are often made at a variety of temperatures. Even if the relative humidity is kept constant, the saturation deficit – that is, the amount by which the

Table 6.3. *Water loss from a variety of arachnids*

Species	Habitat	Weight (g)	Temp. (°C)	Water loss (% body wt/hour at near to 0% RH)	Reference
Scorpions					
Androctonus australis	Desert	7	33	0.032	Cloudsley-Thompson, 1961
Hadrurus hirsutus	Desert	7	33	0.042	Cloudsley-Thompson, 1967
Diplocentrus spitzeri	Montane	0.04–1.9	30	0.12–0.84	Crawford & Wooten, 1973
Centruroides sculpturatus	Desert	?	30	0.091	Hadley, 1970*b*
Solifugids					
Galeodes granti	Desert	7	33	0.090	Cloudsley-Thompson, 1967
Spiders					
Eurypelma sp.	Desert	4–6	33	0.147	Cloudsley-Thompson, 1967
Eurypelma helluo (at 25% RH)	?	?	25	0.600	Herreid, 1969*a*
Dugesiella hentzi	?	10	25–40	0.21–0.27	Stewart & Martin, 1970
Pardosa lapidicina	Ground layer	0.01	20	0.3	Vollmer & MacMahon, 1974
Dolomedes scriptus	Semi-aquatic	0.1	20	1.0	Vollmer & MacMahon, 1974
Uropygi					
Mastigoproctus giganteus	Montane	0.1–6.5	30	0.3–2.2	Ahearn, 1970*a*
Mastigoproctus giganteus	—	4.1–7.5	26	2.27	Crawford & Cloudsley-Thompson, 1971
Acari					
Ixodes ricinus	Humid pasture	0.002	25	0.7–2.1	Lees, 1946*a*, 1947
Hyalomma savignyi	Desert	?	25	0.033	Lees, 1947
Ornithodorus delanoei acinus	Desert	?	25	0.004	Lees, 1947

humidity of the air falls short of saturation – changes with temperature. Permeabilities are therefore sometimes expressed as water loss per unit area per unit time per unit of saturation deficit, at a particular temperature, and some figures are shown in this way in Table 6.4. Even here, although this is a convenient way of comparing different species, rates do not always relate directly to saturation deficit (see, e.g. Warburg, Goldenberg & Ben-Horin, 1980). In addition, animals may lose water through not only the general surface of the cuticle, but also the spiracles of the tracheae or the lung books, so that comparisons of total water loss do not necessarily reflect cuticular permeability.

Taking into account these factors, it is evident that the problem of water loss is a very complicated one, and that we are not in a position to be able to compare a wide variety of arachnids in any detail. But even the 'overall view' expressed as % loss per hour at 0% RH in Table 6.3 shows some interesting points. The lowest rates measured in this way are shown by desert scorpions, with the desert solifugids showing only a slightly greater rate. Some of the ticks have extremely low rates of water loss, while others have relatively high rates. Spiders and Uropygi also show a wide range.

When the rates of water loss are given as permeabilities, i.e. as $\mu g\, cm^{-2}\, h^{-1}\, (mmHg)^{-1}$, as in Table 6.4, the same points are evident. It may be postulated, therefore, that the overall rate of water loss is related more to habitat than to the order in which an arachnid is placed, but such a suggestion must be carefully hedged about with conditions. First, the habitat as defined by such terms as 'desert', 'montane' and so on, reflects only the macrohabitat, while the animals concerned normally live in a much more restricted microhabitat. This has been well illustrated in desert scorpions by the work of

Cloudsley-Thompson (1956). From measurements in Tunisia, he showed that the sand surface may vary in temperature from 0 °C to 45 °C. The sand 3 cm down, however, never reached more than 33.5.°C. Similarly, while the minimum humidity on the surface was 29% RH that 3 cm down never fell below 50%. Since some scorpions burrow, whilst others hide under rocks on the desert surface, the desiccating power of the air to which they are exposed varies tremendously between species. The microclimate is, of course, also defined by behaviour itself: all scorpions are nocturnal and therefore avoid daytime extremes. In many species this behaviour has been shown to be governed by a diurnal rhythm – partly intrinsic, and partly reinforced by environmental stimuli. Other forms of behaviour also influence the temperature of scorpions: in *Opisthophthalmus latimanus* and probably in many other species the phenomena of 'stilting', i.e. of raising the body from the surface of the ground, allows air currents to flow all round the animal. In sunlight this may lead to a reduction in temperature of up to 10 °C, compared with animals resting on the substratum (Alexander & Ewer, 1958). In nature this scorpion normally remains just inside the entrance to its burrow during the daytime, and feeds on other arthropods which approach. If the temperature rises above a lethal value of 40–50 °C, the scorpion retreats further down into its burrow.

At the other end of the temperature scale, many arachnids can withstand surprisingly harsh conditions. Some sand dune spiders, for instance, can survive down to −11 °C in summer, and down to −22 °C in winter. This suggests that as in many other animals some arachnids may form a type of antifreeze (Almquist, 1970).

Another example of differing habitats is shown in the spiders. Three English species of the genus *Amauro-*

Table 6.4. *Water loss from a variety of arachnids, in relation to saturation deficit*

Species	Water loss ($\mu g\, cm^{-2}\, hour^{-1}\, (mm\, Hg)^{-1}$ at 20–30 °C)	Reference
Scorpions		
Pandinus imperator	82	Cloudsley-Thompson, 1959
Androctonus australis	0.4–0.8	Cloudsley-Thompson, 1956
Solifugids		
Galeodes arabs	6.6	Cloudsley-Thompson, 1961
Othoes saharae	11.3	Délye, 1969
Spiders		
Lycosa amentata	32.6	Davies & Edney, 1952
Ciniflo similis	11.0	Cloudsley-Thompson, 1957
Uropygi		
Mastigoproctus giganteus	21.7*	Crawford & Cloudsley-Thompson, 1971
Acari		
Ixodes ricinus	60	Wigglesworth, 1945
Hyalomma savignyi	0.4	Lees, 1947
Ornithodorus delanoei acinus	0.35	Lees, 1947

* The observations of Ahearn (1970*a*) imply a much lower rate than this.

bius are all nocturnal, and each can be found in places defined by a particular relative humidity (Cloudsley-Thompson, 1957). Such preferences indicate the possession of specific receptor mechanisms, and species of *Amaurobius* have been shown to select higher humidities in preference to lower ones – according to Blumenthal (1935), the 'tarsal organ' of spiders is responsible for their ability to sense humidity. Correlations between rates of water loss and habitat have been shown in a number of unrelated spiders by Vollmer & MacMahon (1974). In particular, it was noted that *Dolomedes scriptus*, a semi-aquatic species, has a loss rate of the order of 1%/hour, whereas the groundlayer species *Pardosa lapidicina* has a loss rate of only 0.3%/hour, despite the fact that it is only one-tenth of the weight of *D. scriptus*. In several related species, on the other hand, Engelhardt (1964) showed very similar tolerances and humidity extremes. All these species, in the genus *Trochosa*, lived in similar habitats, and it was in fact difficult to explain the isolation of the species because of their great similarities.

According to Lagerspetz & Jäynäs (1958), the behavioural mechanisms by which spiders select appropriate humidities can be defined mainly in terms of orthokinetic orientation, in combination with klinokineses and klinotaxes. The characteristics of the orthokineses differ from species to species, and from season to season, and undoubtedly define the spider's microhabitat closely. This is well illustrated (see Nørgaard, 1951) by two species of lycosids found in *Sphagnum* bogs: *Pirata piraticus* lives in spaces between the *Sphagnum* stalks where the temperature varies only from 17 to 22 °C, and the relative humidity stays close to 100%. *Lycosa pullata* lives on the surface of the *Sphagnum* where the temperature range is from 6 to 39 °C and the humidity falls to 40% in the day. The temperature preferred by *L. pullata* is 30 to 32 °C. That for males of *P. piraticus* is 20 to 22 °C. For females of this species which are carrying egg cocoons, however, the preferred temperature is 26 to 28 °C: this fits well with the migration of females to the surface of the *Sphagnum* to 'sun' their eggs, and this type of reaction is probably common in spiders which carry their eggs with them. Those that do not do so usually have other mechanisms for maintaining an appropriate temperature for the development of the eggs. A good example of this is shown by *Theridion saxatile* (see Nørgaard, 1956). This species lives in a tube-shaped nest covered with sand grains and suspended by silk threads. Examples of temperatures in and around the nest are shown in Fig. 6.1. The upper normal limit of activity of *T. saxatile* is about 40 °C, and eggs do not hatch when exposed to temperatures higher than this, but the nest may heat up to 45 °C in the daytime. The eggs are therefore taken out of the nest in the day and hung beneath it, to be returned at night.

Control of the temperature and water content of the body of spiders themselves requires a more sophisticated behavioural regime. The burrowing Australian lycosid *Geolycosa godeffroyi* has been investigated by Humphreys (1974, 1975). In this species body tempera-

ture may rise to 40 °C by absorption of radiant heat from the sun, even though the ambient is only 10 °C. Movement into and out of the burrow allows fine control of body temperature (Fig. 6.2), so that the species could be called a 'behavioural homoiotherm' in the same sense as applied to some reptiles. Water must be lost while basking in the sun, but can usually be absorbed from the damp soil in the burrow. When this is not available, movement of the spider at night from the depths of the burrow to the surface allows dew to condense on the spider's body as well as on surrounding soil particles. Given the appropriate behaviour, then, water is always available and temperature can be accurately controlled. Orb-web spiders have also been shown to regulate their body temperature behaviourally. Species of *Argiope* hold the abdomen at right angles to the sun's rays when temperature is low, so that maximum surface area is presented for heat absorption. At midday, the abdomen is pointed directly at the sun to minimise heat absorption, and heat reflection is aided by 'metallic' coloration (Robinson & Robinson, 1978; Tolbert, 1979).

Even when the habitat is defined as a microhabitat, other factors than environmental ones relate to the rate of water loss. For example, the rate of water loss is size-dependent (see Fig. 6.3; Ahearn, 1970a), even within a species. This is likely to be so between different species also, so that it is not really possible to make accurate comparisons between small animals such as ticks and animals 1000 times their weight. Nevertheless, the permeabilities shown in Table 6.4 relate surprisingly well to the rates for percentage water loss shown in Table 6.3. Another problem is given by the variations in results obtained by different workers using the same species.

Fig. 6.1. Temperature conditions in and around the nest of the spider *Theridion saxatile*, in the daytime. The temperature of the nest rises above the lethal for the eggs, and the spider hangs them underneath the nest. Redrawn from Nørgaard (1956).

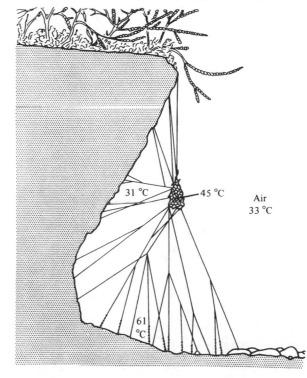

31 °C   45 °C   Air 33 °C

61 °C

Fig. 6.2. Cephalothorax temperatures of the spider *Geolycosa godeffroyi* in relation to ambient temperature. Fine lines show temperature of the cephalothorax, obtained from implanted thermocouples; thick lines show ambient air temperatures; dashed line shows temperature of dead spiders. Records were obtained in Australia: (*a*) in midwinter (June); (*b*) in early autumn (March); (*c*) in late winter (September). Increases in temperature of the body were due to emergence from the burrow and absorption of radiant heat. Redrawn from Humphreys (1974).

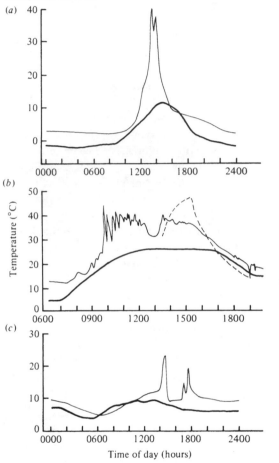

Fig. 6.3. Weight loss during desiccation in the uropygid *Mastigoproctus giganteus*, as a function of body weight. Each point represents a rate for one individual exposed to moving air for 2 hours at 35 °C and approximately 0% RH. Redrawn from Ahearn (1970*a*).

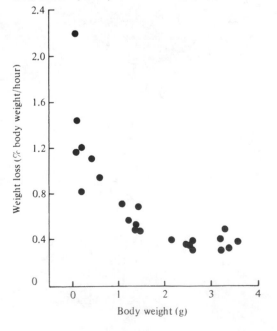

This is well illustrated by the results for *Mastigoproctus giganteus* (Table 6.3). According to Crawford & Cloudsley-Thompson (1971), the rate of water loss from animals of 4–7 g was about 2.2%/hour. According to Ahearn (1970*a*) this rate was found only in small animals (0.1 g), while animals of 6 g lost water at only 0.3%/hour. It seems likely that this difference is due to differences in the degree of abrasion of the cuticle, and to handling of the animals, and this brings us to the whole question of how water is lost, or retained, by the cuticle, and to the exact site of the loss.

The cuticle of ticks such as *Ixodes* consists of a thick endocuticle, with a very thin epicuticle on the external surface. The whole cuticle is traversed by the ducts of dermal glands which open on its surface and appear to secrete a waxy or greasy layer similar to that of insects (Lees, 1946*a*, 1947; Beament, 1949). However, as in insects, much of the wax is found in exceedingly narrow wax canals, and exactly how it is transported to the surface of the cuticle is uncertain (Hadley, 1981). The thin epicuticular wax layer is very delicate, and the experimental handling of ticks vastly increases the rate at which they lose water. This observation suggests that it may be mainly the wax layer which acts as a water-proofing mechanism. Additional evidence for this has come, as in the insects, from observations on the rate of water loss at different temperatures. For instance, if the loss from species of *Ixodes* is measured, there appears to be a transition temperature at 30–35 °C above which the rate increases rapidly (Beament, 1959). By analogy with the insects (see p. 166), this may be thought of as indicating a temperature at which the molecules in a monolayer of wax become disoriented, allowing water to pass through (Beament, 1959). In insects the transition temperatures of whole animals have been shown to co-incide closely with those of isolated waxes. It should be noted, however, that some authors doubt the reality of the transition temperature, and believe that increases in the rate of water loss with temperature really result from a process that is a continuous function of temperature (see discussion by Hadley, 1981). Further evidence for the presence of a waterproofing wax layer has been obtained in the clover mite *Bryobia praetiosa*. In this species treatment with chloroform leads to a higher rate of water loss (Winston & Nelson, 1965). The epicuticular lipids which provide waterproofing in arachnids are similar to those found in insects. In scorpions, however, the lipids show differences such as a higher proportion of sterols (Toolson & Hadley, 1977).

The situation in ticks in also more complex than it would seem at first because of the variation which occurs in the wax layers at different stages of the life cycle. The cattle tick *Boophilus microplus* increases the amount of cuticular wax after it drops off the host (Cherry, 1969). In the same period the rabbit tick *Haemaphysalis leporispalustris* shows a drop in rate of water loss, and a rise in the critical temperature of the cuticular wax (Davies, 1974*a*). All these processes would seem to be advantageous, since the ticks leave a host which guarantees a high humidity and moisture supply, for an environment with lower and more erratic humidity. Other differences between species are more fundamental: most ticks have a wax layer of the type discussed above, but only in *Ornithodorus moubata* has this been shown to be protected by an outer cement layer (Lees, 1947). Added to this, it has been suggested that the low rate of water loss in ticks is not due entirely to the waterproofing bestowed by the wax layers, but also depends upon an active removal of the water from the cuticle (Lees, 1947). This has actually been shown to be so in insects, where the activity of the water in the cuticle is not in equilibrium with that in the haemolymph (Winston & Beament, 1969). In mites, dead animals lose more water than living ones, and this has suggested active water retention in the cuticle, possibly by the epidermal cells (Wharton & Kanungo, 1962; Winston & Nelson, 1965).

Very little is known about possible epicuticular waxes in arachnids other than ticks and scorpions, but if they exist they might explain the results given above for the uropygid *Mastigoproctus giganteus*: the animals showing low rates of water loss were maintained on damp filter paper, while those which lost water rapidly were kept on sand. This would be very likely to cause abrasion of the wax and therefore higher rates of water loss.

The susceptibility of the waxes to handling has meant that, unlike the situation in woodlice, where the waxes are (if present at all) deep in the cuticle (see p. 75), it is difficult to block the respiratory openings in order to measure the effect of this upon water loss. When this is done, as it has been with ticks, the rate of water loss actually rises because of the handling effect (Lees, 1946a). Similar increases in rates of water loss following blockage of the lung books in dead scorpions have been attributed to 'restrictive mechanisms' in the cuticle (Hadley, 1970b), but here again operational damage to the cuticular waxes is likely. Another way of tackling the problem of respiratory water loss has been to use animals treated with carbon dioxide. In spiders this has the effect of opening the spiracles of the lung books, whereas normal resting spiders usually have their spiracles nearly closed. When this treatment was given to species of *Lycosa*, the rate of evaporation increased by nearly 50%, suggesting that a great deal of water may be lost from the lung books (Davies & Edney, 1952). Loss from the tracheae is likely to be small because the surface area of tracheal tubes is only about one-thirtieth of that of the lung books (Davies & Edney, 1952). The function of the spiracles in preventing water loss is in fact not well documented. In some ticks the spiracles are closed by plates, but there is a variety of opinions about whether these have open pores (see, e.g. Browning, 1954; Hinton, 1967).

Nothing is known about the amount of water lost in faeces or with excretory products, but in the ticks at least there is one time in the life cycle when much water is eliminated. Most studies have concentrated on ticks which are isolated from their host, but when they are attached and absorbing blood, they have to cope with a greater inflow of water than is needed. In the argasid ticks, such as *Ornithodorus moubata*, this excess water is removed by the coxal glands, which excrete a fluid slightly hypo-osmotic to the haemolymph of the tick (Boné, 1943; Lees, 1946b; Kaufman, Kaufman & Phillips, 1981). In ixodid ticks excess water is removed by the salivary glands (Kaufman & Phillips, 1973; Hsu & Sauer, 1975). In *Dermacentor andersoni*, for instance, 75% of the water is excreted back into the host in this way, accompanied by sodium and chloride. In this species some fluid is also eliminated through the anus, together with most of the potassium (see summary in Fig. 6.4). In the male the system probably does not involve the salivary glands at all. In most ixodids, however, very little water is excreted through the anus during feeding, and virtually all the excreted sodium, chloride and water is presumably eliminated by the salivary glands (W. R. Kaufman, personal communication).

The water which is lost by all routes must of course be replaced if an animal is to maintain a constant water content. Some arachnids probably never drink, but depend on the water content of their prey. Others such as Amblypygi and Opiliones will drink if water is available, while spiders have been shown to be able to extract water from soil capillaries, even when these exert quite high suction pressures (Parry, 1954). When *Tarentula barbipes* is desiccated, for example, it can drink against a suction pressure of 450 mmHg. Some spiders at least can regulate the amount that they drink in order to restore the body weight to its normal value : a striking parallel to the desert camel investigated by Schmidt-Nielsen (1964). When the American tarantula *Dugesiella hentzi* does this, the water taken up does not enter the haemolymph for some time (Stewart & Martin, 1970), an observation which suggests that the cessation of drinking is probably regulated by stretch receptors in the body wall and not in the haemocoel.

One further mechanism of obtaining water must also be described. This has been best characterised in ticks and mites, and it may possibly occur in other groups such as the spiders (Davies & Edney, 1952). Work on *Ixodes* (Lees, 1946a) has shown that when ticks were desiccated and then placed at 90% RH they showed a gain in weight, and within 2 days had returned to their weight prior to desiccation. In humidities lower than 86% all the ticks lost water, but while some could maintain equilibrium at this humidity, others needed up to 96% RH. In *Acarus siro* water uptake occurred down to humidities as low as 71% (Knülle, 1965), and the equilibrium humidity for *Bryobia praetiosa* may be even lower, at about 50% (Winston, 1967). Fully engorged females of *Ixodes*, on the other hand, could not take up water even from a saturated atmosphere: the uptake process appears to be lost soon after the onset of feeding.

In both ticks and mites the uptake process is thought to involve five stages: first, the secretion of hygroscopic material on to some external surface; second, the hydration of this material from water vapour in the atmosphere; third, swallowing of the hydrated material into the gut; fourth, uptake of both the hygroscopic material and its added water into the haemolymph; and fifth, secretion of the material externally to complete the cycle (see similar conclusions for insects, p. 172).

In ticks (*Amblyomma* spp.) the secretion of the salivary glands contains hygroscopic material, and this is extruded on to the mouthparts. Blocking of the mouth prevents uptake of water vapour (Rudolph & Knülle, 1974, 1978; McMullen, Sauer & Burton, 1976). In mites, where particularly *Dermatophagoides farinae* has been investigated, the supracoxal glands secrete a hygroscopic fluid rich in potassium and chloride (Wharton, 1978; Wharton & Richards, 1978; Arlian & Veselica, 1979, 1981).

6.1.2 Respiration and circulation

As in other terrestrial animals, some compromise must be reached between the necessity for an outer body layer which resists water loss, and a more permeable region through which respiration can take place – the loss of water from the respiratory openings in arachnids

has already been mentioned. The possession of two sets of respiratory organs in arachnids – lung books and tracheae – is a curiosity which must be examined in some detail. The distribution of lung books and tracheae in the various arachnid orders is shown in Table 6.5. The lung books are thought to be homologous with the gill books of the genus *Limulus* because of their similarity of development and their position in the body: they open on the abdomen behind the genital segment (Levi, 1967). Tracheae appear to be of two types: the 'sieve tracheae' are modified lung books (Kästner, 1929), whereas the 'tube tracheae' are new developments, analogous to those of insects and onychophorans. A diagram comparing the arrangement of lung books and tracheae is given in Fig. 6.5. Lung books consist of a series of haemolymph-filled lamellae which project into a restricted air space. This in turn connects with a larger

Table 6.5. *The distribution of lung books and tracheae in arachnids (from Levi, 1967)*

Lung books only	Tracheae only		Lung books and tracheae	No respiratory system
	Sieve tracheae	Tube tracheae		
Scorpiones	Some Araneida	Opiliones	Most Araneida	Palpigradi
Uropygi	Ricinulei	Solifugae		Some Acari
Some Araneida	Pseudoscorpiones	Some Araneida		
Schizomida		Some Acari		
Amblypygi				

Fig. 6.4. Summary of ingestion and elimination of ions and water during feeding in the adult female of the ixodid tick *Dermacentor andersoni*. Light stipple shows the tick; heavy stipple shows the host. Heavy solid arrows denote major routes; fine solid arrows denote minor routes; heavy broken arrow denotes a possible major route. Percentage figures refer to the proportions of the total amount excreted over the complete feeding period. Circled letters indicate the routes: a, the meal is derived from a mixture of whole blood and other tissue fluids; b, sodium and water, but probably a lesser amount of potassium, are transferred from the gut diverticula to the haemolymph; c, sodium (as NaCl), water and some potassium are transferred back to the host in the saliva; d, a small quantity of water evaporates; e, most of the potassium probably passes directly from the gut diverticula to the rectal sac and out through the anus, but alternatively, f, potassium may enter the haemolymph and then be transferred to the faecal material via the excretory tubules. Redrawn from Kaufman & Phillips (1973).

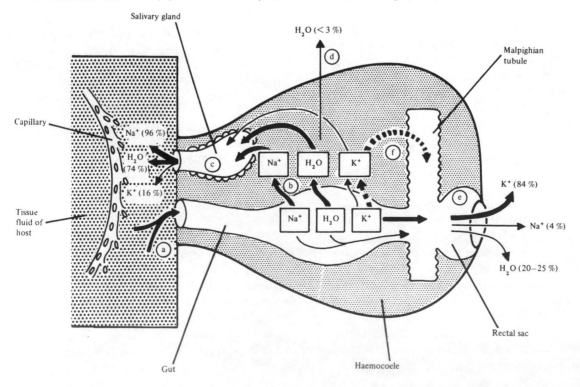

Fig. 6.5. A comparison of the structure of lung books and sieve tracheae in spiders. (*a*) Diagrammatic section showing the ventral median end of a left lung book. Only eight leaves are shown, although the usual number in spiders is more than 50. The haemolymph-filled lamellae are held apart by projections. Air diffuses between them, and this may be aided by some degree of ventilation. (*b*) Diagrammatic section of a sieve trachea. Only the lower half is shown. Two tube tracheae are also shown, coming off the atrium. Redrawn from Levi (1967).

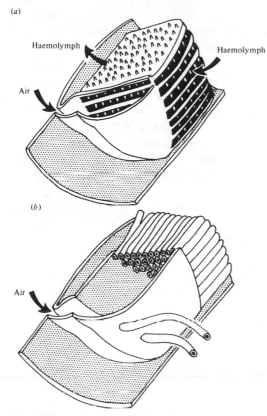

air space, the atrium, which opens to the outside via a spiracle. The lamellae are held apart by chitinous projections (Pohunková, 1969; see Fig. 6.6) which allow air to diffuse between them. The interlamellar space is probably not ventilated, although ventilation of the atrium may occur (Kästner, 1929). However, the possibility that fluctuations in the flow of haemolymph may cause the interlamellar spaces to expand and contract, thereby ventilating them, has been discussed by Anderson & Prestwich (1980). There is little information about the overall functioning of the respiratory system, but at least in spiders, rates of oxygen consumption are low in comparison with other poikilotherms (Anderson & Prestwich, 1982), reflecting a low rate of activity for much of the time. The respiratory pigment is haemocyanin, and this has a high affinity for oxygen: P_{50} is 5 to 37 mmHg in spiders (Angersbach, 1975, 1978; Loewe & Linzen, 1975) and 16.5 mmHg in the scorpion genus *Heterometrus* (Padmanabhanaidu, 1966). This low P_{50} suggests that gaseous exchange in the lung books must be inefficient.

Details of the importance of haemocyanin in the transport of oxygen have been provided for the spider *Eurypelma californicum* (Angersbach, 1978). When at rest, the haemocyanin is only 50% saturated with oxygen even in arterial haemolymph, partly at least because the spiracles are almost closed. However, when active the spiracles open and arterial pO_2 rises. Partly because of a pH drop which initiates a normal Bohr effect, the haemocyanin in oxygenated haemolymph becomes nearly 100% saturated. In active spiders, therefore, 82% of the oxygen carrying capacity of haemocyanin is utilised.

Fig. 6.6. Diagram of the anatomy of a spider lung book. (*a*) Section showing the relationship of stigma, atrium and respiratory leaves. (*b*) Section showing details of the respiratory leaves. Redrawn after Pohunková (1969).

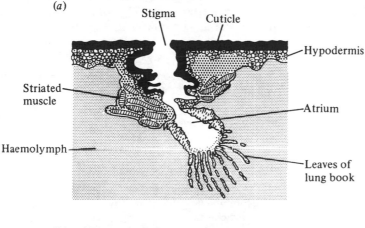

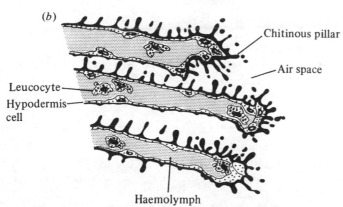

The characteristics of haemocyanin in spiders at rest and during exercise are shown in Fig. 6.7a. The carriage of carbon dioxide is not so well understood, but pCO_2 in *E. californicum* haemolymph is comparable to those of the terrestrial decapods. It may be that pCO_2 is regulated by ventilatory rate, but this has not been proved (Loewe & de Eggert, 1979).

The sieve tracheae, which are developed from lung books, are found mainly in the smaller forms. They are especially characteristic of the smaller spiders, and this is probably because they are more efficient at promoting diffusion of oxygen to the tissues, thereby permitting water loss to be cut down by more frequent closure of the spiracles (Levi, 1967). It has also been suggested that in small spiders these tracheae allow a saving of space, because their development means that the circulatory system need no longer be concerned with respiration and so can be reduced. Some larger spiders also have sieve tracheae, and these are found to a large extent in areas where desiccation is a problem: once again, they probably allow restriction of water loss.

Tube tracheae are found in a variety of groups. They have reached their peak of development in the desert solifugids, where there is a large, well-ventilated

Fig. 6.7. Oxygen dissociation curves in chelicerates. The open arrows show oxygen tensions (pO_2) of venous haemolymph, and filled arrows show oxygen tensions of arterial haemolymph. (a) The spider *Eurypelma californicum*. The dashed line shows the curve with the spider at rest (pH 7.5). The solid line shows the curve after exercise (pH 7.25). At rest the haemocyanin is never more than 50% saturated, and only 47% of the oxygen carrying capacity of the haemolymph is utilised. After activity the haemocyanin is nearly 100% saturated, and 82% of the capacity is utilised. (b) The horseshoe crab *Limulus polyphemus*. The upper curve was obtained at 8 °C, the lower at 24 °C. Both curves are nearly hyperbolic, and the haemocyanin is more than 50% saturated in both venous and aerterial haemolymph. Utilisation of oxygen carrying capacity is only about 20%. (a) From Angersbach (1978); (b) curves from Falkowski (1973), with in-vivo pO_2 values from Mangum, Freadman & Johansen (1975).

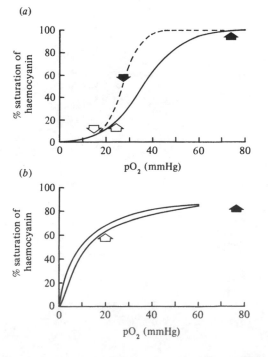

system. This is undoubtedly responsible for the well-known activity and speed of the solifugids, which can run 100 m in 35 s, and regularly move up to 12 km in one night (Junqua, 1966). At the other extreme, a particular variety of trachea is found in some mites: the larger water mites in lakes and ponds do not have tracheae with open stigmata, but small, blind tubes which are separated from the outside by thin regions of cuticle (Mitchell, 1972). Other aquatic and semi-aquatic mites have evolved plastrons which enable them to use a thin film of air as a physical gill (Hinton, 1971; Crowe & Magnus, 1974). This is, however, a development in forms which were originally terrestrial, but have secondarily taken up an aquatic mode of life.

A summary of the respiratory systems of arachnids would be incomplete without mention of the circulatory system. In spiders the heart (which lies dorsally in the opisthosoma) pumps haemolymph to the anterior through the pedicel and into the prosoma. Haemolymph passing back through the pedicel passes through the lung books, but haemolymph from the opisthosoma returns directly to the heart, which therefore contains a mixture of aerated and non-aerated haemolymph. During intense activity the hydrostatic pressure in the prosoma rises, due to the contraction of the prosomal muscles, to as much as 40 cmHg (Parry & Brown, 1959a; Wilson, R. S., 1970; Stewart & Martin, 1974; Anderson & Prestwich, 1975). The pressure in the abdomen rises to a much lower value, and the heart is not strong enough to pump haemolymph into the prosoma under these conditions. During this intense activity, therefore, haemolymph moves back into the opisthosoma and through the lung books, but is not redistributed to the organs in the prosoma (Wilson, R. S., 1970; Wilson & Bullock, 1973). In fact during a struggle the heartbeat becomes very slow and distorted, but it speeds up greatly immediately after activity (Wilson, 1967). It is apparent that spiders could not maintain very high rates of activity for very long because fluid would accumulate in the opisthosoma, and the prosoma would not receive an oxygen supply. Some spiders have better developed muscular systems in the opisthosoma than others, and these may raise the opisthosomal pressure further than in those so far investigated (Wilson, R. S., 1970). This would allow a smaller difference in pressure between opisthosoma and prosoma during intense activity, and the pressure distribution would then resemble that in normal walking. In this case the difference in pressure between prosoma and opisthosoma is not so great, and the heart continues to pump haemolymph into the prosoma (Anderson & Prestwich, 1975).

The details of locomotory mechanisms are discussed later. Here we must note, however, that the utilisation of a hydrostatic system poses severe limitations for spiders' behaviour and modes of life. Presumably similar limitations are also present in other arachnids which depend on lung books. In any case there is obviously a great selection pressure to produce alternative respiratory mechanisms, and this may be why tracheae have evolved in so many groups. The solifugids provide an extreme example of the relative efficiency of tracheae: here the change from lung books to tracheae has allowed both the reduction of the circulatory system and the separation of the prosoma from the opisthosoma by a muscular diaphragm (Kaestner, 1968). This would presumably allow the internal pressure in prosoma and opisthosoma to be independent. The high pressure required in the prosoma would then not impede other physiological processes, and would allow the animals to undergo the sustained activity noted earlier.

The inefficiency of the arachnid lung books together with the limitations imposed on the circulatory system by the use of high hydrostatic pressures in locomotion suggest that the whole mechanism is ill-adapted to terrestrial life. It seems likely, in fact, that it exists only because it was evolved directly from the gill book system of aquatic ancestors. It is most improbable that it would have been evolved directly by any group which originated on land. Such a conclusion weighs heavily in favour of an aquatic origin for the chelicerates, and not an origin in some moist environment on land.

6.1.3 Haemolymph composition and nitrogenous excretion

The role of the haemolymph as an oxygen carrier has been discussed, but not its capacity as a medium for tissues and cells. Figures for ionic composition are available for some scorpions and spiders only, and are shown in Table 6.6. These demonstrate a relatively high osmotic pressure, comparable to that of the terrestrial Crustacea, and to the Insecta. The composition is, however, totally unlike that of most insects because sodium and chloride account for most of the osmotic pressure, and there is evidently no large fraction of small organic molecules. It therefore resembles the rather simple composition shown by terrestrial isopods (see Table 5.4), but has lower proportions of calcium and magnesium. In comparison with the marine *Limulus*, the proportions of ions are again similar, except for magnesium, which *Limulus* retains at a high level. The haemolymph of the genus *Limulus* is fairly typical of primitive marine invertebrates, i.e. it is little modified from sea water. No great changes are needed to derive from it the modern terrestrial arachnid condition: a reduction in osmotic pressure, accompanied by a proportional reduction in the concentration of each ion, and active excretion of magnesium would account for the composition of scorpion and spider haemolymph. This may well suggest that the terrestrial arachnids have in fact been derived from a marine ancestor, since any group which arose directly on land might be expected to have some very significant differences from marine forms. Such a point of view will be taken up again in discussions concerning the insects.

The mechanisms which control the composition of the haemolymph and remove nitrogenous waste have

Table 6.6. *Inorganic composition of some chelicerate haemolymphs*

Species	Osmotic pressure (mOsm)	Na$^+$	K$^+$	Mg^{2+}	Ca^{2+}	Cl$^-$	pH
		(mM/litre)					
Scorpions							
Palamnaeus bengalensis (1)	—	111	4.0	—	2.7	121	6.3
Heterometrus swammerdami (2)	—	150–335	0.5–3.0	5–25	5–10	150–450	—
Heterometrus fulvipes (3)	—	147	0.8	4.9	4.8	166	7.3
Urodacus novaehollandiae (4)	—	185	15	—	10	220	6.3
Leiurus quinquestriatus (5)	—	250	7.7	1.0	5.0	270	7.0
Centruroides gracilis (6)	—	250	7.4	2.0	5.2	260	—
Hadrurus arizonensis (7)	—	353	11.2	1.7	9.8	344	7.2
Paruroctonus mesaensis (7)	—	332	9.2	1.8	8.1	306	7.5
Androctonus australis (15)	480	263	8.0	—	2.4	267	—
Spiders							
Cupiennius salei (8)	—	223	6.8	—	4.0	258	8.1–8.3
Eurypelma hentzi (9)	—	220	5.0	1.1	4.0	—	—
Tegenaria atrica (10)	400	207	9.6	—	—	193	—
Ciniflo similis (11)	460	251	6.0	0.8	7.1	245	—
Dugesiella hentzi (14)	360	—	—	—	—	—	—
Ticks							
Ornithodorus moubata (13)	—	—	—	—	—	159–166	—
Ixodes ricinus (13)	—	—	—	—	—	119–146	—
Xiphosura							
Limulus polyphemus (12)	—	445	12	46	10	514	—

References: 1, Kanungo (1955); 2, Padmanabhanaidu (1962); 3, Padmanabhanaidu (1967); 4, Zwicky (1968); 5, Gilai & Parnas (1970); 6, Bowerman (1972); 7, Bowerman (1976); 8, Loewe, Linzen & von Stackelberg (1970); 9, Rathmeyer (1965); 10, Croghan (1959); 11, F. L. Farr-Cox, personal communication; 12, Robertson (1970); 13, Lees (1946*b*); 14, Stewart & Martin (1970); 15, Bricteux-Grégoire *et al.* (1963).

not been examined in many arachnids. Details of salt and water balance in ticks have already been discussed. In non-parasitic forms little is known about the ability to control even the osmotic pressure of the haemolymph. However, a study of two scorpions by Robertson, Nicolson & Louw (1982) showed that while a mesic species could not control the osmotic pressure of its haemolymph during desiccation, a desert species, *Parabuthus villosus*, could maintain its osmotic pressure at an almost constant level when losing water by evaporation. This recalls a similar ability in some woodlice. It is not known how widespread is this ability, nor what organs are involved. In another desert species, *Paruroctonus aquilonalis*, no capacity for osmoregulation was found (Riddle, Crawford & Zeitone, 1976).

There are four relevant organs which may be involved to different extents in controlling composition of the haemolymph in various groups: salivary glands, coxal glands, excretory tubules and the midgut. The salivary glands are known to be involved in ionic regulation and water excretion in some ticks and have already been mentioned (see p. 113). This leaves as the main excretory organs the coxal glands, and the gut with its associated caeca and tubules. Generally it is thought that the former deals mainly with ionic regulation while the latter eliminates nitrogenous waste, but this is doubtless an oversimplification.

Coxal glands are found in all arachnids except some of the Acari, and mostly open on the coxae at the base of the walking legs. Most groups have only one pair of these glands, opening at the base of leg 1 or leg 3, but some spiders have two pairs. *Limulus* has one pair with four lobes which all empty into one duct opening at the base of leg 3, so that the glands seen in present-day terrestrial forms are probably remnants of serially repeated organs. The situation in *Limulus* is therefore

probably primitive, once again suggesting an initially marine origin for the chelicerates. The structure and function of the coxal glands have been investigated in detail in the Acari and Scorpiones only. In oribatid mites the glands consist of a thin-walled sacculus and a long tubule which is folded back upon itself. In freshwater species this tubule is long, but it is shorter in terrestrial species and saltmarsh forms, and very much reduced in species living in the marine littoral zone (Woodring, 1973). Such circumstantial evidence points to the involvement of the coxal glands in osmoregulation. The elegant work of Lees (1946*b*) has elucidated details for two species of ticks. Here the problems of salt and water balance are associated with the taking of a large blood meal in a relatively short time: the water taken in must be expelled but the blood corpuscles must be retained and digested. Two ticks were investigated: the argasid *Ornithodorus moubata* and the ixodid *Ixodes ricinus*. *O. moubata* feeds rapidly and after 15 min the coxal glands produce a watery fluid. Each coxal gland has a thin-walled sac with muscles attached. The sac opens into a double-looped tubule which eventually leads to the outside (Fig. 6.8). The system is therefore basically very similar to that of oribatid mites. When the tick takes in a blood meal the cuticle stretches and the coxal gland muscles are stimulated to contract. The pressure in the sac is therefore lowered, and fluid moves into it from the haemolymph. The membrane of the sac is permeable to molecules as large as haemoglobin, but bigger proteins do not pass through. By using inulin as a marker, Kaufman, Kaufman & Phillips (1982) have confirmed that the sac is an ultrafiltration chamber, and electron microscopy studies have confirmed that filtration occurs through a basement membrane supported by podocytes (Groepler, 1969; Hecker, Diehl & Aeschlimann, 1969). When the coxal gland muscles relax, the ultrafiltrate

Fig. 6.8. Diagrammatic dorsal view of the right coxal gland of the argasid tick *Ornithodorus moubata*. The filtration sac is shown as if cut away from the posterior region of the gland. Arrows indicate the direction in which fluid flows along the tubule. Coarse stipple shows the first loop of the tubule; fine stipple shows the second loop. Redrawn from Lees (1946*b*).

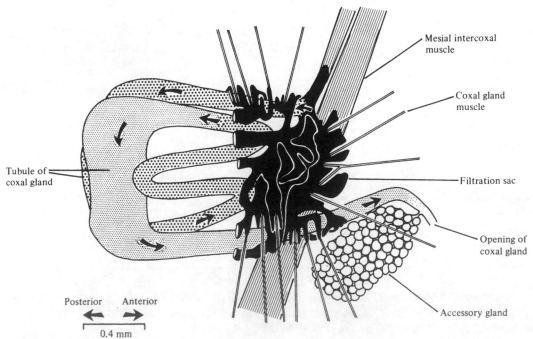

Mesial intercoxal muscle

Coxal gland muscle

Filtration sac

Opening of coxal gland

Tubule of coxal gland

Accessory gland

Posterior   Anterior

0.4 mm

passes into the tubule, where chloride ions are re-absorbed so that the fluid passed out is more dilute than the haemolymph; the ultrastructure of the tubules is typical of reabsorbing epithelia. The speed at which the system acts is remarkable: one animal which weighed 45 mg unfed ingested 257 mg of blood, and liberated 123 mg of coxal gland fluid in 1 hour; of this, 58 mg were produced within 45 s of the first appearance of the fluid.

In *Ixodes ricinus* the situation is quite different. Feeding takes place on the host for up to 8 days, so that part of the excess water is lost by evaporation over this period. There are no coxal glands, and the regulation of salts and water is presumably carried out by the salivary glands as described in another ixodid tick of the genus *Dermacentor* by Kaufman & Phillips (1973).

The coxal glands in other orders have at least some similarities to those of argasid ticks. In scorpions and pseudoscorpions the sacculus consists of a basement membrane supported by podocytes (Rasmont, 1960; Heurtault, 1973), and is obviously a filtration site, but efforts to detect differences in proteins on either side of the membrane have unfortunately not been successful owing to technical difficulties (Rasmont & Rabaey, 1958). The tubules are lined by cells with deep basal infoldings and mitochondria, and must be involved in reabsorbing ions and probably water.

There is little evidence to suggest that the coxal glands are involved in nitrogenous excretion. Instead, various diverticula of the gut produce the substance guanine, which as an excretory product is almost restricted to the arachnids. This substance is even more insoluble than uric acid, and contains one more nitrogen atom per molecule, so it is ideally suited to terrestrial life involving water conservation. There are reports that adenine and hypoxanthine are important nitrogenous waste products in scorpions (Kanungo, Bohidar & Patnaik, 1962; Rao & Gopalakrishnareddy, 1962), but guanine normally accounts for 35 to 94% of the total nitrogen excreted (Haggag & Fouad, 1965; Anderson, 1966; Horne, 1969). Uric acid is important in some species, and may account for up to 29% of the excreted nitrogen (Rao & Gopalakrishnareddy, 1962), but normally it contributes only a few per cent to the total. Variation in excretory product within the arachnids is emphasised by the results of Yokota & Shoemaker (1981) who examined a desert scorpion, *Paruroctonus mesaensis*. In this species no guanine or uric acid is produced, and over 90% of the excreted nitrogen is in the form of xanthine, previously unknown as an excretory product.

The organs which excrete nitrogenous waste vary in the different arachnid groups. In many there is a pair of tubules which branch from the midgut and have been called Malpighian tubules. There is no reason to consider any homology with the Malpighian tubules of insects, however, and it seems more appropriate to call these organs excretory tubules, as have Coons & Axtell (1971) – especially since in the mites these open into the hindgut, and in the spiders they open into a 'cloaca'. The scorpions have two pairs of these excretory tubules, but in some orders (the Opiliones, Palpigradi, Pseudoscorpiones and some Acari) there are none at all. In their case it is likely that the wall of the midgut, or the diverticula of the midgut, have cells which produce the excretory concretions (McEnroe, 1961; Weygoldt, 1969). It has also been suggested that in some scorpions and Uropygi (which *do* have excretory tubules) most guanine is produced by the walls of the midgut itself, and not by the tubules (Horne, 1969). In the mites, however, this is certainly not so, for Coons & Axtell (1971) found the excretory tubules of *Macrocheles muscaedomesticae* to be full of white excretory granules. No granules were present in the tubule cells, and the excretory products are presumably passed into the tubule in soluble form and then precipitated. The tubule cells are like those of many other arthropods in ultrastructure, with basal infoldings associated with mitochondria; but the apical regions of the cells have microlamellae and not the more usual microvilli. It is obvious that much further work on arachnid excretion needs to be done before useful comparisons with the insect systems can be made.

6.1.4 Locomotion

Another possible function of animal body fluids is to act as a hydrostatic skeleton. While such a phenomenon may seem unlikely in animals with a hard exoskeleton, it appears that in the terrestrial arachnids, as already discussed, hydrostatic pressure plays an important part in locomotion. Most terrestrial arachnids (with a few exceptions) have jointed legs with intrinsic flexor muscles but not intrinsic extensor muscles except in the very basal segments. The same can be said of the genus *Limulus*, as well as of some other unrelated groups such as the myriapods and pauropods. It is therefore important to consider how leg extension is achieved. Two viewpoints have been put forward. One is that increased hydrostatic pressure in the prosoma of arachnids extends the legs and provides locomotive force (Parry & Brown, 1959a). The other is that although hydrostatic pressure extends the legs when they are off the ground, locomotive force when they are on the ground is provided by muscular action: depressor muscles at the base of the leg contract, following rotation of the leg, to push down the femur and therefore to straighten out the limb (Manton, 1958a). These views have been put forward as mutually exclusive. The supporting evidence for the two alternatives has also been derived from different sources. Manton (1958a, 1973) has examined in detail the architecture of the leg muscles and joints. Although Parry (1957) has also investigated leg muscles in the spiders, Parry & Brown (1959a,b) have concentrated upon measurements of internal hydrostatic pressures in spiders, and the comparison of these with the pressures necessary for extension as calculated from torque measurements. They have concluded that sufficient pressure is generated in the prosoma to account for the extension of the legs in walking: normal pressure in *Tegenaria* is about 5 cmHg, but this may rise to 40 cmHg in activity. In the jumping spiders such as species of *Sitticus* such a pressure is sufficient to account for jumps of 7 cm by the use of the fourth pair of legs only, and in tropical jumping spiders, working at higher temperatures, jumps of up to 20 cm can be explained. More recent work has confirmed the high hydrostatic pressures found in the prosoma and legs during intense activity, and has shown that the opisthosoma is also maintained at a raised pressure (Wilson, 1962; Stewart & Martin, 1974; Anderson & Prestwich, 1975).

In both theories, locomotor force is achieved in the first half of the backstroke of each leg by the flexion of the leg, as it were 'pulling' the body forward. The flexor muscles are powerful and there is no problem in explaining flexion. The action of the anterior legs in the second half of the backstroke differs from that of the posterior ones, but the latter certainly extend. Manton's (1958a) view is that the depressor muscles at the base of the leg provide locomotor force. If this is so, however, they are working at a tremendous mechanical disadvantage, and they do not seem to be particularly massive to counteract this. The tensions in the various flexor muscles along the leg would also have to be very carefully controlled to prevent the leg buckling instead of straightening. It therefore seems more probable that in spiders the legs are straightened by hydrostatic pressure, and that the depressor muscles are responsible for posture, keeping the body well off the ground. The flexor muscles are large and powerful because on the first half of the backstroke they have to work against the constant high hydrostatic pressure in the legs. Their relaxation in the second half of the backstroke allows leg extension. The details of fluid flow during stepping movements have not been worked out; but if these were so arranged that flexion of one leg effectively pushed fluid into another which was about to extend, the system could be a very efficient one.

Very little is known about leg extension in any other arachnid group except for the scorpions. In these the mechanisms of leg extension are much more varied than in the spiders (Alexander, 1967). The evidence suggests that in some legs hydrostatic pressure is responsible, while in others Manton's (1958a) theory involving extrinsic coxa muscles may apply. While either or both of these mechanisms may be the basic ones, they are aided by the elasticity of the hinge joints – which does not seem to be important in spiders – and by the action of sclerites in the arthrodial membrane. These are bent during leg flexion and then unfold, aiding in leg extension. The hydrostatic pressure system in scorpions also seems to differ drastically from that in spiders. According to Alexander (1967), pressures in the prosoma and opisthosoma are both low, and rise to only about 1 cmHg during activity, while the pressure in the legs may rise to 25 cmHg. If this is correct the source of high pressure in the legs is mysterious, and their hydraulic separation from the prosoma is a most interesting feature.

It is difficult to decide whether the spiders have 'specialised' in hydrostatic pressure and have given up the use of aids such as elasticity at the joints and arthrodial sclerites, or whether these aids are in fact specialisations of the scorpions. More information on other groups, especially perhaps on the very active solifugids, might help to answer this question. At present, however, the

only other information concerns *Limulus* (Ward, 1969). Here the hydrostatic pressures are lower than in spiders, and the depression of the femur is thought to extend the limb, aided by the weight of the leg. There are some strange points concerning *Limulus*, however, which suggest that the mechanism of leg movement is not entirely clear. The hydrostatic pressure in the prosoma is 1.4 cmHg when resting, and 2.4 cmHg when active. When pressure was applied to an isolated leg, approximately 10 cmHg was needed to achieve an extension of about 40°. Prosomal pressures would account for an extension of only 15°, and it is prosomal pressures that must be taken since these are applied at the base of the leg just as the pressures applied *in vitro*. Pressures in the leg itself are not so relevant; but towards the distal end they are approximately 0.5 cmHg. This is a surprising pressure drop considering that the haemocoel of the prosoma and the legs must be directly connected to ensure circulation. It must be concluded that on the basis of the measurements given (Ward, 1969), the legs are not extended entirely by hydrostatic pressure, but that in *Limulus* this is only an accessory mechanism. It may be, indeed, that this method is only effective when the legs are off the substratum. The terrestrial arachnids have increased their dependence upon a hydrostatic pressure system to a marked degree. Presumably this is concerned with the need for supporting the body weight in air without the aid of extensor muscles. The smaller arachnids have used the system effectively, but in comparison with the insect systems which use opposing sets of flexors and extensors their performance is poor. The jumping capacity of salticid spiders compared with that of jumping acridid insects, for instance, is relatively feeble; but most jumping spiders are small, so that their jump is effective relative to their body size (Parry & Brown, 1959b), and this is often the most significant factor. Nevertheless, it appears that the derivation of present-day terrestrial arachnids from aquatic forms which lacked extensor muscles, and the subsequent dependence upon the use of hydrostatic pressure, has severely limited their activity and diversity. This, coupled with its side effects upon respiration discussed above, may go some way towards accounting for the fact that although arachnids are numerous, they are far outnumbered by the insects.

6.1.5 Sensory systems and behaviour

This brings us to the last area of discussion concerning the terrestrial arachnids: their behaviour, and the development of sensory systems and control mechanisms. Much of the behaviour of arachnids is guided by the receptors which detect tactile, vibratory and chemical stimuli. Most have eyes, but the visual sense is really important only in predators which move rapidly such as the hunting spiders (Lycosidae and Salticidae). Since it is not possible – for reasons of lack of information as well as lack of space – to go into all the aspects of arachnid behaviour, we will consider in detail only two such aspects: prey capture and reproductive behaviour. All arachnids, with the exception of a few members of the Acari, are carnivores, and the detection and capture of prey is one of their most important activities. In general, chemical and tactile receptors on the palps and legs are responsible for sensing both prey and predators. For example, in one scorpion, *Paruroctonus mesaensis*, sense organs at the end of the walking legs are used to detect vibrations in the sand. From measurement of the differences in time at which all eight legs are stimulated, the scorpion can detect the direction of prey (Brownell & Farley, 1979*a,b*). In the Amblypygi the tarsi of the front pair of legs are held out in front of the animal. They have many receptors, of many different types: contact chemoreceptors, mechanoreceptors and probably hygroreceptors, together with others of unknown function (Foelix, Chu-Wang & Beck, 1975). In spiders both the mechanoreceptors and chemoreceptors are modified hairs, and resemble those found in other terrestrial arthropods (see, e.g. Foelix & Chu-Wang, 1973; Harris & Mill, 1973). Chemoreceptors, for instance, are curved, blunt-tipped hairs which have a subterminal pore well supplied with nerve dendrites. These are kept moist by a secreted fluid which may also be essential in trapping the stimulatory substance. In the solifugids more complex chemoreceptors called racquet organs are found on the hind legs, where they pick up scent from the ground. They resemble a row of simple chemoreceptors joined together, and open to the outside by a slit instead of a pore (Brownell & Farley, 1974). Not a great deal seems to be known about how these sensory systems are actually used by arachnids. Some observations on scorpions have shown that besides contact chemoreceptors, they have distance chemoreceptors on the palps and can use these to detect prey before contact is made (Abushama, 1964). For an attack to follow, however, the eyes are necessary, so that the sense organs of various types are probably used in sequence.

The same may be true for those spiders which use a web to catch prey. Here the prey is detected when it is caught in the web, and the spider approaches to investigate; but subsequent behaviour differs greatly depending upon the type of animal caught (Walcott, 1963). There is little doubt that spiders can perceive both airborne sounds and substrate-borne vibrations, but to what degree they can distinguish between the two, the means by which they are perceived and the uses to which such perceptions are put are questions that are at present in dispute. The situation for other arachnids is even more obscure. Perhaps it would be best first of all to describe briefly some of the ways in which webs are used to trap prey, since these are very diverse, and involve different combinations of senses. Excellent detailed accounts can be found in Bristowe (1971). In some species, such as *Atypus affinis* (Atypidae), the spider waits inside a sealed silk tube and strikes through this tube when insect prey lands upon it. In this case vibration and tactile senses are involved. In others such as species of *Segestria* (Dysderidae) the burrow is lined with silk from which radiate long 'fishing' lines. When these are moved the spider emerges to investigate, and here the only necessary stimulus is the vibration of the thread. Many spiders, however, spin complex webs – the tangles of the Pholcidae, the circular or orb webs of the Araneidae

and the sheet-like webs of the Linyphiidae to mention just three. When the prey touches or is trapped in these webs, many possible stimuli are produced, and the spider could theoretically use sound, vibration, smell, sight or touch to locate it. Most of the work on vibration receptors and the possible sense of hearing has been concerned with this type of spider. Two types of sense organ may be involved: the long sensitive hairs or trichobothria, which are often found in large numbers on the body surface; and the lyriform organs (Salpeter & Walcott, 1960), which are slits in the cuticle linked to sense cells capable of detecting strains in the cuticle.

The long hairs of web-spinning spiders respond to air movements and to tactile stimulation, but can also respond to airborne sound when the sound intensity is high (Frings & Frings, 1966). In general such stimulation produces a defensive reaction, and it may be that the normal use is for the detection of flying predators. It appears, however, that discrimination between air movement and sound would not be possible, so that the situation is not entirely clear.

The lyriform organs generally function as proprioceptors (Pringle, 1955), detecting the position of leg joints. When the legs are raised off the ground, the lyriform organs of the tarsal/metatarsal joint respond to airborne sound, presumably because the leg tips vibrate with the sound waves (Walcott & van der Kloot, 1959). When the legs are on the substratum, they can detect vibrations; when on a web, for instance, the legs vibrate with the silk strands. These responses to vibrations usually result in an attack response (Walcott & van der Kloot, 1959; Frings & Frings, 1966). The importance of various stimuli that come through the web has been investigated (e.g. Liesenfeld, 1961), but the specific stimuli causing attack are not known. It has been suggested that the fast transient changes in tension of the silk threads when they snap or change position may be involved (Parry, 1965).

Very little work has been done on vibration senses in other arachnid groups. Trichobothria are similar in mites and scorpions (Hoffman, 1965; Haupt & Coineau, 1975). Lyriform organs or their homologues have been shown to be proprioceptive in Amblypygi and scorpions (Pringle, 1955; Barth & Wadepuhl, 1975). Many animals in these orders can produce sounds by stridulation, so that there is at least a possibility of acoustic communication. In some lycosid spiders the males produce a drumming on the substratum with their palps, and these vibrations are picked up and can cause a response in other males and in females (Rovner, 1967). This instance appears to be the only one in which intraspecific communication by acoustic means has been shown. Other spiders can produce sounds by different means: species of *Anyphaena* can vibrate the opisthosoma and the front legs to make a buzzing noise (Bristowe, 1971); and those of the spiny orb-weaver *Micrathena* use their hind legs and the cover of the lung books to stridulate (Hinton & Wilson, 1970). This latter noise appears to be a defensive reaction. Many scorpions can stridulate using a variety of different organs (Alexander, 1958), and here again the sounds do not seem to be used in intraspecific

communication, but may be defensive. Some of the
Amblypygi can stridulate (Shear, 1970), using spines on
the inner faces of the chelicerae, but the function of the
sounds produced is not known.

   While the web spinners use mainly tactile, chemical
and vibratory senses, the Salticidae and Lycosidae have
well-developed eyes. Normally there are eight simple eyes:
six of these are like those of other arachnids, but the
two anterior ones, or principal eyes, are more complex.
The six lateral eyes resemble the eyes of pycnogonids
to an amazing degree, as will be discussed later. They
consist of a cuticular lens with a cellular vitreous layer
beneath, bordering on the retinal cells (Baccetti & Bedini,
1964; see Fig. 6.9). These have their light-sensitive ends,
or rhabdomeres, internal to the cell bodies, and are
backed by a light-reflecting tapetum. Behind this again is
a pigment layer. This type of eye can detect movement,
but has no form vision. The spiders use these lateral eyes
to detect objects which move into the visual field, and
when they are observed, the principal eyes, the head,
or the whole body, are moved to bring the object into
the field of view of the principal eyes (Land, 1972;
Duelli, 1978). These principal eyes are modified in
many ways (see Fig. 6.10). They are long and tubular,
with a cone-shaped retina consisting of cells with their

photoreceptive areas in front of, and not behind, the cell
bodies (Baccetti & Bedini, 1964; Melamed & Trujillo-
Cenóz, 1966). The retinae can be moved by a series of
muscles, and indeed when an object is being scrutinised
they scan continuously from side to side, and rotate
about the axis of the eye (Land, 1969a,b). These movements probably allow the spider to align orientated rows
of photoreceptor units with diagonals in the image, and

Fig. 6.10. The eyes of a salticid spider *Metaphiddipus aeneolus*.
(a) Diagrammatic frontal section of the cephalothorax, showing
the positions of the eyes and their fields of view in the
horizontal plane. (b) Frontal (horizontal) section of the retina
of the right antero-median (principal) eye. The section is taken
close to the centre of the retina and shows the four layers of
receptor endings (1–4). In layers 1, 2 and 3 the presumed
receptive part of each receptor is the straight terminal portion;
in layer 4 it is the terminal ovoid swelling. From Land (1969a).

(a)

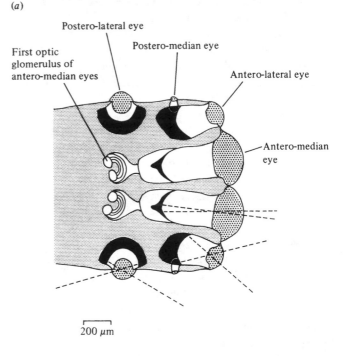

200 μm

(b)

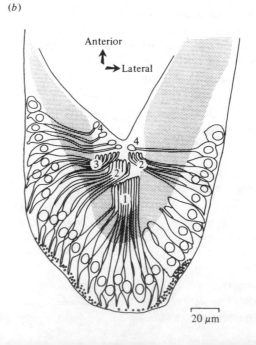

20 μm

Fig. 6.9. The structure of simple eyes in chelicerates. (a) Section
of a postero-lateral eye from a lycosid spider *Arctosa variana*.
The lens is omitted. (b) Section through the eye of a pycnogonid.
(a) From Baccetti & Bedini (1964); (b) after King (1973).

(a)

(b)

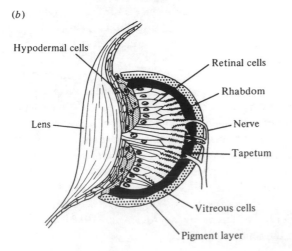

in so doing certain features of the image are 'recognised'. By some complex mechanism working on these lines, salticids can recognise other members of the species, and can differentiate them from other objects such as prey. Such recognition is aided by the fact that they have colour vision: the elements of the retina are placed in four rows at different distances from the lens, and in the planes of focus of light of different wavelengths (Land, 1969*a*). The spectral sensitivities of salticid principal eyes show cells with peaks at 580 nm (yellow), 520–540 nm (green), and 480–500 nm (blue) (DeVoe, 1975; Yamashita & Takeda, 1976), which presumably correspond to the three layers most distant from the lens. There are also cells which are ultraviolet receptors, and these probably constitute the most anterior retinal layer. However, some lycosids can also detect the plane of polarisation of light (Magni *et al.*, 1965), and the anterior retinal cells are thought also to be involved in this process. Certainly the lycosid *Arctosa* can orientate by the position of the sun and by the polarised light pattern of the sky (Baccetti & Bedini, 1964).

In distinct contrast to these specialised spiders, the scorpions have poor vision and eyes which can probably only distinguish between blue/green and ultraviolet (Machan, 1968). Their exact role in scorpion behaviour has not been worked out. Work on their ultrastructure has revealed the fact that the rhabdomeres of the retinular cells in *Euscorpius carpathicus* are united into compound rhabdoms (Bedini, 1967), i.e. they resemble the compound eyes of other arthropods except that they lack the external facets to the eye. The significance of this will be discussed later (see p. 125). An interesting variant in photoreception is the location of a neural photoreceptor in the scorpion *Heterometrus*: here part of the ventral nerve cord in the metasoma is sensitive to light (Geethabali & Rao, 1973).

As with the techniques involved in the capture of prey, the senses involved in courtship and reproduction differ between web-spinning spiders and hunting species. In both cases the male has to search for the female, but in the web spinners he uses mainly tactile and chemical stimuli, and he locates first the web and then the female; while in the hunting species the male uses mainly the visual sense to search for the female. In other arachnids tactile and chemical stimuli are probably the main guides. When male and female have met, a process of copulation unique among the arthropods ensues. In some groups such as the scorpions the male deposits a spermatophore and then guides the female to it; but in most groups he has at some time previously produced a small quantity of sperm and then picked it up on the highly specialised tips of a pair of appendages. In some groups these are the chelicerae or walking legs, but in spiders the pedipalps are used. At copulation the tip of the appropriate appendage is inserted into the female opening. Thus there is no juxtaposition of the male and female reproductive openings as in most animals. In the jumping and hunting forms the courtship displays which precede copulation have become extremely complex. Crane (1949) has investigated the evolution of display mechanisms in some salticids, and has distinguished

running types, hoppers, and those intermediate between the two forms. The runners use mainly chemical senses during courtship. The intermediate species use distance chemoreceptors and some visual cues. The hoppers have high visual acuity and hardly use the chemical senses. This sequence has suggested that the development of visual acuity has occurred in parallel with the capacity for jumping. The display motions are very varied, as can be seen from Fig. 6.11, and in highly vision-dependent species serve as interspecific barriers. The situation of the most 'advanced' terrestrial forms being highly vision-dependent therefore parallels that in crabs (see p. 105). Once again it seems to indicate that the importance of the visual senses often develops more on land than in water; but this suggestion must be tempered with caution more than it was in the Crustacea, since we know little about courtship even in *Limulus* and the pycnogonids, and nothing at all about courtship in the direct aquatic ancestors of arachnids.

6.2   Xiphosurans, eurypterids and pycnogonids

Several points so far considered have suggested an aquatic origin for the arachnids: the inefficiency of the lung book system, with its limitations on other physiological processes, together with the development of tracheae in the very active forms such as the solifugids; haemolymph composition and its resemblance to that of *Limulus* which in turn has a composition typical of primitive marine invertebrates, and not of an animal which has returned to the sea; and the use of hydrostatic pressure to extend the legs, a property required because of derivation from aquatic ancestors without extensor muscles. These points do not, however, suggest which if any of the modern arachnid groups is the most primitive, nor do they tell us very much about the supposed marine ancestors. A brief examination of the groups thought to be at least related to these aquatic ancestors is therefore appropriate here. Of the eurypterids we know very little except for details of external morphology. They are found as fossils from the Cambrian to the Permian. Early forms were primarily marine, but later specimens were common in fresh water with insects (Fage, 1949). Their fossil record overlaps that of the scorpions, which have been found as early as the Silurian (Savory, 1964), but there is little evidence to support the theory of Versluys & Demoll (1923) that it was the scorpions which gave rise to the eurypterids. Fage (1949) has pointed out some of the misconceptions involved in this theory. We may add to these the point that Versluys & Demoll arranged the arachnids in an evolutionary series with the solifugids being the most primitive, and the scorpions being derived via the remaining orders, finally to give rise to the eurypterids and xiphosurans. It has already been shown that in many ways the solifugids are a highly specialised group adapted for fast running, and with an extensive tracheal system which is a secondary development. They could not possibly have been ancestral to the remaining orders. It has also been pointed out that instead of arranging the orders in a linear series, it is probably more appropriate to consider each as having a 'mosaic' of primitive and advanced characters.

The question of the origin of scorpions from the eurypterids may be taken more seriously. The external body forms of some species of the two groups are extremely similar, as shown by a number of authorities, including Fage (1949) and Versluys & Demoll (1923). The embryology of the lung books of scorpions shows many similarities to that of the gill books of *Limulus*, and there appears to be general agreement that there is great affinity between the eurypterids and *Limulus*. There seems, then, to be no objection to the suggestion of a eurypterid origin for the scorpions. The origin of

Fig. 6.11. Examples of display motions in salticid spiders. (*a*), (*b*), (*c*) *Corythalia xanthopa*. (*a*) Shows a threat display, (*b*) and (*c*) show courtship displays. (*d*), (*e*) *Corythalia fulgipedia*. (*d*) Shows threat, (*e*) shows courtship. Redrawn from Crane (1949).

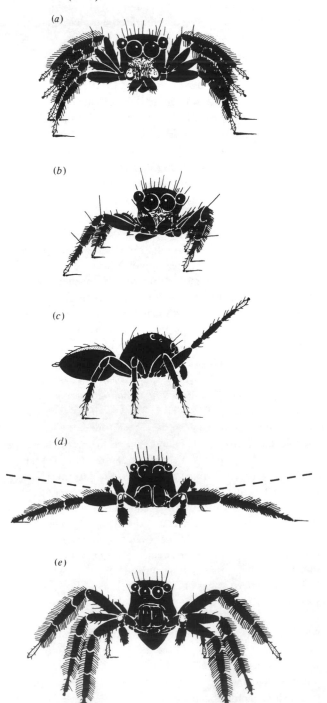

(*a*)

(*b*)

(*c*)

(*d*)

(*e*)

other arachnid orders from the scorpions, however, is very difficult to accept. We shall return to this problem later.

For the moment, let us look at the xiphosurans and pycnogonids, to see what light they may throw upon arachnid ancestry. Both these groups are completely marine. While there is no suggestion that either were direct ancestors of the terrestrial arachnids, they do allow us some insight into some of the probable characters of the early aquatic chelicerates.

Haemolymph composition of *Limulus* has already been discussed: it is typical of primitively marine invertebrates which perform little ionic regulation and shows no signs of derivation from former terrestrial ancestors. No analyses of pycnogonid haemolymph are available. *Limulus* has one pair of coxal glands with four lobes (Fage, 1949) and one may speculate that these may be involved in producing urine while retaining metabolites, but nothing is actually known of their function, apart from the fact that urine is similar in inorganic composition to haemolymph, with a lower pH (Mangum *et al.*, 1976). Pycnogonids have no such glands.

The respiratory system of *Limulus* bears a striking resemblance to that of spiders (Kästner, 1929). The gill books are five modified posterior appendages: each consists of numerous haemolymph-filled lamellae attached to, and protected by, a gill operculum. The lamellae are therefore exposed to circulation of water when the gill opercula are moved. From this system that of spiders may be derived by enclosing the lamellae inside a cavity, and in the ontogeny of spider lung books this process appears to occur. In *Limulus* the haemolymph contains haemocyanin, of which the oxygen dissociation curve has been determined by Redfield & Ingalls (1933) and by Falkowski (1973) (Fig. 6.7). Measurements of oxygen tension *in vivo* have been made by both Falkowski (1973) and Mangum, Freadman & Johansen (1975), but the two sets of measurements differ considerably. According to Mangum *et al.*, their figures reflect normal respiration under water, while Falkowski's figures show a situation with hypoventilation and reduced oxygen pressures. Mangum *et al.*'s figures are therefore shown in Fig. 6.7. When *Limulus* is out of water, or when it burrows, its ventilation rate is reduced, and internal oxygen pressures fall to 24 mmHg (post-branchial) and 10 mmHg (pre-branchial). In this situation the haemocyanin carries a much greater percentage of the total oxygen than normally, partly because the loading and unloading tensions now fall on the steeper part of the curve, and partly because the curve itself is made even steeper by a reverse Bohr effect. Mangum *et al.* (1975) have suggested that the reversed Bohr effect, which is relatively uncommon in animals, and is not known in arachnids, is actually an adaptation to such periods of low internal oxygen tension. Since *Limulus* may spend up to 12 hours buried in the sand, this could be a very important factor (Eldredge, 1970). The only major similarity between the characteristics of haemocyanin in *Limulus* and arachnids is the low P_{50} (Fig. 6.7), which probably reflects the inefficiency of gaseous exchange in both lung books and gill books.

In contrast to the xiphosurans, the pycnogonids have no respiratory pigment. They are mostly small, with a large surface area, and in general are not very active. They have no specialised respiratory organs, but nothing is known about whether these have been lost during their evolutionary history, or whether the group evolved before the advent of gill books.

The possible hydrostatic function of the haemolymph in *Limulus* has been discussed, and it has been pointed out that the question of whether *Limulus* uses hydrostatic pressure to extend its legs remains to be finally proved or disproved. The lack of extensor muscles in the legs of primitive chelicerates certainly imposed a number of limitations upon terrestrial descendants, however, as discussed above. Pycnogonids show a slightly different situation, since they have joints with condyles and not simple hinges. Manton (1978) has shown that the pycnogonid leg joints have more in common with those of arachnids than with those of any other arthropods. Most of these joints have both flexor and extensor muscles (see Fage, 1949). The second tibia/tarsus joint has no extensor muscles, however, and extension is considered to be a passive reaction (see Morgan, 1971). The possibility of the combination of direct muscular action and hydrostatic pressure has apparently not been suggested as a mechanism for limb extension in pycnogonids. Manton (1978) pointed out that no pressure-isolating mechanisms have been described from the base of pycnogonid legs, and that in their absence an increase in general hydrostatic pressure could only favour leg extension at the expense of hindering flexion of other legs. However, an observation by Prell (1910) that the bristles on the leg of *Nymphon mixtus* become outspread during the swimming powerstroke is reminiscent of the similar observation on spiders during activity made by Parry & Brown (1959b). It may be that this effect is actually a passive reaction against the water, but the possibility of raised hydrostatic pressures in pycnogonids during movement should be investigated.

Coming now to sense organs, we can consider two types. First of all, the fine structure of chemoreceptors has been examined. In contrast to those of terrestrial arachnids, the chemoreceptors in *Limulus* are not found in fine hairs, but in the large spines on the chelae and gnathobases. Although the organisation of the receptors is very similar to that of arachnids, the sense cells are therefore buried in the cuticle of the spines, and only the dendrites come near to the surface of the cuticle, being exposed in the usual fluid-filled pit (Hayes, 1971). The receptors are sensitive to many amino acids at concentrations as low as 0.01 M, and to extracts of clams and fish (Wyse, 1971): stimulation of the receptors results in opening of the chelae, and they are presumably responsible for short-range detection of food. In this they differ from the chemoreceptors of spiders, where distance chemoreception is involved. This distance chemoreception in spiders can therefore be seen as a development which occurred by positioning the sense cells on protruding hairs on the legs, instead of having them on the organs and mouthparts involved in catching prey. Such a development would be called for in response to a change

in diet from relatively immobile molluscs and annelids in an aquatic environment to very mobile insects on land.

There are five sets of photoreceptors in *Limulus*, but three are rudimentary, leaving only two as the major pairs: the centrally placed ocelli and the lateral compound eyes (see Fahrenbach, 1975). Both types have cuticular lenses, and the general structure of the ocelli is very similar to that of spider eyes. The retinular cells have microvillar rhabdomeres found distal to the cell bodies, as in the principal eyes of spiders. In the ocelli these retinular cells are sandwiched between guanophores. It is supposed that the ocelli are not image forming, but have some directional sensitivity because of the lens. The eyes of pycnogonids are also similar to those of arachnids (Jarvis & King, 1973; King, 1973), but the rhabdomeres are proximal to the cell bodies as they are in the lateral eyes of spiders. The ultrastructural organisation of arachnid eyes is, in fact, practically indistinguishable from that of pycnogonids (Fig. 6.9).

In the compound eyes of *Limulus* the lens has no external facets but is produced internally into a number of cuticular cones. Beneath each of these lies a group of retinular cells arranged so that they form the segments of a cylindrical ommatidium. The rhabdomeres of adjacent retinular cells are in close contact within each ommatidium, producing a stellate appearance in cross-section. There is therefore a superficial resemblance to the compound eye of insects and crustaceans, but the detailed organisation is sufficiently different to rule out any homology. The resemblance to scorpion eyes is striking.

The receptor cells in *Limulus* eyes show a variety of spectral responses. Most cells are maximally sensitive at 520–530 nm, but some are sensitive also to ultraviolet, while some in the ocelli are responsive only to ultraviolet. There seems, therefore, to be the same diversity of receptors as in the lateral eyes of spiders, and it is only the jumping spiders which are known to have further receptors sensitive to other wavelengths. Unfortunately the work on *Limulus* eyes is mostly confined to structure and basic physiology, and little is known of their uses to the animal. It is possible that the ultraviolet receptors are of use in detecting shallow water (Lall & Chapman, 1973), and that the ocelli are responsible for this, but the uses of the compound eyes are not understood. This makes it difficult even to speculate upon the selection pressures under which compound eyes have evolved, or to come to any conclusions about the question of why most terrestrial arachnids have only simple eyes. One possibility is that compound eyes developed in *Limulus* and possibly in the eurypterids when these animals achieved a large size and consequently came to possess large eyes. In other words, small non-image-forming eyes could be efficient with random placement of photoreceptors, but with larger eyes the receptors had to be organised in groups if any overall visual advantage was to be achieved. This idea is supported by the work of Bedini (1967) which showed that in scorpions, the largest of the terrestrial arachnids, the eyes are in fact arranged with compound rhabdoms. These may have been inherited from a large ancestor such as the eurypterids, or may have independently developed as the scorpions reached a large size. Since most other terrestrial arachnids have not evolved large photoreceptors, no such compound organisation has been required: in general, the arachnids are not a 'visually oriented' group as compared to, say, the insects, which have developed large, compound, image-forming eyes.

6.3 The origin of terrestrial chelicerates

This brief summary of some of the characteristics of the marine chelicerates has shown that they possess many of the features currently found in the terrestrial arachnids. The marine ancestors of the arachnids therefore probably also possessed these characteristics. This does not, of course, answer the question of how such ancestors actually gave rise to the terrestrial forms. Savory (1971) has suggested that since the marine ancestors undoubtedly consisted of a variety of types, several of these might have colonised land independently and in different ways. In particular he envisaged groups such as the scorpions being derived from eurypterid types, and being preadapted to resist desiccation. Their later colonisation of desert areas would therefore be a likely development. Other groups such as the palpigrades, Schizomida, Uropygi and Amblypygi are normally found in cryptozoic habitats in the tropics – in forest humus, under stones, and so on. These, Savory suggested, probably arose from ancestors which were not well armoured, and moved directly from the sea to the terrestrial cryptozoic niche. While the details of Savory's scheme are debatable, the suggestion that several orders colonised land independently does help to explain why it is so difficult to relate one order of arachnids to another. Both Kraus (1976) and Manton (1978) agreed that several chelicerate groups invaded the land independently at different times. If this suggestion is correct, we would expect to find many differences in physiological detail between the different orders, depending upon the route taken when moving on to land. Such detail is not available, but some differences between the scorpions and the spiders – in two different groups according to Savory – may be pointed out. First of all, it seems that in both groups leg extension probably involves hydrostatic pressure; but this is achieved in the spiders by a general rise in pressure, and in scorpions by a rise in the legs only. Such a difference implies considerable differences in circulatory and respiratory physiology which should be investigated. The eyes are organised

into compound rhabdoms in scorpions, and not, as far as is known, in other arachnids: further work might suggest whether this is due to a difference in origin, or to a generally larger size in the scorpions. In particular, an investigation of eyes in large species from a variety of orders might indicate whether the compound organisation is related to increase in size. Haemolymph composition in general shows remarkable similarity in the two groups (Table 6.6), but it is noticeable that magnesium concentrations are lower in spiders than in scorpions. It would be interesting to know if this is a general difference between the 'cryptozoic complex' and the scorpion line, or if it is in some way related to the greater activity of spiders. Haemolymph pH of the only spider investigated is over 8.0, while in scorpions values usually lie around 7.0. Once again the significance of this difference is not understood, but it must imply some large physiological differences. When such questions have been further investigated, and physiological studies have spread to other orders, there may be much more to say about the origins of the present-day arachnid orders.

One question which has not so far been raised is the relative times of origin of the terrestrial arachnids and of the terrestrial insects. At the present time these two animal groups are closely linked in many food webs, since the arachnids feed almost exclusively upon insects. The arachnids are thought to have produced all their known orders before the Carboniferous period (Savory, 1971), and the fossil record shows the scorpions present as early as the Silurian, with Acari and possibly Araneida present in the Devonian (Størmer, 1955). The insects, in contrast, have not been found as fossils earlier than the middle Devonian, and even these forms are primitive wingless Collembola (Moore, Lalicker & Fischer, 1952). The winged insects have not been found before the Carboniferous period. The early scorpions therefore presumably preyed on groups such as the millipedes, which also appeared in the Silurian. It seems likely that the great explosion of arachnid orders which took place in the Carboniferous was due to the radiation of the insects, providing suitable prey, and that without the growth of winged insects, the arachnids would never have become such an important component of terrestrial ecosystems. This, of course, is in contrast to the insect–myriapod group, which has produced very many vegetable feeders, and therefore often forms the first stage in terrestrial food chains. Their movement on to land is presumably related to the development of terrestrial vegetation, as will be discussed in Chapter 10.

7

Onychophorans and myriapods

'The hatching of a Millipede brings curious
   things to light . . .
An Arthropod with yolky eggs, whatever be
   its name,
Must own aquatic ancestors that once laid
   smaller eggs,
And hatched as tiny larval forms with many
   fewer legs.'

From 'The millipede's egg-tooth' (Larval forms
and other zoological verses) by Walter Garstang,
1951, Blackwell, Oxford.

The origins of the onychophorans, myriapods and insects
are shrouded in even more obscurity than those of other
terrestrial arthropods, since these groups have no primitively aquatic relatives at the present day. Nor, indeed,
were there any certain aquatic relatives in the past. The
Cambrian fossils of *Aysheaia* are conjectured to be marine
forms of the Protonychophora (Walcott, 1931), but even
if this is so, the origin of the group is not clear. There has
been much re-interpretation of the structure of *Aysheaia*
(Hutchinson, 1931; Cave & Simonetta, 1975; Whittington
1978, 1979), and it seems unlikely that members of this
genus had antennae. Its true relationships are obscure
(Bergström, 1979). The most primitive myriapods were
the Archipolypoda and Arthropleurida, found as far back
as the Upper Silurian. The morphology of few of these
forms is well known, yet as Hoffman (1969) has re-
marked they may well lie at the base of the radiation
of all terrestrial mandibulate arthropods. Consequently,
there is a variety of views concerning the classification
of these animals. Some authors believe that the pauropods, symphylans, centipedes and millipedes should be
retained as one class, the Myriapoda, and that this is
closely related to the Onychophora (Manton, 1964,
1972). Manton has argued that the myriapods and the
onychophorans have been derived from an aquatic
ancestor of polychaete type, and therefore had a
marine origin. She also indicated a relationship with
the insects, and as described earlier (Chapter 5) regarded
the onychophora–myriapod–insect group as a major sub-

division of the arthropods, the uniramians, equivalent
in status to the crustaceans and chelicerates. An
alternative view is to retain the pauropods, symphylans,
centipedes and millipedes as separate classes, and not
necessarily to link them closely with the onychophorans.
Many authors would retain the Onychophora as a
separate phylum, related to, but not included in, the
Arthropoda. Since some authors would also suggest a
purely terrestrial origin for the insects, they would presumably consider the questions of how and why these
animals colonised the land to be irrelevant (Hinton,
1977a). In view of this complex situation, it is best, as
with the chelicerates, to consider the biology of the
various groups to start with, and to raise the problem
of their origin afterwards. A tentative classification of
Onychophora and Myriapoda is given in Table 7.1.

Table 7.1. *The classification of Onychophora and
Myriapoda*

Taxon	Genera
Onychophora	*Peripatus, Peripatopsis*
Pauropoda	*Pauropus*
Symphyla	*Scutigerella*
Chilopoda	
Subclass Epimorpha	
Order Geophilomorpha	*Geophilus*
Order Scolopendromorpha	*Scolopendra*
Subclass Anamorpha	
Order Lithobiomorpha	*Lithobius*
Order Scutigeromorpha	*Scutigera*
Diplopoda	
Subclass Pselaphognatha	*Polyxenus*
Subclass Chilognatha	
Order Oniscomorpha (Glomerida)	*Glomeris*
Order Limacomorpha (Glomeridesmida)	*Glomeridesmus*
Order Colobognatha (≡ superorder?)	*Polyzonium*
Order Nematophora (Chordeumida)	*Lysiopetalum*
Order Polydesmida	*Polydesmus*
Order Iuliformia *sensu lato*	*Iulus*

7.1 The onychophorans

These animals, of which an example is shown in Fig. 7.1, are found in the tropics and south temperate regions, in rain forests and similar humid habitats, but they do not all need a warm climate: in New Zealand, some species live in areas where snow lies in winter. While most species are thought to be primarily forest dwellers, *Peripatus acacioi* from Brazil is today a member of the soil fauna (Lavallard *et al.*, 1975). It is found in areas which used to be covered by forest, but now have only herbaceous cover. During the wet summer it is found near the soil surface, but it retreats to galleries deep in the soil during the dry winter – much in the fashion of earthworms, as discussed earlier (Chapter 3). The air temperatures in this part of Brazil are moderate because of the altitude (over 1100 m), with a mean maximum temperature of 22.9 °C and a mean minimum of 13.3 °C. Within the soil, the variation is very much less than this (see Fig. 7.2), and *P. acacioi* cannot withstand fluctuations of even 6 °C between day and night. The associated fauna consists of earthworms, planarians, opilionids, scorpions, myriapods and cockroaches.

Onychophorans have a thin cuticle, flexible like that of insect caterpillars, and maintain their shape by using a hydrostatic skeleton: as their speed increases so their turgor rises, and they increase in length (Manton, 1950). They have legs which are conical projections from the body, bearing claws, and are operated by extrinsic limb muscles. These muscles provide most of the motive force, while the body wall muscles are used for maintaining rigidity (Manton, 1950). The movement of onychophorans is thus in some ways similar to that of the arthropods, and not to the annelids. Nevertheless,

the onychophorans can change their shape drastically, and frequently do so to pass through cracks and crevices. Indeed, they may be considered highly specialised in order to be able to live in their environment of leaf litter and rotting logs.

According to Manton (1950, 1952*a*), members of the genus *Peripatopsis* are able to vary their gait widely, depending upon the speed required. It has been suggested by Manton that such an ability is primitive, and that the gaits of the onychophorans could have provided a common origin for all the more specialised types of arthropodan gait. However that may be, the variation in movement available to *Peripatopsis* is evidently an adaptation to its habitat, where slow crawling through decaying wood must be replaced by fast movement in the relatively desiccating environment outside. Changes in gait are permitted by changes in body shape not available to arthropods with a rigid exoskeleton.

As with many other members of the cryptic fauna, onychophorans living in forest emerge into the macroenvironment at night. Their activity is governed by a diurnal rhythm (Alexander, 1957), which continues even if the animals are left in constant darkness. Their activity is also influenced by the humidity of the air: at 98% RH the activity of *Peripatopsis moseleyi* is minimal, and it becomes more active both in 100% saturated atmospheres and in humidities lower than 98% (Bursell & Ewer, 1950). The effect of this orthokinesis is, in general, to bring the animals into areas of 98% RH. Two reactions – avoidance of low RH and avoidance of saturation – are mediated by two sets of receptors: the first by those on the general body surface, and the second by those on the antennae.

The matter of water balance, as may be predicted from the above reactions, is critical to onychophorans. Species of *Peripatopsis* lose water at about 40 times the rate of a caterpillar (Manton & Ramsay, 1937): at 30 °C,

Fig. 7.1. A New Zealand species of *Peripatus*, *Peripatus novaezealandiae*, showing movement of the legs during walking. Body length 5 cm.

in moving air, *Peripatopsis sedgwicki* lost 39% of its original weight in 30 min. The major part of this loss is thought to occur from the spiracles, since these have no closing muscles, and the number of tracheae in one animal is enormous: tracheae open from the base of pits in the cuticle, 80 μm deep and 14 μm wide, each pit separated from the next by only 80 μm (Manton & Ramsay, 1937). The remainder of the cuticle, although only about 1 μm thick, bears small papillae and is hydrofuge. There are no pore canals or dermal glands as in insects, and the cuticle may be secreted in regions of complex interdigitation between the epidermal cells (Lavallard, 1965). The cuticle contains chitin and protein, but the non-wettability is probably ensured by the micropapillae acting as a plastron, since there appears to be no wax layer like that of insects or arachnids (Morrison, 1946; Robson, 1964). Onychophorans are, therefore, strictly limited to a humid environment.

Besides being lost by evaporation, water is lost as urine from the so-called nephridia – segmental organs opening at the base of each leg. Since these may all be discharged simultaneously, as much as 10 μl may be produced by a large specimen of *Peripatopsis moseleyi*

(Manton & Heatley, 1937). Little is known of the function of these nephridia, but they are probably osmoregulatory organs. In *P. moseleyi* they produce a fluid slightly hypo-osmotic to the haemolymph (176 mOsm compared with 196 mOsm for haemolymph – Picken, 1936). Picken believed that this fluid could initially be produced by ultrafiltration, and the structure of the sacculus suggests that it is a filtration site (Gabe, 1957). Seifert & Rosenberg (1976) have shown that the sacculus contains podocytes as in arthropod coxal glands. These authors and Seifert (1980) have suggested that the podocytes have evolved from cells termed nephrocytes, but the widespread occurrence of podocytes in phyla where cells with the unique structure of insect nephrocytes are unknown does not support this suggestion. The nephridia show some variation in structure in different segments (Fig. 7.3), but typical ones have a ciliated region leading from the sacculus to a long duct. This duct is subdivided into regions with different structure: one has cells with apical microvilli, one has cells full of mitochondria associated with basal infoldings, and another contains cells which secrete mucus (Gabe, 1957; Eakin, 1964; Storch, Ruhberg & Alberti, 1978). The duct is therefore

Fig. 7.2. Daily temperature regimes in the habitat of the Brazilian *Peripatus acacioi*, which lives in soil on the sites of former forests. In winter this species burrows to a depth of up to 1 m, but in summer it may be only a few cm below the surface. Closed circles, air temperature; open circles, soil surface temperature; closed triangles, temperature at a depth of 20 cm; open triangles, temperature at a depth of 40 cm; closed squares, temperature at a depth of 50 cm; open squares, temperature at a depth of 60 cm. (a) Winter; (b) summer. From Lavallard *et al.* (1975).

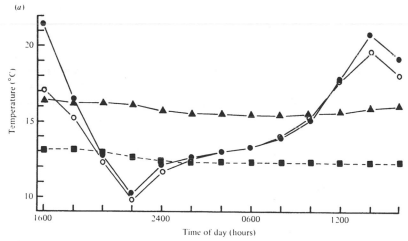

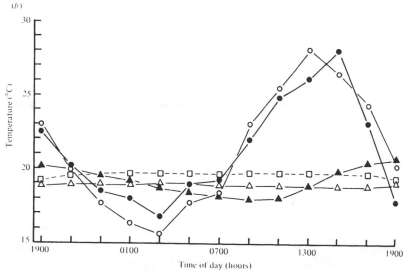

probably responsible for reabsorption of salts and other substances. The salivary ducts are probably modified nephridia, with a much greater capacity for mucus secretion. The production of hypo-osmotic urine parallels the situation in terrestrial cyclophorid snails which live in similar environments, where humidity is always high, and water gain rather than water loss is the major problem (see Chapter 4).

Fluid balance is, of course, also involved in the feeding of onychophorans. *Peripatopsis moseleyi* and other species can 'spit' slime from their oral papillae up to a distance of 45 cm. This is used to entangle large prey, e.g. crickets or woodlice, and may sometimes be used in defence (Alexander, 1957). When the prey has been immobilised, *P. moseleyi* approaches it, applies the oral lips and with the jaws slices a hole in the body wall. Salivary secretions are injected and the whole of the inside of the animal is then sucked out (Manton & Heatley, 1937). The rate of water loss in the subsequent elimination of faeces is unknown, but it is thought that at least some loss is avoided because onychophorans are uricotelic. Uric acid crystals are formed between the layers of the thin peritrophic membrane which lines the gut, and once every 24 hours the membrane and associated uric acid are passed out (Manton & Heatley, 1937). Uric acid may constitute as much as 50% of the dry weight excreted, and may amount to 0.1 mg/day in specimens weighing 300 mg. The possibility that nitrogen is excreted as gaseous ammonia has not been investigated.

Onychophorans also drink liquid water, and can take up water from damp substrates. At the base of the legs are coxal sacs, and in some species such as *Opisthopatus cinctipes* these can be everted and absorb water (Alexander & Ewer, 1955). Other species,

such as *Peripatopsis moseleyi*, also have coxal sacs, but these are not eversible, nor so well developed. This species cannot take up water as rapidly as *O. cinctipes*.

The ways in which all these mechanisms of water loss and gain are balanced are not known, but the measurements of haemolymph composition given by various authorities are remarkably constant, and all indicate an osmotic pressure in the region of 200 mOsm (see Table 7.2). This osmotic pressure is accounted for almost entirely by sodium and chloride (Robson, Lockwood & Ralph, 1966; Campiglia, 1976). The concentrations of other anions and cations are low, and it might seem unlikely that large amounts of amino acids could be present as they are in myriapods and insects. Nevertheless, amino acids have been reported from one species in concentrations very similar to those found in myriapods (Sundara Rajulu & Manavalaramanujam, 1972; see also Table 7.2). This points to a functional, if not an evolutionary, relationship to the primitive insects and myriapods.

Another process in which water loss is usually minimised in terrestrial animals is fertilisation. Two descriptions are available for onychophorans. In *Peripatus acacioi* fertilisation appears to take place only once, spermatophores probably being placed directly into the vagina of the female by the male. Spermatozoa are stored in copulatory vesicles, and after copulation are sufficient to supply the female for the remainder of her life (Lavallard & Campiglia, 1975). In contrast, *Peripatopsis* possesses no such copulatory vesicles, and the males deposit spermatophores on the body of the females. The spermatozoa then penetrate the body wall and make their way to the ovary (Manton, 1938). Subsequent development varies a great deal be-

Fig. 7.3. Diagrammatic drawings of segmental glands in *Peripatopsis capensis*. The glands have been modified as salivary glands in the head segments (*a*), but otherwise are probably excretory or osmoregulatory in function. (*b*) Glands in

segments bearing legs 1–3; (*c*) glands in segments bearing legs 4 and 5; (*d*) glands in remaining segments. Redrawn from Gabe (1957).

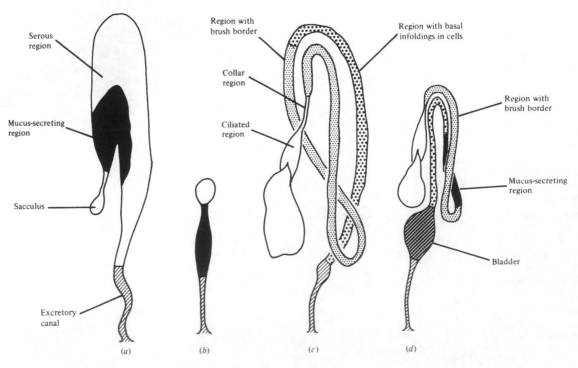

tween species. In most, the eggs are retained in the uterus and live young are produced, but in the Australian genera *Ooperipatus* and *Symperipatus* the female lays the eggs in moist soil and the young develop with no parental care.

The characters discussed so far are fairly strictly related to the environment in which onychophorans live; but much interest has also surrounded the onychophorans because of their mixture of annelid and arthropod characteristics, and it is appropriate here to spend a short time examining some of these. On the annelid side, we have the serial repetition of segments and organ systems, including the nephridia. Although the latter do not at all resemble annelid nephridia and conform in outline to the arthropod plan, they have 'normal' cilia, while arthropods possess cilia only of very modified form. Much discussion has taken place concerning the development of nephridia and their possible annelidan/arthropodan character (see Gabe, 1957; Seifert & Rosenberg, 1976). Gabe concluded that they were essentially annelidan, but in the adult state there is no doubt that they act as arthropodan excretory organs: fluid from the haemocoel is filtered into the sacculus and then modified as it passes down the tubule. This is in contrast to most annelid nephridia, where fluid is thought to enter from the coelomic body cavity via the nephrostome, without filtration. Yet it is simple to derive one system from the other, since in the annelid system we may regard filtration as occurring from the blood vessels into the coelom.

When this coelom is reduced, as it is in Onychophora and Arthropoda, filtration still occurs into the coelom, but this is represented only by the coelomic vesicle, or sacculus. The difference between annelid and arthropod excretory organs may therefore be thought of as simply one of degree; and the Onychophora may therefore in this one characteristic be said merely to be *nearer* the arthropods than the annelids.

Another character of annelid type is the eye structure. The general structure of the eyes is rather simple: they lie at the base of the antennae and each consists of a large lens with a simple pigment cup and retina behind it (Dakin, 1921). In contrast to arthropod eyes, where the lens is cuticular, the lens in onychophorans is separate from the cuticle. The cells of the retina form rhabdomeres as in arthropod eyes, but also contain rudimentary cilia (Eakin & Westfall, 1965). The eyes therefore show a complex mixture of arthropod and annelid features. Some other sense organs have been described (Storch & Ruhberg, 1977) which are probably mechanoreceptors and olfactory organs, but no sense cells with cuticular pores such as those found in the organs of Tömösvary, so characteristic of myriapods and apterygote insects, have been reported.

The fine structure and physiology of neuromuscular synapses in *Peripatus dominicae* have been investigated by Hoyle & Castilo (1979) and Hoyle & Williams (1980). They found that the ultrastructure and function of the neuromuscular synapses were very different from those of

Table 7.2. *Haemolymph composition of Onychophora and Myriapoda*

Species	OP (mOsm)	Na$^+$	K$^+$	Ca^{2+}	Mg^{2+}	Cl$^-$	PO$_4^{-3}$	HCO$_3^-$	Amino acids*	pH	Reference
					(mM/litre)						
Onychophora											
Peripatopsis moseleyi	–	93.0	4.2	5.3	1.2	71.9	–	–	–	7.3	Robson *et al.*, 1966
Peripatus acacioi	180	92.9	3.4	3.4	0.5	89.1	4.5	5.9	–	7.1	Campiglia, 1976
Peripatopsis sp.	200	–	–	–	–	–	–	–	–	–	Picken, 1936
Eoperipatus weldoni	–	–	–	–	–	–	–	–	21.8	–	Sundara Rajulu & Manavalaramanujam, 1972
Diplopoda											
Pachydesmus crassicutis	151–165	66	5.7	–	–	38	–	–	–	–	Woodring, 1974 and Stewart & Woodring, 1973
Glomeris marginata	158	58.5	4.9	40.7	2.9	53.0	–	20.0	24.7	8.3	Farquharson, 1974*a*
Orthoporus texicolens	175	–	–	–	–	–	–	–	–	–	Stewart & Woodring, 1973
Orthoporus ornatus	204	94.3	6.5	9.3	8.0	59.7	–	–	4.1	–	Pugach & Crawford, 1978
Iulus scandinavicus	236	70	18	–	–	61	–	–	–	–	Sutcliffe, 1963
Spirostreptus asthenes	–	–	–	–	–	–	–	–	20–22	–	Sundara Rajulu, 1970*a*
Jonespeltis splendidus	–	–	–	–	–	–	–	–	4.3–9.0	–	Naire & Prabhu, 1971
Chilopoda											
Lithobius sp.	318	157	11	–	–	154	–	–	–	–	Sutcliffe, 1963
Lithobius forficatus	370	190	6	–	–	175	–	–	–	–	Wenning, 1978
Cormocephalus rubriceps	379	214.7	4.9	4.8	1.6	112.9	0.7	–	45.1	–	Bedford & Leader, 1975
Ethmostigmus spinosus	–	–	–	–	–	–	–	–	19	–	Sundara Rajulu, 1970*a*
Haplophilus subterraneus	438	205	–	–	–	–	–	–	–	–	Binyon & Lewis, 1963
Strigamia maritima	462	274	–	–	–	–	–	–	–	–	Binyon & Lewis, 1963
Hydroschendyla submarina	498	283	–	–	–	–	–	–	–	–	Binyon & Lewis, 1963

* Amino acid values are given as mM/litre after conversion as in Sutcliffe (1963) where necessary.
OP = osmotic pressure.

either annelids or arthropods. The muscles of *P. dominicae* are controlled by summation of the activity of many small motor neurones, and there is no apparent inhibitory innervation, a feature often found in annelids and arthropods. Hoyle & Castilo (1979) therefore argued that there is little evidence for a close relationship of the onychophorans with either annelids or arthropods. The structure of the muscle fibres in onychophorans is also different from that of both annelids and arthropods (Lanzavecchia & Camatini, 1980).

In summary, arthropod features are, perhaps, more obvious than annelid ones in the onychophorans. The chitinous cuticle, method of locomotion and presence of tracheae have already been discussed. The whole body plan is organised on the basis of a haemocoel, with a typical dorsal arthropodan heart operating by the use of ostia. Manton's (1964, 1977) claims that the Onychophora should be placed near the Arthropoda are therefore well founded. What is much more difficult to decide is the group's exact status and relationship to the Insecta and Myriapoda. We shall be in a better position to discuss these relationships when we have examined the myriapods in more detail.

7.2 The myriapods

The myriapods have been held by some to be an unnatural but 'convenient' group, and by others (e.g. Manton, 1964) to form a natural unit. Certainly the Symphyla, Pauropoda, Chilopoda and Diplopoda are groups which have much in common in their general organisation, and they can usefully be considered together. The Symphyla and Pauropoda have been least studied. Both contain very small, cryptozoic animals, unable to withstand desiccation. The Symphyla have many features in common with the primitive apterygote insects, and are therefore of great phylogenetic interest. The Pauropoda are the smallest of the myriapods and many of their characteristics may be related to this small size. In contrast, both the Chilopoda and the Diplopoda have produced large forms, although even these are mostly limited to humid microenvironments. With many exceptions, the chilopods or centipedes are essentially specialised for running, whereas the diplopods or millipedes are adapted for burrowing. These two modes of life have probably been responsible for many of the structural differences between the two groups.

7.2.1 The pauropods

None of these animals is longer than 2 mm. All are without eyes and live in cryptic habitats such as moist logs, leaf litter and soil. They have nine pairs of walking legs, but only five dorsal sclerites. Together with the way in which the musculature is arranged, these help to prevent the body from undulating laterally during walking (Manton, 1966). Most pauropods can move rapidly, but since they have relatively long legs they are unstable on flat surfaces. In their normal habitat, however, this cannot be a problem. Pauropods are extremely intolerant of desiccation, and can survive only for a matter of minutes at room humidity (Manton, 1966). This must be due partly to their small size; it may also be due to a lack of

any epicuticular wax layer, but this point has never been investigated. Most species do not have tracheae, so that water loss must occur directly through the cuticle, but a few forms have spiracles which open on the coxae of the walking legs. In these animals the tracheae are short, except for the first pair which supply the head.

Pauropod populations may be dense, although they are not often noticed on account of the small size of the individuals. Over $500/m^2$ have been recorded (Starling, 1944), and their density is thought to be controlled mainly by temperature and humidity. In summary pauropods represent the epitome of the cryptozoic existence – they are completely dependent upon a moist microhabitat of even temperature; and like other animals in this environment, they show few adaptations to terrestrial life.

7.2.2 The symphylans

The largest of these animals is *Scutigerella immaculata*, which grows to 8 mm in length. Unlike pauropods, symphylans have more tergites than legs, and this allows the animals to flex in any plane. Consequently they can twist and turn, and follow complex crevices in any direction. Symphylans live in soil, leaf litter, and even in the marine intertidal zone, always in moist conditions. They cannot tolerate desiccation, partly, as with pauropods, because of their small size. *S. immaculata* has no external wax layer in the cuticle, but an Indian species, *Polyxenella krishnani,* has been shown to have an external layer consisting of lipid (Krishnan & Sundara Rajulu, 1964). Whether this should be considered homologous with the epicuticle of insects is not clear, but similar external lipid layers have been shown in chilopods and diplopods (Blower, 1951). What effect such lipid layers have in retarding water loss is unknown; but when the surface layers of the soil dry out, symphylans move downwards, as do the other soil inhabitants. Conversely, some species at least can move upwards in the soil as the water table rises during flooding. Even if they are submerged, each animal retains an air bubble around itself, and can live in this way for several weeks (Michelbacher, 1949).

The regulation of water balance thus depends mainly on behaviour. Symphylans have organs on the head which can probably detect humidity changes. These are the organs of Tömösvary, which consist of cuticular invaginations filled with an anastomosing network of cuticular tubules. The tubules are occupied by the ciliary processes of sense cells, which are exposed to the exterior by minute pores (Haupt, 1971). Nothing is known of how efficiently the behavioural mechanisms help to maintain a constant haemolymph composition, nor is anything known about the constituents of the haemolymph. Species of *Scutigerella* have maxillary glands which presumably act as osmoregulatory organs. These glands resemble excretory glands in other myriapods, and consist of a sacculus where an ultrafiltrate is produced and a tubule with ultrastructure typical of organs which reabsorb salts (Haupt, 1969b, 1976). There is one pair of Malpighian tubules, so that the pattern typical of other myriapods probably also applies in this case: osmoregulation is dealt with by the maxillary glands, while nitrogen excretion is confined to the Malpighian tubules.

Symphylans have only a single pair of tracheae, with spiracles opening on the head. The importance of tracheal and cutaneous respiration has never been investigated.

Reproduction involves indirect fertilisation as in most myriapods. The males produce spermatophores at the tip of stalks attached to the substratum. The females bite the spermatophores off and sperm travel from the pre-oral cavity to the spermathecae. Eggs are later laid in the soil and the female stays with them, but it is not known what parental care is provided.

Members of the Symphyla have a number of features in common with the apterygote insects, including the presence of styli, coxal sacs which may function in the uptake of water (Tiegs, 1940) and spinnerets. Together with some similarities in development and in head structure these factors have suggested that the symphylans have descended from the ancestors of insects, although there are differences, such as the structure and mode of operation of the mandibles (Manton, 1964; Kaestner, 1968). One of the objections to this idea has been that all insects are opisthogoneate, whereas in symphylans the genital openings are on the third segment of the trunk. Tiegs (1940) has argued that the opisthogoneate condition is secondary, and Krishnan & Sundara Rajulu (1964) have described an opisthogoneate symphylan. Therefore there seems little reason to doubt a close affinity between the Symphyla and the apterygote insects (Gupta, 1980), although Manton (1980) has concluded that the Symphyla cannot be *ancestral* to the hexapod classes.

7.2.3 The chilopods

The centipedes are easily distinguished from other myriapods by their prominent poison claws. Their functional morphology and general biology have been described in detail by Lewis (1981). There are two main subclasses: in the Epimorpha the young hatch with the same number of segments as the adults, while in the Anamorpha the young develop further segments at each moult, until they reach maturity. Within the Epimorpha, members of the order Geophilomorpha are long, very thin forms burrowing in soil – an unusual habitat for centipedes. Most centipedes are found in cryptic habitats such as under logs, stones or bark. This cryptic fauna consists of the Scolopendromorpha (Epimorpha) and the Lithobiomorpha and Scutigeromorpha (Anamorpha). None of these orders can be thought of as having given rise to the others, since all of them show a mosaic of primitive and specialised characters. The Scutigeromorpha are perhaps the most specialised for an active predatory life, with extremely long legs for fast running, compound eyes and tracheal lungs.

Surprisingly little work has been aimed at defining the habitats of centipedes. On a world scale they are common from temperate to tropical latitudes, and the largest species are found in the tropics. Many remain in humid dark places in the daytime and emerge to feed only at night (Cloudsley-Thompson, 1968). In some species at least, this behaviour is governed by a diurnal rhythm which persists even in continuous darkness (Cloudsley-Thompson, 1959; Cloudsley-Thompson & Crawford, 1970). The Scutigeromorpha form an outstanding exception. Sinclair (1895) recorded *Scutigera* in Malta actively hunting for insects in hot sunshine. Although little has been recorded about the ecology of *Scutigera*, it seems to be able to exploit desiccating conditions. Centipedes are widely distributed, from humid forest zones to relatively dry areas, but because of their habits few species are ever exposed for long to desiccating environments. The food consists mainly of small arthropods, but some species also eat molluscs, oligochaetes and, at some times of the year, leaf litter (Lewis, 1965). Some species such as the geophilomorph *Strigamia maritima* have colonised the marine littoral zone (Lewis, 1961). This species lives in rock crevices and in shingle banks, at a fairly high tidal level, and is rarely covered by sea water. Nevertheless, its food consists of such marine animals as *Balanus*, *Littorina saxatilis* and *Orchestia*, so that it forms a part of the littoral food web. Another geophilomorph, *Hydroschendyla submarina*, is found as far down the shore as mean tide level (Binyon & Lewis, 1963).

Since most centipedes are found in moist habitats it is not surprising to find that they have a high rate of water loss. Unfortunately, the available information on this subject is far from satisfactory. Various authorities (e.g. Edney, 1957; Cloudsley-Thompson, 1968) state that the cuticle has no waterproofing wax layer, and certainly the characteristics of water loss are not like those of insects or chelicerates; but Blower (1951) described a thin external sudanophil layer on the cuticle in *Lithobius*, probably secreted via canals by epidermal gland cells. Blower also found that there were lipoids within the main layers of the cuticle. These lipoids may aid in waterproofing to some degree, but they are not arranged as they are in insects, because there is neither a 'critical temperature' at which water loss suddenly increases, nor even an exponential increase in water loss with increasing temperature. It may be that they are more important in making the cuticle hydrofuge. Curves showing rates of water loss in two species are given in Fig. 7.4, and from these it may appear that water is lost more readily from *Lithobius* than from *Scolopendra*, which tends to live in drier habitats (Cloudsley-Thompson & Crawford, 1970). It is not possible to compare the figures, however, because of the different experimental conditions. Neither curve suggests a situation similar to that in insects; but even in insects and mites, it has been stressed by Berridge (1970) that the epidermis underlying the cuticle must in some way be responsible for reducing water loss. How this might be achieved remains entirely a matter for conjecture.

Much water must be lost through the spiracles, since these have no closing mechanism (see p. 135), but no measurements have been made to show what proportion is lost by this route. Experiments by Curry (1974) suggest that a large proportion may be lost through the general cuticle. Some control of water loss is probably

exerted by the excretory organs. These are found in
the head of Anamorpha but not Epimorpha, as part of
a segmental series of glands, some of which are involved
in feeding and digestion (see discussion by Lewis, 1981).
The paired maxillary glands of Anamorpha consist of
a sacculus and a long tubule (Bennett & Manton, 1962),
and in fact represent two pairs of glands that have fused
(Fahlander, 1938). Their structure is very similar to that
of excretory organs in other terrestrial arthropods (Gabe,
1967b; Rosenberg, 1979), and suggests ultrafiltration
in the sacculus, followed by reabsorption in the tubule.
Three centipede orders have organs called coxal glands,
but these are not part of any segmental series. They are
lined by cuticular intima and are not homologous with
coxal glands in other groups. These glands open on the
coxa of some of the legs. They show a structure typical
of transporting epithelia (Rosenberg & Seifert, 1977;
Rosenberg, 1982), and may be involved either in water
conservation, or in the uptake of salts and water from
the environment.

The work of Wenning (1978, 1980) on *Lithobius
forficatus* has emphasised that because centipedes live
mostly in damp habitats their major problem in osmotic
terms is the removal of excess water. In *L. forficatus* the
Malpighian tubules produce a potassium-rich fluid slightly
hyperosmotic to haemolymph, as in many insects, and
overall ion and water balance is to a large extent controlled by the hindgut. This is bordered by a transporting
epithelium, but far from drying the faeces and producing
hyperosmotic fluid, its contents are hypo-osmotic to the
haemolymph: their osmotic pressure is about 200 mOsm,
compared with about 370 mOsm in the haemolymph.
Excess water is therefore excreted by the gut, and ions
retained, because physiologically the animals have the
same problems as those in fresh water.

The details of the ways in which the hindgut and
maxillary glands are combined in their activity to control

Fig. 7.4. Rates of water loss from myriapods. In no case is there
any evidence of a 'critical temperature' which might indicate
the presence of cuticular lipids. Filled circles, a centipede,
Lithobius sp. The animals were dead, and after their spiracles
were blocked they were exposed to dry moving air. Data from
Mead-Briggs (1956). Open circles, another centipede, *Scolopendra*
sp., and triangles, a millipede, *Paradesmus* sp. Both these animals
were alive and were exposed to dry still air. Data from Cloudsley-
Thompson & Crawford (1970) and Cloudsley-Thompson (1950).

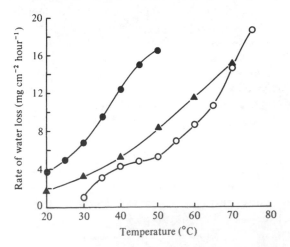

haemolymph composition are unknown, but some
measurements of typical haemolymph composition have
been made (Table 7.2). There is little obvious difference
between the total osmotic pressure in species from
different environments: even the marine *Hydroschendyla
submarina* is only slightly more concentrated than the
terrestrial *Haplophilus subterraneus*. However, the permeability to water of both *Strigamia maritima* and *Hydroschendyla submarina* is less than that of *Haplophilus
subterraneus*, suggesting that the marine forms have
found some way of cutting down exchange rates with
the external environment. The salivary glands in marine
forms are much larger than in terrestrial relatives, and it
may be that these glands are responsible for the removal
of excess salt taken in from the surrounding water and
from food (Binyon & Lewis, 1963). If this is so, it would
be a close parallel with the mechanisms found in marine
teleosts, birds and reptiles.

The osmotic pressures of centipede haemolymph
are higher than those found in onychophoran haemolymph (Table 7.2), and are not entirely accounted for
by inorganic ions. Some, at least, of this deficit is made
up by amino acids which reach concentrations found
in insects (Tables 8.2 and 8.5). Indeed, the total composition of the haemolymph resembles that of the
'primitive' haemolymph type in insects (Sutcliffe, 1963).
The accumulation of amino acids thus forms yet another
link between the chilopods and the insects. It does not
appear to be correlated with present-day habitats in any
way, but could be a feature inherited by both myriapods
and insects from a common ancestor. The significance of
the amino acids is hard to assess. It may be that their
presence is in some way to do with the abandonment of
the haemocoel as a carrier of respiratory pigments, and
the adoption of a tracheal system. In this context an
examination of *Scutigera*, which does use the haemolymph for oxygen transport, would be most interesting.
On the above suggestion, we should expect *Peripatus*
to have significant levels of amino acids, and the one
measurement that has been made suggests that indeed
this is so (see p. 130). Further speculation must wait
upon more knowledge of the source, fate and function
of these amino acids.

Most nitrogen is excreted from the body in
the form of ammonia: in *Lithobius forficatus* 50% of
the total nitrogen excreted is in this form. A further
1–12.5% is in the form of uric acid, which is produced
in the Malpighian tubules, and passes out in the faeces
(Bennett & Manton, 1962; Horne, 1969). Both Anamorpha
and Epimorpha possess one pair of these tubules, but
their histological appearance is different in different
species. In *Scolopendra* the tubules have a brush border
along their entire length, and crystals of urate can be
seen in them. In *L. forficatus* there are three distinct
regions in the tubules, and no crystals can be seen when
studied using the light microscope (Bertheau, 1971),
although small excretory spherules are visible when
studied using the electron microscope (Füller, 1966).
The cells making up the tubules are typical of the
transporting type, having basal infoldings associated
with mitochondria, as well as apical microvilli. These

microvilli, however, are not like those of insects, which have mitochondria within them (Füller, 1966). During normal activity it is unlikely that the uric acid secreted by the tubules is of great significance, but it may be that during aestivation it becomes more important. The fact that centipedes are primarily ammonotelic indicates their limitation to humid microhabitats, and shows an obvious parallel to the isopods.

The respiratory systems of chilopods show similarities to those of insects, since they consist of spirally thickened tracheae opening through spiracles. The spiracles, however, have no true closing mechanisms (Curry, 1974). In most species there is no muscle associated with the spiracle (Dubuisson, 1928; Füller, 1960), but in the Scolopendromorpha, muscles insert on the tracheae just inside the spiracle (see Fig. 7.5): these may serve both to open the spiracle and to expand the atrium, thus drawing in more air. There is no specific closing muscle. The linings of the atria are covered in small projections of the cuticle (see Figs. 7.5, 7.6), which are thought to act very much as do insect plastrons: they permit the entry of air to the trachea in waterlogged conditions. They also ensure that air can enter even when the spiracular lips are pressed together, so that it follows that there is no way of entirely preventing water loss through the spiracles. Nevertheless, the spiracular openings in many species are partly obstructed by cuticular lappets of a variety of types (Verhoeff, 1941; Fig. 7.6), which presumably hinder water loss as well as preventing the entry of foreign bodies. In geophilomorphs there is some correlation between the extent of these lappets and resistance to desiccation (Lewis, 1963).

The details of methods by which centipedes ventilate their tracheae are not at all clear, but the observations of Dubuisson (1928) showed that in geophilomorphs the elongations and contractions of the body during movement flush air through the system. These forms have long, tube-like tracheae joining one spiracular atrium to another (Manton, 1965; Fig. 7.6), so that air can be moved from one part of the body to another. This may be an adaptation concerned with the burrowing habit, allowing air to be passed to segments in which the spiracles are blocked by external soil. Air is also probably moved through the tracheae by alternate compressions and relaxations of the spiracular atria, caused by the pulsations of the heart (Dubuisson, 1928). As a further adaptation, it has been noted that in the marine littoral *Mixophilus* the tracheal tubes themselves pulsate rhythmically, especially after periods of submergence (Sundara Rajulu, 1970*b*). The scolopendromorphs also have long tracheae connecting atria in different segments, but these are flimsy and could not act as efficient conducting channels. They may, possibly, provide temporary oxygen stores (Manton, 1965). Lithobiomorphs, in contrast, have not such long tracheae, and in the scutigeromorphs the system is different again. Here, the tracheae do not supply the tissues directly, but run to the pericardium. It is thus the body fluid which acts as an oxygen carrier, as in chelicerates, and a protein fitting the description of haemocyanin has now been recognised in at least one species of *Scutigera* (Sundara Rajulu, 1969). *Scutigera* has dorsal spiracles, unlike the lateral ones in other chilopods, with tracheae radiating out in a fan shape to contact the pericardium (Dubuisson,

Fig. 7.5. Spiracular openings in scolopendromorph chilopods, showing examples of degrees to which the spiracles are closed off from the outside. (*a*) Surface view of spiracle in *Cormocephalus*; (*b*) diagrammatic section through a spiracle in *Cormocephalus*; (*c*) surface view of spiracle in *Otocryptops*; (*d*) section showing the muscle which can open the spiracle in *Cryptops*. (*a*), (*b*), (*c*) redrawn from Verhoeff (1941); (*d*) redrawn from Füller (1960).

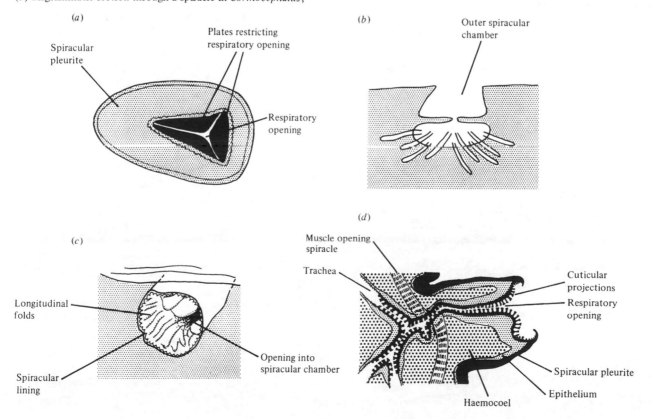

1928). The heart therefore causes oscillations of air in these tracheae, rather as in the atria of geophilomorphs. Air is also apparently sucked into the spiracles, presumably by solution at the pericardial end. This system of respiration must be thought of as helping to adapt *Scutigera* to its fast-running mode of life, and may suggest that the normal tracheal system supplying the tissues of most chilopods is far from efficient. If this is so the chilopod tracheal systems differ distinctly from those found in other terrestrial arthropods. The tracheae of insects have allowed the development of very active forms, and the fastest chelicerates, the solifugids, have replaced the traditional chelicerate system of book lungs and oxygen transport in the blood with large tracheae (see p. 116). Why the direct tracheal system should have been replaced in *Scutigera* therefore deserves further investigation.

The locomotion of chilopods has been examined in depth by Manton (e.g. Manton, 1977). Of the four orders, the most primitive is probably the Geophilomorpha. It is possible that a form like a rather short geophilomorph without any burrowing specialisations gave rise to the four present-day orders (Manton, 1952*b*). Modern geophilomorphs are very long and thin, and can burrow by elongating and contracting their segments rather in the fashion of earthworms. They have up to 170 pairs of legs and are slow movers: their maximum recorded speed is 23 mm/s. In general, however, chilopods have specialised in increased speed rather than in burrowing or in 'pushing power'. Their legs swing about a vertical axis, allowing a rapid but rather weak move-

ment (Manton, 1958*b*). Their cuticle is made rigid by sclerotisation, and therefore remains light in comparison with the calcified skeletons of diplopods. Both the Scolopendromorpha and the Lithobiomorpha are relatively heavy bodied, but are faster than the geophilomorphs. In the scolopendromorphs the duration of the backstroke – i.e. the propulsive stroke – of the legs is equal in length to that of the forward stroke, and speeds up to 80 mm/s can be reached. In the lithobiomorphs the backstroke is short relative to the forward stroke, and with increasing leg length to help, these animals can run at 280 mm/s. In conjunction with this gait, lithobiomorphs have reduced the number of leg pairs to 15 (Manton, 1952*b*), and have developed tergites which alternate down the body from long to short, thus helping to reduce lateral undulations. The scutigeromorphs, however, are the speed specialists. They have greatly elongated legs, and these differ in length down the body so that each can take its maximum stride without becoming entangled with another. *Scutigera coleoptrata* can run at 420 mm/s and possibly faster. The gait is such that the backstroke is of very short duration compared to the forward stroke, so that the animals have negligible pushing power. This has, as it were, been sacrificed for greater speed. The coxae are enlarged, and powerful extrinsic muscles attach to them to allow the legs to rotate as they move: this again lengthens the stride. Many of the aspects of adaptive radiation in chilopods can therefore be seen as linked to a general tendency to produce forms which move rapidly as opposed to the burrowing diplopods, and

Fig. 7.6. Spiracular structure and tracheae in geophilomorph chilopods. (*a*) Section through the spiracle of *Scolioplanes*. The cuticular lappets or 'pillars' prevent the spiracle from closing completely, but probably also restrict water loss. They also prevent the spiracles from being blocked by water films. (*b*) The tracheal system of *Orya*. The spiracles are joined by a complex system of tracheae which may ensure an air supply to the tissues even if spiracles are blocked during burrowing. Small branching tracheae are shown on one side only. (*c*) Section through the spiracular atrium of *Geophilus*, showing cuticular projections which prevent collapse of the atrium and may function as a 'plastron in reverse', holding an air film when the soil is waterlogged. (*a*) From Füller (1960); (*b*), (*c*) from Manton (1965).

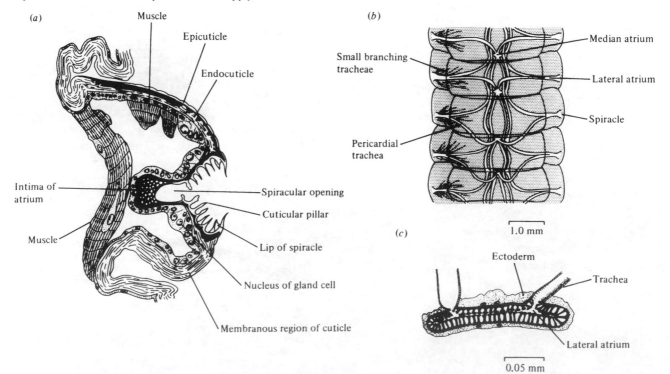

this may be thought of as the retention of early habits, in which the chilopod ancestors tended to run round obstacles, while the ancestors of the diplopods burrowed through or under them (Manton, 1952a).

These differences in speed and method of locomotion are, of course, reflected in the animals' general behaviour. Thus the slow geophilomorphs and scolopendromorphs tend to seize prey when they come upon it, but do not actively pursue it; whereas the lithobiomorphs and scutigeromorphs pounce on spiders and flies, using their poison claws to bite the prey and paralyse it (Manton, 1965). Correspondingly, the vision of scutigeromorphs is good: *Scutigera forceps* has compound eyes with 200 ommatidia, and a Japanese species, *Thereuopoda clunifera*, has 600 (Hanström, 1934). Most chilopods, however, have only simple ocelli, with a structure similar to that of other arthropods (Bedini, 1968; Bähr, 1974), and forms such as *Lithobius* must rely upon these to mediate simple negative phototaxis (Görner, 1959). Other reactions have been investigated further in *Lithobius forficatus* than in other chilopods. This species exhibits positive thigmotaxis, and it responds to odours, probably using chemoreceptors on the antennae (Meske, 1961): sensillae with terminal pores are presumed to be combined contact chemoreceptors and mechanoreceptors (Keil, 1976). The organs of Tömösvary (see Fig. 7.7), like those of the Symphyla (p. 132), may be hygroreceptors: *L. forficatus* can orientate itself in a humidity gradient, but this ability disappears when the organs of Tömösvary are blocked (Tichy, 1973). Many of these reactions, of course, suit

L. forficatus for a cryptozoic life. One final reaction which has been demonstrated is that to sound (Meske, 1960), and this ties in with a few scattered observations suggesting that some centipedes can also produce sounds. Attention to this has been drawn by Cloudsley-Thompson (1960). The anal legs of some scolopendromorphs and scutigeromorphs normally point backwards and do not touch the substrate. These legs are often autotomised when the animals are disturbed, and after autotomy they produce a 'creaking' noise, presumably as a way of distracting predators. In other species these anal legs may be rapidly vibrated, producing a 'rustling' sound (Lawrence, 1953).

The complex behaviour patterns of centipedes are best exemplified by considering some of the details of reproduction. Most centipedes, like many chelicerates, have indirect fertilisation without copulation, although copulation could have been the primitive method of sperm transfer (Brunhuber, 1969), and some centipedes do copulate (Klingel, 1962). The initial indirect methods of sperm transfer probably involved the deposition of sperm droplets in humid sites in soil, and subsequent collection of these by the females. In geophilomorphs the male spins a small web on which sperm is deposited. There is no courtship behaviour, but when the females have laid eggs, these are brooded. A good example of this is given by the marine littoral form *Strigamia maritima*, in which the females aggregate about 30 cm below the surface of the shingle high up on the beach, and hollow out small cavities (Lewis, 1961). In scolopendromorphs some very simple courtship occurs: when the

Fig. 7.7. Diagrammatic sections through parts of the organs of Tömösvary, which are unique to myriapods and apterygote insects and may therefore indicate some phylogenetic relationship between the two groups. The organs may be hygroreceptors or chemoreceptors. (*a*) A diplopod, *Glomeris marginata*, redrawn from Bedini & Mirolli (1967). (*b*) A chilopod, *Lithobius forficatus*, redrawn from Tichy (1973).

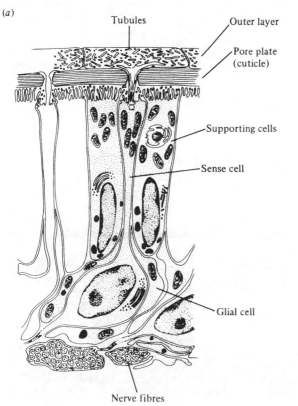

(*a*)
Tubules
Outer layer
Pore plate (cuticle)
Supporting cells
Sense cell
Glial cell
Nerve fibres

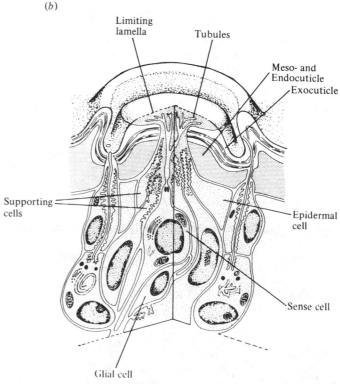

(*b*)
Limiting lamella
Tubules
Meso- and Endocuticle
Exocuticle
Supporting cells
Epidermal cell
Sense cell
Glial cell

male and female initially meet, they react defensively. After this they use the antennae to tap the posterior part of the partner's body. The male spins a pad of threads, and deposits a single spermatophore which is later taken up by the female through the genital opening. Once again, the female broods the eggs in a small subterranean chamber, grooming them periodically (Brunhuber, 1970). In the Lithobiomorphs courtship is slightly more complicated (Brunhuber, 1969), and a complex spermatophore is deposited on a delicate web (see discussion by Lewis, 1965). Eggs in this case are laid singly in the soil, and there is no brooding. Finally, in some scutigeromorphs a complex courtship pattern has evolved, and the male, after depositing a spermatophore directly on the ground, guides the female to it – somewhat as in primitive insects such as the Thysanura (Brunhuber, 1969). While these different mechanisms cannot be interpreted as a direct evolutionary sequence, they do reinforce the suggestion by Manton (1952*b*) that some form near to the geophilomorphs probably gave rise to all four present-day orders. They also demonstrate the increasing degree of communication between individuals required in the scutigeromorphs. Thus in very many ways it is the scutigeromorphs which have become most advanced and, to a greater extent than the other orders, the most adapted to life outside the cryptozoic niche.

7.2.4 The diplopods

There are approximately 7500 described species of millipedes, compared with about 2800 centipedes. All are characterised by the presence of diplosegments, each bearing two pairs of legs – a feature thought to

be of importance in increasing the ability to burrow (Manton, 1954). Also probably in this connection, most millipedes have a heavily calcified cuticle. At least in some soils, millipedes are important in the breakdown of leaf litter, and may have as great an effect on soil structure as do earthworms (Kevan, 1962), although this is not true in at least one calcareous region, where diplopods were not important in decomposition (Dunger & Steinmetzger, 1981).

The members of the subclass Pselaphognatha are atypical, being soft bodied and not having diplosegments along the entire length of the body. The major subclass, the Chilognatha, consists of a number of orders (see Table 7.1) with different body forms but a remarkably uniform basic structure. The two largest of these are the Iuliformia and the Polydesmida. The Iuliformia are now often split into a number of separate orders, but in this account all these will be treated as one since very little is known of differences in physiology between the recent subdivisions. These 'Iuliformia' are long and cylindrical (Fig. 7.8), and mostly burrow by pushing directly into the substratum; they are common in soil. The Polydesmida or 'flat-backed millipedes' (Fig. 7.9) have wide lateral keels and burrow by inserting their anterior ends into cracks and then pushing upwards with their dorsal surface to widen the crack. They are very common in leaf litter where this method of burrowing is ideal. Members of a third order, the Oniscomorpha, are very short and can roll up into a ball, woodlouse fashion. The Limacomorpha are long and flattened, and cannot roll into a ball. The Colobognatha also have lateral keels, and are much more tolerant of arid conditions than the other

Fig. 7.8. A giant iuliform millipede from the forest floor of Papua New Guinea. This species was active during the day, crawling on the surface of the litter. Length approximately 15 cm.

millipedes. Finally, the Nematomorpha includes a group, the Lysiopetaloidea, which can run rapidly and is made up of scavengers or predators.

Most information is available for the Pselaphognatha, the Iuliformia and the Polydesmida, and discussion will be centred upon these three groups. Of the Pselaphognatha, the best-known species is *Polyxenus lagurus* (Schömann, 1956). This lives mainly under the bark of trees, and is usually to be found there, in leaf litter on the ground, or in crevices of walls and rocks. On sunny days it is active on the bark surface where it feeds on single-celled algae such as *Pleurococcus*. This diurnal activity is somewhat surprising since it has a thin cuticle, but authorities differ on whether or not *Polyxenus lagurus* can withstand desiccation (Manton, 1957; Seifert, 1967), and there have been no studies on rates of water loss. However, the preferred humidity lies between 72 and 86% RH (Schömann, 1956), so that it is probably more resistant to desiccation than many other millipedes. Its major structural adaptations appear to be concerned with living in crevices and walking on smooth surfaces. For its life in crevices it is equipped with small cuticular projections (Seifert, 1967) and large, air-filled spines (Manton, 1957) which ensure that an envelope of air is maintained around the body even if the crevice is flooded. The tarsal claws have specialised adhesive lappets which allow the animals to run upside down on smooth surfaces. According to Manton (1957) this also is an adaptation to crevice life, but it seems more likely to allow the species to run over the surface of trees with smooth bark where the algal food supply is abundant. From many considerations the

Pselaphognatha thus appear rather specialised, and cannot be primitive forms. Particularly compelling evidence against a primitive position is the presence of at least some diplosegments. If it is true that these are primarily a burrowing specialisation, their presence in the non-burrowing *Polyxenus* must mean that the latter was derived from a burrowing ancestor. We must, however, note that in fact opinions differ concerning the uniqueness of the diplosegments to the burrowing diplopods (see p. 145).

In contrast to the Pselaphognatha, most of the Chilognatha spend their time burrowing, in regions of high humidity and in a protected environment. They are probably essentially animals of the forest floor (Blower, 1955), but in spite of this many species are found on bare ground in savanna (Lewis, 1971*a,b*), and in this environment may be subjected at times to desiccation or to flooding. While intrinsic rhythms ensure that some species are only active at night when humidities are high (Paulpandian, 1965), others have no such activity rhythm (Cloudsley-Thompson, 1951) or are active during the day (Dwarakanath & Job, 1965), and some are found in the open when conditions are hot and dry (Crawford, 1972). Factors which control water loss and gain are therefore of major importance.

Some information concerning rates of water loss in a variety of species is given in Table 7.3 (see also O'Neill, 1969*b*). There appears to be little relationship between rate of water loss and taxonomic grouping, except that only the Iuliformia contain species capable of restricting water loss to any degree. The desert millipede *Orthoporus ornatus* loses water, in terms of % body weight per hour,

Fig. 7.9. A polydesmid millipede from the forest floor of Papua New Guinea. This species was found within the litter layer, where it could easily move between the flattened leaf surfaces. Length approximately 7 cm.

at about the same rate as many chelicerates (solifugids, scorpions, spiders; see Table 6.3); but most other millipedes lose water at rates which are an order of magnitude greater than this. The structure of the cuticle and the tracheal openings must govern this rate of loss, but in most species it is not clear whether most water evaporates through the cuticle or through the spiracles.

The cuticle is usually thick and calcified, but is penetrated by pore canals and dermal glands, and these glands produce a thin film of lipoid on the surface which renders it hydrofuge (Blower, 1951, 1955). Lipoid also impregnates the exocuticle. In polydesmids, abrasion of the surface did not influence water loss (Cloudsley-Thompson, 1950; Stewart & Woodring, 1973), but extraction with hot chloroform increased it, suggesting that perhaps only the deeper lipoid was involved. No critical temperature was observed (Cloudsley-Thompson, 1950; see Fig. 7.4), again pointing away from the importance of a wax layer; nor was this found in *Glomeris marginata* (Edney, 1951). In iuliforms similar findings apply (Crawford, 1972; Stewart & Woodring, 1973), but in the desert form *Orthoporus ornatus* weight loss doubled after death, implicating restriction of water loss either by the epidermal cells or by spiracle closure. When the spiracles were blocked experimentally the rate of water loss in a polydesmid, *Pachydesmus* sp.,

was reduced to 8% of the original rate, and in an iulid, *Orthoporus* sp., the rate was halved (Stewart & Woodring, 1973). However, it appears that these species had no capacity to close the spiracles themselves, because treatment with carbon dioxide (which opens arthropod spiracles) caused no increase in the rate of water loss.

This brings us to the much debated question of whether there are *any* millipedes which have mechanisms to effect spiracle closure. In spite of statements that these exist, there appears to be little evidence to prove their existence, and in the two species discussed above there can be no direct closing mechanism. There are, however, indirect mechanisms, involving coiling of the body and, in *Glomeris marginata*, rolling up or conglobation. This may be very efficient, but there appears to be no study showing that this is so. Coiling is not effective in some forms, such as *Pachydesmus*, but in *Orthoporus texicolens* the posterior edge of each segment covers the spiracles on the next segment when the animal coils up. In this position the coxae also cover the spiracles on the next anterior segment, and water loss is significantly reduced (Stewart & Woodring, 1973). In another species of *Orthoporus*, however, coiling is only effective in reducing water loss at high temperatures (Crawford, 1972).

Table 7.3. *Rates of water loss in myriapods*

Species	Temperature (°C)	Rates of water loss		Reference
		(%/hour)	(mg cm^{-2} hour^{-1})	
Diplopoda				
'Iulida' (*sensu lato*)				
Ophistreptus sp.	30	—	7.2	Cloudsley-Thompson, 1959
Tachypodoiulus niger	30	—	4.3	Cloudsley-Thompson, 1959
Schizophyllum sabulosum	26	0.6–4.8	—	Perttunen, 1953
Iulus terrestris	26	5.5	—	Perttunen, 1953
Spirostreptus assiniensis	24	0.5	—	Toye, 1966
Orthoporus ornatus	30	0.08–0.24	0.09–0.25	Crawford, 1972
Orthoporus texicolens	24	*c.* 0.1*	—	Stewart & Woodring, 1973
Polydesmida				
Paradesmus (= *Orthomorpha*) *gracilis*	26	3.5–4.7	—	Perttunen, 1953
Paradesmus (= *Orthomorpha*) *gracilis*	30	—	3.0	Cloudsley-Thompson, 1950
Oxydesmus platycerus	30	—	3.8	Cloudsley-Thompson, 1959
Oxydesmus sp.	24	1.1	—	Toye, 1966
Habrodesmus falx	24	2.3	—	Toye, 1966
Pachydesmus crassicutis	24	*c.* 1.6*	—	Stewart & Woodring, 1973
Glomerida				
Glomeris marginata	26	—	4.0	Edney, 1951
Chilopoda				
Scolopendromorpha				
Scolopendra sp.	30	—	1.08	Cloudsley-Thompson & Crawford, 1970
Lithobiomorpha				
Lithobius forficatus	30	—	6.8	Mead-Briggs, 1956

All values are given for dry, moving air, except for those marked with an asterisk, which refer to still air.

In view of the lack of control of the tracheal openings, diplopod behaviour must be most important in regulating water exchange (see, e.g. Baker, 1980). The intrinsic rhythms ensuring nocturnal activity in some species, together with coiling behaviour, have already been mentioned. One species of iuliform, *Narceus americanus*, shows increased activity when first subjected to desiccation, but after about 4 hours it becomes immobile and shows coiling (O'Neill, 1969*a*). Other species show a variety of reactions to desiccation, including intense activity followed by pressing the ventral side of the body closely against the floor when resting (Toye, 1966). The variety of responses has been emphasised by Perttunen (1953) who showed that some species prefer lower humidities than others when in a normal state; but when desiccated, all species move into humid regions. In *Schizophyllum sabulosum* this response appears to be mediated by receptors on the body surface, whereas the movement of hydrated animals to low humidities is controlled through receptors on the antennae (Perttunen, 1955): a situation exactly like that of the onychophorans (see p. 128). The receptors on the body may be the organs of Tömösvary (see Fig. 7.7), by analogy with the conclusions for Symphyla and Chilopoda, but the possible olfactory function of these organs has also been discussed (Bedini & Mirolli, 1967).

In contrast to the many studies made of loss of water from millipedes, very few have concerned the uptake of water, or the reactions to flooding. It is known that some species can take up water through the cuticle (O'Neill, 1969*a*), while the desert form *Orthoporus ornatus* can extrude rectal tissue through the anus and absorb water from damp substrates, and can also drink (Crawford, 1972). Crawford (1978, 1980) also suggested that it may be able to take up water from the soil during dormancy. Some species, such as those in the Lysiopetaloidea (a suborder of the Nematophora) have coxal sacs at the base of some of the legs. These have been shown experimentally to take up water from damp surfaces (Manton, 1958*c*). There must also be reactions to immersion in water, but although such situations must frequently occur in nature they have not apparently been described.

Physiological mechanisms of regulating body water content and haemolymph composition involve two systems: the maxillary glands, and the Malpighian tubules in combination with the gut. Of the maxillary glands, nothing is known but the structure (Fahlander, 1939; Hubert, 1970; El-Hifnawi & Seifert, 1971; El-Hifnawi, 1973). They consist of a simple pair of glands opening in the head region, each having a sacculus lined by podocytes, leading into a labyrinth or duct lined with typical reabsorptive cells. The sacculus is attached by tendons to muscles, and can apparently be stretched, suggesting that there is a mechanism of ultrafiltration very like that proposed for the Acari (see p. 118). The system is plainly organised for an osmoregulatory function.

Concerning the second system, some investigations have been made of both structure and function (Hubert, 1970; Farquharson, 1974*a,b,c*). There is a single pair of Malpighian tubules which open at the junction of the

midgut and the hindgut. The tubules pass forward and then fold back along the length of the body in *Polyxenus*, *Polydesmus* and the Nematophora, and do not obviously consist of different regions. In the Iuliformia and in *Glomeris* they are folded three times on themselves and consist of regions with differing structure. The proximal region secretes a fluid when maintained *in vitro*, and this requires the presence of sodium for its production. It therefore probably depends primarily on the secretion of sodium as in some, but not the majority of, insect Malpighian tubules. This proximal section is permeable to very large molecules which probably pass through the intercellular spaces (Johnson & Riegel, 1977*a,b*). The function of the distal section is not known, but a resorptive role has been postulated. It is probable, however, that the major site of reabsorption of salts and water is the hindgut or the rectum. Both these regions have many projecting folds, well supplied with tracheae (Miley, 1930), and with typical transporting epithelia (Schlüter, 1980*a,b*). The isolated hindgut of *Orthoporus ornatus* can transport sodium from the lumen to the haemolymph, and potassium in the reverse direction (Moffett, 1975), but since no net solute transport has been observed, the overall function of the hindgut in reabsorbing the water and ions contributed by the Malpighian tubules remains unknown. However, the gut seems to act as a general reservoir of water, allowing *O. ornatus* to regulate its haemolymph composition by taking up appropriate amounts to counteract desiccation (Crawford, 1978).

Some figures for the haemolymph composition of diplopods are given in Table 7.2. The osmotic pressure is more in line with that of Onychophora than with that of the Chilopoda, being rather low for terrestrial arthropods; amino acids form a significant proportion of the constituents, as in Chilopoda. Otherwise the only remarkable item is the high level of calcium reported for *Glomeris marginata*. In this species the Malpighian tubules produce a fluid with very much lower calcium concentration, suggesting that much is bound in the haemolymph (Farquharson, 1974*b*); and indeed, it is unlikely that such high concentrations would remain in true solution. If this high calcium level is typical of diplopods, it may reveal an interesting correlation with the heavily calcified exoskeleton.

Since renal and osmoregulatory organs are traditionally associated with nitrogen excretion, this subject may be considered next. It is probable that ammonia is the main end product of nitrogen catabolism. In *Cylindroiulus londinensis* it accounts for 20 to 87% of the non-protein nitrogen excreted (Bennett, 1971). Some uric acid is excreted, but authors disagree as to the amount involved. According to Bennett, the range is 2.8–6% of the non-protein nitrogen, while Hubert & Razet (1965) give a figure of 70%. It is possible that the proportions of ammonia and uric acid vary depending upon the state of the animals: during starvation the percentage of uric acid increases, probably because of the increased breakdown of tissue constituents. This parallels the increase of uric acid production shown by centipedes during aestivation (see p. 135). Much of the non-protein nitrogen excreted has not been accounted for, but amino acids certainly make up part of this (Bennett, 1971). The general picture appears to be similar in a variety of other species. In *Orthoporus*, *Pachydesmus* and *Glomeris* twice as much ammonia as uric acid is excreted (Hubert & Razet, 1965; Stewart & Woodring, 1973). It is assumed that the nitrogen is lost mainly via the faeces, and the possibility of evaporative loss or loss through the maxillary glands has not been investigated. If ammonia is lost as a gas to any extent, the significance of the uric acid excreted would of course decrease even further. Uric acid is excreted in solution by the Malpighian tubules (Hubert, 1979, 1980) but the site of ammonia production has not been identified.

Perhaps the greatest amount of work on diplopods to date has involved their great variety of life styles in relation to their methods of locomotion. As with the chilopods, these have been extensively investigated by Manton (1952*b*, 1954, 1957, 1958*c*, 1961). The most widespread adaptation is to a burrowing habit, and this reaches its culmination in the Iuliformia. In these animals the power for head-on burrowing is achieved by the presence of many short diplosegments which cannot telescope: these allow for many legs per unit length. It was suggested by Manton that normal segments would not allow such close packing of legs because of the space required for intersegmental muscles as well as extrinsic leg muscles. The iuliforms are circular in cross-section, with fused tergites and sternites for rigidity, and the legs inserted midventrally. This gives them maximum length without projecting beyond the outline of the body. The legs move with a 'bottom gear' gait, i.e. the backstroke is very long in duration compared to the forward stroke, and there are consequently more legs pushing at any one time. The structure of the diplosegments allows each to move relative to the next by 'ball and socket' joints. Rotation is therefore possible, but telescoping – which would impede burrowing – is prevented. The legs articulate with the fixed sternites of the diplosegments, and the extrinsic leg muscles are inserted on the tracheal pouches which project into the body cavity from the ventral surface. This gives a powerful leg action since larger muscles can be accommodated in the body cavity than in the legs. The legs swing about a horizontal axis instead of a vertical axis as in the chilopods. The coxa is not rotated relative to the body, as it is in chilopods, where it is an adaptation to allow a longer stride and hence more rapid running. The organisation in diplopods produces strong joints useful in burrowing. The stance of diplopods also preserves a ventral air space supplying the spiracles, and ensuring that these do not become blocked.

In the Polydesmida a wide lateral keel protects the legs, and the animals burrow by wedging open cracks and crevices. Some polydesmids, however, can also run quite fast, and their gait is more a 'middle gear' than low gear – i.e. the backstroke is of about the same duration as the forward stroke. Some have also adopted the conglobating habit and can curl up as effectively as woodlice and the glomerids. Conglobation is presumably a means both of increasing protection from predators

and decreasing water loss, but involves so much enlargement of the lateral keels of the body that the glomerids at least are not good burrowers, and are limited to relatively loose leaf litter.

Many of the Nematophora have a similar body plan to that of the Polydesmida, and are good burrowers. Members of one suborder, the Lysiopetaloidea, are, however, not able to burrow and have become rapid runners. In this they parallel the Pselaphognatha, and the two groups therefore provide interesting comparisons with the rapid-running chilopods. The pselaphognath *Polyxenus*, for example, shows elongation of the body during running, and an active rotation of the leg to ensure an increased swing. The lysiopetaloids show a 'top gear' gait, and have become rock climbers and are carnivorous. According to Manton (1958*c*) these two groups are the most specialised diplopods which have moved furthest away from basic diplopod habits and morphology.

The Colobognatha present yet another line of locomotory evolution. In contrast to the Iuliformia, the body rings of colobognaths can telescope, because there is an extensive area of arthrodial membrane between diplosegments. The body length can therefore be altered radically as it is in chilopods. This property is not, however, an adaptation for rapid running but a further adaptation for burrowing. The trunk muscles as well as the leg muscles exert force: a segment with its legs not touching the ground is drawn forward and then put down, in a manner reminiscent of earthworm locomotion. The body can also form a hump when this occurs, because the sternites move further forward than the tergites. In this way the colobognaths can exert more force in burrowing than any other diplopod. They are not good at burrowing head on into soil, partly because they would telescope, and partly because in many species the legs project laterally beyond the protective keel; but they are very adept at opening cracks in a manner analogous to that of the polydesmids.

A variety of detailed comparisons between the diplopod groups and the chilopods may be made, but here there is room to consider only the structure and function of the leg joints. The base of the leg in most chilopods is seated in a large area of arthrodial membrane so that the angle of swing of the leg can be altered in running. In diplopods the legs articulate with rigid sternites so that there is little flexibility of movement, but the possibility of developing greater power. This adaptation of structure to function is nicely emphasised by the situation in the diplopod *Polyxenus*, which is soft bodied and can run actively. In this case the leg base has come to resemble that in chilopods (see Fig. 7.10). Within the legs of diplopods there may be up to three 'pivot joints' – that is, joints requiring opposing levator and depressor muscles. Such joints are mechanically strong and suited to burrowing. In chilopods there is only one such joint, and even that is a rather weaker version. Hinge joints, however, are present in both diplopods and chilopods. These are flexed by muscles, but have no intrinsic extensor muscles. According to Manton (1958*a*) they are extended during stepping by extrinsic retractor muscles, but when off the ground by hydrostatic pressure. There are no measurements of the hydrostatic pressure in relation to movement, presumably because of the small size of diplopod limbs. Measurements of the general haemolymph pressure in the iuliform *Cingalobolus* are very low – 1.7–4.9 mm saline (Sundara Rajulu, 1971). The pressure in the legs has not been investigated, nor has the pressure necessary for leg extension been evaluated. The whole importance of hydrostatic pressure in myriapod locomotion is, indeed, an open field.

Since many diplopods are restricted to dark, humid microhabitats, it is perhaps not surprising that their sense organs are not very well developed. The organs of Tömösvary have already been mentioned. There are no faceted eyes, but ocelli are aggregated into rows to form eye fields behind the bases of the antennae. The ocelli are not very different from those of chilopods in general structure (Bedini, 1970), but details differ between the Pselaphognatha and Chilognatha (Spies, 1981). In the Pselaphognatha each ocellus is probably derived from a single ommatidium, whereas in the Chilognatha the ocellus has arisen by the fusion and rearrangement of several ommatidia. *Polyxenus* at least can probably perceive outlines and contrasts by using its ocelli (Schöman, 1956), and the restriction of this ability in most diplopods is presumably related to life in a cryptozoic habitat (Spies, 1981).

Sensory hairs acting as mechanoreceptors are present, and resemble those in chilopods (Tichy, 1975). The presence of pheromone glands (Juberthie-Jupeau, 1976) is interesting in emphasising the communication which must occur between the opposite sexes. Such communication (see Haacker, 1974) initially involves

mechanical signals such as tapping with the antennae, drumming on the ground, or stridulating: in many species longitudinal ribs on the male's posterior walking legs are rubbed against the pygidium, and the female responds – although the receptor organs have not been identified. Chemical signals from glands such as the coxal glands or postgonopodial glands then ensure correct positioning of the female for copulation. Whether distance pheromones are used – for example in the initial meeting of the sexes – is not known. In many ways it is the general points concerning reproduction and life cycles which help to give a picture of how these animals are adapted to the environments in which they live. It is appropriate, therefore, to end this section on diplopods with some account of life cycles.

Excellent descriptions of the life histories of several tropical species of polydesmids in Nigeria have been given by Lewis (1971a,b). In these savanna forms the adults are active on the soil surface, so that it is the adult stage which is responsible for dispersal of the species. Copulation, which in most millipedes involves the use of specialised gonopods developed from walking legs, occurs above ground. Sperm is transferred to the reproductive opening of the female, so that fertilisation is internal. In order to lay eggs the females burrow into the soil and construct a special chamber. Here the young hatch before the onset of the dry season. These young, but not usually the adults, survive through the dry season: they construct 'moulting chambers' beneath the soil, and in these are protected from desiccation. After several moults they emerge at the start of the wet season as adults or subadults. This sequence is typical of a number of species,

Fig. 7.10. Diagram to show convergent organisation of the leg joints in a diplopod and a chilopod, both adapted for fast running and not for powerful movements. (a) The diplopod *Polyxenus lagurus*; (b) the chilopod *Lithobius forficatus*. In each, the coxa is greatly widened into a plate, and is set into a large area of arthrodial membrane. The only firm union between the coxa and the more distal parts of the leg is the Y-shaped sclerotisation marked in black, which crosses the arthrodial membrane. This allows the angle of swing of the leg to be altered when running. Redrawn from Manton (1958b).

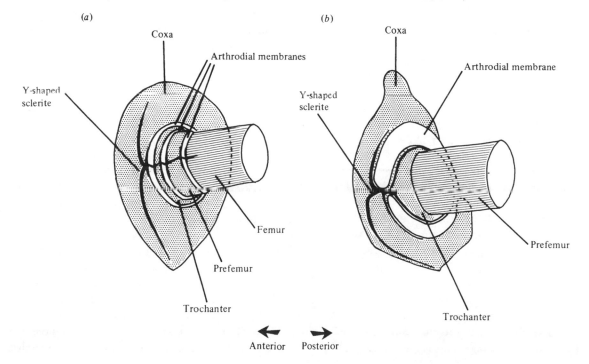

(a)

(b)

but contrasts with life cycles in temperate zones, where it is normally the adults which overwinter. The ability of the juvenile stages of tropical species to build moulting chambers is probably, therefore, the most important factor allowing particular species to survive in areas with marked dry seasons. Distribution of various species in the temperate regions is also related to reproductive strategies. Dunger & Steinmetzger (1981) found that moist regions contained in the main semelparous species – those whose individuals breed once and then die. In drier regions most species were interoparous, i.e. individuals bred more than once. This may be explained in terms of juvenile survivorship. In moist regions a semelparous species can maintain a population because few juveniles die from desiccation stress. In drier regions many juveniles are killed by desiccation stress, and it is necessary for the adults to be able to reproduce several times to ensure continuation of the population. The fact that juveniles are actually less resistant to desiccation than the adults reinforces the conclusion that the general behaviour and life cycles of millipedes are probably just as important in their survival as physiological factors governing rates of water loss.

7.3 The origins of onychophorans and myriapods

Any coherent theory of the origin of the myriapods and their present-day occupation of cryptozoic terrestrial niches must take into account several points: the anatomy and habitats of the earliest myriapods, the Archipolypoda of the Palaeozoic; the position of the Onychophora, and of the fossil *Aysheaia*; the relationships, if any, between the Pauropoda, Symphyla, Chilopoda and Diplopoda; and the relationship of the four myriapod groups to the hexapods. Discussion of each of these topics must to a very large degree be speculative, but some attempt at such speculation must nevertheless be made.

The morphology of the Archipolypoda has been reviewed by Kraus (1974). Comparison of forms from the Upper Carboniferous with present-day species suggests that in the Palaeozoic, eyes were generally larger and were composite, many species had long and complex spines, and the body sternites and tergites were free, not fused. The large compound eyes, with up to 1000 units, contrast markedly with those of modern diplopods in which the maximum number of ocelli is 45. The modern reduction is probably to be associated with the adoption of a cryptozoic habit (Spies, 1981). The structure of forms from periods earlier than the Carboniferous is not so well known, but Størmer (1976) has described a centipede-like form, *Eoarthropleura devonica*, from the Lower Devonian. This had lateral keels on the tergites, as in many later forms. It lived in sheltered, vegetation-rich swamps, probably in brackish water. From the characteristics of Palaeozoic forms, both Kraus and Størmer have concluded that they were not burrowers but lived above ground, feeding on plant material. The tendency for myriapods to become cryptozoic may therefore have been a later, secondary development, possibly related to some change such as the emergence

of insects as the dominant terrestrial animals. If this is so, we must seek a very different origin for the myriapods to the origin postulated in Chapter 6 for many of the terrestrial chelicerates: the movement from water to land seems, in the case of the myriapods, not to have been concerned initially with colonisation of cryptozoic niches, but with movement following the evolution of large land plants suitable for food. However, it should be noted that Manton (1980) emphasised two basic differences between the Palaeozoic forms and those of the present day. She pointed out that the lack of spiracles and the different organisation of the sclerites may suggest lack of any close relationship with modern myriapods.

It has been suggested that several modern groups of diplopods had already become differentiated by the late Carboniferous (Kraus, 1974) and the earliest fossil myriapods date from as far back as the Upper Silurian (Hoffman, 1969). Myriapod ancestors therefore appeared on land before the great radiation of terrestrial plants, which took place in the Devonian and Carboniferous. True forests almost certainly did not arise until the Carboniferous, so that it is unlikely that the Onychophora could have lived on land before then: as described above, the Onychophora are specialised for life in forests, and are highly dependent upon the microclimates found there. Terrestrial Onychophora are therefore unlikely to have given rise to the first terrestrial myriapods. Rather they should be regarded as a 'grade' of organisation showing possible ways in which forms intermediate between soft-bodied annelid types and hard-cuticled arthropods could have evolved. If this is the case, many of the characters found in the Onychophora, such as the tracheae and the haemocoel, must have developed independently from those of the true arthropods. Major evidence which might militate against such reasoning would be the discovery of undoubted marine ancestors of the Onychophora in the early Palaeozoic: hence the interest in *Aysheaia*. Although the structure of this fossil is now quite well known (Whittington, 1978), the resemblance to present-day Onychophora is very possibly entirely superficial: the phylogenetically 'important' characters of Onychophora are, after all, almost entirely internal. For the present, then, it seems both safest and most sensible to regard the onychophorans not as arthropods, not as ancestors of the myriapods, but as a separate phylum in their own right.

There is still considerable discussion as to whether the Pauropoda, Symphyla, Chilopoda and Diplopoda form a natural group (see, e.g. Dohle, 1974). Both Manton (1977) and Demange (1967) have concluded that a natural grouping does exist, but for entirely different reasons. Manton has investigated the functional morphology in relation to movement in great detail, and claimed that all the groups apart from the diplopods have simple segments, and that the diplo-segments of diplopods are purely an adaptation to burrowing. Demange (1967) has examined the comparative anatomy of muscular systems in diplopods and chilopods, and their morphogenesis. From these studies he concluded that the tergites of chilopods are arranged in blocks of two – a macro- and a microtergite, linked together by a set of muscles. The underlying segments to a large degree are effectively equivalent to the diplosegments of the diplopods, although not used for the same mechanical purposes. The 'anomaly' of the seventh and eighth segments of chilopods, where two macrotergites are juxtaposed, is explained by the suppression, during development, of a microtergite. In diplopods a similar transition point occurs posterior to the fourth diplosegment, since it is only in diplosegments five and beyond that there are two pairs of legs, and the repugnatorial glands. This is taken as evidence that the first eight metameres of chilopods and diplopods form a specific tagma, the thorax. The organisation of both groups may, therefore, be more similar than has been thought. Manton (1974) did not accept Demange's theory, but produced no detailed critique of it, and for those who are not myriapod specialists the position is somewhat bewildering, with important concepts appearing to rest upon the minutiae of anatomical investigations. The point to emerge, however, is that the organisation of chilopods and diplopods is very similar, and some close link between the two groups must be admitted. Recent studies which show the similarities in organic composition of the haemolymph of one species of diplopod and one chilopod may add further evidence (Sundara Rajulu, 1974), although this must be very tentative until further species are examined. The relationship may also be extended to the pauropods and symphylans, although there is less evidence. The presence of the unique organs of Tömösvary in Symphyla, Chilopoda and Diplopoda (see Fig. 7.7), for instance, may be taken to indicate a relationship between these groups. Organs with similar fine structure, called the 'pseudoculus', occur in Pauropoda and in some of the apterygote insects, the Protura. In other apterygotes, the Collembola, organs found in a similar position on the head are called 'postantennal organs', but again they have a comparable organisation. Some authors would argue from the morphological similarities of these organs that some homology between the myriapods and the primitive insects is indicated (François, 1969). Others (e.g. Haupt, 1973, 1980) are more cautious, pointing out that the structure may reflect above all the function of the organ, and that even in Crustacea there are organs which show some similarities. As in many other areas of biology, it is difficult to be sure that convergent or parallel evolution has not occurred. Nevertheless, in this case the fine structure is strikingly similar in many details in members of the various groups, and this adds to other arguments which suggest that the four myriapod classes are related and should be regarded as a natural group. It may also support suggestions of the possible relationship of the myriapod groups to the hexapods, or at least to the apterygote hexapods. A discussion of this is more appropriate in the next chapter, and here it may be sufficient to suggest that there appear at least some reasons to suppose that myriapods and insects arose from the same or similar ancestors. As far back as the Carboniferous, and even in the Devonian, numerous animals with hard exoskeletons evolved from aquatic forms, became air breathing, and lived on terrestrial vegetation (Størmer, 1976). From these varied ancestral stocks arose the modern Onychophora, the Myriapoda and the Insecta. Because of a number of adaptations to terrestrial existence arising in the hexapods, it is these forms which have eventually come to make up at least 90% of the total number of present-day terrestrial animal species.

Hexapods

'. . . whether multitudes of those other little creatures that are found to inhabit the Water for some time, do not, at certain times, take wing and fly into the Air, others dive and hide themselves in the Earth, and so contribute to the increase both of the one and the other Element.'

From 'Of the water-insect or gnat' in *Micrographia* by Robert Hooke, 1665, Royal Society, London.

At the present time, hexapods (i.e. apterygote and pterygote insects) make up approximately 70% of extant animal species. They form an essentially terrestrial group but have recolonised fresh water very successfully, and marine environments to some extent. Yet their origins are far from clear, and discussion of their possible ancestors has occupied many zoologists over the past century. Much of this discussion hinges upon consideration of the apterygote orders, and their relationship to the pterygotes. In order to be able to place the terrestrial adaptations of the pterygotes in context, the apterygotes will therefore be considered first. A brief description will allow the possible relationships of myriapods, apterygotes and pterygotes to be reviewed before proceeding with an analysis of some of the ways in which pterygotes are adapted to life on land. A brief classification of hexapods is given in Table 8.1.

8.1 The apterygotes

It is generally accepted that there are four present-day orders of wingless insects, although some authorities raise these to the rank of classes, and some would divide the Thysanura. All the apterygotes are thought to be primitively wingless. The general biology of the Thysanura, Diplura, Protura and Collembola is described in turn.

8.1.1 The Thysanura

The bristle-tails and silverfish can be recognised by their two many-segmented cerci and a segmented median posterior process, which give them their characteristic 'bristle' tails. They also have many-segmented antennae, but only the basal segment of these has its own intrinsic

muscles, in contrast to the Diplura. The mouthparts are not enclosed in the head capsule, and are not highly specialised. Some authorities (e.g. Hennig, 1969, 1981; Kristensen, 1975, 1981) would split the order into two, because the members of the superfamily Machiloidea (bristle-tails such as *Petrobius*) differ in a number of ways from those of the Lepismatoidea (silverfish such as *Lepisma*). The lepismatoids have many specialised features in common with the pterygotes (see Richards & Davies, 1977), whereas the machiloids are somewhat more primitive.

Machiloidea (Archaeognatha)

Although members of this group are found in a variety of habitats, even including dry forests, the best known are marine supra-littoral forms such as those in the genus *Petrobius*. *Petrobius brevistylis*, for example, lives in rock crevices above high water mark, where the accompanying fauna is mainly of terrestrial origin: the eggs of *P. brevistylis* are eaten by beetle larvae, while adults are preyed upon by spiders such as *Amaurobius* and *Zygiella* (Delany, 1959). *Petrobius maritimus*, while generally living in similar habitats, is also found some distance inland where prevailing winds cause maritime conditions to extend (Davies & Richardson, 1970). The relationship to the terrestrial, rather than the marine, environment is also reflected in haemolymph composition: the osmotic pressure of the haemolymph in *P. maritimus* is only 420 mOsm, less than half the value for normal sea water (Lockwood & Croghan, 1959). The haemolymph has a relatively primitive composition compared to most pterygotes, since sodium, potassium and chloride almost entirely account for the osmotic pressure (see Table 8.2). This composition is presumably regulated by the maxillary (labial) kidneys which have a sacculus, labyrinth and excretory canal similar to those of other arthropods (see Gabe, 1967a). The fine structure of these organs suggests the normal processes of filtration followed by reabsorption (Fain-Maurel & Cassier, 1971): the sacculus is lined by podocytes, and the cells of the labyrinth and excretory canal have extensive basal infoldings associated with mitochondria.

The excretory complex normally associated with insects, i.e. Malpighian tubules and a rectal reabsorbing system, has been investigated by Gabe, Cassier & Fain-Maurel (1973). In *Petrobius maritimus* the Malpighian tubules are small, and the cells, while they have some microvilli, do not have a typical brush border. Alkaline phosphatase, normally present in pterygote tubules, was not found. In the hindgut there are longitudinal folds, but no rectal papillae, and it is suggested that the system of Malpighian tubules and hindgut is not as important in salt and water balance as are the labial kidneys. Excretory spherules are found in the cells of the midgut, some of them accumulating calcium. These spherules are liberated when the cells degenerate (Fain-Maurel, Cassier & Alibert, 1973). No concretions are found in the Malpighian tubules; this may indicate that the latter are not as important in nitrogen excretion as is the epithelium of the gut – a feature to be noted in other apterygote groups. Thus it may be that in the mechanisms involving regulation of salts, water and nitrogenous end products, the machiloids differ radically from the pterygotes. As will be discussed later, the change from emphasis on labial kidneys to the use of Malpighian tubules may originally have been related to increasing size.

In the supra-littoral habitat, water supply must sometimes be critical, and it is perhaps not surprising that machilids have specialised organs for water absorption. They do not drink, but use abdominal vesicles which they can evert and apply to the substratum. *Petrobius brevistylis*, for instance, has 22 vesicles (Houlihan, 1976). When the animals have been dehydrated and are then placed on a damp substratum, these vesicles are everted and water is absorbed into them. In this way the normal weight can be regained in a short time. Water is generally taken up only from fluids which are hypo-osmotic to the haemolymph, but an active process is nevertheless involved (Houlihan, 1976), and the fine structure of the sacs shows a typical transporting epithelium (Bitsch & Palévody, 1973). The oxygen consumption of *P. brevistylis* rises while water uptake occurs, due both to the extra use of abdominal muscles to maintain an altered stance and to the employment of extra energy in active transport (Houlihan, 1977). There is no evidence that *P. brevistylis* can take up water vapour from the air.

In considering the relationships of the apterygotes, a great deal of evidence has been based upon feeding and locomotory mechanisms. *Petrobius* is probably an omnivore, but appears to feed largely on unicellular algae, and other vegetable matter such as lichens and mosses (Davies & Richardson, 1970; Manton, 1972). Its mandibles are used to scratch and not to bite: they project well below the head capsule and operate by rotating on a dorso-ventral axis. The mandibular muscles insert on the endoskeleton of the head, a system of tentorial apodemes which is equivalent to the tentorium of pterygotes, but which is here partly mobile (Manton, 1964). While the

Table 8.1. *Classification of the Hexapods referred to in the text (from Richards & Davies, 1977; Kristensen, 1975)*

Taxon	Genera
Apterygotes	
Order Thysanura	
Superfamily Machiloidea (= Archaeognatha)	*Machilis, Petrobius*
Superfamily Lepismatoidea (= Zygentoma)	*Thermobia, Lepisma*
Order Diplura	*Campodea, Heterojapyx*
Order Protura	*Acerentomon, Eosentomon*
Order Collembola	
Suborder Arthropleona	*Podura, Anurida, Folsomia, Isotoma*
Suborder Symphypleona	*Sminthurus, Bourletiella*
Pterygotes (selected orders only)	
Exopterygotes	
Palaeopteran orders	
Order Odonata	*Aeshna*
Orthopteroid orders	
Order Orthoptera	*Schistocerca, Locusta*
Order Phasmida	*Dixippus (Carausius)*
Order Dictyoptera	*Periplaneta, Arenivaga* ·
Hemipteroid orders	
Order Siphunculata	*Haematopinus*
Order Hemiptera	*Rhodnius*
Endopterygotes	
Order Coleoptera	*Eleodes, Tenebrio*
Panorpoid orders	
Order Diptera	*Calliphora, Sarcophaga, Glossina*
Order Lepidoptera	*Manduca, Papilio*
Order Hymenoptera	*Bombus, Perga*

mandibles operate, saliva is poured on to the food, and the resultant fluid mass is sucked up into the oesophagus. The organisation of the mandibles is relatively unspecialised, and of a type which could have given rise to the entognathous mouthparts of Collembola and Diplura, and also to the biting ability of the pterygotes (Manton, 1964).

Petrobius can walk with a gait similar to that of pterygotes, and can also jump, in which case legs on opposite sides of the body move in phase (Manton, 1972),

indicating the flexibility of the neural mechanisms co-ordinating leg movements. When jumping, *P. maritimus* can reach a speed of 210 mm/s. There is also a 'high jumping' escape reaction, in which the abdominal muscles, as well as the legs, provide motive power. The rapid tail beat of this jump occurs too quickly to be an ordinary muscle contraction: two abdominal muscle systems contract in opposition, and when one relaxes, the stored energy is suddenly released (Evans, 1975). During normal locomotion, however, the mechanism

Table 8.2. *Composition of the haemolymph of some apterygotes*

Species	Osmotic pressure (mOsm)	Na$^+$	K$^+$	Cl$^-$	Amino acids (mM/litre)	Reference
		(mM/litre)				
Thysanura Machiloidea						
Petrobius maritimus	421	208	5.8	194	–	Lockwood & Croghan, 1959
Petrobius brevistylis	463	–	–	–	–	Houlihan, 1976
Thysanura Lepismatoidea						
Thermobia domestica	–	219	15.7	268	14.3	Okasha, 1973
Collembola						
Isotoma viridis						
in fresh water	292	–	–	–	–	Weigmann, 1973
in 50% sea water (OP 372 mOsm)	388					
Folsomia sexoculata						
in 50% sea water (OP 372 mOsm)	394	–	–	–	–	Weigmann, 1973
Archisotoma pulchella						
in fresh water	510	–	–	–	–	Weigmann, 1973
in sea water (OP 898 mOsm)	911					

OP = osmotic pressure.

Fig. 8.1. A forest-dwelling machiloid thysanuran, *Nesomachilis* sp., from New Zealand. This mottled brown specimen is excellently camouflaged on the forest floor. Body length approximately 1.5 cm.

by which the legs articulate with the body is necessarily an important consideration. There is, in *Petrobius,* only one point of articulation: the anterior region of the coxa articulates with a pleurite, which in turn is supported by the tergite. Manton (1972) has emphasised that this type of articulation differs from that of other arthropods. The significance of this point will be discussed later.

Besides the maritime species of machilid, there are many species which normally live inland (Fig. 8.1). Although one species, *Dilta hibernica,* is found mainly associated with moss (Sturm, 1955), most live in rocky areas, or, like *Machiloides delanyi,* under rocks in natural forest. The latter habitat, although drier than the litter layer, is nonetheless a stable environment (Heeg, 1969). The distribution of *M. delanyi* in South Africa coincides with that of indigenous forests and their remnants: this is the area where at least 1 cm of rain falls each month. The rate of water loss in *M. delanyi* is low (Heeg, 1967a; Table 8.3), but the barrier to water loss is not understood, since neither abrasion of the cuticle nor treatment with carbon dioxide increases it, and there is no evidence of a critical temperature. There is, therefore, no evidence for the presence of epicuticular wax layers as there is in the lepismatoids (see p. 150). The rate of loss increases markedly after death (Fig. 8.2) suggesting an active process of water retention. As in *Petrobius,* the coxal vesicles in both *Machiloides delanyi* and *Machilis helleri* take up water from damp substrates (Heeg, 1967a; Weyda, 1974). *Machiloides delanyi* avoids light and high temperatures, and when faced with a range of humidities, chooses an RH of 70–85% (Heeg, 1967b).

It is in inland species also that some of the complex behaviour, as exemplified in reproduction, has been best described (Sturm, 1955). In *Machilis germanica,* for instance, the sequence of procedures is as follows. After a short initial series of 'display' movements, the male secretes a thread from the genital appendages and attaches it to the ground. He then draws out the other end of the thread, meanwhile attaching sperm to it. The abdomen and the thread are then raised and the female is manoeuvred into the angle between them. The female finally takes up sperm into her ovipositors (see Fig. 8.3). The procedure is reminiscent of that shown by scutigeromorph centipedes, in which it is thought to have evolved in stages from the deposition of sperm drops in humid soil (see p. 137).

Lepismatoidea (Zygentoma)

The silverfish (*Lepisma*) and the firebrat (*Thermobia*) are, unlike the machiloids, well known from their presence in buildings. Some of their general ecology in human habitations has been described by Sahrhage (1954). They are believed to have their origin in deserts or semi-deserts, but little is known about their natural role in these habitats. One lepismatoid which is known as a desert form is *Ctenolepisma terebrans,* living at the base of dunes in the Namib desert of southwest Africa (Edney, 1971a). These forms are therefore very interesting indeed in relation to the movement of primitive insects out of the cryptozoic niche. Lepismatoids have to a great extent completed this movement, and can provide evidence about the mechanisms involved.

In Europe, *Lepisma* and *Thermobia* thrive at temperatures well above the normal ambient, the optimum being approximately 37 °C. The development, hatching and maturation of *Thermobia domestica* all depend upon critical conditions of humidity and temperature (Sweetman, 1938). The optimum conditions for development (see Fig. 8.4), i.e. about 37 °C and 84% RH, are similar but not identical to the conditions under which the adults are induced to lay eggs; but *T. domestica* can maintain a flourishing population when the RH is as low as 35–40%, providing that it rises for short periods to 60–70% (Sahrhage, 1954).

Table 8.3. *Rate of water loss in some apterygotes*

| Species | Temperature (°C) | Wet weight (mg) | Rate of water loss | | Reference |
			(% body wt/ hour)	(μg cm^{-2} hour^{-1} (mmHg)$^{-1}$)	
Thysanura Machiloidea					
Machiloides delanyi	–	–	0.28–0.38	–	Heeg, 1967a
Thysanura Lepismatoidea					
Thermobia domestica	20	30	0.23–1.47	–	Beament, Noble-Nesbitt & Watson, 1964
Ctenolepisma terebrans	27	8.7	0.23	0.68	Edney, 1971a
Ctenolepisma longicaudata	–	–	0.18–0.26	–	Heeg, 1967a
Collembola					
Seira domestica	20	0.45	0.61	3.67	Vannier, 1973b
Allacma fusca	23	0.08	0.73	5.4	Vannier, 1973a
Orchesella villosa	18	1.6	11–15	–	Vannier & Verhoef, 1978
Tomocerus minor	20	1.16	64	517	Vannier, 1977
Tomocerus problematicus	20	1.25	89	701	Vannier, 1977

All figures refer to living specimens except those for *Thermobia domestica.*

These humidity effects introduce the subject of water economy, and the studies which have shown that lepismatoids can take up water from air which is not saturated with water vapour. It has been shown by Beament, Noble-Nesbitt & Watson (1964) that the permeability to water of the epidermis of *T. domestica* is very low – of the order of 0.004 to 0.026 mg (mmHg)$^{-1}$ hour^{-1} for a 30 mg animal. This low permeability is probably conferred by an epicuticular grease, protected

externally by an external lamina of some sort. In *Ctenolepisma longicaudata* there is no critical temperature, but the rate of water loss rises after death as it does in machiloids, suggesting an active mechanism for the retention of water (Heeg, 1967a; Fig. 8.2). Added to the low rate at which water is lost in lepismatoids is the capacity of living animals to absorb water from relative humidities well below saturation (e.g. Heeg, 1967a; Edney, 1971a). It has now been established (Noble-Nesbitt, 1970, 1975) that in *T. domestica* this uptake occurs in the rectum, and not over the general body surface. The major factor governing water uptake appears to be the volume of the body and not its water content (Okasha, 1971, 1972), and the rates therefore alter with nutritional state and with different stages in the moult cycle. The mechanisms of water absorption have not been elucidated, but the structure of the hindgut is informative (Noirot & Noirot-Timothée, 1971a,b). The anterior regions of the hindgut do not contain rectal papillae as in the pterygotes, but the cells have narrow extracellular channels between them and may be responsible for the first stage in the drying of faeces. The posterior part of the hindgut, the anal sac, has a much folded epithelium in which a thin cuticle overlies cellular lamellae containing enormous numbers of mitochondria. It seems likely that this anal sac is responsible

Fig. 8.2. Water loss in two species of Thysanura. Circles show the machiloid *Machiloides delanyi*. Triangles show the lepismatoid *Ctenolepisma longicaudata*. Filled symbols represent dead animals, while open symbols show live ones. Loss rates are very much faster in dead specimens, showing the importance of active (but unidentified) mechanisms of retaining water. From Heeg (1967a).

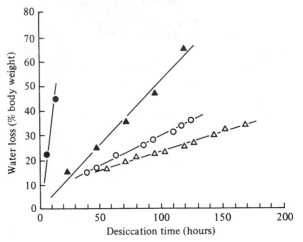

Fig. 8.3. The mating position in the thysanuran *Machilis germanica*. (a) Shows the positions of the two individuals from above. (b) Shows the same individuals from the side. The male attaches the sperm droplets to the thread, and then manoeuvres the female into the angle between himself and the thread. The female then takes up the sperm into her ovipositors. In the diagram the male is stippled and the female is white. Redrawn from Sturm (1955).

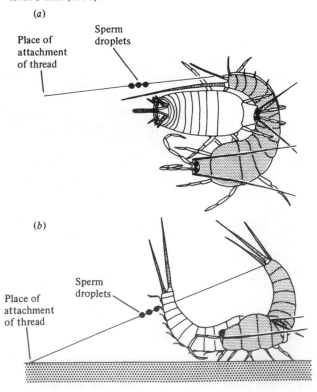

Fig. 8.4. The responses of the lepismatoid thysanuran *Thermobia domestica* to temperature and humidity. (a) The effects on maturation of the nymphs. Heavy stipple shows up to 100% maturation; light stipple shows up to 75% maturation; outside this zone, no nymphs matured. (b) The oviposition responses of the adults. Heavy stipple shows up to 95 eggs laid per adult; light stipple shows up to 33 eggs laid per adult; outside this zone, no eggs were laid. Redrawn after Sweetman (1938).

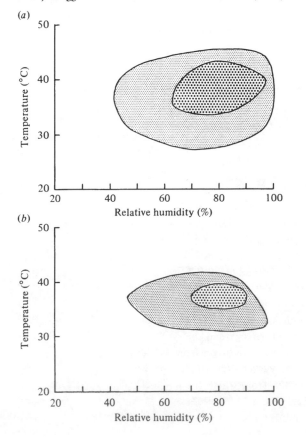

for the second stage in the drying of faeces, and for absorbing water vapour from the air. Noble-Nesbitt (1978) has suggested that some organic molecule which absorbs water vapour from the air, such as a mucopolysaccharide, may be involved in the uptake process.

The haemolymph of *Thermobia domestica*, like that of *Petrobius*, has a composition mainly accounted for by sodium, potassium and chloride, but it also contains some amino acids (see Table 8.2). During desiccation the water content of the whole animal falls, but the composition of the haemolymph remains the same (Okasha, 1973). Since relatively little sodium is eliminated in the faecal pellets, and the total sodium in the body remains approximately constant during desiccation, it follows that some salts must be transferred from the haemolymph and stored temporarily in another tissue. In pterygotes this storage occurs in the fat body, and it seems likely that this tissue has the same function in *Thermobia*. The importance of the labial kidneys in the Lepismatoidea has not been assessed, but their structure indicates that, while they are less well developed than in the Machiloidea (Gabe, 1967a), they are nevertheless osmoregulatory organs. In *Lepisma* (Haupt, 1969a) a typical podocyte-lined sacculus is present, and the labyrinth has cells with basal infoldings and mitochondria.

To summarise water balance in the lepismatoids, it can be said that they have efficient mechanisms for uptake of water and for prevention of water loss. They have means of regulating haemolymph volume and composition which resemble those found in the pterygotes. As will be seen, this situation contrasts strongly with that in most other apterygotes, and since the lepismatoids are really the only apterygotes to have moved out of the cryptozoic habitat, it seems reasonable to invoke causality to explain the difference: above all else, it is the mechanisms of water economy which have allowed the lepismatoids to colonise environments as harsh as deserts.

Turning now to the factors which might elucidate the relationships of lepismatoids, we must consider the mechanisms of feeding and of locomotion. Unlike the machiloid *Petrobius*, *Ctenolepisma ciliata* can eat dry grain, and employs a strong biting action with its mandibles. Such an adduction–abduction motion could have been derived from the rolling action shown in *Petrobius* by the development of a long hinge (Manton, 1964). In connection with these hinged mandibles, the endoskeleton of the head forms a fused unit in *Ctenolepisma*: it seems that movement of the tentorial apodemes is associated with the capacity to protrude the mouthparts, and this does not occur with mandibles which are hinged at their base and bite against each other. Manton (1964) viewed the development of transverse-biting mandibles in *Ctenolepisma* – and in the pterygotes – as essentially different from that in myriapods such as the Symphyla. She therefore regarded the symphylan mandibles as having a separate origin from those of the pterygotes and apterygotes. Nevertheless, the anterior tentorial apodemes, from which the mandibular muscles operate, are regarded as structures which are homologous in myriapods and apterygotes, and there seems little objec-

tion to the suggestion that, although a symphylan jaw mechanism could not have given rise to an apterygote one, some common ancestor with one pair of mobile tentorial apodemes could have produced, on the one hand, the symphylans with their fairly free mandibular articulation, and on the other, the apterygotes with two pairs of tentorial apodemes and hinged mandibles.

The movement of Lepismatoidea has been investigated by Manton (1972) and compared with that of the Machiloidea. Unstable gaits have been developed by *Lepisma* as well as by *Petrobius*, and the animals can, by using these essentially 'jumping' gaits, move very rapidly. There is, however, no bipedal jumping in *Lepisma*. The legs in *Lepisma* have coxae which are always backwardly directed, and work horizontally in a deep groove between the sternites and the paratergal lobe. The stepping movement is unique in that most movement involves only the distal part of the leg. It is the rapid stepping of this telopod which allows an increase in speed: the insect may be thought of as progressing in a series of rapid one-legged hops, and the system is evidently more specialised than that of the Machiloidea. The operation of the legs almost entirely beneath the body presumably aids locomotion in enclosed places.

Some of the behaviour of *Thermobia* has been investigated in relation to mating (Sweetman, 1938). As in *Petrobius*, the male displays to the female. He touches the female's head with his mouthparts, curls the tip of his abdomen up and turns sideways to the female. This and a variety of movements are repeated many times before the male deposits a spermatophore on the substratum in front of the female, and touches her. She then straddles the spermatophore and consumes it by biting it into fragments: a process strongly reminiscent of that found in the Symphyla.

A number of different sense organs must be employed in this complicated reproductive behaviour. The compound eyes are reduced, however, compared to those of *Petrobius*, and ocelli are for the most part absent in Lepismatoidea. There is no homologue of the post-antennal organ which is present in Protura and Collembola. Tactile receptors and chemoreceptors are abundant (Slifer & Sekhon, 1970), however, and *Ctenolepisma* has many trichobothria – the long slender hairs which respond to air movements. Although these sense organs are better known in arachnids, they are also found in Collembola and Diplura, and in some pterygotes.

8.1.2 The relationships of Thysanura, myriapods and pterygotes

The Thysanura, and particularly the Lepismatoidea, are widely regarded as the most closely related to the pterygotes of all the apterygote orders. Attention has also been drawn at times to the similarities between some of the apterygotes and the Symphyla among the myriapods. It is necessary, therefore, although it must involve mainly speculation, to discuss briefly the current views about the relation of the Thysanura to the myriapods on the one hand, and to the pterygotes on the other.

The essential similarity between the Thysanura and pterygotes is the three-segmented thorax bearing six locomotory limbs, together with an abdomen of eleven segments. Because of these characters, there is no doubt that the two groups are related, but it is more difficult to define the exact nature of the relationship. The view of Imms (1936) was that forms like the thysanurans might have given rise to the pterygotes. Manton (1972), after examining the articulation of the legs and the methods of movement, believed that while pterygotes and apterygotes could have arisen from a common ancestor, they could not have been derived one from another. Indeed, Manton concluded that the structural and functional differences between the individual apterygote orders and between the apterygotes and the pterygotes were so great that all five groups must have evolved independently and in parallel from a soft-bodied lobopod ancestor. Manton's views have been reached almost entirely from consideration of the functional morphology of the mandibles and the systems involved in movement. A further hypothesis has been advanced by Hennig (1965, 1969, 1981) and Kristensen (1975, 1981), in which many more characters have been considered, and some attempt has been made to distinguish between characters which are useful in determining relationships and those which are not. In particular, we may note that the presence of unique characters in one group is not considered good evidence for suggesting that the group is unrelated to some other group: rather the approach is positive, and the specialised characters which are held in common between two groups are used to assess the possibility that they had a common ancestor. The specialised characters are termed synapomorphies, and contrast with primitive characters, or symplesiomorphies, which do not necessarily indicate a close relationship. The difficulty of distinguishing synapomorphies from symplesiomorphies remains, of course. The conclusion of Hennig, using these criteria, is that present-day Thysanura and pterygotes could indeed have been derived from a common *sclerotised* ancestor. Since Manton herself has shown the adaptability of exoskeletal articulations and their attached musculature, it seems reasonable to adopt the theory of Hennig, at least as a working hypothesis. One may note in support of this the suggestion (Manton, 1964) that the jaw mechanism of the thysanuran *Petrobius* is sufficiently generalised to have provided an origin for the pterygote jaws.

Next, one must raise the question of the identity and origins of any ancestor which might have given rise both to the Thysanura and to the pterygotes. There seems little doubt that none of the other three apterygote orders could be invoked, since they must all be regarded as, in some sense, offshoots from the main line of insect evolution. Nevertheless, these three groups must be to some degree related to the supposed ancestor, and they will be treated later. Since the Myriapoda are probably the most closely related to the hexapods of all other arthropods (Manton, 1973), it is natural to turn to them for evidence. Imms (1936) envisaged the derivation of the Protura from the Symphyla, and from them in turn the Thysanura and the pterygotes; but this theory must now be discarded because the Protura, Diplura and

Collembola have all been recognised as rather specialised entognathous groups. The similarities between the Symphyla and the Thysanura remain, however. It is difficult to decide which of these characters indicate relationship (i.e. are synapomorphic), but the number of them is impressive: the cephalic excretory organs are almost identical; both have Malpighian tubules; both have abdominal sacs for water absorption; both have cerci; although most Symphyla are progoneate, some are opisthogoneate and the opisthogoneate condition is found in Thysanura; the mode of sperm transfer and associated behaviour is very similar. Against these points it must be said that the articulation of the legs with the body is different in the two groups: Symphyla have coxo-pleural and coxo-sternal articulations, whereas thysanuran coxae articulate only with the pleurite. Manton (1972) apparently regarded these conditions as inflexible, and believed that it was impossible to conceive of viable functional intermediates, despite the adaptive flexibility she has shown in other systems. Another difference is found in the mandibles and their operation: in Symphyla the mandibles are made up of two segments and there is only one pair of tentorial apodemes to support the mandibular muscles; in Thysanura the mandibles are unjointed and are supported from two pairs of tentorial apodemes. Again, these systems were regarded as inflexible by Manton (1973), and it must be noted that Hennig (1981) regarded most of the similarities between Symphyla and Thysanura as symplesiomorphies. As Kristensen (1975, 1981) has argued, however, it is very difficult to deny the possibility of a functional sclerotised ancestor which could have given rise to both present-day conditions. While not supposing, therefore, that the Symphyla themselves gave rise to the Thysanura, we may conclude that they have probably retained many of the characteristics shown by such an ancestor. In particular, this ancestor is likely to have had many more than three pairs of legs, was a small cryptozoic form, and was sclerotised. The possible origins of such an ancestor will be touched upon later.

8.1.3 The Diplura

The Diplura are mostly small forms, the campodeids having two terminal segmented cerci, the japygids with strong terminal forceps which are used to catch prey (Richards & Davies, 1977). *Heterojapyx soulei* reaches a length of 5 cm. Diplura live in cryptozoic situations where at least some of them are territorial (Pagés, 1967). They are characterised by the possession of intrinsic muscles in each segment of the antenna, allowing the flagella to telescope when exploring the environment (Manton, 1972). They have no eyes, and lack some of the other organs usually associated with insects, such as Malpighian tubules. The labial kidneys show the typical arthropod excretory structures: a sacculus lined by podocytes, and an excretory tubule lined by reabsorbing cells (François, 1972). No information is available about haemolymph composition or its regulation. Water can be taken up by eversible abdominal sacs resembling those of the Thysanura (Drummond,

1953). In *Campodea staphylinus* these vesicles have a specialised transporting epithelium, so that active transport is probably concerned in this uptake of fluid, as in other apterygotes (Eisenbeis, 1976a; Weyda, 1976). The cuticle of the vesicles is thin – only 0.15 μm in places – and lacks endocuticle (Weyda, 1980).

The mouthparts of Diplura are entognathous – that is to say, to a great extent enclosed within the head capsule. The head structures concerned with feeding can easily be derived from a generalised type such as *Petrobius* (Manton, 1964), but here the mandibles are protrusible, and they can reach out and grasp and cut relatively small objects. Yet most Diplura are omnivorous or carnivorous (Manton, 1972; C. Bareth, personal communication), eating particularly insect larvae and other apterygotes. Although it seems difficult to explain how this is achieved by such relatively weak mouthparts, it must be noted that many of the prey species have a soft cuticle.

The Diplura move around in their habitats of soil and leaf litter by using existing crevices, and not by active burrowing. In this they parallel the Symphyla, and like these they also have extremely flexible bodies (Manton, 1972). Some of the Diplura can run rapidly: species of *Campodea* can reach 54 mm/s, even with a body length of only 4.5 mm. Presumably as an adaptation to crevice living, the legs project laterally from the body: the limb base is therefore at least superficially like that of chilopods. In the Diplura the coxae have strong articulations with the sternites, but not with pleurites or tergites.

Reproductive biology of the Diplura has been studied in the Campodeidae by Bareth (1968, 1980). In *Campodea*, sperm transfer is indirect as it is in other apterygotes: the male deposits stalked spermatophores,

very similar to those of Symphyla, without any courtship of, or contact with, the female. The spermatophore of *Campodea remyi* is shown in Fig. 8.5.

Interest has mainly centred upon the Diplura because of their resemblance to the Symphyla (Imms, 1936). As with the Thysanura, there is a long list of similarities. To this list must be added the fact that in both japygid Diplura and Symphyla, the cerci have anal glands which open at the tip; and both Diplura and Symphyla have abdominal styli. There are differences, however: Diplura have no Malpighian tubules, and their mouthparts and leg articulations differ from those of the Symphyla. In general characteristics, though, the number of similarities is striking, and must reinforce the suggestion of a common ancestor for apterygote orders and the Symphyla.

8.1.4 The Protura

The Protura are all minute animals (none is longer than 2.5 mm) found mainly in soil and woodland humus, but also in damp grassland (Nosek, 1973, 1975). They lack a number of insect characters such as antennae, eyes, ocelli, cerci and typical Malpighian tubules. The absence of some of these characters may be due to the small size of the animals, but others have probably been reduced in relation to the cryptozoic habit. The mouthparts are entognathous, but it is probable that this state has been reached independently from that in Diplura and Collembola (Manton, 1964, 1973; but see also Tuxen, 1970, for the opposite view). At least some species eat fungal hyphae. Nothing is known about haemolymph composition and its regulation or about the ultrastructure of the excretory organs. One interesting feature is the presence of sense organs called pseudoculi on the head. These are

Fig. 8.5. Spermatophore of the dipluran *Campodea remyi*. (a) The whole spermatophore showing two sperm bundles enclosed. The sperm droplet is supported by a peduncle raising it off the ground by about 100 μm. (b) One sperm bundle unrolled in saline, showing the spiral filament and attached spermatozoa. Redrawn after Bareth (1968).

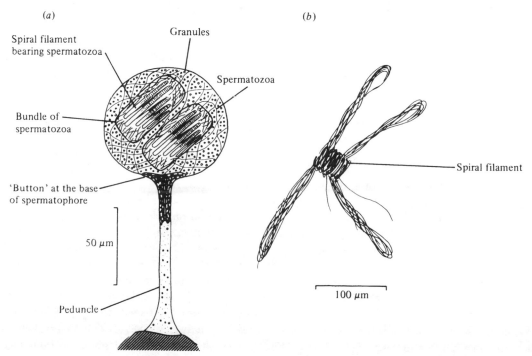

(a)

Spiral filament bearing spermatozoa

Granules

Spermatozoa

Bundle of spermatozoa

'Button' at the base of spermatophore

50 μm

Peduncle

(b)

Spiral filament

100 μm

very like the postantennal organs of Collembola, and the organs of Tömösvary found in the myriapods (François, 1969; Bedini & Tongiorgi, 1971; Haupt, 1972). These organs are not necessarily homologous in all groups, since similar types are found in some of the pterygotes, where they are believed to be chemoreceptors. Their presence in proturans and myriapods may, however, be another suggestion of common ancestry.

Protura do not move rapidly, and often walk using only the second and third pair of legs. The first pair are held out anteriorly and probably act as sense organs (Manton, 1972). This type of behavioural change has been commented upon in chelicerates, but in that case, of course, it still allows hexapodous movement. The explanation for lack of antennae and their replacement by walking legs in Protura is not evident. The walking legs are directed horizontally from the body as in Diplura, but the coxae here articulate both with a sternite and a pleurite. They therefore show yet another alternative from the other apterygotes in this character. Manton (1972, 1973) felt that this was sufficient evidence to show that they have evolved from a soft-bodied lobopod ancestor independently from all other orders; but once again, the general similarity of characters between Protura and the other apterygotes does not support such a contention. In particular, possession of such sclerotised structures as abdominal styli and the pseudoculus suggests a common origin of Protura and other insects from a sclerotised ancestor. Unfortunately nothing is known about any further characteristics such as the reproductive habits of the Protura.

8.1.5 The Collembola

Of all the apterygotes, the Collembola perhaps show the most differences from the pterygotes. In particular, the collembolan abdomen has only six segments instead of the pterygote eleven. The most obvious of the abdominal appendages is the unique forked springing organ or furca on segment 4. This organ can be rapidly extended, producing the well-known jump. According to Manton (1972), this is brought about by an increase in the hydrostatic pressure of the haemolymph, and much of the rest of the body has been modified to withstand the sudden large changes in hydrostatic pressure involved in the jumping mechanism. Christian (1979), however, suggested that hydrostatic pressure does not in fact play a large part in jumping. He was of the opinion that direct muscular action provides the main propulsive force, because punctured animals could still jump effectively.

Collembola are found in a great variety of environments. Many are found in soil, on the soil surface, or in litter. Others live on damp wood or under the bark of trees. A few, such as *Sminthurus*, feed on the leaves of green plants and are therefore well known as agricultural pests. A significant number have also recolonised the marine intertidal zone, where some, such as the common species *Anurida maritima* (Fig. 8.6), have an activity rhythm synchronised with the tide (Foster & Moreton, 1981). This allows the animals to be active at low water, and to retire underground before the tide rises. Excellent

examples of variation in tolerance of salt water are given by Weigmann (1973), who examined collembolan zonation on saltmarshes. Some species, such as *Archisotoma pulchella*, can live only in the salty soils of the *Puccinellia* zone, while others, such as *Isotoma viridis*, can tolerate both salty and salt-free soils. *I. viridis* lives in the *Festuca* zone. In contrast to these two species, most Collembola, e.g. *Isotoma notabilis*, live in salt-free environments, and will not tolerate the presence of salt.

To some extent these different tolerances are reflected in the osmoregulatory capacities of the different species. In fresh water, the species so far investigated have haemolymph with osmotic pressures ranging from 290–510 mOsm (Weigmann, 1973; Table 8.2). The marine species *Archisotoma pulchella* is unique among arthropods in regulating hyperosmotically at all concentrations below 100% sea water (Fig. 8.7). This property presumably represents the complete adaptation of a primitively terrestrial form to the marine environment. The brackish water species *Isotoma viridis* regulates much more in the fashion of normal brackish water arthropods. Unfortunately no investigations have yet been made on exclusively freshwater collembolans which might be considered to be living in environments more similar to those of the primitive ancestors. The colonisation of marine habitats by Collembola is presumably to be compared with the similar trends found in the Thysanura Machiloidea, and there have been no suggestions that either is anything but a secondary movement by terrestrial stock.

Terrestrial species of Collembola are greatly dependent upon environmental humidity and temperature. In habitats where seasonal drought occurs, the adults of many species die and the species survives only in the egg state (Poinsot-Balaguer, 1976). A few species, such as *Folsomides variabilis* and *Brachystomella parva*, can apparently withstand massive dehydration, and can survive by the process of anhydrobiosis or cryptobiosis. Such a phenomenon is for the most part restricted to some pterygote larvae and a number of small phyla already discussed. On the other hand, it must be emphasised that not all collembolan eggs are resistant to desiccation: many require humidities of 95–100% if they are to survive and hatch (Choudhuri & Bhattacharyya, 1975). Water loss from the adults also varies greatly between species. Some comparative figures are given in Table 8.3. Of those species which live in humid microhabitats, *Tomocerus minor*, for instance, loses approximately 0.03 mg min^{-1} mg^{-1} dry weight, when exposed to 0% RH and 18 °C, whereas *Orchesella villosa* of the same size loses only 0.01 mg min^{-1} mg^{-1} (Vannier & Verhoef, 1978). Species which live in desiccating conditions, however, lose water at only approximately one-hundredth of this rate although they are smaller. One example is the symphypleonan *Allacma fusca*. This species, when subjected to similar conditions of desiccation as those given above, loses water at a low but constant rate (Vannier, 1973a; Betsch & Vannier, 1977). In this it contrasts with most other small arthropods, in which water loss is usually initially high and then declines. Fig. 8.8 shows the response to desiccation, from which

Fig. 8.6. A common marine collembolan, *Anurida maritima*, on the intertidal mud of a saltmarsh. This species has an intrinsic activity rhythm which ensures that it emerges to feed at low tide, and hides away in crevices before the tide covers the marsh. Maximum body length approximately 3 mm.

Fig. 8.7. Osmoregulation in two species of saltmarsh Collembola. (*a*) Shows the brackish water species *Isotoma viridis*. (*b*) Shows the marine species *Archisotoma pulchella*. The left-hand diagrams show osmoregulation curves, i.e. the osmotic pressure (as depression of freezing point, Δt) of the haemolymph plotted against that of the medium. Both species regulate hyperosmotically, but *A. pulchella* is hyperosmotic at all concentrations less than that of sea water. The right-hand diagrams show rates of change of the freezing point of the haemolymph when animals are placed in different salinities. The arrows indicate salinities to which the animals were moved. Rates of change in both species are fairly slow, considering their size. Redrawn from Weigmann (1973).

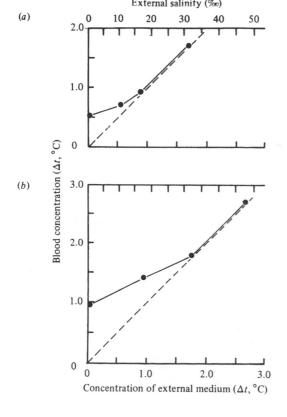

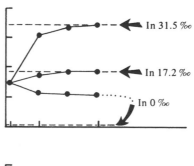

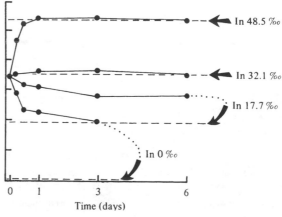

it is clear that the ability to maintain a low rate of transpiration develops only after the first moult. In the juvenile phase which follows, the cuticle thickens, tracheae ramify further through the body, and an epi-cuticular film appears, containing lipoproteins. Experimental removal of this film vastly increased the rate of transpiration. Comparison of further species from dry habitats with those from moist leaf litter has shown that the ability to maintain a low rate of water loss is found only in those from desiccating environments (Verhoef & Witteveen, 1980; Verhoef, 1981). Another example of one of these xeric collembolans is the arthropleuran *Seira domestica*. This species has a rate of water loss of only 0.1 μg min^{-1} mg^{-1} dry weight (Vannier, 1973*b*). The rate can be maintained except during the moult, and even this is very rapid and is concluded in 4 hours.

The factors which affect the rate of transpiration in a cryptozoic species, *Podura aquatica*, have been

described by Noble-Nesbitt (1963*a,b*). In this species the cuticle over most of the body is thin and consists only of endocuticle and epicuticle. The epicuticle is reinforced by layers of wax and cement on numerous tubercles, and these confer upon the animal the property of a hydrofuge cuticle. This is most important in that it maintains an air layer around the body even under water, yet allows gas exchange to continue across the thin areas of cuticle. The wax also slightly reduces the rate of transpiration, but it appears that this is incidental to its primary function. This interesting point suggests that in the primitive insects, epicuticular waxes were developed to produce a hydrofuge surface: the later development in pterygote insects and a few xeric collembolans of an impermeable layer which cut down transpiration was probably a secondary function.

Given that most collembolan cuticles are relatively permeable to water, the ability to osmoregulate suggests

Fig. 8.8. Transpiration from the symphypleonan collembolan *Allacma fusca*. (*a*) Shows a first stage larva (wet weight 52 μg). (*b*) Shows a second stage larva (wet weight 83 μg). (*c*) Shows an adult female (wet weight 1.06 mg). Stipple indicates the water loss due to production of faeces, and the end of the defaecation period is indicated by the open arrow. The closed arrows mark

the end of the 'plateau' period, during which weight loss is kept at a fairly constant level. The first stage larva has no capacity to regulate its rate of water loss, but both the second stage larva and the adult maintain a very steady rate of weight loss until a critical point is reached (i.e. the solid arrows), when the rate rises. Redrawn after Betsch & Vannier (1977).

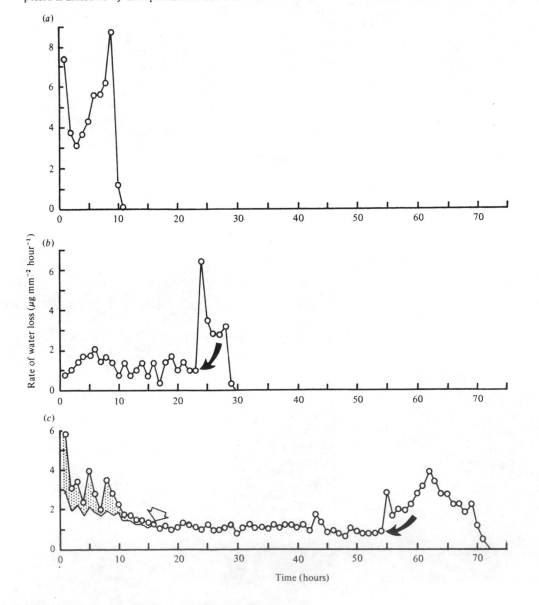

the presence of organs which can efficiently control haemolymph composition. Three such systems have been described, although the relative importance of each and the mechanisms by which they are regulated are unknown. First, there are the labial kidneys which have been described in *Onychiurus* by Altner (1968) and in *Orchesella* by Verhoef *et al.* (1979). They have the typical sacculus, made up of podocytes, and a tubule lined by reabsorptive cells. These kidneys are very like those of other apterygotes, and presumably work on the principle of filtration followed by reabsorption. Nothing is known about the composition of the urine produced. Second, there is a reabsorptive epithelium in the rectum, although this is not developed into rectal pads as it is in pterygotes (Verhoef *et al.*, 1979). Third, there is the organ known as the ventral tube – an appendage of the first abdominal segment. At the tip of this tube are vesicles which can be everted by hydrostatic pressure, much as are the coxal vesicles of Machiloidea, Diplura, some myriapods and Onychophora. Indeed, since the ventral tube is believed to be homologous with the coxae of abdominal appendages (Denis, 1949; Eisenbeis, 1976*b*), the vesicles may be considered to be homologous with those of Diplura and Machiloidea. The vesicles in Collembola are concerned with the uptake of water and salts – not from liquid water, from which the animals drink, but from the capillary spaces of damp substrates (Nutman, 1941). Sodium ions exchange across the epithelium of the sacs, and since potassium does *not* exchange with external potassium, it is probable that at least some of the sodium which exchanges is actively absorbed (Noble-Nesbitt, 1963*c*). High concentrations of chloride have been localised in both the epithelium and the cuticle (Eisenbeis & Wichard, 1975*a*). The fine structure of the sacs is typical of a transporting epi- thelium – the cells have many basal and lateral infoldings associated with mitochondria (Eisenbeis, 1974; Eisenbeis & Wichard, 1975*b*). The structure of the epithelium appears to change when animals are placed on substrates of different salinities: in highly saline media, the apical brush border is reduced and the number of mitochondria diminishes (Eisenbeis & Wichard, 1977). Eisenbeis (1982) has shown that the rate of water absorption diminishes as salinity increases, again pointing to an active uptake mechanism.

Collembola have no Malpighian tubules, and it is thought that the midgut and the fat body are probably responsible for nitrogenous excretion (Feustel, 1958). The fat body accumulates urate concretions throughout the life of the animal. The midgut epithelium also con- tains concretions, and some of these are expelled when individual cells break down (Humbert, 1974). Collembola also undergo a curious process of replacement of the midgut epithelium at ecdysis, and many concretions are therefore excreted at this time. There appears to be no account of the overall significance of various nitrogenous end products in Collembola.

Only some members of the symphypleonan Collembola have tracheae, so that for the most part respiration must take place by diffusion through the cuticle as mentioned above. Rates of oxygen consump-

tion vary between species, and within species they vary with activity, temperature and state of nutrition (Zinkler, 1966). Since the exchange of gases through the cuticle must be accompanied by water loss, it is not surprising that Collembola are restricted, during activity, to damp environments.

The entognathous mouthparts of Collembola are adapted for biting, and the food ranges from fungal mycelia and decayed plant material to small arthropods, dead or alive (Manton, 1972). From her study of the structure and function of arthropod mandibles, Manton (1964) concluded that the entognathous condition of collembolan mouthparts has been derived independently from that of the Diplura and Protura. She therefore did not attach any phylogenetic importance to the group 'Entognatha'. In contrast, many other authorities such as Tuxen (1970), Hennig (1969, 1981), Boudreaux (1979*a,b*) and Kristensen (1981) have continued to re- gard the Entognatha as a sister group of the Ectognatha (Thysanura plus Pterygota).

Whether or not the Entognatha is regarded as a monophyletic group, several differences distinguish the Collembola from the Diplura and Protura. Although the normal walking gait is comparable to that in other hexapods, most Collembola have the ability to jump. As mentioned above, the use of high hydrostatic pressure for this jump may have greatly influenced body structure (Manton, 1972): the thin arthrodial membranes normally present at joints are lacking, because they would balloon out under high pressure. The legs therefore have no sclerotised articulation with the body, but are suspended by an internal system of muscles and tendons. From this characteristic form, a few non-jumping forms which live deep in the soil have been derived by degeneration of the jumping organ. These forms have secondarily re- gained body flexibility with thin arthrodial membranes, and consequently can penetrate easily through soil crevices. Manton's theory now needs to be reinvestigated following the suggestion of Christian (1979) that hydro- static pressure is not necessary for the jump.

Much of the life of collembolans is spent in cryptozoic habitats. Sense organs for vision are therefore not well developed, although ocelli are present; these may have been derived by reduction from more typical insect ommatidia (Paulus, 1974). Many other types of sense organ are present (see, e.g. Altner & Ernst, 1974), but the most interesting structure is the postantennal organ (Altner & Thies, 1976). This organ is similar in structure to the organs of Tömösvary in myriapods, and the pseudoculus of Protura: it consists of a pore plate with sensory cells beneath. As with the other pore plate sense organs, the possibility of convergence must be borne in mind, since quite similar structures are found in widely divergent groups such as the Crustacea and the Pterygota. The fact remains, however, that this structure could be yet another factor in common between aptery- gotes and myriapods. Unfortunately the exact function of the organ, whether as chemoreceptor or hygro- receptor, is unknown.

In most Collembola complex behaviour involving interactions between the sexes is unknown. Since many species tend to live in aggregations, the random deposition of spermatophores presumably allows a high enough density to bring about fertilisation of the females. In many Symphypleona, which tend to live on or above the soil surface, reproductive behaviour is reminiscent of that in other apterygotes (Bretfeld, 1970). In some genera males display to the females, but in *Bourletiella* a complex organ is developed on the abdomen of the male, and this is presented towards the female in courtship behaviour. Spermatophores are deposited immediately in front of the female, who then takes them up through the genital orifice (Raynal, 1973; Klaver, 1975). This behaviour seems, at least in part, to be linked to life in a relatively desiccating habitat, as compared with life in the soil itself, and Betsch-Pinot (1980) has emphasised the transition in reproductive behaviour within the Symphypleona from those species living beneath the litter to those living in drier grassland habitats. In *Sminthurus aureus*, which lives in forest leaf litter, the male deposits spermatophores in the absence of the female, who therefore finds them by chance or is attracted to them chemically. In *Sminthurus viridis*, found in grassland, the male may deposit spermatophores with the female absent, but these are ignored when the female comes across them. The female picks up spermatophores *only* if they are deposited in her presence by the male after a primitive courtship display (Betsch-Pinot, 1976). The possibility of the spermatophore drying up is therefore considerably reduced in the grassland species.

8.1.6 The 'Entognatha' and the origin of the hexapods

Although it is apparent that the Diplura, Protura and Collembola (referred to loosely as the Entognatha) are all rather specialised orders, each possibly having reached its entognathous condition independently, a number of common features emerge from the foregoing account. As pointed out above, there are many authors who believe that the features in common within the group indicate a common origin (see, e.g. Lauterbach, 1972; Kristensen, 1981). All three orders have well-developed labial kidneys which are very similar to those in Thysanura, and which are at least partly responsible for ion and water regulation. All three orders also have eversible abdominal sacs like those of the Machiloidea, which take up water, although it must be admitted that the homology of these is not entirely certain. None of the three orders has Malpighian tubules, although Protura have large papillae at the junction of the midgut and the hindgut with many of the characteristics of Malpighian tubules (Dallai, 1976). It appears that the epithelium of the midgut is responsible for at least some of the forms of excretion. In this they differ from Thysanura and from the pterygotes. The interesting possibility arises therefore, that *if* all the hexapods are ultimately to be derived from one ancestral type, this may not have had Malpighian tubules. In turn, this would suggest that the Malpighian tubules of myriapods and of the Thysanura and pterygotes have been independently evolved. On this basis it would be likely that significant differences in structure and function of the Malpighian tubules would be found between the two groups, and

the few studies on myriapods (see p. 141) certainly suggest that this is so.

Another similarity between the Protura, Diplura and Collembola is the mode of sperm transfer: in all, spermatophores are produced and transferred in ways similar to those found in the Thysanura and the Symphyla (Schaller, 1971). The presence of spermatophores, which in this case are little more than sperm drops supported by a peduncle, is a characteristic of almost all small cryptozoic arthropods, and is probably a primitive feature which arose when animals moved into the interstitial habitat, and long before they became terrestrial (Schaller, 1979). In the cryptozoic habitat on land, discharge of gametes into the water was no longer possible, and the use of stalked spermatophores provided a pre-adaptation allowing the possibility of sperm transfer within a humid environment, initially without complex behavioural adaptations. As already described, however, more complex behavioural sequences constituting courtship have been evolved, even in Collembola that have partly emerged from the cryptozoic environment. In some active myriapods such as *Scutigera*, and in most pterygotes, the development of courtship mechanisms and internal fertilisation has replaced the use of sessile spermatophores, allowing reproduction in desiccating environments.

There are naturally many characters in the Entognatha in which only two orders resemble each other. The peculiar cephalic sense organs are an obvious example, and their resemblance to the organs of Tömösvary in the myriapods has already been mentioned. Here once again the possibility of convergence must be considered, especially in the light of the suggestion that there is probably only a limited number of ways of producing, say, a chemoreceptor in a sclerotised animal (Altner & Thies, 1976).

The general structure of the cuticle does not seem to be particularly helpful in comparing the apterygote orders, but the position in the Collembola does suggest that epicuticular waxes first arose in small arthropods as a means of producing a hydrofuge layer. Only when tracheae had developed, as in some symphypleonan collembolans, would it have been profitable to spread the epicuticular wax layer over the whole of the body surface (see Ghilarov, 1959). In present-day hexapods, at least, this has occurred mainly in the larger forms. Presumably it was this increase in size which also made the use of the midgut inadequate for excretion, and provided the impetus for the development of Malpighian tubules. At the same time, the labial kidneys ceased to function in osmoregulation, and the hindgut assumed an overall importance in the reabsorption of salts and water. The emancipation from cryptozoic habitats allowed by all these developments had many other results. The transfer of sperm droplets or delicate spermatophores, for instance, became impossible in unsaturated atmospheres, and with increasing development of courtship behaviour, direct copulation became possible, and is the rule in pterygotes. The use of abdominal sacs for water uptake was lost, probably because an overall waterproofing of the integument was adopted: it is striking, for instance, that in *Petrobius*, which has abdominal sacs, the integument is quite permeable to water, whereas in *Lepisma*, which has no sacs, the integument has a very low permeability to water. These developments, allowing or being imposed by the colonisation of unsaturated atmospheres, may be envisaged as producing 'emancipated apterygotes', and in turn allowed the development of new structures and adaptations to produce the pterygotes.

Consideration of the Entognatha also allows us to add to the brief speculations already made upon the origins of the ancestral forms which produced both the hexapods and the myriapods. It has been suggested (p. 152) that these forms were multilegged, cryptozoic and sclerotised. They evidently had labial kidneys but probably no Malpighian tubules. They had eversible abdominal sacs for water absorption. Among their sense organs they probably had a humidity receptor (the pseudoculus, postantennal organ or organ of Tömösvary), so important for cryptozoic forms. The osmotic pressure of their haemolymph was high – of the order of half that of sea water – and the haemolymph consisted mainly of inorganic salts, suggesting a primitively marine origin. The geological period in which these forms flourished cannot be fixed, but must have been pre-Devonian, since typical Collembola were in existence by that time (Delamare Deboutteville & Massoud, 1967; Wootton, 1981). Indeed, since the archipolypod myriapods were in existence in the Silurian, the origin of their ancestors probably lies in the Cambrian period. Further speculations about the origin of hexapods and myriapods are probably fruitless, but one possibility concerning their habitats may be discussed. The significance of cryptozoic habitats in the life styles of both myriapods and apterygotes has been emphasised. To some extent this habitat grades into the interstitial habitat of seas and fresh waters. If the ancestral insect–myriapods were indeed small, as supposed, one possibility is therefore that they lived in interstitial habitats: such a life would to some extent preadapt them for life in litter. This route has been proposed by Ghilarov (1959) who has pointed out the transitional nature of the surface layers of the soil, relative to aquatic and truly terrestrial existence. The idea was expanded by Vannier (1973c, 1978) to the concept of a 'porosphere', consisting of all solids which contain internal surfaces: muds, sands, sedimentary rocks, animal and vegetable debris, soil and leaf litter. The habitats of the porosphere coincide roughly with other authors' concepts of the cryptozoic niche. However, the radiation of ancestral insect–myriapods within at least the cryptozoic leaf litter habitats would not have been possible to any great degree until the radiation of the terrestrial plants had occurred: for this reason it is the Carboniferous which saw the sudden appearance of large terrestrial arthropods. Naturally it is these which are preserved as fossils, but it is probably from the smaller forms that the ancestral stocks arose. It is, of course, necessary that most reconstructions of phylogeny depend heavily upon living forms and well-preserved fossils; but in the case of the terrestrial arthropods, it may be more realistic to appreciate that the primitive forms were probably insignificant; we shall be extremely fortunate if any of them have been preserved as fossils.

8.2   The pterygotes

In comparison with the apterygote orders, most of the pterygotes are essentially *not* cryptozoic: many can tolerate desiccating conditions and are often exposed to the conditions of the macroclimate. Their other essential difference, discounting secondary reductions, is the possession of wings, which has allowed them to utilise the air, again exposing them to conditions very different from those experienced by most apterygotes. With these two major changes from the apterygotes, winged insects have become the most diverse and abundant group of animals on earth: approximately three quarters of a million species have been described, in nearly 1000 families; and this may be only a fraction of the true number. Such diversity makes it impossible to cover the insects in great detail, and it will be the aim of this section to discuss only the mechanisms by which pterygotes have left the cryptozoic niche, and some of the ways in which they have become adapted to life in dry habitats.

8.2.1 Insect habitats

It is therefore appropriate at this point to consider some of the characteristics of the macroclimate, and the ways in which these vary in places likely to be inhabited by insects. Those insects that fly are probably exposed to the climate as measured by normal meterological stations; but for most of the time, insects are usually confined to a layer near to the ground, or to the vegetation. The climate of this layer has been discussed in detail by Geiger (1961), and has been reviewed with reference to insects by Cloudsley-Thompson (1962). The factors which govern temperature and humidity near the ground are complex. In terms of temperature, for instance, the radiation balance, the rate of evaporation, the flow of heat from the ground to the soil surface, the movement of air from neighbouring areas and precipitation are all important. Many of these will vary with height above the ground, type of soil, cover by different vegetation types and time of day. Two examples will suffice to show some of these effects. First, some measurements obtained in the Sonoran Desert of North America are shown in Fig. 8.9a. At a height of 15 cm above the ground the temperature reached approximately 30 °C. The temperature of the air nearer the ground was higher, and that of the soil reached 55 °C (Patten & Smith, 1975). The times and durations of these temperature maxima also varied at different vertical heights. In contrast to this situation on open ground, soil in the shade at the same site was cooler than the air, and this point is reinforced in a second example, taken from a temperate meadow (Fig. 8.9b). In this case, with vegetation reaching up to about 50 cm, the maximum temperature was recorded at about 30 cm above the ground, while the soil itself was cooler by several degrees (Waterhouse, 1955). This second example also shows that humidity changes drastically over the vertical profile, and the drying power of the air, as reflected by saturation deficit, is very much less at the soil surface than near the top of the vegetation. The effect of vegetation is to reduce the amount of radiation which reaches ground level, and this can easily be appreciated when it is noted that the total surface of vegetation growing in a meadow is some 20 to 40 times the area of the ground on which it grows (Geiger, 1961). Vegetation also reduces eddy diffusion and lowers wind speed, and since water is transpired by plants, humidities are almost always higher under vegetation than on bare soil. From these brief points, it can be seen that in fact to talk of the macro-climate is seldom relevant unless the insect under consideration is actually flying. Even then, the characteristics of the physical environment are not the only ones to be taken into consideration, because the size, structure and other physical characteristics of the animals themselves have to be taken into account.

Parry (1951) and Digby (1955) have investigated some of these problems, especially with reference to direct sunlight, by the use of models. Parry used blackened brass spheres as models of small insects, and investigated the effect of height above ground on temperature in direct sunshine. The equilibrium temperature of these models increased nearer to the ground, because of the increase in air temperature and the decrease in wind velocity, without change in the net radiation load. The latter factor may itself be greatly affected by the colour, shape and orientation of the body concerned, and is therefore unlikely in fact to be constant even for

Fig. 8.9. Climatic conditions near the surface of the ground. (a) Diurnal fluctuations in air and soil temperatures for a day in March 1973 in the Sonoran Desert, Arizona. Solid line, air in the open at 15 cm height. Alternate dashes and dots, air in the open at 1.5 cm height. Filled circles, air at 1.5 cm height under the plant *Cercidium microphyllum*. Long dashes, soil in the open at 1.5 cm depth. Short dashes, soil under *C. microphyllum* at 1.5 cm depth. Greatest fluctuations in temperature are seen at the soil surface and in the air immediately above it. Increasing height and addition of shade generally reduce the extremes, but it is noticeable that air under the plant becomes hotter in the day and cooler at night than soil in the same area. (b) Temperature, humidity and wind profiles in grass during June, in a meadow in Scotland containing clover to a height of 10 cm. Conditions were sunny after rain on the previous day. Solid line, mean wind speed, with stipple showing fluctuations. Long dashes and circles, saturation deficiency. Short dashes and triangles, air temperature. In this case, maximum temperatures occur well above the ground surface, but below the surface of the vegetation. Both wind speed and saturation deficit rise with height above the ground. (a) Redrawn from Patten & Smith (1975); (b) redrawn from Waterhouse (1955).

(a)

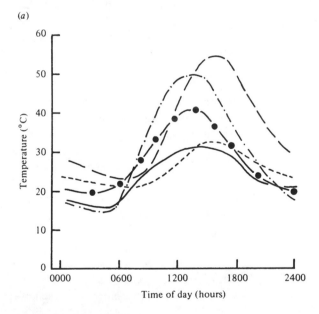

(b)

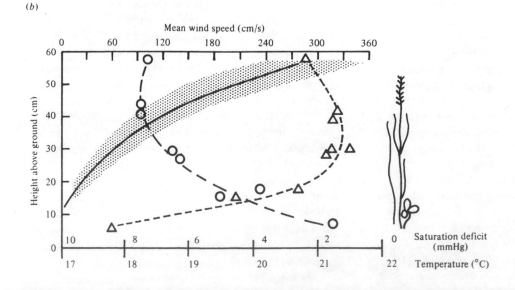

one individual. Besides this, it must be remembered that although the loss of heat by radiation is accounted for in the term 'net radiation', heat may also be lost in large degree by convection. Since the rate of loss by convection varies inversely with linear dimensions, larger bodies have higher equilibrium temperatures than smaller ones – an important point for arthropods living in desert conditions. Loss of heat by conduction was minimal in Parry's experiments, and might be expected to be so in insects because of their small area of contact with the ground. Loss of heat by metabolism and evaporation was also estimated to be small because of the ability of insects to close their spiracles – a point to which we must return when considering respiration. Digby (1955) came to similar conclusions to those of Parry (1951), but also investigated in detail the effects of wind speed on the temperature excess – the amount by which temperature of the models exceeded that of the ambient air. At wind speeds of over 20 cm/s the temperature excess varied inversely with the square root of velocity, but below this figure the relationship broke down. At low wind speeds, therefore, natural convection instead of forced convection was occurring, with the rate of heat loss independent of wind speed. Such conditions would naturally occur very close to the ground.

8.2.2 Temperature regulation

Although treatment of the heat relationships of insects by reference to models is informative in some aspects, it necessarily considers only direct physical effects produced by the environment. In fact, many insects can regulate their body temperatures, either by behavioural or physiological means, or by a mixture of both these. For instance, Edney (1971*b*) has shown that desert tenebrionid beetles can alter their temperature by changing their orientation to the sun, and by moving in or out of shade. Hamilton (1975) has gone further than this and demonstrated that some desert tenebrionids maintain their body temperature between 37 °C and 41 °C by a variety of behavioural means, including climbing on to stones or burrowing. He also pointed out that most desert beetles are black, and that this is presumably to allow maximum heat absorption at low external temperatures, so that the preferred body temperature can be reached early in the day. When the ambient temperature rises, behavioural mechanisms are used to prevent overheating. The details of how these desert tenebrionids control their body temperature vary from species to species (Henwood, 1975). *Onymacris plana*, an inhabitant of sand dunes in the Namib Desert, selects dunes facing in appropriate directions at different times of day. In order to absorb solar radiation in the morning, it moves to east-facing dunes. At midday, it burrows into the loose sand of dune faces to avoid temperatures which would be lethal. In the late afternoon, it emerges to move to west-facing slopes to absorb heat from the sun again. In contrast, *Stenocarpa phalangium* lives in the stony gravel plains between dunes. It cannot burrow at midday, but it climbs the highest stones, elevates its extremely long legs, and

points the white posterior part of its abdomen directly at the sun. In this way it avoids the highest temperatures of the desert surface, and minimises the amount of solar radiation received during the hottest part of the day. Other beetles of arid areas, such as the tiger beetle *Cicindela hybrida*, can also maintain a relatively constant body temperature during activity (Dreisig, 1980). *C. hybrida* begins its daily activity when ambient temperature rises to 19 °C. It maintains a preferred body temperature of 35 °C by basking in the sun at low ambient temperatures and by stilting to keep its body away from the soil surface at high temperatures. Burrowing occurs only when the ambient temperature rises to near lethal levels.

Body temperatures are normally high in a variety of desert insects. The larvae of one desert species, the Australian sawfly *Perga dorsalis*, have been investigated by Seymour (1974). When there is danger of body temperature rising above the critical thermal maximum, the larvae raise their abdomens from the substratum so that more heat is lost by convection. At very high temperatures they exude a watery fluid from the anus and spread it over the body: a rare instance of the use of evaporative cooling. Larvae of some desert Lepidoptera can also alter their body temperatures by behavioural means (Casey, 1976*a*). The larvae of the sphinx moth *Hyles lineata* maintain a relatively constant body temperature by changing their orientation to the sun, and by moving in and out of shade. This behaviour well suits their utilisation of a variety of annual plants which do not give much shade, and where high body temperatures ensure rapid feeding and rapid development during a short growing season. In contrast, the larvae of *Manduca sexta* follow air temperatures closely, but obtain protection from extremes of heat by moving into the dense foliage of the jimson weed, *Datura metalloides*, at midday. *M. sexta* lives on this sparsely distributed perennial, and is adapted more to economical consumption of the whole plant than to rapid consumption of a variety of annual plants.

For insects in flight the whole process of temperature regulation is more complicated. Some of the generalities noted still apply – as, for instance, that the heat lost by evaporation is normally insignificant (Church, 1960*a*). More important, however, is the point that a very great deal of heat is produced by the flight muscles. Unless these are actively cooled, they must, in large insects, heat up to 30 °C or more above ambient temperature during continuous flight. They have, therefore, become adapted to operate at these high temperatures, and as a corollary it follows that the muscles must be warmed up if the insects are to be able to initiate flight (Heinrich, 1975). The two main ways in which this is done are basking and 'shivering'. Good examples of the former are provided by the butterflies, which may sit with their wings outstretched, or may fold their wings together and tilt sideways (Vielmetter, 1958). In either case, the sun's radiation is absorbed, allowing temperature excesses of up to 14 °C (Wasserthal, 1975). The body temperature tends to rise to between 32 and 37 °C, and it is only when it is this high that

Argynnis paphia (the silver-washed fritillary) shows its full range of flight activity, courting behaviour and feeding. *A. paphia* is thus a homoiotherm for much of the time. The mechanism by which radiation is absorbed is not as direct as might be thought. Very little heat is transferred from the wings themselves to the body by the haemolymph or by conduction (Kammer & Bracchi, 1973). Instead, the air underneath the wing bases heats up, thus heating the body directly (Wasserthal, 1975). The colour and morphology of these wing bases is therefore most important. In *Papilio machaon* (the swallowtail), basking continues until the thoracic temperature reaches 28.7 °C – at this temperature flight ensues (Fig. 8.10).

Other examples of basking are found among the dragonflies (Heinrich & Casey, 1978), but many of these heat up their flight muscles by small-amplitude wing-beats – a mechanism analogous to shivering in vertebrates. Similar warm-up is characteristic of the sphinx moths: *Manduca sexta*, for instance, warms up to a thoracic temperature of 40 °C before flying (see Kammer & Heinrich, 1978). In these insects which use shivering, only the thorax heats up, while the temperature of the abdomen remains approximately the same as the ambient temperature. The warm-up is aided by insulation of the thorax, either by hairs or by a thick covering of scales, or, in the case of dragonflies, by a layer of air sacs between the muscles and the cuticle (Church, 1960*b*).

During flight, thoracic temperature in large insects such as sphinx moths, bumblebees and syrphid flies is stabilised over a wide range of ambient temperatures.

During hovering flight, for instance, the eastern tent caterpillar moth, *Malacosoma americanum*, retains a thoracic temperature of between 34 and 41 °C when the ambient temperature varies from 10–30 °C (Casey, 1981). It should be emphasised, though, that not all large insects act as homoiotherms: the desert locust *Schistocerca gregaria* does not maintain a constant temperature in flight; and the temperature of small insects such as *Drosophila* always remains near ambient temperature. For the large regulators the problem in flight is often how to get rid of excess heat produced by the flight muscles. In some, at least, this heat is lost by allowing haemolymph to pass from the thorax to the poorly insulated abdomen, from which heat is then lost by convection. In *Manduca sexta* the abdominal temperature remains at about ambient during warm-up, but rises suddenly at take-off (Heinrich & Bartholomew, 1971). In another sphinx moth, *Hyles lineata*, the thorax in flight remains at 40–42.5 °C, while the abdomen is much cooler, but still up to 6 °C above ambient (Casey, 1976*b*). In this case, some heat may also be lost from the head and legs as well as from the abdomen. In the dragonfly *Aeshna multicolor*, Heinrich & Casey (1978) have shown that haemolymph is shunted to the abdomen when the thorax is artificially heated. The most complete account of the use of haemolymph in temperature control is that given for the bumblebee *Bombus vosnesenskii* by Heinrich (1976). In this species a countercurrent mechanism in the petiole joining thorax and abdomen normally ensures that haemolymph returning to the thorax is heated efficiently and does not cool the flight

Fig. 8.10. Warming rate and body excess temperature in the swallowtail butterfly *Papilio machaon* in relation to different ambient temperatures. The lower threshold for flying activity, 28.7°C, is indicated by the stippled plane. The intersections of the two warming curves with this plane are projected on to the basal plane. These projections therefore show the basking time needed by *P. machaon* at different ambient temperatures before it will fly spontaneously. Redrawn from Wasserthal (1975).

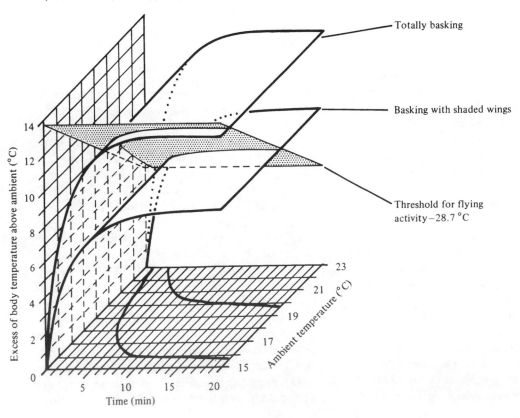

muscles. When overheating becomes a problem, how-ever, haemolymph can be shunted to the abdomen in pulses so that the countercurrent mechanism does not operate. Heat is then lost by convection from the underside of the abdomen, which has little insulation. A further method of getting rid of excess heat produced by the flight muscles is found in the honeybee *Apis mellifera* (Heinrich, 1980a,b): nectar is regurgitated on to the mouthparts and then evaporation cools the head. Because of the circulation of haemolymph this indirectly cools the thorax as well. While for most species a high rate of loss of fluid could soon be fatal, honeybees can afford to use evaporative cooling because they carry large volumes of nectar in their honeycrop.

The energetic cost of thermoregulation in flying dragonflies has been calculated by May (1977), who showed that smaller species of *Microthyria* did not thermoregulate in flight, while larger species such as *Microthyria atra* maintained relatively constant thoracic temperatures over a wide variation in ambient tempera-ture. The energy that would have been needed for the smaller species to thermoregulate would have been several times greater than that expended by *M. atra*. Even in this species more than 30% of the total energy expenditure during flight was used in thermoregulation. The larger species gained the advantage of being active both earlier and later in the day, and therefore at the breeding sites the males avoided aggressive interactions with males of other species, as well as enjoying reduced competition for perches.

A final example of temperature regulation is that of the African dung beetles, which gather dung into balls before rolling it away to bury it prior to consumption. When these beetles are at rest the body temperature is the same as ambient (Bartholomew & Heinrich, 1978). While walking, body temperatures may, or may not, be elevated. Prior to flight, body temperature is greatly elevated, probably by isometric contractions of the metathoracic muscles; extra oxygen is supplied to them by pumping movements of the abdomen which flush out the tracheal trunks. Take-off temperature for large beetles is of the order of 40 °C, and this temperature is maintained in flight. A difference from bees and moths is that the abdomen is nearly at the same temperature as the thorax, since their connection is wide. During other activities, such as ball making and ball rolling, body temperatures were nearly as high as those during flight: this allowed faster collection of dung in a situation of in-tense intra- and interspecific competition (Bartholomew & Heinrich, 1978). In this example, then, the beetles are homoiothermic whenever it is advantageous to be so, and not just in flight, where raised temperatures are, as it were, forced upon them. The contrast to small crypto-zoic insects, in which temperature varies entirely with the environment, could hardly be more extreme.

Little mention has been made so far of heat loss by evaporation, except to say that it is usually minimal. This, in turn, is because the structural and physiological mechanisms which have allowed insects to leave humid microclimates have largely been concerned with reducing water loss. Undoubtedly the two most important of these mechanisms have been the formation of an efficient closing device for the spiracles and the development of a waterproofing layer in the cuticle. We shall take these two factors in turn.

8.2.3 Respiration and associated water loss

There is little evidence that any of the apterygotes have muscles which can close the spiracular openings. Even in the Thysanura, Heeg (1967a) has been unable to demonstrate any active occlusion of the spiracles, although Noble-Nesbitt (quoted by Miller, 1974) has described a mechanism whereby the spiracles of *Thermobia* may close by elasticity. Many spiracles in the pterygotes are covered by sets of interlocking bristles, or sieve plates. These protect terrestrial species against wetting of the spiracles by rain (Hinton, 1969), but do not affect the diffusion of gases. In addition to these protective struc-tures many pterygotes have two sets of muscles for each spiracle: the opener and closer muscles (Miller, 1974). The effect of the closer muscles has often been investi-gated by applying carbon dioxide, since at high concen-trations of this gas the muscles relax and all spiracles will open. In the tsetse fly 10% carbon dioxide is sufficient for this (Bursell, 1957). Under these conditions the flies lose a maximum of $1.0 \, \text{mg fly}^{-1} \, \text{hour}^{-1}$, presumably as water vapour. In contrast, when the spiracles are blocked the loss rate is reduced to $0.09 \, \text{mg fly}^{-1} \, \text{hour}^{-1}$. It has been realised since the early work of Ramsay (1935) that many factors influence the rate of water loss through the spiracles. Purely physical factors such as the wind velocity and direction are certainly important; but perhaps more relevant here are situations in which the degree of open-ing of the spiracles, and hence the rate of loss of water, is altered. For example, both tsetse flies and mosquitoes in which water reserves have been depleted show marked decreases in the amount of spiracular opening at low humidities (see Fig. 8.11; Bursell, 1957; Krafsur, 1971). High temperatures usually cause increased spiracular opening: in the tsetse fly all the spiracles open at 40 °C (Edney & Barrass, 1962). In this case the water lost by evaporation is then sufficient to reduce the body tem-perature by as much as 1.7 °C.

The mechanisms which control the operation of spiracular closing muscles are now reasonably well under-stood (Miller, 1964a,b). In the dragonfly *Ictinogomphus ferox*, for example, desiccation causes an increase in the frequency of motor impulses passed to the closer muscle of spiracle 2, with a consequent increase in the threshold for opening, as well as an increase in the threshold of central responses to such factors as lack of oxygen. A rise in temperature also raises the threshold for opening of spiracle 2. These factors come into play while the dragonfly is perching; but when it flies, the spiracle neurones are 'switched off', because the flight muscles must then be provided with a good oxygen supply, regardless of external conditions. This observation brings out the point that the ability to close the spiracles has enormous consequences for respiratory physiology; and of necessity, there must be some compromise between closing the spiracles in order to diminish water loss and opening them to gain oxygen. Part, at least, of this compromise appears to be brought about in some species by intermittent opening of the spiracles. In the pupae of many insects found in dry environments, of which the best known is the diapausing pupa of the silkworm *Hyalophora cecropia*, the spiracles open briefly for approximately 30 min every few hours in order to eliminate the accumulated carbon dioxide. Oxygen, in contrast, is taken up continuously even when the spiracles are closed, since in fact they do not close very tightly. Thus as oxygen is used up by the tissues, a partial vacuum develops in the tracheae, and air is sucked through the spiracular openings (Buck, 1958, 1962). This partial vacuum occurs because the carbon dioxide generated by metabolism is not immediately released to the gaseous phase, but can be banked as bicarbonate for long periods. Meanwhile, the slow inward flow of air through the spiracles prevents any outward diffusion of water vapour (or of carbon dioxide). This process of 'passive suction ventilation' therefore allows minimal rates of water loss over long periods. At intervals, however, the carbon dioxide and bicarbonate rise to such concentrations that they trigger the opening of the spiracles. Because of the large concentration gradient, carbon dioxide then diffuses out rapidly, so that the stored bicarbonate produces gaseous carbon dioxide which in turn diffuses out until the store is removed. Water loss must also occur at this time, but is

relatively small because of the restricted time of spiracle opening. In the pupa of the silkworm *Hyalophora cecropia* water lost in the opening 'bursts' is approximately equal in amount to water produced metabolically, so that the pupa can remain in water balance over very long times (Kanwisher, 1966).

Many insects do not store carbon dioxide and release it intermittently, and for these the mechanisms of reducing water loss are less well understood. Here it is appropriate to discuss briefly the relative importance of diffusion, convection and ventilation in tracheal gas exchange. Classically, diffusion has been regarded as the only method by which oxygen moves along the tracheoles to the cells. Brocas & Cherruault (1973) suggested, however, that a convection current caused by the diffusion of nitrogen into the haemolymph at the distal end could accelerate the inward movement of oxygen, particularly during activity. Active ventilation cannot occur within the tracheoles, but can produce bulk movement of air in the tracheae and tracheal trunks. The more efficient is this ventilation, however, the higher will be the rate of water loss (see Loveridge, 1968). An example has been provided by the work of Wasserthal (1980, 1981) on a butterfly, *Papilio machaon*, and a moth, *Attacus atlas*. In both species, haemolymph periodically oscillates from thorax to abdomen, causing a decrease in thoracic fluid volume, and hence expansion of the

Fig. 8.11. The effects of different relative humidities on the tsetse fly *Glossina morsitans*. (*a*) Rates of water loss under different experimental conditions: E1, E2 and E3 show the flies in air, one, two and three days after emergence. Closed circles show the flies with added carbon dioxide in the atmosphere. Closed triangles show flies with blocked spiracles. Since carbon dioxide opens the spiracles, their partial closure is

probably responsible for the low rates of water loss from the flies. Even so, most of the water lost *is* lost through the spiracles, as shown by the blocking experiment. (*b*) The degree of spiracular closure at the different emergence stages. A lower percentage closure in E1 correlates well with the higher rate of water loss in E1 shown in the left-hand diagram. Redrawn from Bursell (1957).

(*a*)

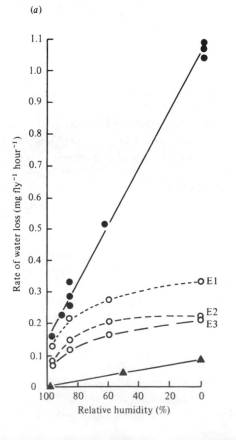

(*b*)

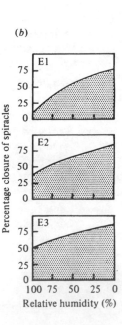

tracheal trunks, and inspiration. When the haemolymph passes back into the thorax, warm moist air is expired from the spiracles. Active muscular ventilation of the tracheal trunks in the abdomen of such flying insects as bees and moths is also common, and must entail a high rate of water loss: because air leaves at a high temperature, it must have a high water content.

One possible way of reducing water loss during ventilation is to use one-way ventilation of the tracheal trunks instead of tidal flow through each spiracle. The locust *Schistocerca gregaria* provides a good example. Although during flight the air supply for the flight muscles is tidal, entering and leaving through spiracles 2 and 3, the air supply to the brain is supplied through spiracle 1, but leaves through abdominal spiracles 5–10 (Miller, 1960). This means that air leaves the body at a temperature of approximately 20 °C instead of the thoracic temperature of about 30 °C. Although it is still saturated with water vapour, its water content is therefore much less. It is also possible that in some actively ventilating insects the air in the tracheal trunks will not always be completely saturated, since the first few ventilatory strokes may clear the trunks of saturated air (Hamilton, 1964). Water loss would therefore not be as high as when saturated air is expelled.

The phenomen of unidirectional airflow within the tracheae has also been found in desert tenebrionid beetles. Air taken into the thorax of these beetles leaves from the abdomen and is collected in a special cavity under the fused elytra before emerging to the exterior (Fig. 8.12; Ahearn, 1970*b*). The purpose of this flow and the reason for its association with the subelytral cavity are, however, far from clear. For one thing, in most black tenebrionids the air in the subelytral cavity is actually hotter than that in the thorax (Hadley, 1970*a*), so that air passing out by this route would actually contain more water vapour than if it had emerged directly from the thorax. A very few species of desert tenebrionids have white elytra, and in their

Fig. 8.12. Diagram of the tenebrionid beetle *Eleodes armata* showing the subelytral cavity and probable avenues of respiratory water loss. Heavy arrows from the thorax indicate spiracular water loss. These spiracles do not open to the air but to a thoracic air space, from which water passes out through the intersegmental areas. Light arrows show losses from the abdominal spiracles. These open into a much larger cavity, the subelytral cavity, which opens to the atmosphere through a single orifice above the anus. Redrawn from Ahearn (1970*b*).

Subelytral cavity

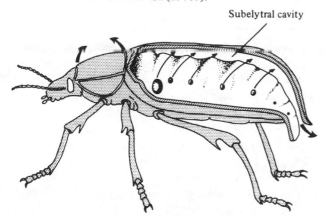

case the temperature of the abdomen is lower than that of the thorax by several degrees (Edney, 1971*b*). In black species the elytra certainly do reduce water loss from the abdomen, since their removal or partial removal increases water loss rates (Cloudsley-Thompson, 1964; Ahearn & Hadley, 1969); but whether this is because the subelytral cavity has some specific function, or whether it merely reduces eddy currents in the spiracular openings, is not known. One possibility, which has not been investigated, is that there is a water-absorbing organ situated within the subelytral cavity. The presence of such an organ would then form a very efficient means of water conservation, since it would reduce the water lost in expired air. Cloudsley-Thompson (1964) found that the relative humidity within the cavity was equal to that of the ambient air, and was not saturated as tracheal air would presumably be, so that the possibility of water absorption within the cavity would repay investigation.

In summarising the extent to which water is lost from insect spiracles, it can be said that at least in many species found in dry habitats the rate of loss is extremely small, except during violent activity or at very high temperatures. This is nicely exemplified in results obtained by Ahearn (1970*b*) for the desert tenebrionid beetle *Eleodes armata*. At 25 °C respiratory water losses account for only about 3% of total transpiration. At 35 °C this percentage rises to 21%, and at 40 °C loss through the spiracles is 36% of transpiration (i.e. disregarding losses such as the ejection of quinones from the abdomen, faeces, etc.). From these figures it is evident that loss through the integument itself, although low in comparison with many animals, often represents one of the major hazards leading to desiccation.

8.2.4 The integument and associated water loss

Despite the many investigations into the properties of insect integuments, there has not as yet been complete agreement about the mechanisms which restrict water loss from the body surface. The possible sites at which control may be exercised have been admirably discussed by Berridge (1970), who pointed out that whereas most studies have considered the properties of the cuticle, the properties of the underlying epidermal cells may be equally important (Fig. 8.13). If, for instance, the apical plasma membrane of these cells constitutes an important permeability barrier, then the water content of the cuticle will tend to equilibrate with the surrounding air, and the permeability of the epicuticle will not necessarily play such an important role in reducing water loss. The majority of available evidence, however, suggests that the major permeability barrier lies in the epicuticle. At this point it is appropriate to give a very brief introduction to the organisation of the integument, although excellent detailed accounts will be found in Neville (1975), Locke (1974) and Edney (1977). The epicuticle is a thin outer layer which covers the main mass of procuticle. This procuticle is made up of a complex of chitin and protein, with chitin crystallites laid down in laminae. Pore canals from the epidermal cells run right through the procuticle. The outer part of the procuticle is hardened by quinone tanning, and is called the exocuticle. The inner part remains flexible and is called the endocuticle. The outer layer, or epicuticle, has a composite structure (Locke, 1974). Its inner side consists of a 'cuticulin' layer, which marks the termination of the pore canals, although it is permeated by the wax canals which run outwards from the tips of the pore canals. Outside the cuticulin layer is a layer of wax, and outside this again is a layer of cement. In many insects this cement layer acts as a protective varnish, while in others it may act as a sponge to retain mobile lipids from the wax layer (Locke, 1974). There is general agreement that the waterproofing properties lie in the wax layer, and perhaps the most widely accepted theory concerning this has been provided by Beament (e.g. 1964, 1965). According to this theory the wax nearest to the cuticulin layer is organised as a monolayer, while the remainder is not well ordered, and forms a much thicker layer. The putative monolayer will reduce water loss because water molecules will be unable to pass between the closely packed wax molecules. According to Locke (1974) the wax canals penetrate even this monolayer. The canals are supposed to be lipid–water liquid crystals, whose

Fig. 8.13. Diagram to show the locations of possible mechanisms which could be responsible for regulating water loss from the integument of arthropods. (a) The sites of the three major permeability barriers: 1, the basal membrane; 2, the apical membrane; 3, the cuticle. (b)–(e) Show how the activity of water in the cuticular compartment can be altered by varying the permeability of barriers 2 and 3, or by the imposition of a water pump (P). Degree of stippling represents water activity. (b) Shows the main resistance in the epicuticle: water activity in the endocuticle equals that in the epidermal cells and haemolymph. (c) Shows the resistance shared between apical membrane and cuticle: water activity in the endocuticle is less than that in the cells, and may approach that of the air. (d) Shows a water pump moving water from the endocuticle to the cell with the same result as in (c). (e) Shows the water pump able to reduce water activity in the cuticle below that of the air, so that water vapour can be absorbed. From Berridge (1970).

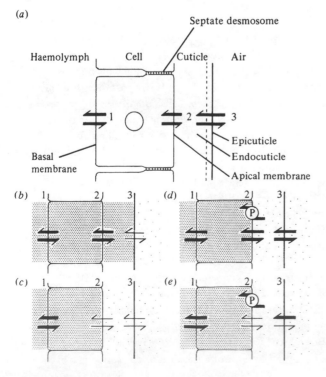

normal function is to form a route by which wax can be transported to the outside of the cuticle.

Based on this monolayer theory, there are two suggestions as to how water may at times penetrate through the cuticle. As in other terrestrial arthropods, such as the arachnids, the phenomenon of a transition or critical temperature is commonly found: at a particular temperature for any cuticle, the permeability to water suddenly increases by a factor of approximately five, although it should be noted that there are authorities who doubt the reality of these transition phenomena (see e.g. Hackman, 1974; Edney, 1977; Hadley, 1981), and even some desert insects show no critical temperature (Vannier & Ghabbour, 1983). The monolayer theory may explain the transition *either* as a phase change of the oriented monolayer such that more space is available between the wax molecules for outward diffusion of water (Beament, 1964), *or* as a phase change of the lipid–water liquid crystals of the wax canals, such that water can diffuse directly through these canals instead of being restricted by the monolayer which normally covers their openings (Beament, 1965; Locke, 1974). In either case, the major boundary to water loss above the transition temperature is the layer of relatively unordered lipid external to the monolayer.

The evidence for the monolayer comes from various sources and appears to be convincing, but as pointed out by Edney (1977) it is not direct evidence. The agreement in transition temperatures between artificial membranes with wax monolayers, and insect cuticles from which the waxes were derived, is very good (Beament, 1964). It has also been shown that the quantity of grease taken from a supposed monolayer on the aquatic beetle *Gyrinus natator*, when allowed to form an artificial monolayer, covers approximately the same area as that of the beetle's body surface (Beament, 1964). Such lines of evidence, and those deriving from the effects of abrasives, have been accepted by many workers (e.g. Ebeling, 1974). Others, such as Hackman (1974), did not believe that the lipid constituents isolated from insect cuticles would form appropriate waterproofing monolayers. Edney (1977) apparently favoured the view put forward for ticks by Davies (1974*b*) that transpiration rates are not controlled by an oriented layer but by bulk packing of lipids in the wax layer. Since Beament (1964) has shown that at least in *G. natator* there is only sufficient lipid to account for a monolayer, this 'bulk' theory is unlikely to be valid for a wide range of insects. The evidence of Davies (1974*b*) for ticks, along with that of others for insects, does, however, show that the rate of transpiration may vary immensely with life history and stage in the moult or feeding cycle, and this suggests that there may be physiological mechanisms by which it is controlled. Work by Treherne & Willmer (1975) strongly suggests that there is hormonal control of transpiratory water loss in the cockroach. No doubt this mechanism could function by affecting the cuticular lipid directly, but it would seem more likely to affect some property of the epidermal cell membranes. This brings us back once more to the role of the epidermal cells in controlling water loss.

So much effort has been expended on examination of the cuticular structure, and so little on the function of the epidermal cells, that perhaps the latter should be the next area for intensive investigation.

As a result of the mechanisms discussed above, many insects have extremely low rates of transpiration. Some figures are given in Table 8.4, in which loss rates are expressed as permeability. It can be seen that in general the lowest rates are found in desert species, as might be expected. Most of the species labelled 'xeric' have loss rates not much above $20 \, \mu\mathrm{g \, cm}^{-2} \, \mathrm{hour}^{-1}$ $(\mathrm{mmHg})^{-1}$ while those labelled 'mesic' have a range of about 10 to $50 \, \mu\mathrm{g \, cm}^{-2} \, \mathrm{hour}^{-1} \, (\mathrm{mmHg})^{-1}$. 'Hygric' species have loss rates above $50 \, \mu\mathrm{g \, cm}^{-2} \, \mathrm{hour}^{-1} \, (\mathrm{mmHg})^{-1}$. Such categories deserve a word of caution, however. The larvae of *Manduca sexta* provide a good example. Although they live in deserts the larvae of *M. sexta* live in bushes of the jimson weed, as mentioned earlier. The category 'xeric' is, therefore, a gross oversimplification of the situation. The earlier discussion of microclimates should make it evident that it is very difficult indeed to categorise an animal's habitat, and the categories used by Edney (1977) are useful only as very crude descriptions.

The rates of transpiration in insects show a very similar range to those of arachnids (see Table 6.4), with the lower end of the range being very much less than the minimum rates for decapod or isopod crustaceans (see Tables 5.5, 5.9). There is little doubt that these differences reflect the presence of an external wax layer in insects, and its absence in the crustaceans. But it must

also be said that some isopods and myriapods have transpiration rates as low as, or lower than, the majority of mesic insects (see Edney, 1977). Only the true desert insects which are exposed to harsh desert conditions, as well as some of the arachnids, have perfected a really efficient waterproofing layer. It should also be mentioned that some insect larvae produce cases which significantly reduce the rate of water loss. *Tinea pellionella*, for instance, makes a case of silk and food materials, which has a lower permeability to water than the integument (Chauvin, Vannier & Gueguen, 1979).

8.2.5 The regulation of water balance by the Malpighian tubules and gut

In summary so far, we may say that at least some of the pterygotes have greatly reduced the rate of loss of water through the integument and through the spiracles. The other major avenues of loss to be considered are those associated with the excretory and digestive systems. In pterygotes these are closely associated because there are no cephalic excretory glands, and the end products of nitrogenous metabolism as well as the remnants from digestive processes pass through the rectum and are voided from the anus. The processes of osmoregulation, ionic regulation and water balance are also associated with the alimentary canal. Two sets of organs are involved: the Malpighian tubules, which are blind diverticula arising from the junction of the midgut and hindgut, and the epithelium of the rectum which is often modified into rectal glands. The Malpighian tubules produce a fluid which although slightly hypo-

Table 8.4. *Transpiration rates in pterygote insects (from Edney, 1977)*

Species	Habitat	Permeability ($\mu\mathrm{g \, cm}^{-2} \, \mathrm{hour}^{-1}$ $(\mathrm{mmHg})^{-1}$)	Reference
Orthoptera			
Locusta migratoria	Xeric	22	Loveridge, 1968
Dictyoptera			
Periplaneta americana	Mesic	55	Mead-Briggs, 1956
Blatta orientalis	Mesic	48	Mead-Briggs, 1956
Arenivaga investigata	Xeric	12.1	Edney & McFarlane, 1974
Hemiptera			
Rhodnius prolixus	Xeric	12.0	Holdgate & Seal, 1956
Coleoptera			
Tenebrio molitor (larvae)	Xeric	5	Mead-Briggs, 1956
Tenebrio molitor (pupae)	Xeric	1	Holdgate & Seal, 1956
Diptera			
Bibio sp.	Hygric	76	Wigglesworth, 1945
Calliphora erythrocephala	Mesic	51	Mead-Briggs, 1956
Glossina palpalis	Mesic	12	Mead-Briggs, 1956
Glossina morsitans	Mesic–xeric	8	Bursell, 1957
Glossina morsitans (pupae)	Xeric	0.3	Bursell, 1958
Lepidoptera			
Manduca sexta (larvae)	Xeric	39.6	T. M. Casey, unpublished

All figures are for transpiration between 20 and 30 °C.

osmotic to haemolymph and with a higher potassium
content is generally similar to haemolymph in composition. This fluid passes down the tubules and into
the hindgut, where subsequent reabsorption of salts and
water determines the ionic composition and water
content of the faeces. In turn, then, these processes
control the composition of the haemolymph. The review
by Maddrell (1971) has provided an excellent summary
of the mechanisms of insect excretory systems.

The exact nature of the mechanism by which the
fluid is produced within the Malpighian tubules is not
certain, but it involves in most insects the active transport of potassium into the tubular lumen (Ramsay,
1953, 1955b). In some insects sodium may be actively
transported instead of potassium, and Maddrell (1977)
suggested that the pump involved is situated on the
apical membrane of the tubular cells, and that it will
transport either of these ions depending upon which
gains entry to the cells. This itself is decided by the
permeability of the basal membranes of the cells, i.e.
those facing the haemocoel. The active transport of
sodium or potassium is thought to be the primary
factor responsible for the flow of water and other salts
into the lumen, and other molecules move across if they
are small enough to penetrate the epithelium. The structure of the cells is typical of transporting epithelia, in
that the apical membranes are extended into microvilli
(many actually containing within them mitochondria),
while the basal membranes are much infolded and are
also associated with mitochondria. The exact mechanism
by which these cells transport fluid has been in dispute
for some time. The 'three compartment' model of Curran
(1960), or its modification as the 'standing osmotic
gradient' theory of Diamond & Bossert (1967), which
in this application would mean the transport of ions into
the spaces between the apical microvilli, followed by
diffusion of water, has not been supported by measurements which have been made of ionic concentrations in
these regions (Gupta, Hall & Moreton, 1977). Other
hypotheses such as those based on the concept of
electro-osmosis (e.g. Hill, 1975) have yet to be properly
investigated. The possibility that organic molecules
might be involved has not apparently been taken up.

Whatever the details of the mechanism by which
potassium transport produces a flow of fluid into the
tubules, it seems likely that the composition of the fluid
is further modified by other transport systems as well as
by the permeability characteristics of the tubule walls
(Maddrell, 1977). The net result, however, is the movement into the tubules of a fluid not very different in
composition from haemolymph. The rationale behind
this system is presumably similar to that behind the production of an ultrafiltrate: any 'unwanted' substance can
be excreted without the necessity for a specific transport system. There must, however, be transport systems
available to reabsorb the necessary salts, water and
metabolites. It is now believed that at least in some
insects such as Locusta and Calliphora glucose is reabsorbed directly by the Malpighian tubules (Knowles,
1975; Rafaeli-Bernstein & Mordue, 1979). Salts, water
and amino acids are usually reabsorbed in the hindgut,

although in cases where excess water is taken up with
the food, this is passed out in the faeces (see, e.g. Cohen,
March & Pinto, 1981).

Reabsorption in the hindgut has been studied particularly in locusts and blowflies. In Locusta the degree of
reabsorption of water depends upon the hydration state
of the animal (Loveridge, 1974). The water content of
the faeces may be reduced to as little as 39% when the
animals have been fasting in dry air, whereas in humid
conditions with a moist food supply it is of the order of
80%. In Schistocerca and in Calliphora both water and
ions can be absorbed from the rectum against large
gradients (Phillips, 1964, 1969). Normally, the uptake
of water is accompanied by the uptake of ions such as
sodium or potassium; but water uptake can occur without the uptake of salts, and this originally led Phillips to
point out that the possibility of the active transport of
water could not be ruled out. Because of the modified
structure of the rectal epithelia, however, it now appears
that all the uptake phenomena can be explained on the
basis of solute transport followed by the diffusion of
water. Two detailed investigations have been made, and
will be discussed briefly here, while reference to other
systems can be found in Edney (1977). Periplaneta
has been studied by Oschman & Wall (1969), by Wall
& Oschman (1970) and by Wall (1977). The rectal
epithelium is modified into rectal pads with a structure
as shown in Fig. 8.14. The epithelial cells are tall and

Fig. 8.14. The organisation of rectal pads in the cockroach
Periplaneta americana. (a) Rectum as seen in vivo, showing the
position of the pads. (b) Diagram to show the structure of
a rectal pad. It is thought that solutes may be transported
actively into the intercellular channels, creating a high osmotic
pressure. Water would thus move into these spaces from the
rectum, drying the faeces. Solutes could then be taken up by
the cells again and recycled. Redrawn from Wall (1977).

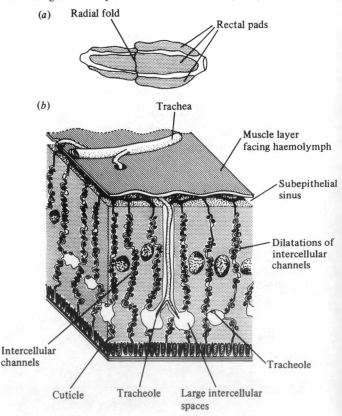

class, the Lissamphibia (see Table 9.3). In support of
a monophyletic origin, Bolt (1977) discussed the characteristics of a Lower Permian labyrinthodont, *Doleserpeton*,
and concluded that it could have been an ancestor of all
three living amphibian orders.

At present it seems impossible to decide between
these various alternative views, and the situation is
complicated by the suggestion that lungfishes, not
rhipidistians, may be the closest fish-like relatives of the
tetrapods (Rosen *et al.*, 1981). Instead of attempting to
select one theory from so many, it may be more profitable to examine the physiology of present-day forms in
the hope of gleaning ideas about how the physiology of
the amphibian ancestors became modified to deal with
life out of water.

Locomotion

Some of the structural modifications which have
allowed the evolution of tetrapod motion from that of
aquatic ancestors have already been briefly described;
but only the study of living forms can make possible
a detailed functional interpretation of these structural
characteristics. Undoubtedly the modern salamanders
show the most primitive locomotory traits found in
modern tetrapods, and an analysis of their methods of
movement gives some insight into how tetrapod locomotion evolved. The question has recently been discussed
by Edwards (1976) who investigated salamander locomotion by using filming techniques. When they are moving
on land, salamanders have three different methods of
locomotion. As they start off, they use what is known
as a 'lateral sequence' walk: in this walk, the footfall of
a hindfoot is followed by that of the ipsilateral forefoot,
then the contralateral hindfoot and lastly the contralateral forefoot. In this way the body is always supported
by three legs which form a stable triangle. A similar
pattern is used by all other tetrapods. As salamanders
speed up, they switch to a gait known as the trot: the
contralateral feet move together, and during most of
each stride there are only two feet on the ground at the
same time. For very rapid escape responses, the limbs are
not used: the animals rest their ventral surface on the
ground and essentially perform swimming motions with
the myotomal musculature.

During the use of these three different ways of
moving, the salamanders employ two different types of
propulsion. The majority of the force producing forward
propulsion is generated by the active protraction and
retraction of the limbs (Fig. 9.8a). This method, the
most common in the tetrapods, is most effective when
the vertebral column is held rigid, as it is at relatively low
speeds during the lateral sequence walk. As a modification of this method, the limbs can be rotated to aid in
their retraction: the retractor muscles insert on projections from the humerus and femur known as the cristae
ventrales. The contraction of the retractors therefore
results in a 'crank' action, and the distal end of the limb
is pulled from a forward-facing position to a backwards-
facing one (Fig. 9.8b). The second method of producing
propulsive force is by the rotation of the pelvic and
pectoral girdles (Fig. 9.8c). These rotate as the body is

Fig. 9.8. Mechanisms of movement in amphibians. Broken lines
represent muscles contracting isometrically. Solid lines with
arrows show muscles contracting isotonically. (*a*) Propulsion by
active limb retraction. Isometric contraction makes the vertebral
column a rigid rod. Thrust from the limbs pushes the body
forward. Propulsion between stages 1 and 3 shown as P_L.
(*b*) Propulsion by limb rotation. The crista ventralis, a process
of the humerus, is shown stippled. When the limb is protracted
(1), the crista ventralis points antero-ventrally. Contraction of
retractor muscles (arrows) brings the crista ventralis to a position
where it points first ventrally (2) and then postero-ventrally (3).
The crista ventralis thus acts as a crank, and the humerus rotates
on its long axis. Protractor muscles (not shown) also insert on
the crista ventralis and reverse the process during protraction of
the limb. (*c*) Propulsion by girdle rotation. Isometric contraction
fixes the limbs relative to the girdles. At time 1, the protracted
limbs are fixed and the axial muscles are just beginning to
contract. This contraction straightens the vertebral column and
then bends it in the opposite direction, rotating the pectoral and
pelvic girdles, and leading to the retraction of the fixed limbs at
times 2 and 3. Propulsion is shown as P_G. Redrawn after Edwards
(1976).

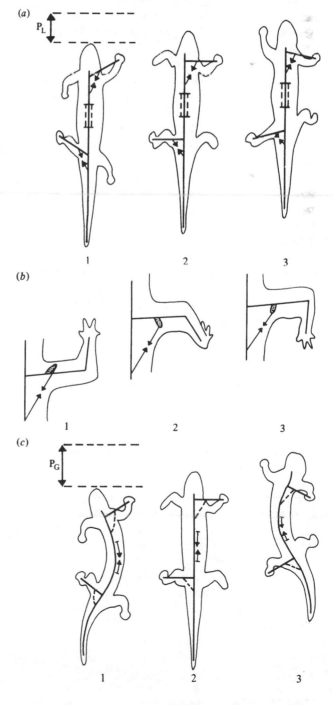

bent laterally, and since the limbs which are not anchored to the ground are rigidly attached to the girdles, they are passively protracted. They are then anchored, and the contralateral legs are protracted by rotation of the girdles in the opposite direction, i.e. by flexing of the whole trunk in the opposite direction. This girdle rotation is of maximum importance during trotting, when it accounts for nearly 20% of the forward propulsive force. Unlike the situation in the walk, salamanders flex their bodies considerably during the trot.

To summarise, the walk and trot can be contrasted as follows. Walking salamanders maintain a fairly rigid vertebral column, and use active limb protraction and retraction to generate up to about 60% of their forward propulsive force. Lateral bending of the body is minimal, and the bend may be described as a standing wave; that is to say, the flexure alternates from one side to the other, but the wave does not travel down the body. In trotting salamanders, in contrast, lateral bending increases and takes the form of a wave which travels down the body, as it does in swimming fish. The importance of active protraction and retraction of the limbs is lessened, while that of girdle rotation is increased. Edwards (1976) interpreted these observations to mean that the rhipidistians and early labyrinthodonts employed the trotting gait as their initial mode of movement when out of water. In support of this hypothesis, it may be suggested that the myotomal muscles of rhipidistians would already have been well adapted to the formation of the travelling wave utilised in the trot. However, some rhipidistians had relatively well-developed limbs, and Rackoff (1980), after an examination of the osteolepiform *Sterropterygion*, concluded that active limb protraction and retraction had evolved before movement on to land. Many of the early labyrinthodonts certainly had short trunks and powerful limbs, and in these some active limb retraction must have occurred, because they possessed cristae ventrales and presumably used the method of limb rotation described above.

The initial method of movement on land was therefore probably an extension of the normal aquatic myotomal mechanism, aided by the use of the limbs as anchoring points, but the labyrinthodonts had already begun to use the limbs themselves for propulsion. The development of a rigid vertebral column, with the major use of active limb protraction and retraction, was a later phenomenon. Other workers have regarded the walk as primitive because of its ubiquity, but from the work of Edwards (1976) it appears that this ubiquity is due to the increased efficiency of the method, and not to its primitiveness. After the evolution of the walk, many other locomotory mechanisms were evolved within the Amphibia. The Apoda lost all the limbs and became adapted to burrowing. Most anurans developed much stronger back legs than front ones, and move by jumping. These later specialisations do not concern us here, however, since the main line of the vertebrates retained combinations of myotomal movement and the lateral sequence walk.

Respiration

The development of limbs which could raise at least the anterior end of the body off the ground when the animal was out of water had repercussions intimately linked with the method of respiration. Schmalhausen (1968) has pointed out that even though the early amphibians possessed lungs, they probably could not inflate them when out of water: without support at the anterior end the weight of the body on the lungs would have prevented this. Once the trunk was supported, however, by the development of limbs braced to the vertebral column by a pectoral girdle, the pressure opposing inflation would have decreased, allowing the lungs to aid dermal and buccal respiration. The present-day amphibians show examples of dermal, buccal and pulmonary respiration, and consideration of this respiratory diversity throws some light upon the problems which faced the primitive amphibious vertebrates.

Both aquatic and terrestrial amphibians show great similarities with air-breathing fish in their overall mechanisms of gas exchange. Air-breathing amphibians obtain 50–78% of their oxygen through the lungs, but lose 57–84% of their carbon dioxide across the skin and gills (Rahn & Howell, 1976). These proportions are very similar to the exchanges found in the air-breathing fish discussed earlier. The presence in amphibians of this 'bimodal' gas exchange and its similarity to exchange processes in fish suggest that early amphibians may have used a similar system (Hughes, 1967). However, if this was the case, early forms must have been very different from the earliest amphibians to be preserved as fossils, since these, the ichthyostegids, were large, probably had a thick layer of scales and possessed well-developed ribs. It has been argued that cutaneous respiration would not have been important in these forms (Gans, 1970*a,b*), and that they had probably replaced the buccal force pump by an aspiration pump. Gans pointed out that such a pump may have been operated either by a lifting movement of the ribs, or by internal shifting of the organs, as happens in present-day crocodiles. In *Ichthyostega* the ribs were broad and heavy, and were probably adapted primarily for giving support to the body, since the vertebral column is not thought to show significant advance from that of the rhipidistians (Thomson & Bossy, 1970). However, since all early tetrapods had two-headed ribs with the heads at an oblique angle (Fig. 9.7), it is possible that aspiration pumps were an early development. Randall, Burggren, Farrell & Haswell (1981) have even suggested that both buccal *and* aspiration ventilation evolved in the aquatic environment, pointing to the situation in an Amazonian teleost *Arapaima gigas* (see Farrell & Randall, 1978). In this species, however, inhalation is brought about not by movements of the ribs, but by a downward movement of the ventral wall of the air bladder, in turn brought about indirectly by arching movements of the body. This unique type of aspiration is actually disadvantageous out of water, because the fish has to struggle each time it breathes. Aspiration as found in the higher tetrapods has evolved from the use of ribs, with their intercostal muscles, and the diaphragm.

The situation in modern amphibians is therefore somewhat anomalous, since they utilise the skin and a buccal pump for respiration. Gans (1970*a,b*) has argued that this is a specialisation developed after the loss of scales and well-formed ribs. Romer (1972) supported these suggestions, and put forward the hypothesis that primitive amphibians probably retained gills which were used for carbon dioxide removal. It must be recognised that the arguments put forward by Gans concerning the 'degenerative' nature of respiration in modern amphibians are based upon the supposition that the earliest terrestrial amphibians were the ichthyostegids. It seems more likely that the first forms to move on to land used extensive vegetation cover, in parallel to almost all other groups moving on to land (see, e.g. Thomson, 1980). They were probably not the large forms which have remained as fossils, but small, fragile species which were unlikely to be preserved. It should also be said that Romer (1958) envisaged *Ichthyostega* as being entirely aquatic.

There is therefore a wide range of possible types to be considered in the water–air transition of the vertebrates. It is interesting to note, for instance, that Cox (1967) was able to make out a plausible case for deriving the present-day amphibians from microsaurian lepospondyls, with their small size and light rib development. Such early amphibians might have retained the buccal pump and cutaneous respiration. However, microsaurs generally had two-headed ribs, and this state could indicate aspiration ventilation. As mentioned above, Randall, Burggren, Farrell & Haswell (1981) believed that aspiration ventilation evolved in aquatic forms. They went so far as to suggest that early air-breathing vertebrates consisted of two separate lineages: one with buccal ventilation, giving rise to the modern-day anurans: and one with aspiratory ventilation, giving rise to reptiles, birds and mammals.

In view of these diverse opinions, it is certainly worthwhile considering the possibility that the bimodal breathing of rhipidistian fishes was directly inherited and utilised by the first amphibians. To do this we must therefore review briefly the mechanisms involved in bimodal breathing in the present-day amphibians. In these the division of gas exchange between lungs and gills/skin does not in fact account for all respiration, because the buccal cavity may also be used. Although in resting anurans, therefore, the constant movement of air into and out of the buccal cavity is thought to serve purely olfactory purposes (Foxon, 1964), it may be that in some forms the buccopharyngeal mucosa is important in releasing carbon dioxide (Hughes, 1967). It is more important in urodeles, and may well have played an important role in early amphibians. However, evidence from the degree of vascularisation suggests that the major exchange of gases occurs across the skin and the lungs. The proportion of total capillaries in these organs varies from species to species, but in anurans the skin always has more than 20% of the total, while the lungs have 30–80% (Foxon, 1964). As one might predict from this variation, the structure of the lungs also varies greatly between species. An example of a well-developed lung is provided by the toad *Bufo marinus* (Smith & Campbell,

1976), in which the main pulmonary cavity is divided up into primary, secondary and tertiary alveoli, with the septa of each being completely covered by a capillary network. The lung epithelium of anurans is lined with surfactant, allowing the alveoli to be kept open (Pattle, 1976).

The lungs are inflated by a buccal force pump in much the same way as in lungfish (see p. 183), but in frogs at least, expiration is brought about not only by the elastic fibres within the lung, but also by the active contraction of the flank muscles (see Foxon, 1964, for discussion). In higher tetrapods this activity of the flank muscles is carried to an extreme whereby the movement of the ribs can cause inspiration as well as expiration. Inspiration in frogs, however, is solely due to the buccal force pump: with the nares open, the buccal floor is lowered. The nares close, the glottis opens, and when the buccal floor is raised, air is forced into the lungs. This process may have to be repeated several times to fill the lungs completely. As noted by Foxon (1964), it may be that the wide, dorso-ventrally flattened heads of air-breathing amphibians developed at least partly in order to increase the size of the forcepump, so that relatively few movements were needed to fill the lungs. This tendency, together with the use of the nares instead of a gulping action with the mouth, points to the much greater sophistication of the pulmonary mechanisms in amphibians compared to those of lungfish. At least some anurans have become so well adapted to air breathing that they suffer respiratory acidosis when they are forced to stay under water (Emilio & Shelton, 1980) – a complete reversal of the situation in fish, which suffer acidosis in air.

Because of this shift in emphasis towards pulmonary breathing, it is important to review in some detail the changes in control of acid–base balance. In lungfish the use of bimodal breathing leads to problems of acid–base balance when the animals are exposed to the air for any length of time: because the gills are the major site of loss of carbon dioxide as a gas, and of bicarbonate regulation in the form of bicarbonate–chloride ion exchange, aerial exposure leads to respiratory acidosis. In amphibians many situations lead to the build-up of carbon dioxide tension in the blood, and this also tends to bring about acidosis; but at least in anurans some control is exercised over this, by a number of mechanisms.

(1) The skin is very well vascularised, as noted above, so that carbon dioxide is lost directly to the air.

(2) The skin is also the site of pumps which exchange chloride for bicarbonate and sodium for hydrogen ions, in just the same way as those in the gills of fish (Maetz, Payan & de Renzis, 1976). Some regulation of pH may therefore occur by uptake of bicarbonate ions and by the direct elimination of hydrogen ions.

(3) In some situations the kidney is able to take part in pH regulation. A good example is shown by *Bufo marinus* (Boutilier *et al.*, 1979*c*). When this toad burrows, its breathing rate decreases and the tension of carbon dioxide in arterial blood rises. The pH falls, but over a period of time the concentration of bicarbonate rises, acting as a buffer, and the original pH is regained.

The extra bicarbonate is probably derived from fixed carbonate in the tissues, but may also partly come from renal reabsorption of bicarbonate ions, or from renal excretion of hydrogen ions. The urinary bladder is important here because it excretes less bicarbonate when blood pCO_2 rises, but more bicarbonate as the level of bicarbonate in the blood increases (Fitzgerrel & Vanatta, 1980). This function of the kidney in acid–base balance is interesting because it has been widely reported that acid–base balance in fishes is effected entirely by exchange at the gills. The involvement of the kidney in amphibians might therefore be thought of as a new departure, except for the fact that recently it has been shown that teleost kidneys *are* involved in hydrogen ion excretion (Wood & Caldwell, 1978). Whether they actually play any part in regulating respiratory acidosis is more dubious (Cameron & Wood, 1978), but at least the basic capacity for pH regulation is there. Its use in amphibians may therefore represent a modification of a very primitive ability by fish kidneys and not a new development.

(4) There is also evidence in *Bufo marinus* that in conditions other than burrowing, a fourth mechanism of acid–base control is present: *B. marinus* can regulate the tension of carbon dioxide in the blood by changing the rate at which it ventilates its lungs (Boutilier *et al.*, 1979*a,b*). During and following exercise this allows it to remove the excess carbon dioxide which cannot escape from the skin (Boutilier, McDonald & Toews, 1980). Although acidosis does occur, to some extent, it is probably mainly due to the accumulation of metabolic acids such as lactate and not directly to accumulation of carbon dioxide; recovery of pH to normal occurs rapidly after exercise ceases. In contrast, urodeles such as *Cryptobranchus*, which have poorly vascularised lungs, are not able to control removal of excess carbon dioxide in this way, and exercise produces in them far more serious and long-lasting acidosis, similar to that seen in amphibious fish: the pH drop is due to build-up of carbon dioxide as well as to that of metabolic acids, and it takes many hours to return to normal (Fig. 9.9). The use of lungs as a means of controlling carbon dioxide levels and pH is therefore a most important feature in Amphibia. It is this control which has been so greatly developed in higher tetrapods that in the mammals it is the major mechanism of respiratory acid–base control.

If present-day amphibians really represent a degenerate group with respiratory mechanisms which have become specialised for amphibious life, as argued by Gans (1970*a*) and Romer (1972), study of their physiology can reveal little about the ways in which amphibians were derived from rhipidistians. However, the similarity of the overall respiratory process in amphibians and in air-breathing fish is very striking, and a case may be made for accepting the idea that cutaneous respiration is indeed a primitive character, and not a recent specialisation. Taking this as a starting point, it can be argued that the development of respiratory mechanisms in Amphibia took place along the following lines. Development of support for the body allowed the lungs to become more important in oxygen uptake, while the gills were

reduced. Carbon dioxide was eliminated through the skin, but the tension of carbon dioxide in the blood increased, and there was initially very little control over it. This rise in pCO_2 in arterial blood allowed the development of internal carbon dioxide receptors instead of the external receptors found in fish. With mechanical improvement in pulmonary breathing and an increase in the rate of pulmonary ventilation, these internal carbon dioxide receptors came to control carbon dioxide tensions by controlling ventilation rates. Short-term control of acid–base balance therefore passed from the gills to the lungs. Initially, longer-term regulation of acid–base balance probably moved from the gills to the skin, linked to the bicarbonate–chloride and hydrogen–sodium exchange pumps. With the skin no longer in contact with water all the time, this longer-term regulation of pH passed partly to the kidney, although the skin retained the ion exchange pumps. In later evolution, when the larger labyrinthodonts evolved with their relatively impermeable skin, it must be assumed that the lungs became totally responsible for short-term acid–base

Fig. 9.9. The characteristics of acidosis following activity in two amphibians. *Cryptobranchus alleganiensis* (filled circles) is a salamander, breathing predominantly through the skin. The toad *Bufo marinus* (open circles) uses both skin and lungs. In both species, a 30-min period of enforced activity (stipple) produced a sharp fall in arterial pH. This was, however, more severe and long-lasting in the skin breather *C. alleganiensis*. Examination of lactate and pCO_2 in the blood shows that in *B. marinus* the metabolic acidosis due to lactate was large but very soon regulated, while respiratory acidosis due to carbon dioxide was small and regulated within 30 min. In *C. alleganiensis*, in contrast, metabolic acidosis from lactate lasted more than 10 hours, and respiratory acidosis from carbon dioxide took about 10 hours to be regulated. The use of the lungs for carbon dioxide elimination in the toad allows it to control respiratory acid–base balance by increasing rate of ventilation, a possibility not open to the skin breather. From Boutilier, McDonald & Toews (1980).

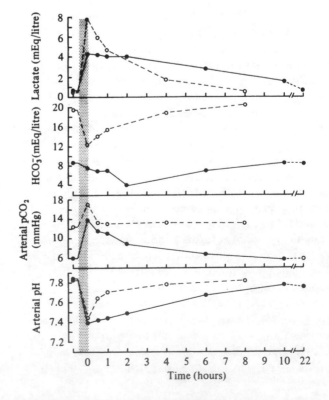

control, while the kidneys took over the entirety of long-term metabolic acid–base balance. This is the pattern in higher tetrapods, and it seems probable that it originated in the amphibian ancestors of the reptiles.

Circulation

One further factor relevant to the development of new respiratory mechanisms which needs to be discussed is the pattern of circulation. The matter of separation of pulmonary and systemic circuits in lungfish has been discussed earlier. In amphibians there is a great deal of evidence to show that although the ventricle is anatomically undivided, the bloodstreams from pulmonary and systemic circuits are functionally separated in the heart (Johansen & Hanson, 1968). In both the salamander *Amphiuma tridactylum* and the toad *Bufo paracnemis* most of the blood which is directed to the systemic circulation is derived from the lungs; that which is pumped into the pulmonary artery, although it contains a substantial proportion of pulmonary venous blood, mainly consists of systemic venous blood. The relevance of this separation is, however, not entirely clear because blood is oxygenated to some degree in the skin capillaries, and this oxygenated blood returns to the right atrium, whereas oxygenated blood from the lungs returns to the left atrium. There is therefore oxygenated blood on *both* sides of the heart. The separation of bloodstreams does not, therefore, ensure an efficient separation of oxygenated and deoxygenated blood. It must be assumed that only with the transfer of *all* respiratory exchange from the skin and gills to the lungs did selection pressure build up sufficiently to bring about complete separation of the pulmonary and systemic circuits. This situation may well have come about in the labyrinthodonts, and it is virtually complete in some reptiles. Complete anatomical separation is found only in birds and mammals.

Water loss

If the early amphibians indeed possessed a permeable skin, it seems likely that they possessed some of the other characteristics associated with a permeable skin in modern amphibians. The location of the ion exchange pumps has already been mentioned in pH regulation. These pumps are also important in salt uptake in a way similar to that of the pumps found in the gills of modern teleosts. Their movement to the skin would have been a logical step, and the later transfer of all osmoregulatory power to the kidneys was presumably coordinated with the development of an impermeable skin in the labyrinthodonts. The initial amphibians with permeable skins probably also used the skin as a site of water uptake. Modern amphibians are able to control the osmotic pressure and ionic composition of the blood by using a combination of the skin, the kidney and the urinary bladder (see Deyrup, 1964; Alvarado, 1979; Koefoed-Johnsen, 1979). Water is continually taken up across the skin when it is available, partly by passive and partly by active processes. However, the high permeability of the skin to water means that in desiccating conditions the animals lose water rapidly. Tables 9.4 and 9.5 show

some examples of rates of loss, and further examples are given by Bentley & Yorio (1979), Canziani & Cannata (1980), Katz & Graham (1980), Warburg & Degani (1979) and Wygoda (1981). With very few exceptions there appear to be no great modifications of the skin to prevent this water loss, even in frogs and toads which live in arid conditions. One of the exceptions, the South American frog *Phyllomedusa sauvagii*, has an extremely low rate of water loss (Table 9.4; Shoemaker, Balding & Ruibal, 1972), but the structural basis of the mechanism by which this low loss is achieved is not known. Generally, the skin of amphibians has a very thin external layer of squamous cells, and the blood capillaries are usually found just below this. Water conservation is therefore mainly ensured by appropriate behaviour patterns rather than by physiological or anatomical adaptations. The kidneys normally produce urine which is hypo-osmotic to the blood, and thereby function as pumps eliminating the excess water entering through the skin. In dehydrating conditions the kidneys cease the production of urine, and water in the urinary bladder is reabsorbed (see, e.g. Balinsky, 1981). The kidneys are therefore of the 'freshwater' type, and cannot produce a hyperosmotic urine, an ability found only in the birds and mammals.

One further point about amphibian skin may be noted. If indeed it was concerned with respiration in early forms it must have remained permeable to water, as it is in Lissamphibia. Very early amphibians must therefore have been confined to damp environments in a similar way to those of the present day. Although the development of thick scale coverings in the labyrinthodonts very probably came about in response to desiccation, and not for any purpose of defence, it should also be noted that thick scales do not necessarily preclude either water loss or a capacity for cutaneous respiration. Both these depend upon the supply of fluid to the surface of the body, the permeability of the scales and the possibility that there are areas where capillaries come very near to the surface. Even with a thick scale covering in places, early amphibians may have had a capacity for cutaneous respiration, and the problem of cutaneous water loss. In this connection, the observations of Bystrow (1947) are relevant. He concluded that while some of the early labyrinthodonts lived in dry environments and had no blood vessels in the dermal bones, others lived in wet habitats and the dermal bones possessed a net of blood vessels which extended into the skin and served in cutaneous respiration.

Nitrogenous excretion

In the section concerning air-breathing fishes it was noted that some shift from ammonia to urea production occurs when the fish are in air, and the inability of both the gills and the kidneys to excrete the urea formed leads to its build-up in the blood. Present-day aquatic amphibians excrete both ammonia and urea. Although they have little ammonia in the blood (Jungreis, 1976), ammonia is the major end product (Goldstein, 1972): the African toad *Xenopus laevis*, for instance, eliminates 70–80% of its nitrogen as ammonia, and the rest as urea.

Some of the ammonia is excreted across the skin, at least in frogs (Frazier & Vanatta, 1980), but most is eliminated through the renal system, and is in fact formed in the kidneys. This is in direct contrast to the situation in fish, in which ammonia is exchanged across the gills. Urea is also excreted through the kidneys by amphibians, and again this is in contrast to the case in fish. During adaptation to saline environments by the

amphibians, urea in the blood is retained, at least in adults (Balinsky, 1981) and is responsible for almost all the observed increase in osmotic pressure. However, urea continues to be excreted in the urine, so that the capacity of the kidneys to excrete urea is well established. This is a marked change in the ability of the kidneys compared with that of fishes. It seems certain that the origin of urea production dates from before the origin of the

Table 9.4. *Evaporative water loss in selected amphibians and reptiles*

Species	Habitat	Weight (g)	Temperature (°C)	Rate of evaporative water loss ($mg\ g^{-1}\ hour^{-1}$)		Reference
				Total	Cutaneous	
Amphibia						
Order Anura						
Rana pipiens	Moist	25.5	20.7	141	140	Spotila & Berman, 1976
Rana temporaria	Moist	24–33	26	14.2–18.9	–	Shoemaker et al., 1972
Bufo cognatus	Fossorial, desert	33–40	26	11.8–12.2	–	Shoemaker et al., 1972
Scaphiopus couchii	Fossorial, desert	24–28	26	13.6–14.2	–	Shoemaker et al., 1972
Phyllomedusa sauvagii	Arborial, semi-arid	23–35	26	0.3–1.6	–	Shoemaker et al., 1972
Order Urodela						
Desmognathus ochrophaeus	Moist	1.1	21.7	656	656	Spotila & Berman, 1976
Reptilia						
Order Chelonia						
Terrapene carolina	Dry woodland	452	20.8	0.5	0.4	Spotila & Berman, 1976
Order Squamata						
Anolis carolinensis	Dry areas	5.8	20.6	1.9	1.3	Spotila & Berman, 1976
Dipsosaurus dorsalis	Desert	22–28	26	0.3–0.4	–	Shoemaker et al., 1972
Uma scoparia	Desert	11–30	26	0.4–0.6	–	Shoemaker et al., 1972

Table 9.5. *Cutaneous permeability in selected amphibians and reptiles*

Species	Habitat	Temperature (°C)	Permeability		Reference
			($mg\ cm^{-2}\ hour^{-1}$)	($mg\ cm^{-2}\ hour^{-1}\ (mmHg)^{-1}$)	
Amphibia					
Order Anura					
Rana pipiens	Moist	20	42	2.41	Spotila & Berman, 1976
Order Urodela					
Desmognathus ochrophaeus	Moist	20	70	4.12	Spotila & Berman, 1976
Reptilia					
Order Crocodilia					
Caiman sclerops	Primarily aquatic	23	1.37	0.07	Bentley & Schmidt-Nielsen, 1966
Order Chelonia					
Pseudemys scripta	Aquatic	23	0.51	–	Bentley & Schmidt-Nielsen, 1966
Terrapene carolina	Forests	23	0.22	0.01	Bentley & Schmidt-Nielsen, 1966
Terrapene carolina	Forests	20	0.54	0.03	Spotila & Berman, 1976
Order Squamata					
Iguana iguana	Tropical terrestrial	23	0.20	–	Bentley & Schmidt-Nielsen, 1966
Sauromalus obesus	Desert	23	0.05	–	Bentley & Schmidt-Nielsen, 1966
Anolis carolinensis	Desert	20	0.18	0.01	Spotila & Berman, 1976
Gecko gecko	–	30	0.1–1.0	–	Zucker & Maderson, 1980
Python sebae	–	16.8	22.1*	–	Hattingh, 1972
Xantusia vigilis	–	25	–	0.002	Mautz, 1980
Lepidophyma smithii	–	20	–	0.01	Mautz, 1980

All values were derived in moving dry air except for the value marked with an asterisk, which was obtained in still air, before equilibrium was established.

Amphibia, but that the transfer of its elimination site from gills to kidney is an amphibian adaptation. This must have been one of the most important early adaptations allowing some emancipation from aquatic life.

As mentioned above, the retention of urea in the body may be useful in osmoregulation as well as in ammonia detoxification. Even on land, some osmoregulatory benefit may be obtained by retaining urea. In the aestivating spadefoot toad *Scaphiopus hammondii*, urea in the blood rises tremendously only when the surrounding soil dries out and the osmotic pressure of the soil water rises (Shoemaker, McClanahan & Ruibal, 1969). The rise in internal osmotic pressure in this case prevents the loss of water to the environment. In spite of this advantage of retaining urea, it is perhaps surprising that in general the ability to convert urea to uric acid has not been evolved within the Amphibia because at least one species of anuran has been shown to excrete uric acid. *Phyllomedusa sauvagii*, an arboreal frog from arid regions of South America, excretes both urea and ammonia in solution, but these are accompanied by semi-solid uric acid (Shoemaker *et al.*, 1972). African fossorial species are also capable of excreting uric acid. The ability is so limited within the Amphibia, however, that it seems to be a recent development by isolated forms adapting to arid conditions. Whether labyrinthodonts evolved such an ability must, of course, remain an open question; but the capacity of some modern anurans to produce uric acid demonstrates the adaptability of amphibian physiology.

Reproductive biology and egg production

It is apparent from the above points that it is very difficult to decide how relevant modern amphibians are in elucidating the characteristics of early forms in the Devonian and Carboniferous. It seems to be agreed, however, that in one respect the present-day representatives give us information about early amphibian physiology. This is the case with the general trends in reproductive biology and egg production. Modern amphibians produce eggs which have no complex mechanisms shielding them from the external environment, and consequently they are not at all resistant to desiccation. The contrast with reptiles is striking, and this is taken up in the next section. Here it is sufficient to summarise some of the basic properties of amphibian eggs, and to contrast them with those of the fishes.

Although the details of early development differ greatly between various fishes and amphibians, there is a striking similarity in the membranes which surround the eggs and separate them from the external environment. In both amphibians and fishes the plasma membrane of the egg is surrounded before oviposition by a membrane called the chorion (variously known in the amphibians as the vitelline membrane and in the fishes as the capsule). At oviposition or fertilisation, which normally occur at the same time, the chorion is pushed out from the surface of the egg by the extrusion immediately beneath it of polysaccharides which attract an osmotic inflow of water from the external environment, and possibly from the egg itself (Smith, 1957;

Salthe, 1965; Blaxter, 1969). This mixture of polysaccharides and water forms the perivitelline fluid which bathes the egg and allows it to rotate. In both fishes and amphibians the perivitelline fluid has an osmotic pressure less than that of the egg, but greater than that of the external water, with a fairly low salt content (Krogh, Schmidt-Nielsen & Zeuthen, 1938; Richards, 1940; Potts & Rudy, 1969). Separating the plasma membrane from the perivitelline fluid in amphibians is a very thin surface coat (known confusingly in fishes as the vitelline membrane) which adds mechanical strength to the egg and prevents it from swelling (Holtfreter, 1943; Løvtrup, 1960). This coat may be a remnant of chorionic material, most of which is pushed out beyond the perivitelline space. Its function appears to be a mechanical one, and osmoregulation takes place at the plasma membrane. The chorion in amphibians is surrounded by several jelly (mucous) coats which are added by the oviducal wall (Lofts, 1974); in fishes such a coat is often lacking, although in demersal species some of the polysaccharide of the perivitelline space may be exuded through pores in the chorion to form an external mucous layer which helps to stick the eggs to the substratum (Harder, 1975). In both fishes and amphibians the chorion acts as the major mechanical barrier to swelling (Berntsson, Haglund & Løvtrup, 1965). It consists of very fine fibres orientated parallel to the surface of the egg (see, e.g. Afzelius, Nicander & Sjöden, 1968). In some species at least, these fibres become orientated at oviposition, and it is at this time that the chorion allows the passage of sperm through it to the egg (Grey, Working & Hedrick, 1977). It could be that this change also affects the permeability characteristics of the chorion.

In none of the eggs just described is there any layer which is specially adapted to prevent water loss. In general the responsibility for regulation of salts and water in the egg seems to lie with the plasma membrane of the egg, although the perivitelline fluid must to some degree act as a buffer between the egg and the outside. The jelly coats of amphibians are extremely hygroscopic, and take up water if it is available; but there is no evidence that they can act as osmotic buffers or that they can prevent the evaporation of water (Salthe & Mecham, 1974). The eggs of amphibians which are laid out of water are therefore provided with further protection such as foam masses (Salthe & Mecham, 1974), and all amphibian eggs laid on land are placed in humid environments which are unlikely to dry out. These characteristics of amphibian eggs by definition restrict their owners to damp environments, or necessitate their return to water to breed. In contrast, the reptiles are defined as having a 'cleidoic' egg which, coupled with the acquisition of internal fertilisation, has allowed them complete freedom from the water. Consideration of this last step in the colonisation of land forms a major part of the next section.

9.4.   The amniotes: emancipation from the water

Two characteristics in particular restrict modern amphibians to life in humid regions or near water. The first is a permeable skin across which water is lost and respiration occurs. The second is the production of a small, relatively unprotected egg which normally develops into an aquatic larva. It may be argued that many fossil amphibians had thick, impermeable, scale-covered skins, so that in the past the major feature which has governed amphibian distribution has been their mode of development. It is believed that at some time in the Carboniferous there arose in the descendants of the amphibians the capacity to produce eggs which were larger and were protected from the environment by a series of membranes, allowing respiration to continue yet reducing the rate of water loss. This type of egg, the amniote egg, revolutionised the capacity of the tetrapods to live on land by completely removing the need for them to breed in or near water. It has been called a cleidoic egg because it consists essentially of a closed box which allows development within it regardless of external conditions. As first defined by Needham (1963), a cleidoic egg allows only *gaseous* exchange with the environment. It will be seen later that by this definition most reptile eggs are far from cleidoic.

Much discussion has centred upon the question of how such an egg evolved, and which fossil forms produced it rather than the unprotected 'anamniote' eggs. Because early eggs have not been found as fossils, however, palaeontologists have had to use other characters with which to define the earliest amniotes. It is therefore appropriate here to consider the whole problem of the origin of the reptiles before going on to deal with the evolution of cleidoic eggs. In particular, this discussion concerns the development of the middle ear, which has played a central part in all considerations of reptilian origins.

9.4.1 The origin of reptiles

The classical idea that seymouriamorphs show characters intermediate between those of amphibians and reptiles and may have given rise to the reptiles (e.g. Watson, 1951) has now been discarded, and *Seymouria* is regarded as a later derivative from primitive anthracosaur stock representing a sterile line which did not produce amniotes (Olson, 1965*b*). At least some of the seymouriamorphs have been shown to have had aquatic larvae, and must therefore be classed as amphibians. The earliest undoubted reptiles were the Captorhinomorpha (order Cotylosauria) from the mid Carboniferous. A hypothesis put forward by Carroll (1970) was that these evolved from small, early anthracosaurs related to such forms as *Gephyrostegus*. However, other authors have contended that reptiles could not have evolved from anthracosaurs. In particular, Panchen (1972, 1975, 1980) has examined the problem, concentrating upon the anatomy of the occipital region of fossil amphibians and early reptiles. The anthracosaurs possessed an indentation at the back of the skull termed the otic notch, which presumably housed the tympanum. Primitive reptiles, on the other hand, had no such notch. The evolution of these reptiles from anthracosaurs would therefore mean that they inherited an otic notch but then reverted to a notchless condition. Alternatively, they might have evolved from some hypothetical 'notchless' anthracosaur. Neither of these ideas seems to be supported by the evidence, and in particular the idea of loss of the otic notch must be seen in the light of its subsequent development – or redevelopment – in later reptiles.

It can be seen that the development of the middle ear and the relationships of its associated structures in amphibians and reptiles is a controversial topic, but one which is crucial in discussion of the origins of reptiles. Ideas about the evolution of the ear in tetrapods have been reviewed by Lombard & Bolt (1979) who contrasted classical theories with an alternative view. According to the classical theory the tympanum developed early in tetrapod history, and the otic notch of the labyrinthodont amphibians, presumably containing the tympanum, provided the starting point for the evolution of the middle ear in reptiles and all higher tetrapods. There are four major problems with this classical view. First, the notchless condition of early reptiles has been mentioned above. Second, the stapes bone in the middle ear, derived from the hyomandibular, bears quite different relationships to the internal branch of the seventh cranial nerve in anuran amphibians and in amniotes, and it is difficult to see how the amniote condition could have evolved from that of amphibians. Third, the stapes is orientated ventro-laterally in rhipidistian fishes, dorso-laterally in labyrinthodonts, ventro-laterally again in primitive reptiles, and directly laterally in the precursors of mammals. It is difficult to envisage these orientations as being sequential. Finally, while the stapes connects directly with the tympanum in amphibians and reptiles, in mammals two more ossicles are interposed between the stapes and tympanum – the quadrate (incus), derived from the back of the skull, and the articular (malleus), from the lower jaw. After detailed consideration of these points, Lombard & Bolt concluded that there was no reason to favour the classical hypothesis of direct homology of the tympanum in all tetrapods over a hypothesis which derived the tympanum independently in three lines: anurans, reptiles and birds, and mammals. Support for this suggestion has been provided by Carroll (1980) who examined the structure and relationships of the hyomandibular (stapes) in fossil rhipidistians, labyrinthodonts and early reptiles. The hyomandibular remained large and was retained as a supporting element for the braincase in early reptiles, and could hardly have been derived from the state in labyrinthodonts, in which it had already been reduced in size and acted as a stapes. If the independent origin of a tympanum in three tetrapod groups is accepted, it follows that labyrinthodonts with an otic notch were not ancestors of the amniotes. As emphasised by Panchen (1972), the phyletic lines leading to labyrinthodont amphibians and captorhinomorph reptiles probably diverged at the level of small prototetrapods. Most authors would presumably relate this small prototetrapod to the rhipidistians, although Rosen *et al.* (1981) would prefer a dipnoan relationship.

The topic is obviously far from being resolved. The general tendency seems to be to push the origin of individual tetrapod groups further back in time, and to derive them relatively independently from aquatic vertebrates, rather than to derive one tetrapod line from another.

9.4.2 The origin of cleidoic eggs

Carroll (1970) pointed out that the maximum size of present-day amphibian eggs is restricted to a diameter of about 9 mm. Larger eggs would have a lower surface area : volume ratio, so that the rate at which oxygen could diffuse into them would be below an acceptable level. The first terrestrial amphibian eggs must therefore have been small. Since in present-day amphibians there is a correlation between egg size and adult size, it follows that the first amphibians to lay their eggs on land must also have been small – probably no bigger than about 10 cm snout to vent. Presumably it was from some such stock of small forms that the ability to secrete protective membranes around the egg arose. This argument fits well with the small size of early captorhinomorph reptiles. Later captorhinomorphs increased in size, and must therefore have developed eggs which had accessory breathing organs and did not rely solely upon diffusion over the whole egg surface.

The amniote egg, with its accessory breathing organ and protective shell, was thus developed in the early reptiles. Whether the capacity to produce this type of egg was developed more than once has been hotly argued. If this were so, it would mean a polyphyletic origin of the reptiles. The arguments have been well summarised by Stahl (1974) and need not be repeated here. The point to be made is that the amniote egg characterises not only reptiles, but also the birds and primitive mammals. Whether or not the reptile and mammal lines evolved entirely separately (see Kemp, 1980), the development in both of the amniote egg is the one great advance which initially allowed the vertebrate classes to expand into the available terrestrial niches.

9.4.3 Structure of the early stages of amniote eggs

Some consideration of the processes involved in the development of the amniotic egg is therefore necessary. Perhaps it is best to begin by describing the egg of present-day reptiles and birds, and to compare this with the eggs of amphibians. Fig. 9.10 shows diagrammatically the organisation of the early stages of some reptilian eggs and that of the eggs of birds. The major differences between all these eggs and those of Amphibia are the existence outside the vitelline membrane of a layer of albumen, and the presence of membranes outside this. All of these are laid down by the oviduct as the egg passes down it. In most snakes and lizards the albumen layer is very thin, and it is surrounded by a fibrous membrane but no calcareous shell. In the crocodiles and the birds, and in many turtles, there is a thick albumen layer surrounded by two fibrous membranes – the inner egg membrane and the outer shell membrane. Outside these again is a calcareous shell. The properties of all these layers are vitally important for the development of the eggs because they determine the rates of diffusion of oxygen into the egg, and those of carbon dioxide and water out of it. Initially, the metabolic rate of amniote eggs is fairly low, but as the embryo develops, the demand for oxygen rises. It is essential, therefore, that the layers are permeable enough to allow oxygen to diffuse in at an adequate rate, while

not permitting sufficient water vapour to diffuse out to cause undue desiccation. During development various structures arise in the embryo which facilitate gaseous exchange with the outside, but it is important to look first at the properties of the inner membranes.

9.4.4 Permeability of egg membranes and shells

The fibrous shell membrane of lizards and snakes (and some turtles) contains crystals of calcium carbonate, but is evidently permeable to water since the eggs swell if they are put in contact with moist substrates. These eggs also lose water by evaporation in humid atmospheres, so that successful development appears to be dependent upon a balance between uptake and evaporation of water (Packard, Tracy & Roth, 1977). The hard-shelled eggs of most crocodiles and turtles can only take up a little water during development, and correspondingly have a low rate of water loss (see review by Packard, Packard & Boardman, 1982). Some of these reptilian eggs are very similar in organisation to those of birds (Packard & Packard, 1980), in which the eggs allow loss of water vapour only, and are therefore truly cleidoic. It is thus apparent that there are great differences between various egg types, and these may reflect to a great extent the environment of the nests in which they are laid (Packard *et al.*, 1979). Most reptilian eggs

Fig. 9.10. Schematic diagram showing the organisation of the eggs of amniotes. (*a*) Aves; (*b*) Chelonia and Crocodilia; (*c*) Lepidosauria. Although a space is shown between the inner egg membrane and the outer shell membrane in the eggs of birds, turtles and crocodiles for clarity of illustration, the egg membranes of turtles and crocodiles are adherent, and those of birds are adherent except in the region of the air space. Redrawn from Packard, Tracy & Roth (1977).

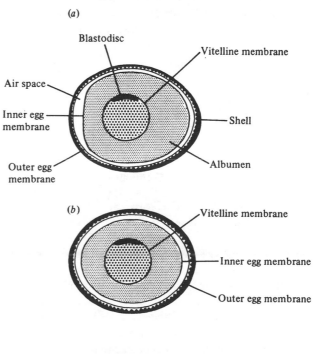

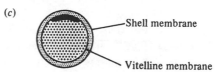

are laid on or under the ground, or in freshly cut vegetation, where the air has a high humidity, and there may also be a supply of water from the surroundings. In contrast, the eggs of birds are subjected, in the main, to dry conditions. All shells have pores in them, and the total pore area is probably connected with nest conditions in a complex way: reptilian eggs need to take up water, but if they are placed in the conditions of reptile nests they may also be subjected to lowered oxygen tension and increased carbon dioxide tension. While developing they therefore need to facilitate the diffusion of gases across the shell as well as increasing the possibility for water entry. Their eggs have a much higher pore area than those of birds, allowing increased diffusion rates for gases. In contrast to this, the nests of birds probably never suffer decreased oxygen tension or increased carbon dioxide tension, and shells with a small pore area are adequate for allowing gaseous exchange, while keeping the evaporative loss of water to a minimum. The size of the pores is related to the geometry of the crystal lattice of the shell (Board & Scott, 1980), and generally increases with increasing egg size (Tullett & Board, 1977). The total pore area is therefore determined by relative changes in the number and density of the pores, and the individual pore sizes.

In both reptiles and birds one of the barriers to oxygen diffusion into the egg is certainly the shell: its limited pore area restricts the movement of gases as discussed above, whereas the underlying shell membrane has no such restrictive effect (Wangensteen, 1972; Lutz *et al.*, 1980). A further barrier to gaseous diffusion is provided by the inner egg membrane plus the cellular membranes of the developing embryo. The net result of these barriers may be summarised for the hen's egg. In 21 days of incubation the embryo takes up 5 litres of oxygen and gives off 4 litres of carbon dioxide and about 10 litres of water vapour (Wangensteen, 1972).

9.4.5 Development of amniote eggs

As the metabolic rate of amniote embryos rises during their development, the limited surface area of the embryo becomes insufficient to deal with gaseous exchange. In spite of the relatively high permeability of the shell, carbon dioxide builds up during development, causing respiratory acidosis (Dawes, 1975; Tazawa, 1980). The problem of nitrogen excretion also increases because although some ammonia may be excreted as a gas, waste nitrogen begins to build up. During development, various cellular membranes grow out from the embryo to produce a complex system of organs which are used in the absorption of nutriment, in respiration, and in nitrogen excretion. Early development of the blastodisc occurs while the yolk is bounded by the vitelline membrane, and separated by this from the external albumen. The vitelline membrane breaks down or is resorbed early on (Romanoff, 1960), however, and the later embryo develops with its cellular surface next to the inner egg membrane.

The yolk sac, like that of many lower vertebrates, grows out initially as a trilaminar membrane composed of endoderm, mesoderm and ectoderm (see, e.g. Luckett,

1976). This membrane is soon split into two by the development of a coelomic space in the mesoderm, so that the definitive yolk sac is composed of only the inner layers – endoderm and mesoderm – which are together called the splanchnopleure (Fig. 9.11*a*). The outer layers enclosing the coelom – the ectoderm and mesoderm, together called the somatopleure – form part of the chorion (which is in no way related to the acellular membrane of the same name in amphibian and fish eggs). The somatopleure expands in area, and folds over the embryo as the coelom increases in volume. The folds then close (see the arrows in Fig. 9.11*a*), defining a fluid-filled cavity known as the amnion which encloses the embryo. By this closure, the chorion also comes to surround the whole embryo. The final membrane to consider is the allantois, which is an outgrowth of the embryonic hindgut. Initially it is used for the excretion of nitrogenous waste produced by the kidney, but as it expands its outer side fuses with the chorion, and the resulting chorioallantoic membrane, which is well vascularised, becomes the site of gaseous exchange with the external environment. It is also a site of active ion absorption, since the avian embryo absorbs its calcium supply during development from the shell (Simkiss, 1980). In viviparous amniotes (some reptiles and mammals) its function is altered, and it forms the foetal part of the placenta.

By these processes the developing embryo comes to be surrounded by the amnion, which acts as a protecting fluid bath, and by a coelomic space lined by the chorioallantoic membrane. The latter comes to lie beneath the inner egg membrane as albumen is reabsorbed in the bird's egg, so that the situation for gaseous exchange becomes as shown in Fig. 9.11*b*. The allantoic artery supplies blood to the capillary network and carbon dioxide comes into equilibrium with that in the air space, or, presumably, with that in the outer shell membrane where no such air space exists. Oxygen does *not* come into equilibrium, however, because of some kind of functional shunt, the details of which are unknown (Wangensteen, 1972). Altogether, this inner barrier accounts for 64% of the resistance to oxygen diffusion, compared with a figure of 36% for the outer shell membrane and shell (Piiper *et al.*, 1980).

9.4.6 Excretion by amniote eggs and adults

While the exchange of respiratory gases allows metabolism to proceed, much of the albumen is resorbed across the chorioallantoic membrane by birds, although in at least some reptiles the albumen is abandoned at hatching (Clark, 1953). The nitrogenous waste produced by metabolism of this protein, and that of the yolk is, perhaps surprisingly, mostly urea in reptile embryos. Ammonia is also produced and excreted as a gas in the early stages of reptile development, but uric acid does not seem to be produced in large amounts until hatching (Packard *et al.*, 1977). Much of the urea which is produced during development is stored in the albumen layer by snakes (Clark, 1953), thus possibly explaining the abandonment of the albumen at hatching. The allantois in reptiles is therefore involved in the excretion of nitrogen but not in its storage. In birds the story is different (see Freeman, 1974), and is best documented in the chick. After about the tenth day of incubation, ammonia, urea and uric acid begin to accumulate rapidly – the uric acid some 10 times faster than either the urea or the ammonia. This uric acid is secreted by the tubules of the mesonephric and metanephric kidneys, passes into the cloaca, and thence into the allantois. The allantoic membranes reabsorb salts and water from the lumen, leaving a solution with a low salt concentration, but high levels of nitrogenous waste. The uric acid is initially present as a colloidal sol, but the action of ammonia is thought to aid in its precipitation as solid ammonium urate. Solid urates accumulate throughout incubation, relieving the embryo of the necessity of storing or excreting urea, either of which require large volumes of water.

In adult birds, reabsorption of salts and water from the urine is mainly carried out by the cloaca, showing an interesting parallel with the situation in insects where the

Malpighian tubules produce an excretory fluid similar to the haemolymph, which is modified by the reabsorption of salts and water in the rectum. The kidneys of some birds are, however, able to produce slightly concentrated urine because they have some kidney tubules with a loop of Henle: these form a countercurrent concentrating mechanism which, together with the variable permeability of the collecting ducts, allows water resorption from the urine leaving the kidney. This mechanism is only well developed in the mammals, probably because both birds and reptiles depend to a great degree upon extra-renal salt-excreting systems. In particular, many species of reptiles and birds have nasal glands known as 'salt glands' (Peaker & Linzell, 1975). These glands secrete strong salt solutions, which allow the animals to conserve water, especially in desert forms and in those living in or near the sea.

The major end product of nitrogenous excretion in adults of both birds and reptiles is uric acid. In many invertebrates, uric acid is excreted almost without water,

Fig. 9.11. The membranes of amniote eggs. (*a*) Generalised diagram of the early foetal membranes in the egg of reptiles and birds. The arrows show the direction of growth of the chorion: the two folds fuse, and the amnion then encloses the embryo. (*b*) Gas exchange across the membranes of the chick embryo.

The section is drawn in the region of the air chamber. The two barriers to diffusion are marked as those between ambient air and the air chamber, and between the air chamber and the blood capillaries. (*a*) Redrawn after Luckett (1976); (*b*) redrawn from Wangensteen (1972).

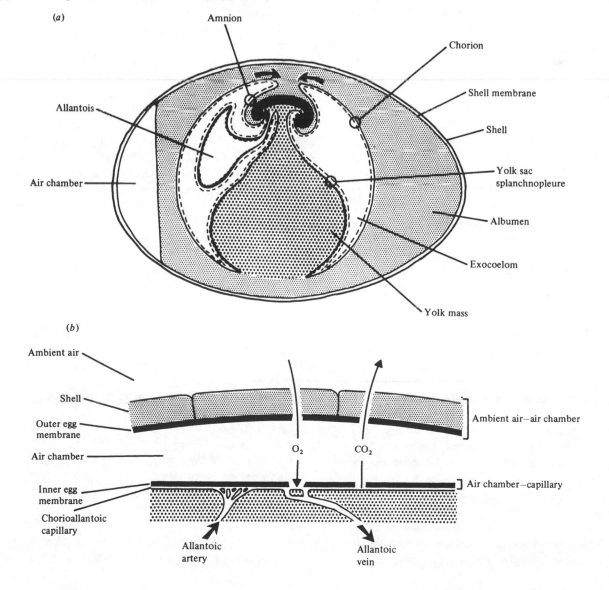

and because of this it can be argued that it aids water conservation: if soluble substances such as urea were excreted, much more water would be lost in the process. In birds, however, uric acid or urate is excreted as a colloidal gel, containing large amounts of water, and Sykes (1971) has pointed out that on average birds lose just as much water by excreting this urate gel as mammals lose by excreting urea in solution. It may be that the use of uric acid in birds was initially evolved to deal with the problem of accumulating nitrogenous waste in the cleiodoic egg, and that the retention of uric acid excretion in adults is to some extent a retention of this embryonic adaptation. Another suggestion has been made by McNabb & McNabb (1975): the excretion of urates may allow cations to be excreted without contributing to the osmotic pressure of the urine, as, for example, sodium urate and calcium urate. The picture is complicated by the situation as just described for the reptiles: these do not have a strictly cleidoic egg, and their embryos produce mainly urea and not uric acid, although *adult* reptiles excrete mainly uric acid. This difference from the birds suggests that the capacity for synthesis of uric acid has evolved several times, and probably for different reasons, in various amniotes. The ability of some of the amphibians to produce uric acid (see p. 201) emphasises the physiological adaptability of vertebrates, and shows that much more information is needed before the situation can be clarified. In particular, investigation of the excretory physiology of reptilian embryos has been much neglected, as pointed out by Packard *et al.* (1977), and further investigations in this field might prove informative.

9.4.7 Origins of amniote egg structures

Few authors have committed themselves on the subject of the origin of the structures within the amniote egg. Presumably the first step was the production of yolky eggs laid on land with a resistant shell membrane (Fisk & Tribe, 1949). These may have had a calcareous shell, but Szarski (1968) has argued that shells were derived *after* the accession of specialised breathing organs. Respiration in the early forms was probably provided for by the blood vessels of the yolk sac. Szarski (1968) envisaged the early replacement of ammonia as an excretory product by urea, which was then stored in the embryonic urinary bladder. The allantois developed from this region, and therefore represents primarily a nitrogen store. The enlargement of the allantois with its own blood vessels provided an additional source of oxygen, which came to replace the yolk sac in importance as larger embryos evolved. Finally, Szarski believed that it was this growing allantois which was responsible for splitting the extra-embryonic wall into somatopleure and splanchnopleure, thus forming the amnion and the chorion. In complete contrast to this view, Luckett (1976) suggested that the accumulation of fluid within the exocoelom promotes the differentiation of the somatopleure and splanchnopleure. Both suggestions are based on the idea of fluid accumulation as a causative factor in differentiation, since the expansion of the allantois has been shown to depend upon the secretion into it of fluid from the mesonephros and

metanephros. However, the allantois is in fact the last of the foetal membranes to differentiate, and the exocoelom has already begun to expand by the time the allantois appears. The origin of the chorion and amnion occurs early in ontogeny, and probably early in phylogeny. Fisk & Tribe (1949) suggested that the amnion and chorion evolved in reptiles because of selective pressure to offset problems of desiccation and adhesion to the egg membrane. This view has been supported by Luckett (1976), who pointed out that the early stages of development in amniotes result in a sinking inwards of the embryo, aiding both in protection of the embryo, and in amnion formation. The origin of the amnion probably therefore predates that of the allantois, and the development of some kind of protective amniotic fold probably came before any definitive split between the somatopleure and the splanchnopleure. By this argument, the amnion was the first membrane to be evolved, and the allantois the last.

9.4.8 Water loss from adult amniotes

At hatching, young reptiles and birds are exposed directly to the nest environment, and shortly afterwards to the general macroclimate. In contrast to amphibians, they can withstand low humidities and lack of water, in the main because their integument has a much lower rate of water loss. Figures for some amphibians and reptiles are given in Tables 9.4 and 9.5. From these it is evident that whether the loss rate is measured as a percentage of body weight, or whether as permeability per unit area, the difference between the two classes is, in general, greater than an order of magnitude. The exception provided by a very small number of anurans such as *Phyllomedusa sauvagii* has received comment above (p. 199). This difference between amphibian and reptilian skin is accounted for by two factors. First, amphibian skin has a large number of glands which open on to its surface, providing a mucous covering and constantly maintaining the surface moist for respiratory purposes. Reptilian skin, in contrast, possesses very few glands, and is dry on the outside: all respiratory processes are undertaken by the lungs. The second difference lies in the thickness and composition of the keratin layers in the skin. The epidermis of vertebrates contains two cell layers. The outer layer, or stratum corneum, has flattened cells containing the protein keratin, while the inner layer, the stratum germinativum, has columnar cells. In amphibians the cells of the stratum germinativum contain filaments of pre-keratin, and as the cells move towards the outside this is converted to α keratin (Parakkal & Matoltsy, 1964). The stratum corneum is formed by a flattening of the cells, and they become filled with α keratin. The cells retain a nucleus but lose all other organelles (see, e.g. Brown & Ilic, 1979). In spite of this keratinisation the stratum corneum in amphibians is highly permeable to water. In reptiles the stratum corneum is much thickened, and the keratin is found in both α and β forms (Spearman, 1966): the Squamata have two distinct layers, one inner layer composed of α keratin, and one outer layer composed of β keratin. The cells are filled with filaments and have no nuclei.

They also contain much bound phospholipid which probably aids in waterproofing. The outer layer of β keratin in the Squamata appears to be mainly for mechanical protection, since its removal or natural absence in aberrant forms (Bennett & Licht, 1975; Zucker & Maderson, 1980) produces no change in the rate of water loss. Damage to the inner layer, however, drastically increases water loss. Either the layers of keratin filaments, or the bound phospholipid, or a combination of the two, produces a very impermeable layer.

In birds the primary form of keratin is the β type, which also forms the feathers. Keratin filaments fill one outer layer of cells. In the mammals a different type of epidermal keratinisation is found (Spearman, 1966): the cells are no longer solid, but are hollow with a layer of α keratin filaments around the periphery. However, the skin is not a permanent barrier to water loss because it is punctuated by the openings of the sweat glands. These allow facultative loss of water as a cooling device, so that the advantages of both evaporative cooling and impermeability to water are present at appropriate times.

9.4.9 Summary of amniote adaptations

The protection of the embryo and the development of waterproofing layers in the adult have allowed the amniotes to become highly successful as terrestrial animals. Many other characters have, of course, contributed to their success, and the functioning of many of the organ systems differs markedly from that of the Amphibia. The development of an impervious skin in amniotes, whether or not it reflects a new or separate development from that of the large labyrinthodonts, has by definition been coupled with changes in respiratory methods. Various points have been mentioned in the section on Amphibia, but may be summarised here. There seems little doubt that some method of aspiration breathing was acquired in the early labyrinthodonts, probably in response to increasing size rather than

directly linked with the development of an impermeable skin. The increase in efficiency of lung ventilation which followed, coupled with the loss of cutaneous respiration, allowed the control of acid–base balance to move from the skin or gills, and in the amniotes the lungs and kidneys are together responsible for all acid–base balance. With the concentration of all respiratory exchange in the lungs, all oxygenated blood returned to the heart from one source, and the selection pressure to divide the blood circulation completely must have increased. Division is not absolute in the reptiles, but higher amniotes have completely divided hearts. The surfactants present in the lungs of all amniotes are capable of maintaining stable bubbles (Pattle, 1976), and their development appears to have been a necessary parallel to the development of an efficient system of very small alveoli, which provide an enormous respiratory surface.

Water conservation in reptiles depends almost entirely upon reabsorption in the cloaca, but in birds and mammals this process has been greatly aided by the presence of long loops of Henle in the kidney. This countercurrent concentrating mechanism initially aided and then took over the function of the salt glands which characterise many reptiles and birds. The mechanism is best developed in desert mammals, which can produce very concentrated urine. The mammalian stock also adopted viviparity instead of egg laying early in their evolution, and the amniote egg in this context became highly adapted to form placental junctions with the parent. Another large step was the adoption of homeothermy by birds and mammals, an advance preceded by the behavioural homeothermy of reptiles, and possibly paralleled by true homeothermy (endothermy) in the dinosaurs. All these developments and many others have aided in amniote success, but in comparison to the major breakthrough of the amniote egg they must be regarded as of secondary importance. The many adaptations of amniotes to terrestrial life are described in detail in standard works, and will not be considered further here.

10

Comparative discussion: the evolution of terrestrial ecosystems

'. . . this desolate and desert earth . . . remaining essentially lifeless for well over five hundred of our million-year periods, in spite of the abundant presence and notable progress of life in the waters. Through all these ages, the lands remained unconquered and must have seemed unconquerable. It is of the invasion of this territory, not only huge in extent but a veritable land of promise from the evolutionary point of view, that this chapter will tell.'

From 'Life conquers the dry land' in *The science of life* by H. G. Wells, J. Huxley & G. P. Wells, 1931, Cassell. London.

In the preceding chapters each animal group has been considered separately. While it is hoped that this approach has allowed an understanding of the ways in which terrestrial forms have evolved in each group, the interactions between different types of animal, and their dependence upon the evolution of a terrestrial flora, have been largely ignored. It is the purpose of this chapter to draw together the conclusions so far put forward, and to consider the evolution of individual animal taxa against the wider background of changing environmental conditions brought about by geological factors, and by the evolution of other elements of the terrestrial fauna and flora.

To accomplish this, the terrestrial groups are first discussed in terms of the routes that they are thought to have taken in their colonisation of the land, and the physiological limitations and advantages that these various routes have conferred. Second, brief consideration is given to those groups which have *not* produced terrestrial forms. Third, the times at which various terrestrial groups evolved is related to changing earth climate, and to the evolution of terrestrial plants. This leads on naturally to the discussion of how successful the various taxa have been in different types of terrestrial habitat. Lastly, the discussion comes full circle with brief consideration of the topic of recolonisation of aquatic environments by animals that are essentially terrestrial.

10.1 Routes on to land and their consequences

From consideration of a number of factors, it has been possible to speculate upon the possible routes by which various animal groups have invaded the land from aquatic environments. A summary of the speculations made in previous chapters is given in Table 10.1, from which it is evident that the two major routes involved have been the direct one, across the marine littoral zone, and an indirect one via fresh water. Individual phyla are by no means limited to one particular route, but in general the Crustacea have tended to move directly across the littoral zone, while many molluscs and annelids have probably passed through an intermediate stage in fresh water. There are also groups that may have had intermediate stages in brackish water or saltmarshes, and further groups which may have moved inland in interstitial water. The use of each of these routes has had very far-reaching consequences, and has contributed largely to the different degrees of adaptation to the terrestrial environment shown by the various groups. These consequences can be considered under a number of headings.

10.1.1 Osmoregulation and water balance

Some representative values of the osmotic pressure of body fluids in terrestrial Nemertinea, Annelida, Mollusca, Crustacea and Vertebrata are given in Table 10.2. The spread of values is, of course, much wider than is evident from the table, but even taking into account the wide variety noted in previous chapters, it can be concluded that the lowest values are found in prosobranch gastropods derived from freshwater ancestors, while higher values are found in gastropods which have been derived from marine ancestors, and in crustaceans. The correlation between osmotic pressure and ancestry is a good one, and has been emphasised in several chapters. It is not a rigid condition, however, for the reason that the osmotic pressure of body fluids is governed by a number of factors, and is therefore not strictly related either to the characteristics of an animal's environment, or to the environment of its ancestors.

Table 10.1. *Possible routes on to land taken by the major animal groups*

Group	Possible route	Page reference
Platyhelminthes	Fresh water	8
Nemertinea	Marine littoral	11
Nematoda	Interstitial	15
Annelida		
Polychaeta	Marine littoral and interstitial	19
Oligochaeta	Fresh water, burrowing	32
Hirudinea	Fresh water	32
Mollusca		
Prosobranchia		
Neritacea	Fresh water	37
Architaenioglossa	Fresh water	48
Littorinacea	Marine littoral: mangroves ?	46
Pulmonata	Salt marshes	62
Crustacea		
Amphipoda	Marine littoral	68
Isopoda	Marine littoral	74
Decapoda		
Astacura	Fresh water	82
Anomura	Marine littoral	87
Brachyura		
Potamonidae, etc.	Fresh water	88
Grapsidae	Marine littoral and fresh water	94
Ocypodidae	Marine littoral	94
Gecarcinidae	Marine/brackish water littoral	102
Chelicerata	Marine littoral	126
Onychophora	?	144
Myriapoda	?	145
Hexapoda	? Marine interstitial	159
Vertebrata	Fresh water	194

Table 10.2. *The osmotic pressures of the blood of some terrestrial animals*

Species	Taxonomic position	Osmotic pressure (mOsm)	Reference	Possible route to land
Nemertinea				
Argonemertes dendyi	(Hoplonemertea, Prosorhochmidae)	145	C. Little, personal observations	Marine littoral
Annelida				
Lumbricus terrestris	(Oligochaeta, Lumbricidae)	165	Ramsay, 1949*a*	Fresh water
Mollusca				
Eutrochatella tankervillei	(Prosobranchia, Helicinidae)	67	Little, 1972	Fresh water
Poteria lineata	(Prosobranchia, Cyclophoridae)	74	Andrews & Little, 1972	Fresh water
Pseudocyclotus laetus	(Prosobranchia, Assimineidae)	103	Little & Andrews, 1977	Brackish water
Helix pomatia	(Pulmonata, Helicidae)	183	Meincke, 1972	Salt marshes
Pomatias elegans	(Prosobranchia, Pomatiasidae)	254	Rumsey, 1972	Marine littoral
Agriolimax reticulatus	(Pulmonata, Limacidae)	345	Bailey, 1971	Salt marshes
Crustacea				
Talitrus sp.	(Amphipoda, Talitridae)	400	C. Little, personal observations	Marine littoral
Holthuisana transversa	(Decapoda Brachyura, Sundathelphusidae)	517	Greenaway & MacMillen, 1978	Fresh water
Porcellio scaber	(Isopoda, Porcellionidae)	700	Barrett, 1972	Marine littoral
Cardisoma armatum	(Decapoda Brachyura, Gecarcinidae)	744	de Leersnyder & Hoestlandt, 1963	Marine littoral
Coenobita brevimanus	(Decapoda, Anomura, Coenobitidae)	800	Gross, 1964*a* (recalculated)	Marine littoral
Vertebrata				
Bufo bufo	(Amphibia, Anura)	205	Schoffeniels & Tercafs, 1965	Fresh water
Bufo americanus	(Amphibia, Anura)	310	Schmid, 1965	Fresh water

All figures are averages for active animals in damp conditions on land, or for equilibrium with fresh water.

For example, while it is true that most marine invertebrates have blood with a high osmotic pressure, it is not true that the blood of all freshwater invertebrates has a *low* osmotic pressure. It is convenient here to consider one group in detail first, and the Crustacea will be used because they demonstrate the physiological differences characteristic of marine and freshwater forms. The Mollusca show comparable physiological variation.

The marine and freshwater routes: crustaceans and molluscs.

The freshwater Crustacea show a wide range of values of osmotic pressure. Freshwater amphipods such as *Gammarus pulex* (274 mOsm – Lockwood, 1961) and isopods such as *Asellus aquaticus* (276 mOsm – Lockwood, 1959a) are at the bottom of the range, far below their marine relatives. This low osmotic pressure was probably necessitated by the small size of the animals, which made the maintenance of a large difference from the external medium too expensive in metabolic terms, in spite of the development of ion uptake mechanisms with high affinity for ions such as sodium (see, e.g. Sutcliffe, 1968, 1974). Freshwater decapods, in contrast, tend to be large and have retained a high osmotic pressure by developing a low permeability to water in addition to efficient ion uptake mechanisms. Thus, for example, *Potamon niloticus* has a value of 492 mOsm (Shaw, 1959), while in *Eriocheir sinensis* osmotic pressure is as high as 577 mOsm (Krogh, 1939).

All terrestrial crustaceans have high values of osmotic pressure, but this may represent either a marine or a freshwater origin. In the amphipods and isopods this is probably due to a direct marine ancestry. We have no evidence that terrestrial forms of either of these groups evolved in fresh water: if they had done so, we might expect osmotic pressures of 270 mOsm or less, based on the values shown by present-day freshwater forms. On the other hand, some terrestrial decapods have a high osmotic pressure because they are derived from freshwater forms which themselves retained a high value. The situation is further complicated because, as pointed out above, many factors are involved in maintaining an animal's osmotic pressure. Overall permeability to salts and water and the ability to absorb ions from the medium are important, especially in relation to size, but in addition to this the function of the kidney in eliminating water while conserving ions may be singled out as a key factor. The large, relatively impermeable freshwater crabs such as *Eriocheir sinensis* and *Potamon niloticus* have kidneys which produce urine iso-osmotic to the blood, but because of the low permeability to water the rate of urine flow is low, and little salt is lost by this route. The smaller crustaceans such as amphipods and isopods actually have a lower permeability to water than the larger crabs, but because of their enormously increased surface : volume ratio they probably all produce hypo-osmotic urine in order to eliminate excess water. This is in contrast to marine forms, which probably produce iso-osmotic urine: although there is no direct evidence for this, brackish water amphipods produce iso-osmotic urine in high salinities (Lockwood, 1961),

and most marine invertebrates are thought not to have the capacity for the production of hypo-osmotic urine. There is little information about urine production in any terrestrial crustaceans, even the larger ones, but on the basis of the above it would be expected that both large and small forms would produce an iso-osmotic urine. The terrestrial crab *Holthuisana transversa* has equal sodium concentrations in blood and urine, so that the urine is probably iso-osmotic with the blood (Greenaway & MacMillen, 1978).

The mechanisms by which terrestrial crustaceans are in fact able to some extent to keep the composition of their body fluids within tolerable limits are not well understood. The problem is essentially one of the supply and loss of water rather than of salt regulation. All the evidence points to the importance of mechanisms which are linked to behaviour, rather than to direct physiological processes as is the case in aquatic forms. Some examples demonstrate this point. The terrestrial crab *H. transversa* is essentially a fossorial species which can withstand extreme losses of water, and which probably stays in water balance by absorbing water that condenses in its burrows (Greenaway & MacMillen, 1978). It has no structural or physiological modifications which allow it to absorb water from the soil. Terrestrial hermit crabs, and some other crabs whose origins were marine, can maintain the osmotic pressure of their body fluids at a preferred level by selecting water of different salinities in the appropriate proportions (Gross, 1964a). In practice this mechanism is operated by the crab spending different periods of time in each salinity. Further behavioural methods of regulating water balance are found in purely terrestrial isopods, as described in Chapter 5: *Porcellio scaber*, like most isopods, remains in the humid environment of leaf litter during the day, but climbs vertically upward at night (Den Boer, 1961), when the environmental air humidity is higher. Few observations have been made of the behaviour of terrestrial amphipods, but for the most part they seem to remain in forest leaf litter, and desiccate rapidly outside this environment. However, like their relatives of the marine littoral zone, they will climb upwards to avoid being submerged (Hurley, 1968).

Coupled with behavioural mechanisms for regulating the water content of the body and the osmotic pressure of the body fluids, most terrestrial crustaceans appear to have a wide tolerance of internal osmotic pressures. This reflects the fact that the behavioural mechanisms are seldom precise enough in nature to be able to control the composition of body fluids to within narrow limits. The origin of this tolerance is of some interest, since many marine crustaceans are stenohaline, and many freshwater species can also tolerate salinities only within a narrow range. Taking the freshwater examples first, the isopod *Asellus aquaticus* presents a typical picture. The normal osmotic pressure of its blood in fresh water is 276 mOsm, and it can withstand up to about 360 mOsm (Lockwood, 1959a). In the freshwater crab *Potamon niloticus* the normal osmotic pressure of the blood is 492 mOsm, and most individuals can withstand internal values of about 550 mOsm; but only a few individuals can tolerate an increase to over 730 mOsm (Shaw, 1959). Generally, freshwater invertebrates regulate their osmotic pressure at approximately the normal level in all salinities up to that at which the environmental equals the internal osmotic pressure. Above this external osmotic pressure, that of the blood starts to rise, and mortality increases. In other words, freshwater invertebrates have homeostatic mechanisms which regulate the internal osmotic pressure within fine limits; but once the external conditions become so altered that the homeostatic mechanisms can no longer cope, the body cannot tolerate the resulting internal changes.

Crustaceans from the marine littoral zone show entirely different responses. The intertidal isopod *Ligia pallasii* can withstand internal osmotic pressures of about 800–1500 mOsm (Wilson, W. J., 1970) for long periods. Similarly, *L. oceanica* can survive with internal osmotic pressures of 800–1700 mOsm (Parry, 1953). Many littoral marine crabs and hermit crabs can tolerate wide ranges of internal osmotic pressures: *Hemigrapsus oregonensis* can tolerate values between 600–1500 mOsm (Gross, 1964*b*) and *Coenobita perlatus* about 800–2000 mOsm (Gross & Holland, 1960). The exceptions to this generalisation are those crabs that have developed excellent powers of osmoregulation, such as *Grapsus grapsus* (Gross, 1964*b*).

The divergence of these two physiological approaches by marine and freshwater crustaceans suggests a possible reason for the lack of terrestrial crustaceans with freshwater origins. The freshwater environment has been associated with excellent control of the osmotic pressure of body fluids, but this good control has in turn eliminated the need for development of tolerance. The system is therefore fundamentally unsuited for the development of a terrestrial existence, in which variable conditions impose a variable water supply. On the other hand, the development of tolerance of widely changing internal osmotic pressures has been widespread in littoral marine crustaceans, and this has preadapted them for life in terrestrial environments, where desiccation, and hence rise in internal osmotic pressure, is effectively the norm. It should also be pointed out that in littoral crustaceans the ability to withstand increasing osmotic pressure is better developed than the ability to withstand dilution – evaporation being more important in the littoral zone than dilution. This may well be correlated with the ability of terrestrial crustaceans to tolerate great increases in blood concentration, and a relative inability to tolerate dilution. Terrestrial amphipods are thought to be killed by osmotic stress when rainwater accumulates into pools (Duncan, 1969), and terrestrial isopods are thought to climb out of their cryptozoic habitat to avoid the osmotic stress produced by flooding.

It must be emphasised that the freshwater 'regulators' and marine 'tolerators' by no means form rigid categories. These are the general trends, but there are exceptions, including the good regulators such as *Grapsus grapsus*, found intertidally, and the good terrestrial tolerator *Holthuisana transversa*, which is of freshwater origin. *H. transversa* can tolerate a change in the osmotic

pressure of its blood from 500 to 980 mOsm (Greenaway & MacMillen, 1978). However, these exceptions merely demonstrate that physiological mechanisms do evolve, and allow animals to invade new habitats, without necessarily involving *pre*adaptation. Indications of this physiological lability are widespread. The case of *Potamon niloticus* showed, for instance, that some individuals could tolerate 75% sea water well, although others could not. If there were a strong selective advantage in this tolerance, no doubt the mean tolerance of the population would change. Similarly, Barnes (1968*b*) has emphasised the variation in osmotic pressure found within a population of the estuarine crab *Australoplax tridentata*, and has pointed out that such variation will have selective advantage under conditions of change. As a generalisation, however, the development of tolerance of change in internal osmotic pressure seen in marine littoral crustaceans seems to have preadapted some of these forms for a terrestrial life. Lack of such preadaptation in freshwater forms has limited the number of terrestrial crustaceans originating from fresh water.

A similar theme may be outlined for the Mollusca. Here the osmotic pressure of the body fluids in terrestrial forms is generally lower than in the Crustacea, although values as high as 400 mOsm have been recorded in active slugs by Bailey (1971). It seems likely that the molluscan epidermis is more permeable to salts and water than the integument of crustaceans, and the general lowering of osmotic pressure probably reflects this. The situation is similar in other soft-bodied invertebrates. However, there are distinctive contrasts between groups of molluscs thought to have reached the land by different routes. The pulmonates, which are postulated to have originated in saltmarshes or similar habitats, retain a relatively high osmotic pressure, paralleled by the pomatiasid prosobranchs which also had a direct marine origin. Neither of these groups produces significantly hypo-osmotic urine. In contrast, helicinid and cyclophorid prosobranchs are derived from freshwater ancestors, have blood with a low osmotic pressure, and produce hypo-osmotic urine. These two groups are confined to tropical forests, probably in the main because they are equipped to deal with dilution but not desiccation, and are intolerant of changes in internal osmotic pressure. As with the Crustacea, therefore, it may be said that the groups derived from marine tolerators have been more successful on land than those derived from freshwater regulators.

The interstitial route: nemertines, oligochaetes and insects

It is instructive to examine the extent to which these generalisations can be applied to other animal groups, although less information is available, and there is no clear series of related animals which have colonised the land by different routes. In the Nemertinea the osmotic pressure of rhynchocoelic fluid has been measured in very few species only. In the terrestrial species *Argonemertes dendyi* it is almost identical to the value found in a freshwater species *Prostoma jenningsi* (see p. 11). Yet the terrestrial *Argonemertes* is postulated to have been derived from marine ancestors. Clearly, this situation does not fit the picture presented above for molluscs and crustaceans. Until further information about salt and water balance is available for a variety of nemertines, no overall explanation can be offered. There is one facet of nemertine organisation, however, which may be linked to the problem. The flame cell system has been developed to an extreme degree in terrestrial and freshwater nemertines, but even in supra-littoral forms such as the New Zealand *Acteonemertes bathamae* the flame cell system is very well developed (Moore, 1973). This suggests that the marine precursors of the terrestrial forms were well supplied with a mechanism for removing water, even without any long period of evolution in fresh water. *A. bathamae* lives hidden in crevices between damp pebbles and stones high on the shore, and these may be subject to drainage of groundwater from the land. A study of this type of environment might be very rewarding in establishing the selection pressures which have brought about the immense development of the excretory system in nemertines.

The characteristics of such transitional environments, including the truly interstitial habitat, are important in considering the evolution of other groups. The arguments presented for molluscs and crustaceans are undoubtedly oversimplified, but apply in many cases because these animals to a great extent live above ground and are subjected to the conditions of the macroenvironment. For the smaller animals, such as nematodes, and for burrowers such as the oligochaete annelids the physical characteristics of soil and crevice habitats must be taken into account. Because sediments act as buffers to changes in salinity of overlying water, the gradations from marine to freshwater soils and marine to terrestrial soils are probably fairly gradual. In neither transition will the selection pressures for tolerance be as great as in animals living on the surface; but in both cases the final osmotic adaptation must be to dilution. It follows, therefore, that animals which are essentially adapted to the soil are likely to have evolved good mechanisms for regulating their salt and water content, in parallel with those shown by freshwater animals. If this line of argument is accepted, it will be apparent that it may be difficult to distinguish between the mechanisms of osmoregulation and water balance in those soil animals colonising land via fresh water, and those colonising via marine intertidal sands.

It has been suggested (p. 32) that since the oligochaetes are today primarily a freshwater group, they have probably invaded the land from fresh water. In spite of this, their body fluids have retained a fairly high osmotic pressure (Table 10.2). The production of hypoosmotic urine presumably helps them to maintain this high osmotic pressure, the mechanisms being typical of freshwater animals. However, there appear to be no available figures for the osmotic pressure of the body fluids of truly freshwater oligochaetes with which to compare the terrestrial forms (see, e.g. Oglesby, 1978). If it should be shown that present-day terrestrial oligochaetes have significantly higher osmotic pressures than freshwater forms, the possibility of a direct origin of

some terrestrial oligochaetes from marine forms might be raised. If this is so, some present-day oligochaetes may be primitively marine, and it should be noted that some tubificids appear to be fully adapted to entirely marine, and even deep water, habitats (e.g. Erséus, 1980; Timm, 1980). It is certainly worth discussing the possibility that modern oligochaetes represent the independent invasion of terrestrial and freshwater soils from marine sediments. Two points concerning the physiology of salt and water balance in terrestrial oligochaetes suggest that while they have often been treated by physiologists almost as freshwater animals, not all their characteristics are typical of inhabitants of fresh water. The first point is their great tolerance of changes in internal osmotic pressure. From the figures given in a comprehensive review by Oglesby (1978), it appears that *Lumbricus terrestris* can tolerate changes in osmotic pressure of coelomic fluid from 150 to approximately 400 mOsm. Such a range is larger than would be expected for a freshwater invertebrate. As argued above, osmotic tolerance is a property usually incurred by animals of marine origin. The second point is one which has not been discussed in the present context. This is the existence in the blood of significant concentrations of amino acids – up to 13.8 mM/litre in the tropical earthworm *Lampito mauritii* (Pampapathi Rao, 1963). Amino acids are found in the blood of a number of other terrestrial invertebrates such as the insects, myriapods and onychophorans, although their function is unknown. Free amino acids are also prominent in the tissues of many marine and brackish water invertebrates. In Crustacea and Mollusca found in brackish water these cellular amino acids change in concentration when the osmotic pressure of the blood changes, thereby reducing the changes in intracellular concentrations of ions. While no explanation is yet available of the role of amino acids in the blood of terrestrial invertebrates, their presence forms a contrast to what is known of freshwater invertebrates. Such a contrast suggests that, whatever their function, the presence of amino acids is more likely to have been inherited from marine than from freshwater forms.

When attempting to use values for the osmotic pressure of body fluids as evidence of ancestry, it is essential to compare animals within related groups, because of the very different structures and mechanisms involved in maintaining body fluid composition. The Insecta, Myriapoda and Onychophora may therefore be considered together, and possibly may be compared with the Chelicerata and the Crustacea. The ranges of osmotic pressure within these groups are (from Tables 6.6, 7.2, 8.2, 8.5): Insecta, 290–510; Onychophora, 180–200; Myriapoda, 150–500; Chelicerata, 360–480 mOsm. Most of these figures, with the exception of those for the Onychophora, are similar to those in the Crustacea given in Table 10.2, and are considerably higher than those in the Mollusca, Annelida, Nemertinea or Amphibia. While it is tempting to assign the cause of these relatively high values to a marine origin, it must be remembered that decapod crustaceans having a freshwater origin retain high values of osmotic pressure. On the other hand, these decapods are relatively large, whereas most Insecta and Myriapoda are small, and if such small forms had evolved in fresh water it seems likely that their osmotic pressure would have dropped radically, as in the smaller freshwater Crustacea. The fact that it has, on the contrary, remained high strongly suggests a marine origin. Values in Insecta and Myriapoda are, in fact, very similar to the osmotic pressures found in Chelicerata, which are believed on other grounds to have a marine origin (see p. 126). If we may extrapolate further from the arguments derived for Crustacea and Mollusca, another point reinforces the suggestion of a marine origin at least for the Insecta. This is the degree of tolerance to dehydration and to changes in composition of the body fluids. Although most insects are thought of as good regulators, many are also extremely tolerant of changes in the osmotic pressure of the blood (see, e.g. Barton-Browne, 1964). Among the Apterygota, some of the Collembola can tolerate a range from about 360 to 1360 mOsm (Weigmann, 1973). Admittedly, these forms may be specialised for life in saltmarshes, and little is known of the tolerance of freshwater and terrestrial soil apterygotes. However, the tolerance of some of the Pterygota is also striking. There is a great deal of variation in the range tolerated by various pterygotes. In some, such as the larvae of *Locusta migratoria*, osmotic pressure varies from a normal value of 338 mOsm to only 378 mOsm after feeding, and rapidly back to normal (Berneys & Chapman, 1974). In *Periplaneta americana* the osmotic pressure is normally well regulated, but extreme values of 245–470 mOsm have been observed (Wall, 1970). *Dysdercus fasciatus*, with a normal osmotic pressure of 270 mOsm in hydration, reached 435 mOsm after desiccation (Berridge, 1965a), and in *Trichostetha fascicularis* desiccation caused a rise from 437 to 742 mOsm (Fielding & Nicholson, 1980). The blowfly *Lucilia cuprina* showed average changes from 336 mOsm immediately after feeding to only 390 mOsm 8 hours later (Barton-Browne & Dudzinski, 1968), but the housefly *Musca domestica* showed changes from 163 to 762 mOsm (Bolwig, 1953). In most of these experiments the lethal change was not really estimated, and many of the species could probably have withstood greater ranges. This is underlined by the ability of the tissues to continue functioning in very wide ranges of osmotic pressure. Good examples are given by the Malpighian tubules: in *M. domestica* these produce excretory fluid (although at varying rates) in media from 110 to 740 mOsm (Dalton & Windmill, 1980). In *Calliphora erythrocephala* the tubules secrete in media from 160 to 1125 mOsm (Berridge, 1968).

These extremes of ability to tolerate changes in osmotic pressure of the blood are so different from those typical of freshwater invertebrates, that their direct origin from freshwater forms seems unlikely. It may, of course, be argued that the tolerance has been developed in response to the conditions of the terrestrial environment, but this again seems unlikely because many insects have also developed systems which in practice maintain the osmotic pressure of the blood at a fairly constant level. The development of tolerance is likely to have taken place at a time when the blood was subjected to large variations in composition. A marine littoral ancestry, rather than a freshwater one, would have provided these conditions. The development of a regulatory system as well as great tolerance may reflect a terrestrial adaptation. But here, in principle, though not in detail, the insects show a close parallel with the oligochaetes. For these, it may be argued that the interstitial or burrowing habit has brought about first the capacity for osmotic tolerance – in situations where osmotic pressure was governed by the influence of the sea and evaporation – and secondly the capacity for production of hypo-osmotic urine, the selection pressure coming from contact with fresh water in soil crevices. Since an interstitial origin for the insects has already been envisaged (p. 159), similar arguments for them may be briefly considered. Thus the insects show, as do the oligochaetes, the development of osmotic tolerance *and* the capacity to produce hypo-osmotic urine. Although it is usually the ability to produce *hyper*-osmotic urine which is highlighted as a terrestrial adaptation in insects, some forms at least can reabsorb ions from the rectum without concomitant reabsorption of water, so that hypo-osmotic urine is excreted. Not all insects can do this – *Schistocerca gregaria*, for example, produces rectal fluid which is slightly hyper-osmotic to the blood even when fed on a water diet (Phillips, 1964). *Calliphora erythrocephala*, however, can produce rectal fluid as dilute as 62 mOsm (Phillips, 1961), and *Dysdercus fasciatus* can produce urine with an osmotic pressure of 94 mOsm (Berridge, 1965a). Research into the ability to produce a hypo-osmotic urine does not seem to have occurred in many cases, and this ability may be more widespread than currently realised. The final parallel with the oligochaetes is the presence of free amino acids in the blood, discussed above. These are present even in apterygotes, and presumably form an ancient characteristic of the group. Since their concentration in the blood does not change during desiccation (Okasha, 1973), at least in *Thermobia domestica*, their present function is presumably not to do with osmoregulation. The parallel with oligochaetes can probably not be taken further at present because of these unknowns, and in any case it must be admitted that while the oligochaetes have remained bound to the soil environment, many insects have become emancipated from the cryptozoic habitat. It might, however, be profitable to investigate further similarities between the oligochaetes and some of the cryptozoic arthropods of the Insecta–Onychophora–Myriapoda line.

The freshwater route: vertebrates

From the discussion given above, it may be concluded that there is at least some evidence to suggest that most terrestrial invertebrate groups have originated directly from the sea. The few groups which are thought to have come from fresh water, such as some of the prosobranch molluscs and some of the decapods, have in general not become very widespread in terrestrial ecosystems. In this they contrast completely with the vertebrates, which are thought to have used a freshwater route to land. It is therefore important at this stage to consider the possible reasons for this difference.

Freshwater amphibians have body fluids with an osmotic pressure of about 180–270 mOsm (e.g. Schmid, 1965; Alvarado, 1979). In general, survival is poor if this osmotic pressure rises, either from desiccation or from increasing salinity of the external medium. Amphibians are similar to freshwater invertebrates in this, and the similarity extends to their renal systems, which produce hypo-osmotic urine, but are unable to produce hyper-osmotic urine. There are very few exceptions to this typical 'freshwater state' in any amphibians, aquatic or terrestrial. One of the most striking is *Rana cancrivora*, a frog which lives in mangrove swamps (Gordon, Schmidt-Nielsen & Kelly, 1961). Adults of this species can tolerate concentrations of up to 80% sea water, and a blood concentration of 280–800 mOsm. Most of the increase above the normal level is due to accumulation of urea, and not to high levels of electrolytes. Probably the build-up of urea allows higher concentrations to be reached than would be tolerated with increasing salt content. This idea is supported by studies of another widespread brackish water amphibian, *Bufo viridis*, which allows sodium and chloride, and not urea, to rise in concentration as osmotic pressure rises. This species can tolerate a rise of only 238 to 528 mOsm (Gordon, 1962). There is an interesting parallel to *R. cancrivora* in a terrestrial desert toad, *Scaphiopus couchii*, which can tolerate changes in blood concentration from 294 to 606 mOsm during desiccation (McClanahan, 1967). Most of this increase is due to accumulated urea.

The ability of a few species of amphibians to tolerate high internal osmotic pressures demonstrates the physiological flexibility of the group; but the fact that the great majority can tolerate only small changes in osmotic pressure of the blood, and especially of its electrolyte content, suggests comparison of the terrestrial species with terrestrial invertebrates of freshwater origin. Like these, most amphibians are limited to areas near water, or to humid microhabitats. However, while there are very few invertebrates of freshwater descent that have moved out of their humid microhabitats, amphibians or their ancestors at some time gave rise to reptilian stock which has colonised the driest of microhabitats. It is important to emphasise the properties which have allowed vertebrates to do this while the invertebrates have been confined to humid niches. The main factors here seem to be a combination of larger size and the development of a great reduction in the permeability of the skin to water. These factors, in spite of the inability of the reptilian kidney to produce hyper-osmotic urine, have allowed the reptiles to become truly terrestrial. The contrast in skin permeability between amphibians and reptiles is strikingly evident from the figures in Table 9.5. The only invertebrate groups which can also withstand desiccating environments are those that are probably of marine origin – the insects and chelicerates. These have developed even better water-proofing systems than the reptiles, and this is presumably associated with their relatively small size, although the waterproofing of insects as discussed earlier (p. 177) probably arose in a different context, to provide a hydro-fuge cuticle. Apart from these three groups – the reptiles,

the insects and the chelicerates – very few terrestrial animal groups have developed waterproofing to a high degree in a wide variety of species. It is true that a few isopods such as *Venezillo arizonicus* have low transpiration rates, and the crab *Holthuisana transversa* probably has a lower transpiration rate than that given for deca-pods in Table 5.9 (MacMillen & Greenaway, 1978), but these are exceptions to the general trend.

Neither the amphibians nor the reptiles has the ability to produce hyperosmotic urine, from which it may be concluded that this capacity arose after the vertebrates became independent of water. This probably contrasts with the situation in insects, where an imperme-able epidermis and hyperosmotic urine were probably early characteristics. Once again, size is important here, because the amphibians were large enough to be able to tolerate long periods away from water, even without a very impermeable skin, whereas the insects, being smaller, must have developed mechanisms for both toler-ance and regulation in order to be able to spend even short periods away from their cryptozoic habitat. According to Maddrell (1981) insects evolved in 'osmoti-cally and ionically stressful environments', and this idea agrees well with the hypothesis put forward here of origination in, or colonisation via, the interstitial en-vironment, where the initial evaporation stress led to development of osmotic regulation. The apparent paral-lels with vertebrates in terms of the ability to live in dry environments therefore disguise basic differences in the origin and function of the physiological systems involved. The vertebrates are basically regulators, with little toler-ance of internal change, having evolved in fresh water. Their development of an impermeable skin, together with their larger size, has, as it were, allowed them to move into desiccating environments *in spite of* their freshwater origin.

The relevance of blood composition

One further consequence of the various routes on to land taken by the different groups must be dis-cussed. This is the proportion of various ions in the blood which contribute to its osmotic pressure. It is perhaps appropriate to begin by mentioning Macallum's (1926) hypothesis that when marine organisms first appeared sea water had an ionic composition far different from that of today, and that many animals retained ions in their body fluids in proportion to this early composition, even when the composition of the sea changed. Macallum's idea now seems unlikely on two grounds: one, that early sea water was probably not very different from that of today (see, e.g. Mackenzie, 1975); and two, that the blood of most marine invertebrates follows closely the composition of sea water into which the animals are placed. Burton (1973), for example, reviewed the factors regulating the composition of both body fluids and tissues. However, a variant of Macallum's hypo-thesis has been suggested by Spaargaren (1978), who attempted to calculate the composition of ancient sea water from the common properties of the blood of present-day marine animals. It can be seen that some of the ideas discussed in the past few pages retain an element of Macallum's original idea. In other words, it is proposed that most animal groups which invade new environments retain at least some of the characteristics of their original blood composition, and some of the methods by which variation in this is controlled or tolerated. In particular, those groups that have produced terrestrial representatives directly from marine ones have retained a relatively high osmotic pressure, while those that have passed through a freshwater stage have a lower osmotic pressure. It has also been asserted, however, that the more 'terrestrial' the species in a group, the more the osmotic pressure tends to rise (Schmid, 1965). No explanation for this is apparent in the majority of cases, although where the rise is extreme (e.g. in the aestivating toad *Scaphiopus couchii*) it may decrease osmotic loss of water to the soil. In *S. couchii* there is a good correlation between rise in the osmotic pressure of the blood and fall in soil moisture tension (McClanahan, 1972).

One of the characteristics of ionic composition of the blood which received early attention is the ratio of the concentration of monovalent cations (Na^+ and K^+) to divalent cations (Ca^{2+} and Mg^{2+}). Florkin (1949) suggested that vertebrates tended to have a monovalent to divalent ratio higher than that of invertebrates, regardless of habitat. Lutz (1969), however, concluded that marine invertebrates have a ratio of approximately 10, while in freshwater invertebrates it is approximately 20. He also suggested that in marine invertebrates chloride concentration is greater than sodium concentration, whereas in freshwater invertebrates the reverse is true. Recent work has allowed these generalisations to be considered in more species, and to be modified. It is useful here to discuss these modifications, and to see how the blood composition of terrestrial invertebrates fits into the general picture. Since most analyses available are for molluscs and for crustaceans, discussion will be concentrated upon these two groups.

Some figures for blood composition in marine, freshwater and terrestrial gastropod molluscs and crustaceans are given in Table 10.3. Marine gastropods have a monovalent : divalent cation ratio of just over 7, which reflects their almost total lack of ionic regulation: the ratio for sea water itself is 7.5. In freshwater gastropods the ratio is little different, although in fresh water itself the ratio is much lower. This reflects the fact that while sodium and potassium are reduced to about 10% of their concentration in marine forms, calcium is retained at 20% of the marine concentration, while magnesium has been reduced to only 5% or less of the concentration in marine forms. Calcium levels are probably maintained in equilibrium with the calcium carbonate of the shell (Potts, 1954a). This trend in freshwater gastropods is continued in terrestrial forms, where similar values for calcium are found in species with a freshwater ancestry, such as species of *Eutrochatella* and *Poteria*. *Pomatias elegans*, a terrestrial species with direct marine ancestry, has a calcium concentration even higher than that of marine species. This introduces another point, namely that some of the calcium present is probably bound to proteins.

Without activities of each ion species, instead of overall concentrations, it is not likely that a coherent picture will emerge. At present, then, it appears that the monovalent : divalent cation ratio betrays little of species ancestry in gastropods. The same may be said for the chloride : sodium ratio. This is linked with the concentration of bicarbonate in the blood, because bicarbonate makes up a large part of the anion fraction in freshwater and terrestrial gastropods, and is in fact relatively constant (Rumsey, 1972). Bicarbonate is involved in equilibrium with the shell, together with calcium. In summary, although the osmotic pressure of gastropod blood to some extent reflects ancestry, there is no evidence as yet that ionic composition is useful in a similar fashion. Ionic composition of marine and freshwater ancestors does not seem to have imposed limitations on terrestrial forms, which have altered compositions fairly freely.

In the Crustacea the situation shows a great scatter of values. A value of the monovalent: divalent cation ratio of about 10 has been used by Lutz (1969) to support a claim for a marine origin in *Sudanonautes africanus*, but since *Gecarcinus lateralis*, which has a very definite marine origin, has a ratio of 19.3, the variation can be seen to be enormous. Two particular problems need to be taken into account at this stage. One has already been mentioned in the context of molluscs: the proportion of bound cations in the measured total must be known before much use can be made of divalent cation levels. The second is the very variable degree of reduction of magnesium found in the blood of marine crustaceans. In both the two crabs shown in Table 10.3 magnesium concentrations are about half those in sea water. However, concentrations vary from about 10% to over 90% of the sea water value in the whole range of marine Crustacea (see figures quoted by Prosser, 1973). It seems likely that individual ion levels are so closely tied to physiological requirements that they can change radically when crustaceans colonise new environments. In contrast, the general level of osmotic pressure tends to be retained from one environment to another.

10.1.2 Respiration

The selective pressures which were responsible for the development of lungs in animals colonising land from fresh water were probably very different from those acting on animals emerging from the sea or brackish water. In fresh water, as shown by some prosobranch gastropods, by the lungfishes and by several groups of teleosts, it was probably the lack of dissolved oxygen in swampy areas which led to the evolution of air breathing. In marine forms, such as many crustaceans, possibly the chelicerates, and the gobioid fishes, the driving force more probably reflected the availability of new food reserves and the possibility of escape from predators. Despite this difference in selective pressures, animals of both freshwater and marine origin have developed some form of internal gas exchanger, or lung, to replace the external gas exchanger, or gill. Marine animals in transition between water and land, such as many decapod crustaceans, frequently possess both structures: in water, the gills provide a very large surface area for

Table 10.3. *Blood composition of selected molluscs and crustaceans*

Species	Na$^+$	K$^+$	Na$^+$ and K$^+$	Ca^{2+}	Mg^{2+}	Ca^{2+} and Mg^{2+}	Cl$^-$	$\frac{\text{Na}^+ \text{ and K}^+}{\text{Ca}^{2+} \text{ and Mg}^{2+}}$	$\frac{\text{Cl}^-}{\text{Na}^+}$	Reference
Mollusca, Gastropoda										
Marine										
Strombus gigas	503	11.1	514.1	11.2	60.2	71.4	577	7.20	1.15	Little, 1967
Nerita fulgurans	484	11.3	495.6	11.2	55.4	66.6	554	7.44	1.12	Little, 1972
Fresh water										
Theodoxus fluviatilis	45.0	2.2	47.2	2.3	2.9	5.2	32.8	9.08	0.73	Little, 1972
Viviparus viviparus	34.0	1.2	35.2	5.7	<0.5	c. 6.0	31.0	c. 5.87	0.91	Little, 1965a
Terrestrial										
Eutrochatella tankervillei	26.5	1.2	27.7	3.2	1.5	4.7	23.7	5.89	0.89	Little, 1972
Maizania wahlbergi	26.0	1.8	27.8	4.5	1.8	6.3	24.5	4.41	0.94	Andrews & Little, 1972
Poteria lineata	31.4	1.8	33.2	5.1	1.5	6.6	25.1	5.03	0.80	Andrews & Little, 1972
Pomatias elegans	110	6.0	116	16.5	2.5	19	106	6.11	0.96	Rumsey, 1972
Crustacea, Isopoda										
Marine										
Ligia oceanica	586	14	600	36	21	57	596	10.53	1.02	Parry, 1953
Fresh water										
Asellus aquaticus	137	7.4	144.4	–	–	–	125	–	0.91	Lockwood, 1959a
Terrestrial										
Oniscus asellus	230	8.2	238.2	16.7	9.1	25.8	236	9.23	1.03	Barrett, 1972
Porcellio scaber	227	7.7	234.7	14.7	10.9	25.6	279	9.17	1.23	Barrett, 1972
Crustacea, Decapoda										
Marine										
Pachygrapsus crassipes	465	12.1	477.1	11.4	29.2	40.6	–	11.75	–	Prosser, Green & Chow, 1955
Carcinus maenas	525	12.7	537.7	14.3	21.2	35.5	502	15.15	0.96	Riegel & Lockwood, 1961
Fresh water										
Paratelphusa hydrodromous	330	6.8	336.8	7.8	7.8	15.6	255	21.58	0.77	Ramamurthi, 1967
Potamon niloticus	259	8.4	267.4	12.7	–	–	242	–	0.93	Shaw, 1959
Terrestrial										
Sudanonautes africanus	207	6.0	213	11.8	10.6	22.4	241	9.51	1.16	Lutz, 1969
Holthuisana transversa	270	6.4	276.4	15.7	4.7	20.4	266	13.55	0.99	Greenaway & MacMillen, 1978
Gecarcinus lateralis	468	12	480	17.3	7.6	24.9	–	19.28	–	Mason, 1970
Sea water	475	10.1	485.1	10.3	54.2	64.5	554	7.52	1.17	Barnes, 1954
Fresh water (soft)	0.25	0.005	0.26	0.07	0.04	0.11	0.23	2.32	0.92	Potts & Parry, 1964
Fresh water (hard)	2.22	1.46	3.68	3.98	1.67	5.65	2.54	0.65	1.14	Potts & Parry, 1964

All concentrations are given as mM/litre.

oxygen uptake, but in air the lung cavity remains patent while the gills tend to collapse. Similarly, some amphibious animals of freshwater swamps, such as ampullariid gastropods and the lungfish, possess both gills and lungs.

There is no hard and fast distinction to be made between the lungs developed by animals of freshwater origin and those of animals coming direct from the sea or brackish water. Some groups of marine origin, such as the pulmonate gastropods, have in some forms developed ventilation lungs, while others, such as the chelicerates, have diffusion lungs. On the other hand, some animals of freshwater origin, such as the cyclophorid gastropods, have diffusion lungs while others, such as the lungfish, have ventilation lungs. There is, however, one point which often distinguishes animals of freshwater and marine origin. Animals emerging from the sea tend to spend relatively long periods out of water because this emersion time is usually governed by, or related to, the tidal cycle. During this time they tend to obtain oxygen by methods which are only slight modifications of those used under water. For example, many of the semiterrestrial crabs pass air through the gill chambers instead of water. Intertidal molluscs with part of the mantle vascularised gain oxygen by diffusion into the mantle cavity. Intertidal isopods gain oxygen by diffusion across the pleopods which often beat in air just as they do in water. All these animals tend to obtain oxygen slowly, but *fairly continuously*. In contrast to this situation on marine shores, animals living in fresh water of low oxygen content have to adopt radically different methods of obtaining oxygen from those employed under water. By definition they must be *intermittent* air breathers, because they must come to the water surface to obtain oxygen. Species of the prosobranch gastropod *Pomacea,* for instance, stretch a flap of mantle skirt through the water meniscus, and then pump air in and out of the lung by movements of the buccal mass (McClary, 1964). In the vertebrates the initial mechanism may have begun by the swallowing of air, as seen today in the South American fish *Hoplosternum thoracatum* (Gee & Graham, 1978). From such beginnings as this came the development of both buccal respiration, as found in many teleosts, and pulmonary respiration, as in the lungfish. Probably because the latter mechanism allowed an ever increasing volume and complexity of respiratory surface, whereas buccal respiration had finite limitations, pulmonary respiration led to the only widespread development of terrestrial vertebrates. It is important here to note that while fairly continuous forms of respiration, including buccal respiration, evolved in fresh water, along with intermittent breathing, the latter has not evolved in marine littoral air breathers. In other words the vertebrate lung evolved in its present form because it arose in fresh water. If terrestrial vertebrates had instead evolved directly from the sea, the mechanism of air breathing would probably have been much more akin to that seen in the semiterrestrial mudskippers, in which air is taken in at the mouth, oxygen is absorbed in the buccal cavity and air is then expelled through the opercular opening.

Although it might be expected that other differences would be linked with these differences in the site

and mechanism of oxygen exchange, the physiology of air breathers seems to be remarkably uniform in a variety of animal groups. Although we know little about the details of gas exchange in the invertebrates, all transitional vertebrates seem to use the skin and gills for carbon dioxide removal, and the lungs for oxygen uptake. A comparative examination of the methods by which invertebrates lose carbon dioxide would be invaluable, since the subject seems to have been neglected. It may well show a different picture from the vertebrates, because there is no doubt that in many cases their oxygen exchange sites are much less efficient than those in the vertebrates, and this is reflected in the very high oxygen affinity of most invertebrate respiratory pigments. In other ways, such as the regulation of acid-base balance, terrestrial invertebrates probably resemble terrestrial vertebrates: blood pCO_2 is higher than in aquatic forms, and the buffering systems need to be more efficient to prevent respiratory acidosis. The invertebrates employ entirely different systems from those of the vertebrates – for example, the calciferous glands are involved in earthworms, and the shell in molluscs – and a comparison of invertebrate with vertebrate mechanisms would repay investigation.

10.1.3 Nitrogenous excretion

The long-held theory that nitrogenous excretory products are related to environment was summarised in Chapter 1: ammonia excretion typifies aquatic animals, whereas urea and purine excretion is typical of terrestrial animals. Having reviewed the evidence for a number of phyla in succeeding chapters, it is evident that this idea must be a gross oversimplification. Let us consider a few examples. Very little is known about excretion in aquatic oligochaetes, but the terrestrial forms excrete both ammonia and urea, with some uric acid (see p. 27). The ratio of urea to ammonia changes with nutritional state, but is probably more related to the control of acid-base balance than to availability of water. In prosobranch gastropods the situation is very confused (see Little, 1981), but even marine forms appear to excrete purines as well as ammonia, and there is little clear relation between habitat and excretory product. Some terrestrial pulmonate gastropods reverse the accepted sequence, and excrete mainly ammonia gas during aestivation, and this process is to some extent linked with control of acid-base balance (Speeg & Campbell, 1968a). The problem in gastropods is aggravated by the fact that very many species accumulate uric acid in the connective tissue surrounding blood vessels, but do not seem to excrete it. Certainly the conventional picture of ammonia excretion in water and purine excretion on land is no longer acceptable in this case. Some of the terrestrial isopod crustaceans also excrete gaseous ammonia, as well as accumulating urea and uric acid. Little is known about the subject in decapod crustaceans, and even in the land crab *Cardisoma*, although it is known that uric acid is accumulated, the mechanisms by which it is excreted have not been investigated. Chelicerates appear to excrete mainly the purine guanine, unlike any other group. Myriapods excrete mainly ammonia, but with some uric acid. The insects probably agree best with early simplified theories, because in general terrestrial forms excrete purines while aquatic larvae excrete ammonia (but see p. 175). The vertebrates do not completely conform to the early ideas. Some air-breathing fish such as *Periophthalmus* and the lungfishes increase urea production in air and ammonia production in water; but both products are formed to some extent in both environments. Smith (1959) concluded that the earliest vertebrates probably eliminated the products of protein catabolism as urea, and certainly the production of urea seems to be a basic feature of all fish. As discussed earlier the origin of urea as an excretory product in aquatic vertebrates has yet to be explained. Even in higher vertebrates, excretory products do not seem to be strictly related to environment. Reptiles, for instance, produce uric acid as adults, but their embryos produce ammonia and urea and not uric acid, despite the problem that their excretory products must, for the most part, accumulate within the eggshell.

It should be evident from these brief examples that the situation is more complex than a simple relationship of excretory product to general environment. First, the categories of 'terrestrial' and 'aquatic' are not particularly meaningful unless broken down into appropriate microhabitats. Thus, woodlice lose ammonia as a gas only when they are in a microhabitat of air saturated with water vapour. This means that they lose ammonia without becoming desiccated. Second, a firm distinction must be made between *formation* of a nitrogenous end product and its *elimination* from the body. *Periophthalmus* forms relatively more urea when it is in air than when it is in water, but it cannot excrete the urea until it returns to water. Third, the biochemistry of nitrogen metabolism does not work in isolation from other biochemical systems. In many cases it has already been shown to be tied to the regulation of blood pH, so that there are many other factors besides the availability of water which might be expected to affect the formation of excretory products. If we add to this the fact that nitrogenous catabolism has to deal with at least two entirely different sources, the purines and pyrimidines as well as the proteins and amino acids, it can be no surprise that there is no simple relationship between product and environment. An animal's nitrogen metabolism can only be understood in the context of its microhabitat, its life cycle, the related biochemical pathways, and its ancestry. As yet there are few if any cases in which we can even begin to assess all these factors.

In view of this, it must be impossible at present to relate the mechanisms of nitrogenous catabolism in terrestrial animals to their routes on to land. It seems very unlikely, in any case, that the nature of the final excretory products should bear any relationship to a marine or a freshwater origin; but it is quite possible that within a single animal group the pathways adopted for producing, say, urea, could differ depending upon the microhabitats used by various ancestors on their route to land. As an example, it can be pointed out that while lungfish produce mainly urea, this is derived from the ornithine cycle; whereas the urea produced by *Periophthalmus* comes from purine breakdown. It is this kind

of difference which should perhaps be investigated in the future.

10.1.4 Reproduction

In terms of reproductive adaptation, semi-terrestrial invertebrates of freshwater origin differ considerably from those of marine origin. In the marine littoral zone, species show a great variety of reproductive methods which presumably reflect responses to emersion and the problems of desiccation, as well as to the availability of a great diversity of habitats. Many of the intertidal gastropods utilise internal fertilisation rather than the dispersal of gametes into the sea. The Littorinacea in particular show both internal fertilisation and a variety of further strategies which concern larval development. In various species of *Littorina* this variety includes the production of pelagic larvae, the attachment of egg masses to rocks or algae, from which the young subsequently hatch as young snails, and the retention of the young in a brood pouch until they can leave as miniature adults. Brooding of the young is also common in isopod and amphipod crustaceans. In decapods the eggs are often retained by the parent for some time, but finally they are released into the sea to become planktonic larvae. The development of ovoviviparity or viviparity can certainly be viewed as a preadaptation to life on land, and the terrestrial isopods and amphipods retain this property, so protecting the eggs and young against desiccation and predators. A marine littoral ancestry thus seems in many cases to have aided the penetration of the littoral ecosystem, although in the species which have continued to produce planktonic larvae, such as the hermit crabs, it has of course limited the animals to a more or less maritime distribution.

The categories of animals which are thought to have invaded land from fresh water have somewhat different reproductive characteristics. Many of them have also developed internal fertilisation of some kind, but brooding the young does not seem to be a common attribute. Even in the freshwater crabs the young, which hatch in the burrow and attach themselves to the female, are soon deposited in fresh water to mature unattended. In most groups fewer and larger eggs are produced, and in those groups which normally have a succession of larval stages the number of these is reduced, as in the freshwater grapsid crabs (see p. 94). These freshwater adaptations are probably concerned with factors such as the provision of the eggs with greater stores of nutriment in an environment where the plankton is either seasonal or scarce, the reduction of the surface area : volume ratio and therefore of the osmotic work necessary, and generally with the preparation of the emerging young for an environment much harsher than that of the sea.

Many terrestrial arthropods, especially those which are cryptozoic, deposit stalked spermatophores or sperm drops, and Schaller (1979) has put forward the hypothesis that this feature probably arose long before such lines became terrestrial, and at a time when they exploited an interstitial habitat (see Chapter 8). It has also been suggested above that most terrestrial arthropods may have had a direct marine origin, so that the habit of producing this type of spermatophore probably arose in the marine interstitial environment.

Among the vertebrates the teleosts exemplify some of the differences between marine and freshwater forms. Most marine teleosts produce enormous numbers of small floating eggs each with a diameter of 0.3–2 mm. Some of the marine littoral forms, and most of the freshwater forms, produce fewer, non-floating eggs, each mostly having a diameter in the range 0.4–7 mm (Marshall, 1965; Blaxter, 1969; Miller, 1979; Fig. 10.1). While the eggs of some freshwater teleosts are therefore just as small as those of marine species, others are very much bigger. This increase in the range of sizes must be seen in the context of the total life history of the fish. For instance, the size of the adult, even within one species, is often related to the size of the eggs it produces (Mann & Mills, 1979). However, the mean egg size of individual species which have the same kind of life history within the same habitat tends to be similar (Miller, 1979). Egg size perhaps relates best to the size of the larva at hatching, and this in turn is dependent among other things upon the size of the food available, and the abundance of competitors and predators (Wootton, 1979). In summary, we may say that it is only in fresh water that many teleosts produce large eggs, and that we may therefore ascribe their production directly or indirectly to some of the characteristics of the freshwater habitat. Marshall (1965) has argued that the presence of tides and currents in inshore waters and rivers may account for the habit of laying large, non-buoyant eggs, since such water move-ments could easily carry floating eggs well away from the appropriate habitats. In fresh water this pressure is probably reinforced by the problems of osmotic regulation inherent in very small embryos because of their relatively large surface area. Linked with these two aspects is the necessity for larvae to hatch at a size at which they can osmoregulate, swim against water currents and obtain food.

Other groups of fish do not show this clearcut relation of egg size to habitat. The elasmobranchs have large eggs, as do the coelacanths (see, e.g. Schultze, 1972), although they are marine. The marine hagfish *Myxine glutinosa* has large eggs (Walvig, 1963), while the freshwater lampreys have eggs only 1 mm in diameter (Piavis, 1971). However, this reinforces the point that comparisons can only be made between similarly sized groups and within phylogenetically related taxa, and that even then they can only be made by taking into account the total life history and reproductive strategies. There is still insufficient information about non-teleost fishes to allow us to form generalisations. It is of interest to note, however, that the lungfish *Protopterus* lays relatively large eggs of 3.5–4 mm diameter (Budgett, 1901).

The amphibians produce eggs very comparable in size to those of the freshwater teleosts. In a series of Australian frogs, Main (1968) showed that adults laying eggs in water produced eggs of 1.1–1.7 mm diameter, despite a body length of only 1.5 to 7 cm. A series of freshwater teleosts with body lengths from 1.8 to 4.6 cm have been found to lay eggs of diameter 0.4–2.1 mm, whereas their estuarine counterparts produced eggs of 0.33–0.45 mm diameter (Miller, 1979). The development of relatively large eggs in freshwater amphibians must therefore be seen to be essentially a freshwater adaptation, allowing the hatching of larvae which can cope with the properties of the freshwater environment. In terrestrial amphibians the egg size is again increased: Australian frogs up to 7 cm in length laying eggs on land showed egg diameters of up to 5.5 mm (Main, 1968). Although the factors which control egg size in amphibians are numerous (see, e.g. Salthe & Mecham, 1974), the increase in size shown by terrestrial forms must be seen as a development from the freshwater egg. It seems unlikely that such an increase would have arisen if the amphibians had evolved from littoral marine forms, because these would have retained small pelagic eggs with pelagic larvae. It is therefore interesting to speculate upon the evolutionary fate of the vertebrates if the tetrapods had evolved in marine and not freshwater conditions. If the above arguments can be sustained, primitive marine littoral tetrapods would have retained pelagic larvae. They might then have developed with some parallels to the semi-terrestrial decapod crustaceans which are mostly limited to maritime existence. It can thus be postulated that it is the evolution of reproductive mechanisms adapted to fresh water which has allowed the vertebrates to become truly terrestrial. The production of larger and larger eggs began as a freshwater adaptation, and by further terrestrial adaptations produced the cleidoic egg. Without a freshwater ancestry such a pattern would have been unlikely to emerge.

Fig. 10.1. The frequency distribution of egg diameters in teleosts from European waters, based on data from 101 marine species (*a*) and 33 freshwater species (*b*). There is considerable overlap between marine and freshwater species, but the trend is for freshwater eggs to be larger. From Wootton (1979).

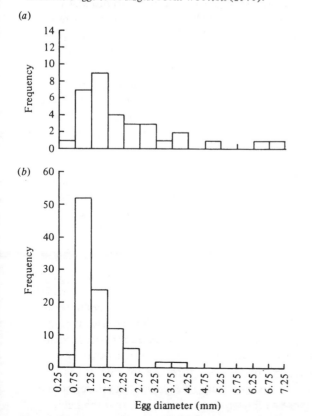

10.1.5 Behaviour and sense organs

In many of the groups of animals with both aquatic and terrestrial members, it is the terrestrial ones which can move most rapidly, and which appear to have the most complex behaviour patterns. In general they have much better visual senses than aquatic species, both in terms of the detection of patterns and shapes and in the ability to detect movement. The amphipod and decapod crustaceans in particular show good examples of these generalisations. The contrast between such a decapod as *Pachygrapsus crassipes*, which hunts for its prey in air and uses visual clues, and the aquatic species which search for food by using chemical senses, is striking. The complex behaviour of the semi-terrestrial fiddler crabs contrasts strongly with low-shore members of the same family, especially in the context of social interactions. It has been suggested that these developments are associated with the very great spatial and temporal variation in physical properties of the terrestrial environment, in comparison with the more homogeneous nature of aquatic environments, both freshwater and marine. A word of caution must temper this suggestion, however, because so much more effort has been put into investigating the behaviour of terrestrial and semi-terrestrial forms than that of aquatic ones that the apparent complexity of terrestrial behaviour may reflect this bias. For instance, it is true that much communication occurs through the visual senses on land, while a great deal of information under water is gained by the chemical senses; but since very little is known about the transfer and reception of chemical stimuli, there is a great temptation to ignore the possible complexities of such systems, whereas the complexities of the visual systems are immediately evident. Until more is known about the behaviour of aquatic animals and the functioning of their sensory systems, their apparent lack of complex behaviour should not be accepted without caution.

Of the terrestrial animals, the behaviour of cryptozoic invertebrates calls for particular comment. Many of the cryptozoic arthropods have developed quite complex behaviour, particularly in relation to reproduction. Emphasis has already been given in this book to the methods by which sperm drops and spermatophores are placed by the male, and the subsequent courtship by which he ensures that these are transferred to the female. In the hexapods, for example, it seems likely that the deposition of sperm drops in the soil is the pattern which preceded the evolution of spermatophores. The evolution of direct copulation may even have been a much later property which arose in association with emergence from the cryptozoic habitat. In the myriapods also, the cryptozoic forms in particular, including pauropods, symphylans and most chilopods, transfer sperm in sperm drops or spermatophores. This use of indirect fertilisation may therefore be seen as a primitive property of the cryptozoic representatives within the insect–onychophoran–myriapod assemblage. Since the property is so widespread, it may be suggested that it probably was present in the initial colonisers of the terrestrial cryptozoic habitat. This in turn may suggest that these early colonisers moved into the cryptozoic habitat directly from the aquatic interstitial habitat, since such reproductive methods could never have survived in animals exposed to the terrestrial macroenvironment. It has already been suggested (p. 159) that the insects may have used this route on to land, coming directly from the sea. Such an idea has also been entertained above for other arthropod groups. The behavioural mechanisms which ensure that individuals remain within the interstitial environment would have then given rise to mechanisms which ensure that on land the animals remained in cryptozoic niches.

While such general traits as the use of indirect fertilisation may reflect an origin in the interstitial environment, it seems unlikely that any major trends in sensory ability or in mechanisms of movement are related to routes on to land. These details of behaviour and sensory perception are very closely related to the physical properties of air, in comparison with those of water, and reflect the life styles of animals in their individual habitats. It has not been possible to detect differences between animals living in similar habitats which are due to differences in ancestral habitat. Thus decapod crustaceans which are active in air show visual senses with properties which are comparable to those of insects (see p. 105). It may be concluded that the routes on to land have had little influence on mechanisms of sensory perception.

10.2 Negative evidence: the groups without terrestrial members

The major groups not represented on land can for the most part be described as filter or plankton feeders. These are all excluded from terrestrial life because, although there is a certain amount of aerial organic matter, and in some seasons a certain amount of aerial plankton, the concentrations of both are too low to support a filter feeding existence. There are therefore no terrestrial hydroids, corals, lophophorates, bivalve molluscs or tunicates.

A second class of animals excluded from land was claimed by Wells, Huxley & Wells (1931) to consist of those animals which had insufficiently developed skeletons to support themselves without the aid of water. To some extent this is the case, particularly in preventing soft-bodied animals of large size from colonising the land. Small, soft-bodied animals such as flatworms, nemertines, slugs and leeches show, however, that on a small scale hydrostatic skeletons can be used on land as well as in water. Some of the earthworms can reach lengths as great as 2 m, without any rigid skeleton, and many other terrestrial animals use hydrostatic support to some degree, either for individual parts of the body or at some particular stage in the life history. Many insect larvae, for example, depend essentially upon the use of a hydrostatic skeleton, and not upon jointed legs and a hard exoskeleton. Lack of either an internal or an external hard skeleton can hardly be claimed to have prevented animals from colonising land, although it has been limiting in the matter of size (see, e.g. Jones, 1978*b*). It must, however, be admitted that those groups with a rigid skeleton have on the whole been more successful on land than those without rigid skeletons.

Two groups remain to be considered: none of the echinoderms is terrestrial, nor are any of the cephalopod molluscs. The echinoderm body plan is unique in many aspects, and is characterised particularly by the water vascular system. Exchange of gases, water, ions and nitrogenous waste occurs partly across the tube feet, which are evaginations of the water vascular system, and partly across a variety of surfaces developed within, or from, the coelom. Although these surfaces differ in different echinoderm groups, all are relatively unspecialised and cater for a number of processes: the active uptake of ions, diffusion of oxygen and carbon dioxide and removal of nitrogenous waste. The echinoderms are therefore typified by the possession of an extremely large area of contact with the external medium, across which exchange occurs, and by the complete lack of any discrete osmoregulatory or excretory organs. This combination, while it has been successful in the sea, is hardly adaptable to other, more variable environments except by major innovation. Since *none* of the classes of echinoderms has produced any members which have moved out of the sea, it must be assumed that the basic echinoderm design is so committed to continuous exchange with the surrounding medium that this, rather than the failings of any particular physiological system, has confined the group to the sea.

The cephalopod molluscs have no such unique design which might prevent their movement on to land. As discussed above, it can hardly be claimed that lack of mechanical support is the barrier. The problem becomes more manageable if we consider separately the reasons why cephalopods used neither the direct route to land nor the route via fresh water. Taking the direct route first, we may start by pointing out that small intertidal octopuses have much in common, in terms of habitat, with the gobies and blennies, and sometimes crawl out of water (Packard, 1972; Johannes, 1981). Yet they have never ventured into modes of life as terrestrial as, say, *Periophthalmus*. Characteristics such as the loss of a protective shell seem unlikely to explain this, because numerous molluscs such as *Onchidium* have become intertidal and retained a soft epidermis. Similarly, any single physiological attribute is unlikely to give a prime explanation because so many other molluscs with the same basic organisation have become terrestrial. The one point which remains is simply that the cephalopods have reached a peak of neuromuscular coordination, but that such a system requires constant conditions to function well. No molluscs have achieved methods of controlling their internal environment to any degree in the face of external changes: the life styles of non-marine forms are characterised by tolerance of internal changes, and perhaps in consequence they have never developed a particularly complex system of sense organs, or of behavioural patterns. It may be the very complexity of cephalopod organisation which has barred their direct entry to the terrestrial ecosystem.

No cephalopods have ever left the sea to enter fresh water. According to Wells (1962), limitations of the physiological processes involved in respiration and excretion have prevented movement out of the sea. Yet as pointed out above, the molluscs as a whole have shown extensive adaptiveness, and the gastropods, as discussed in Chapter 4, have been very successful in many habitats. However, it may be that as with the direct route the physiological systems have not been able to provide a sufficiently constant internal environment to allow the continued function of the sensitive and complex neuromuscular system: while the homeostatic mechanisms of cephalopods control the finer points of the internal environment when the external medium remains fairly constant, they are incapable of widening their limits to deal with gross changes. Added to this presumed physiological inability is an argument based on the point made by Packard (1972). The extensive parallels between cephalopods and fish reflect the intense competition between the two groups. Although the cephalopods arose before fish, the latter have forced the cephalopods out of their adaptive zone and into deeper water. This competition with fish may well also have resulted in their exclusion from fresh water. Certainly in combination with their inability to tolerate changes in their *milieu intérieur*, the competitive action of the vertebrates is likely to have been a major factor in preventing their movement into fresh water.

At this point is is appropriate to consider how the vertebrates managed to move into fresh water, and why it is that their homeostatic systems were adequate to allow this movement. The probable answer to this lies in the degree of development shown by the group when it first moved into estuarine and river systems. Following the argument derived by Berrill (1955) (see p. 178), chordates probably moved into river systems at a relatively small size, and without any especially complex organisation. It is their subsequent developments in fresh water which have produced the complicated machinery of homeostatic mechanisms found in the vertebrates today. Had the group become particularly specialised for a marine life style, as have the cephalopods, it is unlikely that freshwater vertebrates would ever have evolved.

10.3 The co-evolution of fauna and flora, and relationships with climate

The development of terrestrial animals has so far been considered without reference to the evolution of terrestrial vegetation; but because all terrestrial ecosystems depend ultimately upon the fixation of carbon by green plants, it is necessary at this stage to examine how animal–plant interrelationships have influenced the movement of various animal groups on to land. Most of the evidence here is derived from the fossil record, and although this evidence is necessarily incomplete, it gives an indication of the times at which various terrestrial groups made their appearance. Some fossil records are therefore summarised in Figs. 10.2 and 10.3. These figures are much simplified, because they assume that groups have been continuously present between their times of first and last appearance: the fossil record is in fact very patchy for most groups. The figures must also be incomplete because it is likely that groups appeared some time before the earliest period in which they have been recorded as fossils. Nevertheless, the record is probably sufficiently accurate to enable us to relate the emergence of terrestrial faunas and floras. From other evidence, it is also possible to make inferences about climatic conditions at the time and place that these terrestrial ecosystems appeared. These in turn have relevance to present-day distributions of the various faunal assemblages.

10.3.1 The influence of plants on the development of terrestrial faunas

The oldest records of both terrestrial animals and plants come from the Silurian. The first vascular plants, the Rhyniophyta, had erect stems with sporangia at the tip but no leaves. The first terrestrial animals preserved as fossils are thought to be a primitive group of myriapods (Hoffman, 1969). It should be said that these myriapods, described from the Upper Silurian, are somewhat problematic forms, and that their status and environment are in doubt. It must be emphasised, of course, that there may have been other, earlier terrestrial forms which have not been preserved, or have not been found. The Silurian myriapods, or Archipolypoda, have been assigned to the class Diplopoda, and were presumably vegetation feeders. Although the only vascular plants present on land were the Rhyniophyta, there were very probably algae and bryophytes which existed in damp places. Without these the land would almost certainly have been devoid of herbivores. The scorpions found in rocks of the same age are thought to have been aquatic (Kraus, 1976), but show very little difference in structure from later terrestrial forms. Such similarity highlights the problem of deciding which species found as fossils were truly terrestrial, and if they were, under

what ecological conditions they lived. It seems likely that in Silurian times, terrestrial ecosystems were very poorly developed, with vegetation which merely fringed water bodies, and enabled a few species to evolve in its cover.

In Devonian times, terrestrial vegetation cover increased dramatically. The Lycophyta (club-mosses), Sphenophyta (horsetails) and Pteridophyta (ferns) appeared and became abundant, and at the same time the hexapods were present. The chelicerates became terrestrial, and the first amphibians began to utilise the land. The evolution of the ferns and horsetails must have changed terrestrial microclimates from those of the Silurian to a marked degree. By providing shade, lessening the effect of wind and retaining a humid atmosphere, by supplying organic detritus and a direct food supply, they effectively created cryptozoic niches which allowed the evolution of small, desiccation-intolerant forms. Presumably this is the niche which was early exploited by the hexapods, which fed on vegetation and detritus, and were accompanied by the myriapods. The scorpions probably preyed upon these other arthropods, and were the first terrestrial predators. It has been argued that the first amphibians were also small forms which lived in damp vegetation cover, so that the whole of the Devonian terrestrial ecosystems depended closely upon the utilisation of cryptozoic habitats.

The great development of terrestrial forests did not occur until the Carboniferous, when the coniferous trees evolved. These forms, with methods of reproduction which were somewhat independent of moist conditions, and with tall trunks and a variety of leaf

Fig. 10.2. The fossil record of selected animal groups. Black bars show aquatic animals, and stippling shows terrestrial animals. The actual records are more patchy than shown, but discontinuous distributions in time have been joined up for clarity. Information from Termier & Termier (1952); Knight *et al.* (1960); Howell (1962); Harland *et al.* (1967); Glaessner (1969); Hessler (1969).

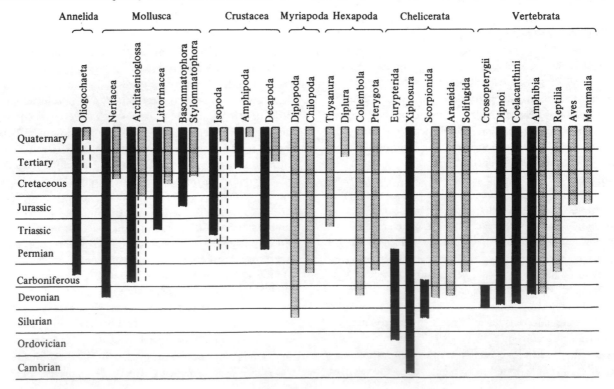

form, produced environmental conditions which were much more constant than those of the Devonian. In these forest conditions the pterygote insects evolved, centipedes appeared, and the reptiles evolved from amphibian stock. The ecosystems of Carboniferous time must therefore have had many parallels with those of today. The vertebrates and insects were the dominant forms on land, living in environments where the conditions were closely controlled by the growth of vegetation. Since vertebrates and insects were the most important herbivores, particular species must have interacted strongly, and the direction of evolution of both groups was related to that of the plants. The insects also produced carnivorous forms such as the dragonflies, but the role of predator was particularly taken up by the chelicerates, which produced many different forms in the period. This early interaction between insects and arachnids has been retained through geological time, and is still one of the dominant characteristics of terrestrial predator–prey relationships. The existence of land snails in the Carboniferous is more doubtful, since the shells of this date may have belonged to freshwater forms (see discussion by Solem (1978) and Fretter (1975)).

During the Permian, Triassic and Jurassic periods the terrestrial faunas which had appeared during the Devonian and Carboniferous showed a great deal of evolution and radiation, but very few new groups appeared. Birds and mammals evolved from reptilian stock in the Jurassic and more types of insects are found progressively during more recent periods, but

the terrestrial snails and crustaceans seem not to have appeared in large numbers until the Cretaceous or Tertiary. This late radiation seems to correlate with the evolution in Cretaceous times of the angiosperms or flowering plants. The first terrestrial Neritacea, Littorinacea and probably the first terrestrial pulmonates all come from the Upper Cretaceous. The first recorded terrestrial Decapoda come from the Tertiary. While there is no direct evidence to suggest why this should be so, the dependence of both groups upon a plentiful calcium supply may be involved. In recent ecosystems the detritus formed in coniferous forests is usually an acid humus because the trees absorb few nutrients from the soil, and leaching produces a resultant downward movement of soluble ions (see, e.g. Eyre, 1968). There are few invertebrates in this habitat except for insects and mites (Edwards, Reichle & Crossley, 1970). It is only in broad-leaved forests that the trees absorb nutrients from deeper layers of the soil and effectively return them to the soil surface via the fall of leaves. In this type of soil many invertebrates are found, including those with a skeleton or shell composed of calcium carbonate. Assuming that these generalities applied in the past, the calcium supply in the humus of gymnosperm forests has always been poor, and this effectively prevented their colonisation by molluscs.

The general features of the insect fauna had already developed great similarities with that of today as early as the Jurassic (Hughes & Smart, 1967). The major exception to this is provided by the Lepidoptera, which feed on nectar, and only arose in conjunction with the evolution in the Cretaceous of the angiosperms. The co-evolution of the insects and plants may, however, be considered in slightly more detail, since this association has been a very important one in the development of terrestrial ecology.

The associations between insects and plants probably evolved gradually from an initial situation in which insects utilised plant debris as food (Hughes & Smart, 1967; Hughes, N. F., 1976). From this initial position, some insects came to feed on living plants, presumably by chance: the greater the 'exposure' of the insect to a plant, the more likely a herbivore–host relationship was to arise (Southwood, 1977). It is appropriate to note, however, that the numbers of insect orders which feed directly on the tissues of higher plants is small compared with those that are scavengers and carnivores (Southwood, 1973). This lack of direct insect herbivory in the taxonomic sense probably derives from several factors. First, the relatively high proportion of nutritionally inadequate material in many plants is a deterrent. Second, the risk of suffering desiccation when feeding on growing leaves is considerably higher than when living in the cryptozoic niche provided by the leaf litter. Third, plants have also developed mechanisms of resistance to insect attack, such as the growth of thickened cuticles and the manufacture of toxic substances such as alkaloids. In reply to this last point, however, many insects have in turn evolved resistance to plant toxins, and some even take them up and utilise them for their own defence (Feeny, 1975).

Fig. 10.3. The fossil record of selected plant groups. Black bars show aquatic plants, stippling shows terrestrial plants. Information from Harland *et al.* (1967). Classification as in Benson (1979).

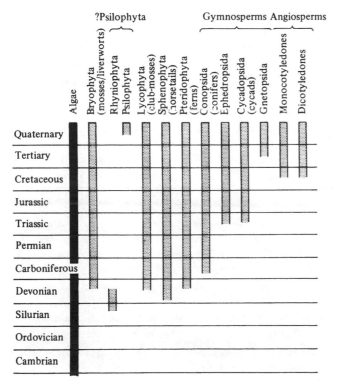

As already pointed out, the dependence of insects on higher plants as a source of food was a major factor in their evolution as terrestrial animals. Of equal significance to terrestrial ecosystems, however, has been the development of mutual interactions involving flowers and their mechanisms for pollination. The first flowers were apparently produced by the Bennettitales (gymnosperms resembling cycads) of the Triassic (Smart & Hughes, 1973). Many insect groups were extant at that time, and certainly one present-day beetle family had already appeared. It may be hypothesised that such crawling insects carried out pollination by seeking food sources within the flowers, although there is no positive evidence of this. Similarly, pollination of angiosperm flowers in the Cretaceous may have been effected by coleopteran feeders, but ants and dipterans were also present by that time (Smart & Hughes, 1973). Since then the co-evolution of flower structures and pollinating insects has been remarkable. Most of the major orders of angiosperms are insect-pollinated to a great extent (Southwood, 1973). The relationships all confer mutual benefit on insect and plant species: the plants provide pollen and nectar from the flowers as food, and the insects, in carrying away pollen, ensure cross-pollination. Diptera, Lepidoptera and Hymenoptera are particularly common as pollinators, and very often the plants attract these insects from a distance by appropriate scents which mimic their normal food, as well as by providing visual cues in the shape of flowers (Yeo, 1973). The specialisations of some flowers such as orchids, which lure insects into copulatory positions because they mimic the female of the species, are extreme. More important for present considerations, however, are the wider ecological effects generated by such mutual associations. An excellent example of these widespread effects has been provided by Gilbert (1975) who studied the feeding and reproductive strategies of the neotropical butterfly genus *Heliconius*. These butterflies lay eggs on the leaves of vines of the genus *Passiflora*, and the larvae then eat the leaves. Relatively few of the eggs and larvae hatch into adults because other insect species such as ants and wasps feed on nectar produced by *Passiflora*, and act to 'defend' the vine against *Heliconius* by eating eggs and parasitising caterpillars. Each species of *Passiflora* also produces its own combination of toxic chemicals, such as glycosides and alkaloids, so that almost every species has only a limited number of species of *Heliconius* which utilises it. The adults of *Heliconius* feed on pollen and nectar of a different genus, *Anguria* (the cucurbit vines). Because of the high nutrient content of this food, and the fact that male *Anguria* plants flower almost continuously, the adult butterflies can live and reproduce for up to six months. In this time, they ensure cross-pollination of the *Anguria* vines. The interactions between the three species are thus extremely complex. In particular, they have led to the presence of relatively few individuals of the plant species, which are widely scattered, but there are many other consequences which are indicated in Fig. 10.4. Although this example is clearly a rather special case, it shows that mutual interactions can be just as significant as predator–prey relationships. The insect–angiosperm interaction has been of very great importance in the development of terrestrial ecosystems.

10.3.2 The influence of climate

From the foregoing discussion it appears that many species of terrestrial animals have evolved strong relationships with terrestrial plants. The microclimate in which the animals live is often very greatly determined, or at least modified, by the vegetation cover. The macroclimate, however, is mainly determined by external physical processes. At the present time the macroclimate of the tropics supports a far greater diversity of terrestrial animal species than do temperate latitudes. This increased diversity is probably related to the more constant climatic conditions in the tropics: the change of temperature over the seasons is often slight or non-existent, while rainfall in at least some regions remains high throughout the year. Especially within the regime of tropical rain forest, conditions are therefore nearly constant and do not impose problems of desiccation. This constancy contrasts strongly with the seasonal climates of temperate latitudes where conditions during part of the year are such as to enforce inactivity or hibernation for extended periods, or migration to more favourable regions.

It is exceedingly difficult to extrapolate from the present distribution of earth climate back into the past. For a number of reasons both the general level of climate and its distribution over the earth are thought to have changed markedly over geological time. These changes are linked with the theory of continental drift, and it is appropriate here to summarise some of the ways in which the movement of the continents may change earth climate. First, if continents move from one latitude to another, and if climate varies with latitude, individual continental climates will change to reflect the new latitude. Second, if a continent with mountainous regions drifts into the region of the poles, ice will accumulate on it, and the growth of glaciers will increase the albedo of the earth – more sunlight will be reflected and less absorbed, and the general temperature of the earth's surface will decrease. Third, the changes in continental configuration will alter the flow of ocean currents and winds, and may therefore indirectly change the degree of latitudinal variation in climate. Fourth, if continents join together their climate will become more 'continental', i.e. it will become more extreme in its variations. If, on the other hand, continents break up, the terrestrial climates will become more mild and equable. Unfortunately there is not yet agreement over the details of past continental movements, so that there can be no agreement over the details of climatic changes; but the outlines of continental movement seem to be fairly clear so that some attempt can be made to interpret the generalities of past climate. The following interpretations are based essentially upon discussion by Pearson (1978) and on the volume edited by Wilson (1976), although it should be accepted that other discussions (e.g. Tarling & Runcorn, 1973) have produced different conclusions. During much of the Palaeozoic many of the continents appear to have been separated from the others.

In the Devonian the land masses were some way south of the North Pole so that the latter was not subject to their influence (Valentine, 1973). The Devonian flora was cosmopolitan (Chaloner & Lacey, 1973), and it may be suggested that not only was the climate of the earth relatively mild, with warm water currents circulating round the continents, but there was also little latitudinal variation in climate. It was in this climatic regime that the species diversity of marine invertebrates living on the world's continental shelves reached a peak (Valentine & Moores, 1976), and in the same period terrestrial vegetation and fauna became well established for the first time. There can be no doubt that these two developments were connected. In the period when the continents were coming together, but had not yet fused, extensive areas of continental shelf with divergent characteristics, and possessing faunas which had been separated for very long periods, were juxtaposed. Large shallow areas would have provided ideal opportunities for air breathing to evolve, both in marine and fresh waters. At the same time plants with aerial leaves would have experienced ideal conditions, and truly terrestrial plants soon followed.

The early Carboniferous period which followed the Devonian also showed a general uniformity of world floras (Chaloner & Lacey, 1973). The luxuriant coal swamp vegetation found in North America and Europe, both of which were at the time near the equator, pro-

duced ideal microclimatic conditions for the evolution of both arthropods and vertebrates. Towards the end of the Carboniferous, Europe and America became joined to the southern continent of Gondwanaland, thus disrupting the flow of water currents and moisture laden winds. Latitudinal variation began to build up, and the southern part of Gondwanaland was probably glaciated. The change of world climate towards a more continental one increased in the Permian, with the joining of all the continents to form Pangaea. The Permian itself was characterised by the extensive spread of deserts, presumably with extremes of climate, and it is noticeable that during this time no new terrestrial plant or animal groups were recorded. The diversity of marine animals decreased, presumably due to the fusion of continental coastlines and the obliteration of the large areas of shallow continental shelf (Valentine & Moores, 1976).

In summary of the development of terrestrial ecosystems in the Palaeozoic, it may be said that they evolved and diversified only in times of widespread mild or tropical climate. The two major groups of terrestrial animals, the vertebrates and the insects, emerged and radiated during these conditions. The present-day distribution of both these groups suggests that they have retained their basically tropical adaptations, and that world distribution can be envisaged in terms of a colonisation of temperate zones from tropical ones. If this is

Fig. 10.4. Pathways by which the mutual interactions between butterflies (*Heliconius* spp.) and vines (*Anguria* spp. and *Passiflora* spp.) influence individual, population and community features. Adult *Heliconius* feed on the nectar of *Anguria* while larvae feed on the leaves of *Passiflora*. Dashed lines show features which have most probably evolved in the context of mutualism. From Gilbert (1975).

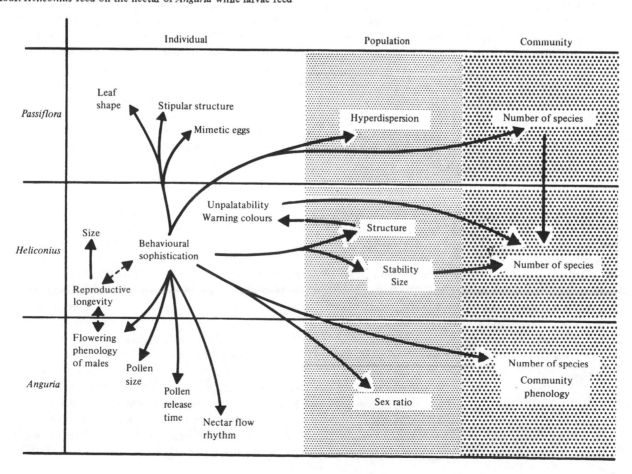

really the case, it is evident that the study of tropical animals in transition between water and land at the present time could reveal much about the ways in which early terrestrial animals evolved.

The Mesozoic period was characterised by the break-up of Pangaea (Dietz & Holden, 1976), and the consequent return of milder conditions and decreasing latitudinal variation in climate. The mild conditions of the Jurassic and Cretaceous periods saw the development of organisms which form the essential parts of present-day ecosystems: the birds and mammals appeared, the land snails evolved or became common, and the angiosperms came to dominate the vegetation. In general terms, then, it can be said that groups of terrestrial organisms have only evolved or radiated extensively during conditions of relatively high and constant temperature. High humidities have also been needed by most animal groups, at least during times of transition between water and land. These have been provided by the plant cover: initially by the gymnosperms, but later increasingly by the angiosperms, which have themselves developed mechanisms of conserving water and of reproducing with only minimal supplies of free water (see, e.g. Delevoryas, 1977). It would thus be of the greatest interest to pinpoint the exact region in which angiosperms originated. So far this has not proved possible (Schuster, 1976), although many palaeobotanists believe that they evolved in the tropics (Brenner, 1976). Since proof of a tropical origin for the angiosperms would add weight to the idea of a tropical origin for many terrestrial animals, it is to be hoped that new approaches will clarify the situation.

The evolution of one particular angiosperm group, and its effects upon the colonisation of land by animals, can be considered a little more closely. This group comprises the tropical salt-tolerant trees known as mangroves. At the present time mangrove communities are dominated by trees from four families which vary in significance in different longitudes: Rhizophoraceae (including *Rhizophora*), Avicenniaceae (including *Avicennia*), Sonneratiacea (including *Sonneratia*) and Combretacea (including *Laguncularia*). These trees are all characterised by their salt tolerance, and they form the dominant vegetation along protected shorelines throughout the tropics. As mentioned in previous chapters, the microenvironment that they provide forms the basis for a community of animals which are air breathing for part of the tidal cycle. The possible involvement of the mangrove habitat in the evolution of terrestrial faunas must now be briefly discussed. The first point to settle is the time at which mangrove communities first appeared. From the fossil record (Chesters, Gnauck & Hughes, 1967), we find that the Combretaceae were present in Upper Cretaceous times, although the earliest records for the other families are no earlier than the Tertiary. The present distribution suggests, taking into account what is known of continental drift, that the four families were extant and spreading between Upper Cretaceous and Lower Miocene times (Walsh, 1974). From Fig. 10.2, it can be seen that most of the terrestrial snails and the terrestrial decapods appeared either in the Cretaceous or the Tertiary. Considering the abundance of these two groups in present-day

mangrove swamps, it seems extremely likely that those of their representatives which moved on to land directly from the sea came through the intermediate stage of the mangroves.

10.4 The success of different animal taxa in a variety of terrestrial habitats

Throughout preceding chapters the restrictions of particular animal groups to certain habitats has been emphasised. It is appropriate now to attempt to draw some of this information together, and to consider the relative success that different taxa have achieved in a variety of different terrestrial conditions. In order to make comparisons we shall be forced to consider only gross regional habitats or biomes, although within these the microhabitats, and hence the fauna, show a great diversity. Even without this inherent variation it is not easy to decide upon any one measure of success for an animal group. Many definitions of success might be used, including perhaps persistence through time, but the three approaches taken here will consider abundance as numbers of individuals, biomass and numbers of species. In ecological terms the best measure of 'importance' in the ecosystem might be gained by considering the metabolic rate of each group, but since so few figures are available for this, only passing reference will be made to it.

In order to narrow down some of the problems of comparing a wide range of habitat types, we shall begin with a discussion of the invertebrate soil fauna in different biomes. This is in most respects equivalent to the cryptozoic fauna, and includes categories A, B and C described in Chapter 1 (Table 1.3): aquatic animals, those requiring constant humidity and those requiring high humidity for activity but able to withstand some desiccation. Some figures for abundance (numbers $/m^2$) and biomass (live weight, g/m^2) have been assembled in Figs. 10.5 and 10.6 but it must be noted that these can only be taken as a general guide. The problems in accumulating such figures are enormous. First of all, the fauna of the soil proper grades into that of the litter, and this in turn grades into the fauna of vegetation above. Second, various methods used to extract the fauna have very different efficiencies, and figures for the smaller arthropods especially tend to be unreliable. When it comes to converting figures for abundance into biomass, many assumptions have to be made, all involving inaccuracies, and while some workers favour reporting biomass as dry weight, others prefer live or wet weight. The results in Fig. 10.6 are given as live weight because most of the results have been expressed in this way. The variations in any one biome are enormous. Geographical variation may be considerable over very short distances, due for instance to changes in pH or calcium availability, which particularly affect the distribution of animals with calcareous skeletons. Over longer distances, variation may be extreme, and the north temperate and south temperate zones show great differences in their characteristic faunas: for instance, while south temperate zones have abundant terrestrial amphipod populations, amphipods are not found on land in the north temperate zone. Seasonal changes also may involve orders of magnitude

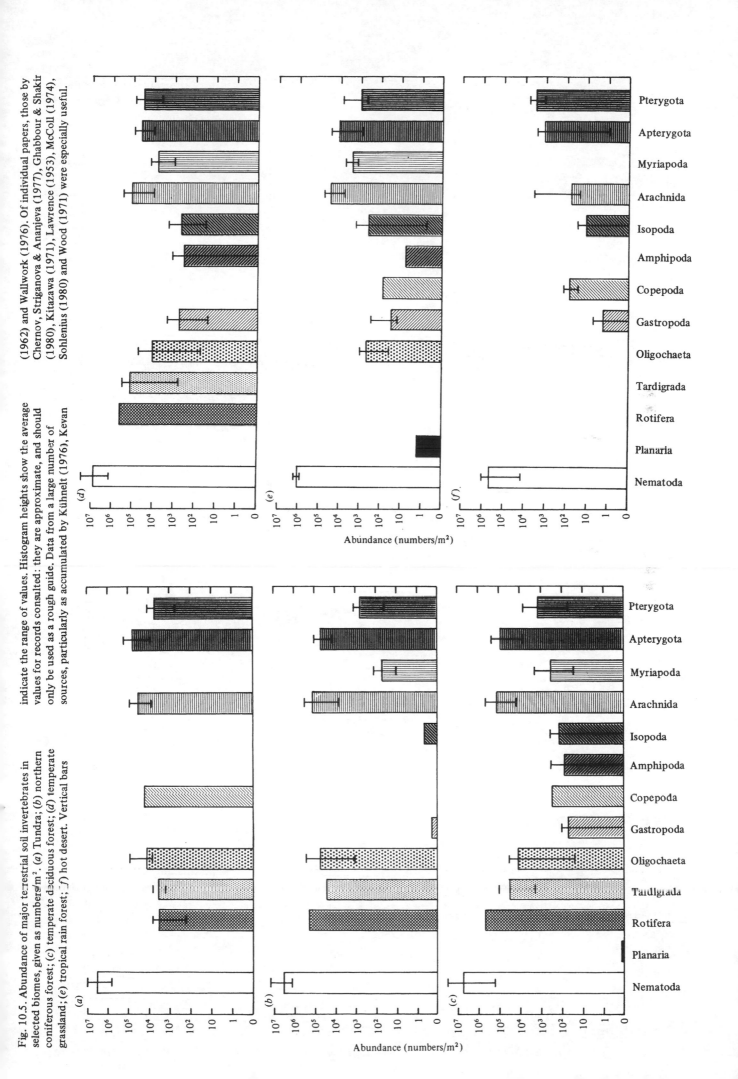

Fig. 10.5. Abundance of major terrestrial soil invertebrates in selected biomes, given as numbers/m². (a) Tundra; (b) northern coniferous forest; (c) temperate deciduous forest; (d) temperate grassland; (e) tropical rain forest; (f) hot desert. Vertical bars indicate the range of values for records consulted: they are approximate, and should only be used as a rough guide. Data from a large number of sources, particularly as accumulated by Kühnelt (1976), Kevan (1962) and Wallwork (1976). Of individual papers, those by Chernov, Striganova & Ananjeva (1977), Ghabbour & Shakir (1980), Kitazawa (1971), Lawrence (1953), McColl (1974), Sohlenius (1980) and Wood (1971) were especially useful.

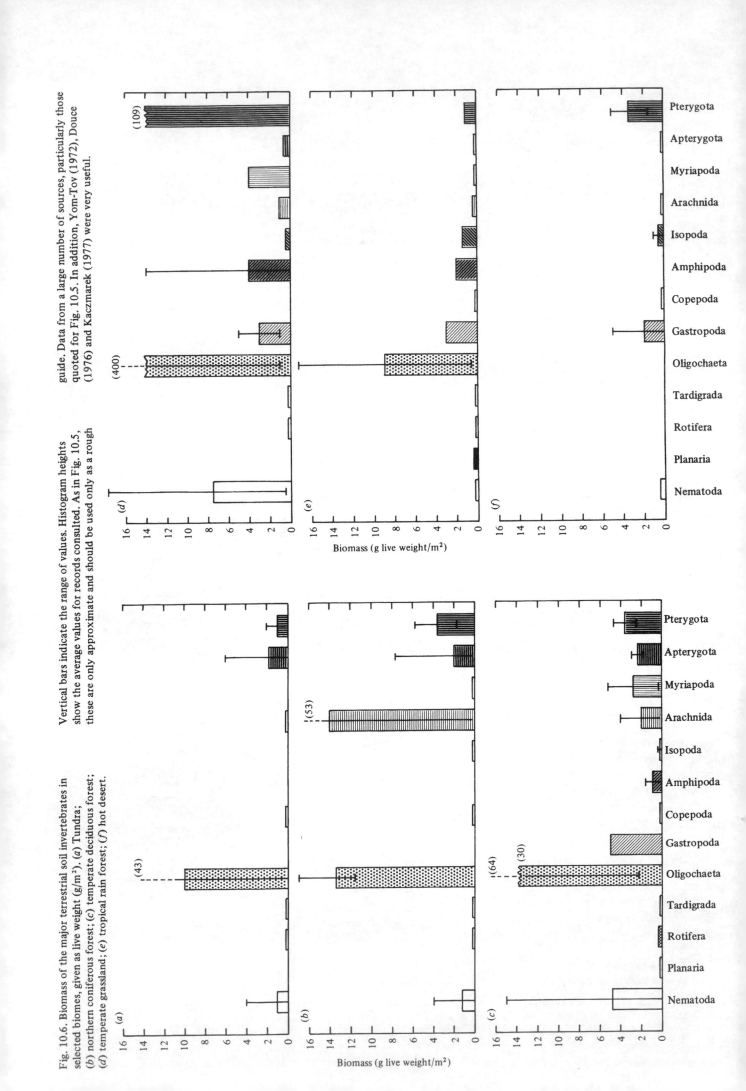

Fig. 10.6. Biomass of the major terrestrial soil invertebrates in selected biomes, given as live weight (g/m²). (a) Tundra; (b) northern coniferous forest; (c) temperate deciduous forest; (d) temperate grassland; (e) tropical rain forest; (f) hot desert.

Vertical bars indicate the range of values. Histogram heights show the average values for records consulted. As in Fig. 10.5, these are only approximate and should be used only as a rough guide. Data from a large number of sources, particularly those quoted for Fig. 10.5. In addition, Yom-Tov (1972), Douce (1976) and Kaczmarek (1977) were very useful.

Biomass (g live weight/m²)

in differences of abundance. For all these reasons the data presented in Figs. 10.5 and 10.6 must be treated with caution, but in spite of their inherent unreliability some generalisations appear.

10.4.1 Abundance

In all the biomes represented in Fig. 10.5 the nematodes are the most abundant taxon. Although their density is higher in temperate zones than in high or low latitudes, they have spread over *all* land masses. The development of the facility of cryptobiosis, and their existence essentially in the water film between soil particles, has made them eminently successful on land. Other small forms which show cryptobiosis, the tardigrades and rotifers, have also been successful, and are widespread and found in large numbers. Of the larger invertebrates, the oligochaetes are found abundantly in all habitats except hot deserts. In view of their susceptibility to desiccation, this may seem surprising, but many of them utilise the damp habitat of the soil as a haven and emerge from it to feed only when external conditions are appropriate. The gastropod molluscs show quite a different trend in their distribution. They have failed to colonise high latitudes but are abundant in temperate zones, and have also managed to penetrate into hot deserts. Their ability to aestivate inside their protective shell and to emerge rapidly when conditions ameliorate has obviously contributed to their success in such a wide range of habitats. Among the arthropods there are some surprising distributions. While copepods are often thought of as being essentially aquatic, they and cladocerans have also colonised the soil water effectively in high latitudes, and to some extent in temperate zones. Harpacticoids are even present in Australian arid soils, together with some cladocerans (Wood, 1971). These forms can probably undergo a cryptobiotic existence like the tardigrades and rotifers, which accounts for their otherwise anomalous existence in dry environments. The amphipods show a distribution the converse of that shown by the copepods: they are abundant in tropical forests and stretch into the south temperate zone, even in grasslands, but are completely absent from deserts. Unfortunately very little is known about their biology and the properties which enable them to withstand the seasonal changes of the temperate climate. The myriapods parallel their distribution to some degree, but are also common in the north temperate zone as well as in the south. Isopod crustaceans are similar to the molluscs in their distribution, being abundant in temperate and tropical regions, and being able to tolerate desert conditions to some degree. Unlike the desert snails, however, desert woodlice live in burrows and can retire into these to escape the rigours of the macroclimate. The high numbers of arachnids are mainly due to the abundance of the acarines. Although most are small in size, the numbers of mites are high from tundra to desert. Similarly, the apterygotes are represented mainly by one group, the collembolans, which are common in all the habitats considered. Both have excellent osmoregulatory and excretory mechanisms, and are also good tolerators. Finally, the pterygote insects are also ubiquitous,

although in terms of numbers they fall slightly below those of the apterygotes; this, however, is related to their larger size, as will become apparent from a consideration of biomass.

One generalisation which springs from this brief account is the greater diversity of taxa in temperate forests, temperate grasslands and tropical rain forest than in other biomes. The wide variety of taxa in tropical rain forest is perhaps to be expected, given the constancy of conditions within these forests, and the possibility that many groups of animals moved on to land in the tropics. It is only here that terrestrial flatworms and leeches are common, both dependent upon the relatively constant high temperature and humidity. The diversity in temperate zones reflects the protective influence of the soil habitat, and its relative constancy in comparison with the macroclimate. As will be seen when species numbers are discussed, the diversity within many groups declines from the tropics towards higher latitudes.

10.4.2 Biomass

The relatively uniform representation of many animal groups in the soils of temperate and tropical regions, when considered in terms of numbers, contrasts strongly with the picture presented by the distribution of biomass (Fig. 10.6). From this it is apparent that a very few groups make up the majority of the soil biomass. Overall, the most widespread dominants are the oligochaetes. Fig. 10.6 hides the fact that Enchytraeidae are abundant in cold acid soils, while Lumbricidae dominate in warmer, more alkaline conditions. In the tropics other families dominate. Still it can be said that it is the oligochaete organisation which dominates in the soils of the world, with the exception of hot deserts. As already pointed out in Chapter 3, earthworms are very important in forming and maintaining soil structure, and in initiating the breakdown of plant material to form detritus which is then utilised by other soil animals. In specific regions other taxa can show high biomasses. Mites are important in northern regions, and in temperate grassland both amphipods and pterygote insects reach very high levels. In these cases the energy flow through the soil system may depend more upon these other groups, since they have higher metabolic rates than oligochaetes. In some temperate regions the biomass of nematodes is also high, and with their enormous numbers they must influence soil biology to a great degree. The large biomass of pterygote insects in temperate grassland reflects the high numbers of adults which live in the above-ground vegetation and have larvae which develop in the soil. In hot deserts, pterygotes have a much lower biomass, but nevertheless this is higher than that of any other taxon. The situation is quite different from temperate grasslands, however, because the main biomass of desert insects is made up by adult forms such as the burrowing cockroaches and the tenebrionid beetles.

10.4.3 Species numbers

Although for many groups both total population densities and total biomass are similar in tundra and temperate soils, it has been seen that several taxa, such as the isopods, amphipods, myriapods and gastropods, do not penetrate into the cold tundra regions. Nevertheless, the taxa which *do* occur in the tundra show quite a large number of species: MacLean *et al.* (1977) reported 20 species of Collembola, 40 Acari and 15 Enchytraeidae in the tundra of Alaska, and many temperate forest soils show similar species numbers. In the polar deserts further north and south, species numbers decline, as exemplified by Fig. 10.7 (Tilbrook, 1967): in the maritime Antarctic the number of species of Acari and Collembola are reduced to very low levels. This reduction in diversity must be ascribed to the harsh physical conditions in these areas, and to the uniformity of the environment. There is a parallel here with other 'stress' environments such as estuaries: these have a small number of species, but a high biomass and number of individuals because the sediments are uniform and physical conditions are extreme. Few species can survive in these conditions, but for those that can, the food supplies allow tremendous population sizes.

At the other extreme of climatic conditions a reduction in the number of species is also found in the soils of tropical rain forest and of hot deserts. In the deserts, harsh physical conditions are responsible, as well as the restricted number of niches. Animals here are primarily governed by the physical conditions, and there is little room for a variety of biological types. It is impressive that only the pterygote insects show much diversity of types in the soils of hot deserts, with the cockroaches and tenebrionid beetles being the main representatives. The only woodlice that are also found in deserts seem to have emulated the insects in the production of an impervious cuticle (see Chapter 5).

The variety of species in tropical rain forest requires quite separate consideration. Here it is necessary to be particularly careful in defining soil fauna,

Fig. 10.7. A diagram to illustrate lowering of the numbers of species with increasing latitude. White histograms show Acari, black histograms show Collembola. For both, numbers of species are lower in Antarctic habitats than in temperate regions. From Tilbrook (1967).

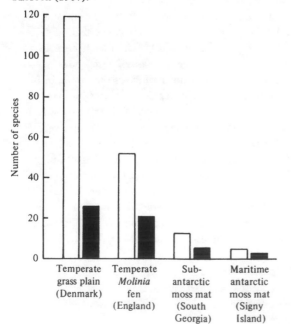

as opposed to the fauna living above ground level. The fauna of the soil *sensu stricto*, i.e. not including the fauna of litter or of ground-layer vegetation, is often depleted in tropical rain forest. Both numbers and diversity are low probably because the high rainfall causes rapid leaching, and with the high temperatures the rate of decay of organic detritus is high. The supply of nutrients in the soil is therefore poor, and the soil structure tends to be rather uniform. In the litter layer and above, in contrast, diversity may be high. Wallwork (1976), for instance, quoted the much greater number of species of cryptostigmatic mites in leaf litter from rain forest in Peru than in temperate forest. Williams (1941) recorded more than 200 species of invertebrates from the leaf litter of rain forest in Panama. Not only was the number of insect and mite species high, but groups uncommon in temperate forests, such as leeches and flatworms, were abundant. This high diversity reflects the enormous diversity of tree species in tropical rain forests: individual species seldom form pure stands like those of temperate forests, and the number of species in any one area is very great. Consequently, the mosaic of different leaf types on the forest floor is extremely complex. In addition to this factor, the category of 'soil fauna' spreads much higher vertically than in temperate forests: with the high humidity many trees are covered in dense epiphytic growths, which themselves collect and form soil. Many species of animals such as prosobranch gastropods, which in general are litter species, are consequently to be found some distance from the forest floor. The number of available niches is thus enormous, and with the constancy of physical conditions which is typical of the tropical rain forest, it is hardly surprising that this habitat today harbours the richest animal diversity of any area.

The diversity of species in leaf litter and above it leads on to a more general consideration of the diversity of animals above the level of the soil. Principally these animals are the ones that can withstand the general macroclimate, as opposed to the cryptozoic niches provided by soil and leaf litter. Some approximate numbers are given in Table 10.4. There is no doubt that overall the insects show the greatest number of species. Present numbers are not known, but with the number described having nearly reached a million, the true total may be as high as 4 to 10 million (Matthews, 1976). Most of these are terrestrial, or have at least one terrestrial stage. The arachnids have produced the next highest number of species. Once again, the total is not known accurately. There are about 32 000 species of Araneae, and 25 000 described species of Acari, but this latter figure is undoubtedly nowhere near the true total. The terrestrial molluscs have about 25 000 and the terrestrial vertebrates at present about 20 000 species, and come third in the order of species abundance. The myriapods have about half this number, distributed between the Chilopoda (3000 species), the Diplopoda (more than 7500), the Symphyla (120) and the Pauropoda (380).

Other terrestrial groups have only relatively small numbers of species. The oligochaetes, although so important in the soil in terms of biomass, energy flow and soil structure, have only 2000 species. Of these the Enchytraeidae have 260 species and the Lumbricidae 160, while most species belong to the tropical earthworm family Megascolecidae (1300). Terrestrial isopod species number about 1000, mostly in the suborder Oniscoidea, while there are only about 50 terrestrial Amphipoda. The flatworms, in the families Geoplanidae (mostly in South America) and Bipaliidae (mostly in the Indo-Pacific) have not been monographed in recent years, and the estimate of 500 is probably some way below the real total. Considering their limitation to humid microenvironments, and their predatory life style, they have been very successful – more so than the Onychophora and Nemertinea, which have similar habitat requirements and are also predators.

The factors controlling these various degrees of success on land have been debated many times, as have the various measures which should be used in judging success. If we take as a measure of success the diversity of different niches that are occupied by a group, then the number of species in the group can probably be justified as giving a reasonable index. However, it would also be appropriate to include some measure of the diversity of nutritional mechanisms used by each group, i.e. whether they contain species which are herbivores, carnivores, parasites, etc. Using these two points as a guide, the insects, with by far the greatest number of species, and with a range of nutritional types spanning every conceivable method from predator and herbivore to parasite and symbiont, must be rated as the most successful group. Their occupation of all habitats from cold to hot climates, and from deserts to rain forest, reinforces this view. In comparison, the only groups which rival the insects in terms of species numbers are other arthropods, vertebrates and molluscs. The terres-

trial molluscs, however, have almost entirely remained herbivores and detritivores. Carnivorous snails such as the New Zealand *Paryphanta*, which eats earthworms, are exceedingly few, and there are also only a few species of carnivorous slugs (Solem, 1974). Because of this, and their dependence upon high humidity for activity, the number of niches occupied has been relatively small. Among the arthropods, the myriapods consist of both herbivorous and carnivorous forms, and have produced both slow-moving and very fast-moving species. Their tolerance of physical extremes is, however, limited, and they have not spread out of the cryptozoic habitat nor out of tropical and temperate latitudes.

The arachnids surely qualify as successful competitors of the insects. They are found from the poles to equatorial deserts, have a multitude of herbivorous and carnivorous forms, and a high degree of physiological resistance to, and tolerance of, the physical environment. The insect–arachnid interdependence has been emphasised earlier, and the rise of the predatory Araneae was probably consequent upon the rise of the insects as prey. The Acari have become both plant and animal feeders, and have representatives in most environments. Most have remained small, and have exploited appropriate microhabitats: they have become enormously abundant, but have in the main become specialised as soil forms, or as parasites on either plants or animals.

The main rivals to the insects in terms of terrestrial success are the vertebrates. Although they do not compare with the insects in terms of number of species, their large size has made their effect upon the environment of a comparable order. Like the insects, terrestrial vertebrates inhabit regions from the polar deserts to equatorial rain forest, and have diversified as a variety of herbivores and carnivores. Their diversity of life styles is, however,

Table 10.4. *Estimated numbers of terrestrial species in selected animal taxa*

Taxon	Number of terrestrial species	Reference
Tricladida (Platyhelminthes)	500	Graff, 1912–1917; Froehlich, 1954
Nemertinea	20	Gibson, 1972
Gastropoda		
Prosobranchia	4 000	Morton, 1979, after Winckworth
Pulmonata	20 500	Solem, 1978
Oligochaeta	2 000	Stephenson, 1930
Onychophora	70	Barnes, 1980
Isopoda	1 000	Waterman & Chace, 1960
Amphipoda	50	Hurley, 1968
Arachnida	>64 000	Barnes, 1980
Myriapoda	11 000	Barnes, 1980
Hexapoda	>750 000	Barnes, 1980; Matthews, 1976
Amphibia (including freshwater and amphibious forms)	2 000	Young, 1950
Reptilia	5 000	Porter, 1972
Aves	9 000	Johnson *et al.*, 1977
Mammalia	4 000	Gunderson, 1976

Numbers have been rounded up from the reports quoted, and are only approximate.

not nearly as great as that shown by insects. Partly, at least, this can be attributed to their greater size, since there are many more microhabitats available to the smaller insects. Most vertebrates live in the macrohabitat and thus cannot experience this diversity of niches. There are several further factors which govern the relative diversity of insects and vertebrates (Matthews, 1976). Their great similarity lies in the comparability of their physiological complexity and efficiency: members of both groups exhibit well-developed homeostatic mechanisms which maintain the characteristics of their internal environment at a constant level in the face of wide external changes. As can be seen from preceding chapters, there is probably no other group with such good regulatory mechanisms, although the arachnids have in some respects produced comparable systems. There are, however, also two basic differences between insects and vertebrates. The first concerns generation times: vertebrates typically have a long life span in which the process of learning has evolved as a major strategy to deal with environmental change, whereas in insects, life spans are typically short, so that individuals do not experience enough of the irregular short-term fluctuations which might initiate the learning process (Slobodkin & Rapoport, 1974). In place of learning, individual insects are provided with innate responses; but insect populations respond by developing isolated genetic units, or species, as a way of adapting to environmental diversity in time and space. In other words, speciation is, for insects, a rapid and adaptive strategy, so that the numbers of extant insects at any one time are enormous.

The second major point to be made about differences between insects and vertebrates is the relative importance of the co-evolutionary interactions in which the two groups are involved. The co-evolution of plants and insects has already been discussed (see p. 223), and the complex interactions of insects with angiosperms emphasised. The importance of plant–herbivore interactions in general as diversity-generating mechanisms in terrestrial ecosystems has been stressed by Ehrlich & Raven (1964). These interactions have triggered more diversity in insects than they have in vertebrates because of the difference in size of the individuals of the two taxa: the interactions of insects with plants are intimately concerned with leaf and flower structure, whereas the vertebrates are mainly macrofeeders, so that detailed plant structure seldom relates to particular herbivore species. Vertebrates therefore tend to eat a great variety of plant species, whereas insects are often restricted to a very few, and the number of insect herbivore species bears a close relation to the number of available plants. Diversity in the vertebrates probably has much to do with their co-evolution with the insects: the insectivores and insect-eating bats comprise at least a fifth of terrestrial mammal species, and many of these have specific dependence upon a limited number of insect species. In contrast, the rodents, containing over 40% of the mammals, have a diet based on both insects and plants, especially seeds; but in very few cases are there examples of specific relationships with individual plant species. Similarly in the birds, many species are seed eaters, and

aid in species dispersal, but most feed on a wide variety of species so that there are few examples of true co-evolution.

In contrast to the diversifying influence produced by the close relations between insects and plants, the effect of detritus in this direction has been minimal, both for insects and for other animals. Partly this is probably because detritus food chains are based more upon the consumption of bacteria and fungi than they are upon direct digestion of plant remains, so that even with many different plant species contributing to the leaf litter, the available energy is present in a fairly uniform state. Partly, also, the relationships which characterise interactions between animals and living plants involve a great variety of different defence strategies by different plant species, whereas no such mechanisms exist once the plant is dead and being formed into detritus. The result has been that the diversity of detritus-feeding invertebrates has remained relatively low. Oligochaetes, while important in energy flow, have only about 2000 present-day species. Even the prosobranch gastropods, which are primarily detritivores, but exist in a much wider variety of niches than the oligochaetes, have only produced about 4000 species. Physical restriction to one range of habitats is obviously also important, as shown by a comparison of isopods and amphipods: isopods, which in some cases can withstand quite dry conditions, have about 1000 terrestrial species, whereas amphipods, which are restricted much more by problems of desiccation, have only 50. No doubt there are many other additional factors in all the cases that have been considered. In particular, the time that various groups have had on land should be emphasised. As far as available evidence goes, amphipods and oligochaetes may be recent inhabitants of the land, whereas the insects and vertebrates have had terrestrial forms since the Palaeozoic (Fig. 10.2). At least in insects it appears that the longer a group has been exposed to particular host plants, the greater is the number of species utilising the plants (Southwood, 1973). This effect may also occur in a much more general sense, so that the longer a group has been exposed to terrestrial habitats, the greater the number of terrestrial species which has evolved. Although some authors have argued that these effects are secondary to those of co-evolution (e.g. Matthews, 1976), they could be appropriate in making comparisons between groups of recent and ancient terrestrial origin. Thus the myriapods, although basically cryptozoic and to a great degree detritivorous, have at present about 11 000 species – very much greater than, say, the isopods. Such differences might be partly attributed to the more ancient origin of terrestrial myriapods. However, so many other factors are important in determining the adaptive radiation shown by various animal groups that such an argument can probably not be widely applied.

10.5 The recolonisation of aquatic habitats

It is appropriate to conclude by examining those animal groups which have reversed the processes of terrestrial colonisation discussed in this book, and have produced secondarily aquatic species. This phenomenon

is probably widespread. It has been described or postulated in nemertines, pulmonate molluscs, arachnids,
insects and vertebrates, but occurs to some degree in
many other groups. Here the main discussion will concern the pulmonate molluscs and the insects, and
secondarily aquatic vertebrates will not be considered
because the subject is covered in many text books.
Before examining insects and molluscs in some detail,
we must briefly review the attempts made at aquatic
recolonisation by other invertebrates.

10.5.1 Nemertines

Nemertines typically form a marine group, and
freshwater species are rare. Six freshwater genera have
been described, three of which are heteronemerteans
and contain only one species each. The other three are
hoplonemerteans: in the family Tetrastemmidae is the
genus *Prostoma* with five or six species, while the family
Prosorhochmidae contains the two genera *Potamonemertes* and *Campbellonemertes*, each with one species
(Gibson & Moore, 1976). *Prostoma* has colonised
fresh water via estuaries (see Chapter 2), but *Potamonemertes* and *Campbellonemertes* show very great similarity
to the supra-littoral and terrestrial genera within the
family: in common with these genera (*Acteonemertes*
and *Geonemertes*), they have enormous rhynchocoels,
protonephridial systems with very numerous excretory
pores, a thick dermis with well-developed parenchyma,
and many other features (Moore & Gibson, 1973). On
this basis it seems likely that both *Potamonemertes* and
Campbellonemertes have re-invaded fresh water from the
land: a cryptozoic existence has allowed sufficient contact with hypo-osmotic stress for the development of
an excretory system which is adequate to deal with
conditions in fresh water. It is noticeable, however, that
while the records for the genus *Prostoma* are worldwide,
those for *Potamonemertes* and *Campbellonemertes* are
restricted to New Zealand. This suggests that for the
nemertines, recolonisation of fresh water from the land
has been much less successful than direct colonisation
via the estuarine habitat.

10.5.2 Myriapods and arachnids

Many arthropod lines of terrestrial stock, besides
the insects, have representatives in the aquatic habitats,
both marine and fresh water. Because they are basically
airbreathing in design, however, much of this movement
back to water can be thought of as 'tolerance of submersion' rather than true adaptation to an aquatic
habitat. For example, there are many species of marine
intertidal centipedes (Roth & Brown, 1976), but these
are mainly active only in air. Since they feed on marine
intertidal invertebrates, they are indeed part of the
marine ecosystem, but they retain the basic terrestrial
air-breathing mechanisms. Symphyla, Pauropoda and
Diplopoda have also been recorded intertidally, but
generally the myriapods are not common on marine
shores, and are not found in fresh water. Arachnids,
on the other hand, have many species which are found
on marine shores, as well as many that live in fresh water.
They have evolved several respiratory mechanisms which

allow them to live completely submerged (see Chapter 6),
and of all the primarily terrestrial arthropod groups, they
have become most completely readapted to aquatic life.
Among other arachnid groups the spiders have many
intertidal forms (Roth & Brown, 1976), but few which
have become greatly adapted to life in fresh water.
Argyroneta aquatica, the water spider, is the only one
that lives entirely under water, by constructing its own
'diving bell' from air bubbles brought down from the
surface (Bristowe, 1971). Palpigrades are common at the
top of marine shores, and scorpions are sometimes found
on the shore (Roth & Brown, 1976), but they show little
specific readaptation to an aquatic environment. Apart
from the Acari, arachnids in general may tolerate submersion, but are not active except when they are exposed
to the air.

10.5.3 Insects

Out of the nearly one million described species of
insects, 25 000–30 000 species are aquatic in at least one
stage of their life history. Most of these are connected
with fresh waters, and only a small number are marine.
Cheng (1976) estimated marine species to be of the
order of hundreds, but if those with intertidal larvae are
included, the number is still probably only of the order
of thousands. Many entomologists have sought for an
explanation of these very small proportions. A brief
survey of the types of insects which have become
aquatic, and some of their adaptations, will clarify
the situation. It is convenient to consider freshwater
species first, and then to move on to marine species.

The adaptations of freshwater insects have long
interested entomologists (see, e.g. Miall, 1895). Many
orders of insects are common in fresh water, but mainly
in the juvenile stages; adults have only rarely adapted
completely to an aquatic life (Imms, 1947). Of the
Diptera, chironomid larvae and mosquito larvae are
almost ubiquitous. The Trichoptera or caddis worms are
usually abundant, as are the nymphs of Ephemeroptera
and Odonata with their various gill mechanisms. Only in
the Coleoptera and Hemiptera are the adults commonly
found in fresh water. Beetles such as *Gyrinus natator*
are mainly associated with the water surface, while
others such as species of *Dytiscus* spend more time
under water. A similar diversity is found in the Hemiptera, of which the water skaters utilise the water surface
while the water boatmen live under water. In all cases
the adults have a close association with the water surface:
if they do not actually live on the surface film, they need
to penetrate through it from below to obtain their air
supply, since there are very few adult species capable
of obtaining a continuous air supply under water.
Aquatic insects have been very successful in the diversity of mechanisms used to obtain oxygen (Hinton,
1976). Some visit the surface at intervals, or have long
spiracles which reach to the surface and penetrate the
surface film. Others have hydrophilic cuticles, usually
with tracheal gills, and absorb their oxygen directly from
the water. Species associated with plants may have
pointed spiracles which they thrust into the plant tissues
to tap intercellular air spaces. Finally, many have evolved

a permanent physical gill or plastron into which oxygen diffuses from the surrounding water. With this variety of respiratory methods, freshwater insects have been able to colonise a wide variety of types of habitat from the still waters of ponds and lakes to fast-flowing streams.

One other major adaptation that freshwater insects have evolved is a method of regulating the osmotic pressure of the body fluids in the face of the continual osmotic inflow of water and loss of salts by diffusion: problems only infrequently encountered by terrestrial insects, which are more often exposed to desiccating conditions. Osmotic problems are probably slight for freshwater adults, because most of these are at least partly surrounded by an air film; but in larvae with hydrophilic cuticles, complex mechanisms have developed. These are based upon ion pumps which are located in special structures such as the rectal gills of dragonflies and the anal papillae of mosquito larvae.

The adaptations pointed out above have allowed freshwater insects to evolve many species. Nevertheless, this number is so much lower than the number of terrestrial species that one is justified in asking why there are not still more freshwater forms. One of the answers may be that although the freshwater insects have coped very well with the physiological problems of respiration and salt and water balance, the reasons for insect diversity on land are based upon co-evolution with plant species, coupled with the capacity for aerial flight. In water, the advantages of flight are lost – indeed, wings are a positive disadvantage – and the diversity of angiosperms is very much reduced compared to that on land. The factors that have made the insects so successful on land have thus resulted in their not being so well adapted for entering fresh waters. On the other hand, it is possible to take an entirely different view of the situation. It must be pointed out that the area of fresh waters at the present time is only about 5 million km^2 (Russell-Hunter, 1978), whereas the total area of land is about 144 million km^2 (Sverdrup, Johnson & Fleming, 1942). On a proportional basis, one would expect about 30 times as many insect species on land as in freshwater. With the numbers described at present being about 30 000 in fresh water and nearly a million total, a ratio of approximately 1 : 30 is found. For fresh waters, then, it could be argued that there is *no* lack of insect species, and that the numbers are approximately as would be predicted.

In the marine intertidal zone, insect species are much less common than in fresh water, and on the open sea the only forms are the hemipteran species of the genus *Halobates* (see Andersen & Polhemus, 1976). Some of these live in flocks miles from land (Foster & Treherne, 1980). On the shore, the Diptera, Coleoptera and Hemiptera have the greatest number of species (Fig. 10.8). In particular, species numbers are higher in saltmarshes than in rocky and sandy intertidal shores (Foster & Treherne, 1976). From this it seems unlikely that high salinity by itself is a major factor limiting the numbers of marine species. Indeed, terrestrial insects are to some extent preadapted for the colonisation of marine environments by their impermeable cuticles and their ability to osmoregulate by reabsorbing water from the

rectum so that rectal fluid becomes hyperosmotic to the blood (Foster & Treherne, 1976). A similar argument may be put forward for respiratory problems: the variety and efficiency of respiratory mechanisms found in freshwater insects should act as preadaptations for life in the sea. However, more than half of the species of insects found in the marine intertidal zone have come not from fresh water, but directly from the land (Hinton, 1976). A spectacular example is the intertidal earwig *Anisolabis littorea* (Fig. 10.9). With the diversity of physiological adaptation shown by the insects it is in any case hard to believe that one or two physiological problems have been the cause of the small numbers of marine insects (cf. Cheng, 1976). Rather, some basic differences between the marine habitat and the freshwater/terrestrial habitats must be involved.

Two key factors must be considered. The first derives from the point originally made by Buxton (1926) that phytophagous insects attempting to colonise the sea would have to change their diet from angiosperms to algae because of the paucity of marine angiosperms. This is an unlikely change, however, because algae are nutritionally poorer than angiosperms, and very few species of benthic marine animals are direct herbivores (see Fenchel & Jørgensen, 1977). Most marine benthic secondary producers are detritivores and, as pointed out in an earlier section, even detritivore systems on land show a much reduced species diversity compared with herbivore-based systems. The small number of marine angiosperms is therefore probably the most significant factor limiting the numbers of marine phytophagous insects, and this in turn will limit the insect species of higher trophic levels. The relatively higher number of insect species on marine saltmarshes in comparison with other intertidal habitats supports this hypothesis, since the biomass of saltmarsh vegetation is made up mainly of angiosperms.

The second factor limiting the number of marine insect species is the nature of the air/water interface. It has been emphasised above that many freshwater insects depend upon the water surface for movement and for their oxygen supply: many also have to pass through this interface to reach the adult environment. It should also be noted that most insects are small enough and light enough to be under the control of the tensional forces

Fig. 10.8. The composition of saltmarsh insect faunas. Total number of species 308. Diptera, Coleoptera and Hemiptera together comprise nearly 80% of saltmarsh insect species. From Foster & Treherne (1976).

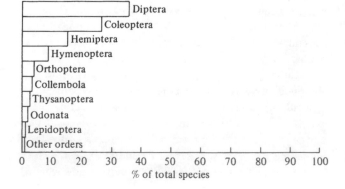

at the water surface. In fresh waters the water surface is often still, and when moving it seldom rises and falls vertically to any extent over short time periods. In the sea the rise and fall of the tide over vertical distances of several metres occurs regularly. The degree of wave action is also much greater in most marine littoral habitats, where the shore is not protected by angiosperm growth, than on freshwater shores, where angiosperms usually provide sheltered niches at the water's edge. The exception in marine shores is provided by saltmarshes, which because of their situation and their angiosperm cover, provided sheltered niches. This combination of shelter from wave action and the presence of angiosperms probably encourages high species numbers. On other marine shores small animals are subjected to violent wave action when covered by water, and to the forces of surface tension as the tide leaves them uncovered (Hinton, 1976). As a consequence, the insects that have colonised the marine intertidal zone tend to remain inactive at high tide, and to regain activity when the substrate remains damp but without water cover. They have in effect retained their terrestrial habitats, and have not, in many cases, become truly aquatic.

No doubt there are many additional factors which reinforce the effect of the two points made above. The respiratory and osmoregulatory difficulties can hardly make the transition to marine life any easier. One final argument may also be advanced, however, which concerns biological interactions rather than physical limitations. Usinger (1957) suggested that in any movement towards colonisation of the sea, the insects would have to compete with the crustaceans which have already occupied many of the available niches. Being small, they would

also have had to counter the predation of numerous intertidal fish. Such competitive effects, when reinforcing the two main factors of lack of angiosperms and the influence of a fluctuating air/sea interface, may have been decisive in limiting the numbers of marine insects to their present low level.

10.5.4 Molluscs

Unlike the situation in aquatic insects, the origin of aquatic pulmonate gastropods is not entirely clear. The present freshwater pulmonates appear to be secondarily readapted to an aquatic life, and can be arranged in artificial series showing the degree of this readaptation (Russell-Hunter, 1978). None has a true gill, or ctenidium and its absence, coupled with the development of a vascularised mantle cavity, is thought to be a characteristic terrestrial adaptation. It has also been suggested, however (Fretter, 1975), that the gill may have been lost as an adaptation to small size. However the loss occurred, many present-day Basommatophora, such as *Lymnaea*, remain primarily air breathers. Others, such as the ancylid limpets, have almost no mantle cavity but possess neomorphic gill lobes. Between these extremes are several types which breathe air to some extent, but can also utilise dissolved oxygen – some via the skin, and others via secondary gills.

Most of the very primitive pulmonates, such as the families Ellobiidae, Otinidae, Amphibolidae and Siphonariidae, form a contrast to those so far discussed, and are found in the marine littoral zone, although the Chilinidae are freshwater forms. Some members of these families breathe air all the time, but many can breathe using the lung filled with either water or air. Several

Fig. 10.9. *Anisolabis littorea*, a marine intertidal earwig from New Zealand, partly hidden in a crevice in driftwood. Most intertidal insects are derived, like this one, from terrestrial ancestors. Body length (including the characteristic unequal forceps) 2.5 cm.

species produce free-swimming veligers, showing that their ancestors were never truly terrestrial, and the Amphibolidae retain an operculum. There seems little doubt that the pulmonates first arose in the marine intertidal or supra-tidal zone. It is important to attempt a reconstruction of the ways in which adaptive radiation produced from these first air breathers a series of types in various habitats, although it must be realised that with only present-day species to use as guides such an attempt must be tentative.

Of the present-day primitive Basommatophora, the Ellobiidae are by far the most diverse in terms of types of habitat occupied, as discussed in Chapter 4, and they help to envisage pulmonate origins and radiations. It seems likely that early supra-tidal snails such as these developed a vascularised mantle cavity for air breathing, although the possibility that the lung is a secondary development has also been discussed (see Fretter, 1975). From present-day distributions, saltmarshes or mangroves seem the most likely place of origin (Morton, 1955*a*), and from there different groups moved into a variety of habitats. Some early pulmonates moved on to estuarine mud flats, and are represented by the present-day Amphibolidae. Others, represented today by Siphonariidae, Trimusculidae (=Gadiniidae), Otinidae and some Ellobiidae, colonised rocky shores and formed secondarily aquatic marine groups. Further forms moved above the supra-littoral zone on to land, but remained strictly within the cryptozoic habitats of forests. Present-day Ellobiidae such as *Pythia* are restricted to coastal forests, but the small *Carychium* is found inland. From small cryptozoic Basommatophora evolved both the Stylommatophora, with their reproductive specialisations and further adaptations for terrestrial life, and many groups which invaded fresh water. These are now represented by such genera as *Lymnaea*, *Ancylus*, *Planorbis*, etc. From this brief account, it can be seen that the recolonisation of fresh water, and that of the marine zones, took place as quite separate events in pulmonate evolution. It is convenient to consider the freshwater forms before those that have moved back into the sea.

Freshwater pulmonates comprise approximately 148 genera and subgenera, in comparison with 82 marine Basommatophora, and 2184 terrestrial Stylommatophora (Taylor & Sohl, 1962). The number of species is uncertain, partly at least because there is a great deal of variation between individuals, and in older treatments some of these variants were termed species. For example, over 1000 species assigned to the genus *Lymnaea* have recently been reduced to about 40 (see Hyman, 1967). According to Russell-Hunter (1964, 1978) this variability within a single species, coupled with a small number of species, can be related to the transitory nature of freshwater habitats. The impermanence of freshwater bodies was not emphasised when considering the evolution of freshwater insects, but most of the present large freshwater lakes have existed for no more than some thousands of years – Lake Ontario for 8000, and Loch Lomond for 11 000 (Russell-Hunter, 1978). Smaller lakes have an even more recent origin. It is interesting to compare the effects of this impermanence upon freshwater insects

and pulmonate gastropods. Insects, with their short life cycles, tend to speciate rapidly, so that physiological adaptations soon become incorporated into new taxa. Molluscs, on the other hand, have relatively long life cycles – generally much greater than one year – and individuals adapt to changing environments with flexible physiological responses. These might well lead to speciation, except that transfer of individuals between isolated environments usually occurs fairly frequently by passive dispersal, so that sufficient gene exchange occurs to prevent it (Russell-Hunter, 1978). This passive dispersal is a well-known phenomenon in freshwater molluscs, and is brought about usually by the transfer of aquatic weed – with molluscs attached – on the feet of birds. Only when bodies of water are isolated for long periods of time is there the opportunity for formation of new species, and this can be seen by examining the molluscan fauna of such ancient lakes as Lake Tanganyika and Lake Baikal (Boss, 1978). In these lakes, which have ages of 10^5 years or more, endemic pulmonate species occur quite commonly.

The major readaptation of Basommatophora to fresh water has been the development of respiratory systems allowing uptake of oxygen either at the water/air interface, or directly from the water. Snails such as *Lymnaea peregra* and *Physa fontinalis* have a vascularised lung but no form of gill. They come to the surface to breathe air in shallow water, but in deeper water the lung may be filled with water (Hunter, 1953). Even further variety is produced by those living continually submerged, but having a gas bubble in the lung. The origins and function of the gas in this situation are unknown. In all situations cutaneous respiration is important, and in *P. fontinalis* the mantle can be spread over the shell, so enlarging the cutaneous surface area (Fig. 10.10). The

Fig. 10.10. The expanded mantle margin in a freshwater pulmonate, *Physa fontinalis*. Mantle tissues, which increase the surface area for cutaneous respiration, are shown stippled. The small arrows indicate ciliary currents on a patch of epithelium posterior to the anus which aid in rejection of faeces. *P. fontinalis* breathes air as well as using the skin, and comes to the surface periodically to refill the lung. (*a*), (*b*) Lateral views; (*c*) dorsal view. Redrawn from Hunter (1953).

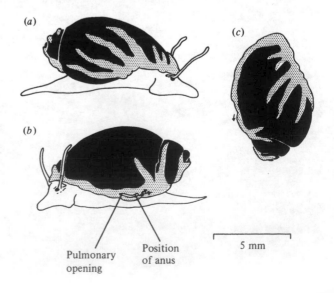

(*a*)

(*c*)

(*b*)

Pulmonary opening

Position of anus

5 mm

Planorbidae and the freshwater limpets possess secondary gills, developed from the mantle, but lying outside the mantle cavity (Fig. 10.11). Unlike true ctenidia, these gills are not well ciliated (Russell-Hunter, 1978). The relative importance of lungs and gills was discussed when comparing *Lymnaea* and *Planorbis* (p. 52).

Fig. 10.11. Structure of the neomorphic gill in the freshwater limpet *Laevapex fuscus*. (*a*) Dorsal view, the tip of the gill being anterior. (*b*) Lateral view, showing folds extending both dorsally and ventrally from the margin. *L. fuscus* and the other freshwater limpets do not come to the surface to breathe air, but obtain sufficient oxygen through the skin and the neomorphic gills. Redrawn from Basch (1959).

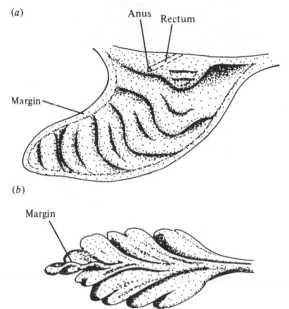

The freshwater limpets are so well adapted to a freshwater existence that they do not migrate to the water surface to take in air. A combination of cutaneous uptake of oxygen and uptake through the neomorphic gill is sufficient for their respiratory needs. These limpets are the most structurally specialised of the Basommatophora, indicating their further divergence from primitive forms than the lung breathers. The mantle cavity is almost completely lost, the reproductive organs show a wide range of variation, and the central nervous system is concentrated in a ring of ganglia around the oesophagus (Hubendick, 1978).

The three families of pulmonates found on marine shores, apart from the Ellobiidae and Otinidae, are the Amphibolidae, Siphonariidae and Trimusculidae (=Gadiniidae). Members of the Amphibolidae are restricted to the Indò-Pacific, and are found on estuarine mud flats. They are detritus feeders, moving across the flats and producing a continuous faecal string (see Fig. 10.12). The lung is a vascularised mantle cavity and may contain either air or water (Fretter, 1975), but appears to have undergone little structural readaptation to an aquatic environment. Despite this apparent lack of structural change, the New Zealand species *Amphibola crenata* can regulate its rate of oxygen utilisation in the face of enormous environmental changes: the rate stays effectively unchanged from that in air when the snail is submerged in either fresh water or sea water. Stabilisation of the uptake rate is apparently effected by the properties of the haemocyanin (Wells & Shumway, 1980). Besides being able to tolerate low oxygen tensions, it can also withstand a wide range of salinity

Fig. 10.12. *Amphibola crenata*, a pulmonate from estuarine sand and mud flats in New Zealand. This species is active in air, producing a continuous faecal string which trails behind the animal. The pneumostome is here shown open. This specimen is beginning to burrow by pushing the edge of the shell into the sand. Shell diameter 2 cm.

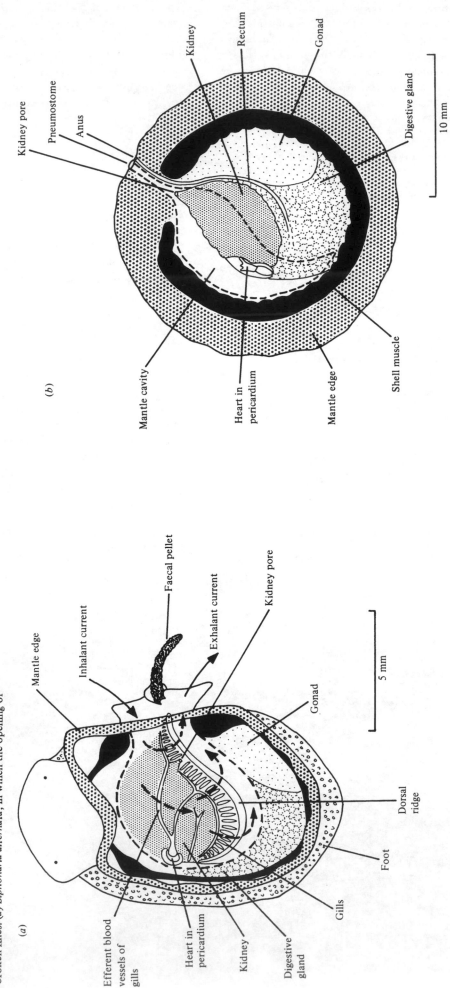

Fig. 10.13. Structure of the mantle cavity in two marine pulmonate limpets. Both are viewed from the dorsal surface after removal of the shell, with the cut shell muscles shown in black. The extent of the mantle cavity is indicated by heavy broken lines. (a) *Siphonaria alternata*, in which the opening of the mantle cavity to the outside is broad, with separate anterior (inhalant) and posterior (exhalant) apertures. Inhalant and exhalant currents are shown by solid arrows. The course of water inside the mantle cavity, where it flows across the secondary gill leaflets, is shown by broken arrows. (b) *Trimusculus (Gadinia) reticulatus*, shown without the head exposed. The pneumostome is small and not divided, and there are no secondary gills. Redrawn from Yonge (1952, 1958).

and temperature, and its range is probably restricted by biological rather than physical factors (Shumway & Marsden, 1982).

The Trimusculidae are limpets found on rocky shores. *Trimusculus reticulatus* lives on the Pacific coast of North America, where it is found in damp regions of the shore under overhangs (Yonge, 1958). The mantle cavity is vascularised and connects to the outside by a narrow pneumostome (Fig. 10.13*b*). There is no secondary gill, but like *Amphibola crenata*, *T. reticulatus* can breathe air or sea water.

The Siphonariidae are also rocky shore limpets, but have a much wider distribution than the Trimusculidae. Species of *Siphonaria* have a wide opening from the mantle cavity to the exterior, and are rare amongst aquatic pulmonates in having a secondary gill inside it (Fig. 10.13*a*). The gill is made up of triangular leaflets but these do not have the cilial arrangements found in ctenidia (Yonge, 1952). Instead, the respiratory current is provided by ciliary bands on the dorsal and ventral surface of the mantle cavity. Although *Siphonaria alternata* was claimed by Yonge (1952) to breathe only water, other species of *Siphonaria* are capable of breathing either air or water (Fretter, 1975; Wells & Wong, 1978). When breathing water, the inhalant current enters through the anterior part of the siphon, and passes across the gill. Exhalant water leaves through the posterior part

of the siphon (Figs. 10.13*a*, 10.14). When breathing air, *Siphonaria marza* utilises only the anterior part of the mantle cavity, and air is pumped in and out by rhythmical movements of the dorsal body wall (Fretter, 1975). In *Siphonaria zelandica*, however, the lung is not ventilated (Wells & Wong, 1978), and gaseous exchange must occur entirely by diffusion. Further differences between species are seen in feeding ecology, since *S. alternata* has been reported to feed mainly when covered by the tide, while *S. zelandica* is apparently adapted for an active life at low tide. These activity patterns may, however, be more flexible in reality, because Cook & Cook (1978) have shown that different feeding patterns in relation to tidal cover can be found within separate populations of the same species of *Siphonaria*.

The organisation of the Siphonariidae has led to a greater degree of success, in terms of species numbers, than has that of the Trimusculidae: while there are only seven species of *Trimusculus,* there are 61 species of Siphonariidae (see Yonge, 1958). At least part of this success may be attributed to the redevelopment of a pallial gill, and the physiological flexibility which allows at least some species to breathe both in air and in water.

The marine pulmonates may be compared with the freshwater forms in their readaptation to an aquatic existence. Many, including *Amphibola crenata*, retain a

Fig. 10.14. A marine pulmonate limpet, *Siphonaria zelandica*, from New Zealand, respiring under water. The surface of the irregular shell is covered by a growth of *Enteromorpha*. The animal is moving from left to right, the inhalant opening being indicated by a straight arrow and the exhalant opening by a curved arrow. Shell length 15 mm.

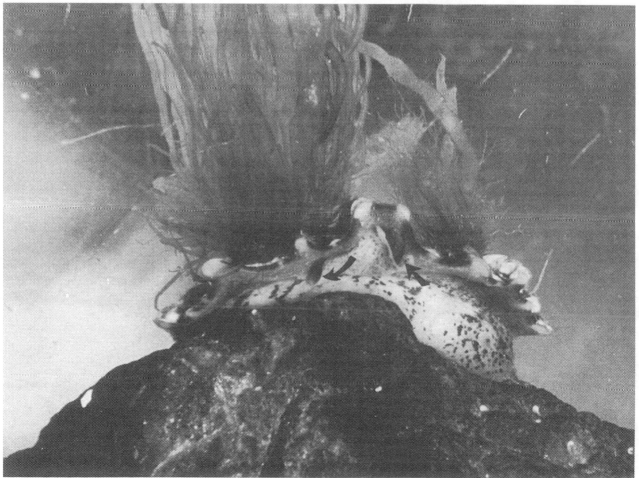

free-swimming veliger stage (Fretter, 1975; Pilkington & Pilkington, 1982), in contrast to freshwater forms, suggesting that the ancestors of present-day marine pulmonates never moved further on to land than the high tidal levels occupied at present by some species of Ellobiidae. All the ancestral pulmonate lines, however, both marine and freshwater, completely lost the ctenidium and breathed air using a vascularised mantle cavity. In later marine and freshwater species secondary gills have been evolved independently. Several species in both habitats have also evolved the ability to fill the lung with either air or water. The physiological flexibility of the freshwater pulmonates has enabled them to colonise most types of freshwater habitat, and with 148 genera they have become a successful and diverse group. Marine pulmonates comprise only 82 genera, of which most are Ellobiidae (55) or Siphonariidae (20). It is perhaps not surprising that the greatest number should be found in the Ellobiidae: these are essentially cryptozoic forms of small size, living in detritus-rich habitats, probably with relatively little competition from the prosobranchs. The Siphonariidae, in contrast, feed on the diatom floras of tropical rocky shores among intense competition with prosobranch limpets (e.g. Black, 1979). Their readaptation to such a species-rich environment has been surprisingly effective. In comparison with the recolonisation of the intertidal zone by insects, which have essentially retained their terrestrial mode of life and consequently are not prominent members of the shore fauna, the secondarily marine forms of the Siphonariidae provide one of the most striking examples of recolonisation of the sea by a primarily air-breathing invertebrate group.

References

van Aardt, W. J. (1968). Quantitative aspects of the water balance in *Lymnaea stagnalis* (L.). *Netherlands J. Zool.* **18**, 253-312.

Abdel Magid, A. M., Vokac, Z. & Ahmed, N. El Din. (1970). Respiratory function of the swim-bladders of the primitive fish *Polypterus senegalus. J. exp. Biol.* **52**, 27-37.

Abele, L. G. & Means, D. B. (1977). *Sesarma jarvisi* and *Sesarma cookei*: montane, terrestrial crabs in Jamaica (Decapoda). *Crustaceana* **32**, 91-3.

Abele, L. G., Robinson, M. H. & Robinson, B. (1973). Observations on sound production by two species of crabs from Panama (Decapoda, Gecarcinidae and Pseudothelphusidae). *Crustaceana* **25**, 147-52.

Abushama, F. T. (1964). On the behaviour and sensory physiology of the scorpion *Leiurus quinquestriatus* (H. & E.). *Anim. Behav.* **12**, 140-53.

Adolph, E. F. (1927). The regulation of volume and concentration in the body fluids of earthworms. *J. exp. Zool.* **47**, 31-62.

Afzelius, B. A., Nicander, L. & Sjöden, I. (1968). Fine structure of egg envelopes and the activation changes of cortical alveoli in the river lamprey, *Lampetra fluviatilis. J. embryol. exp. Morph.* **19**, 311-18.

Ahearn, G. A. (1970*a*). Water balance in the whipscorpion, *Mastigoproctus giganteus* (Lucas) (Arachnida, Uropygi). *Comp. Biochem. Physiol.* **35**, 339-53.

Ahearn, G. A. (1970*b*). The control of water loss in desert tenebrionid beetles. *J. exp. Biol.* **53**, 573-95.

Ahearn, G. A. & Hadley, N. F. (1969). The effects of temperature and humidity on water loss in two desert tenebrionid beetles, *Eleodes armata* and *Cryptoglossa verrucosa. Comp. Biochem. Physiol.* **30**, 739-49.

Akerlund, G. (1974). Oxygen consumption in relation to environmental oxygen concentrations in the ampullariid snail *Marisa cornuarietis* (L.). *Comp. Biochem. Physiol.* **47A**, 1076-75.

Alexander, A. J. (1957). Notes on onychophoran behaviour. *Annls Natal Mus.* **14**, 35-43.

Alexander, A. J. (1958). On the stridulation of scorpions. *Behaviour* **12**, 339-52.

Alexander, A. J. (1967). Problems of limb extension in the scorpion, *Opisthophthalmus latimanus* Koch. *Trans. roy. Soc. S. Afr.* **37**, 165-81.

Alexander, A. J. & Ewer, D. W. (1955). A note on the function of the eversible sacs of the onychophoran, *Opisthopatus cinctipes* Purcell. *Annls Natal Mus.* **13**, 217-22.

Alexander, A. J. & Ewer, D. W. (1958). Temperature adaptive behaviour in the scorpion, *Opisthophthalmus latimanus* Koch. *J. exp. Biol.* **35**, 349-59.

Alexander, S. J. & Ewer, D. W. (1969). A comparative study of some aspects of the biology and ecology of *Sesarma catenata* Ort. and *Cyclograpsus punctatus* M. Edw., with additional observations on *Sesarma meinerti* De Man. *Zool. Africana* **4**, 1-35.

Almquist, S. (1970). Thermal tolerances and preferences of some dune-living spiders. *Oikos* **21**, 230-6.

Alt, J. M., Stolte, H., Eisenbach, G. M. & Walvig, F. (1981). Renal electrolyte and fluid excretion in the Atlantic hagfish *Myxine glutinosa. J. exp. Biol.* **91**, 323-30.

Altevogt, R. (1957). Untersuchungen zur Biologie, Okologie und Physiologie indischer Winkerkrabben. *Z. Morph. Okol. Tiere* **46**, 1-110.

Altner, H. (1968). Die Ultrastruktur der Labialnephridien von *Onychiurus quadriocellatus* (Collembola). *J. Ultrastruct. Res.* **24**, 349-66.

Altner, H. & Ernst, K.-D. (1974). Struktureigentümlichkeiten antennaler Sensillen bodenlebender Collembolen. *Pedobiologia* **14**, 118-22.

Altner, H. & Thies, G. (1976). The postantennal organ: a specialised unicellular sensory input to the protocerebrum in apterygotan insects (Collembola). *Cell Tiss. Res.* **167**, 97-110.

Alvarado, R. H. (1979). Amphibians. In *Comparative physiology of osmoregulation in animals*, vol. I, pp. 261-303, ed. G. M. Maloiy. Academic Press, London.

Andersen, N. M. & Polhemus, J. T. (1976). Water-striders (Hemiptera: Gerridae, Veliidae, etc.). In *Marine insects*, pp. 187-224, ed. L. Cheng. North-Holland Publishing Co., Amsterdam.

Anderson, D. T. (1973). *Embryology and phylogeny in annelids and arthropods.* Pergamon Press, Oxford.

Anderson, D. T. (1979). Embryos, fate maps, and the phylogeny of arthropods. In *Arthropod phylogeny*, pp. 59-105, ed. A. P. Gupta. Van Nostrand Reinhold Co., New York.

Anderson, J. F. (1966). The excreta of spiders. *Comp. Biochem. Physiol.* **17**, 973-82.

Anderson, J. F. & Prestwich, K. N. (1975). The fluid pressure pumps of spiders (Chelicerata, Araneae). *Z. Morph. Tiere* **81**, 257-77.

Anderson, J. F. & Prestwich, K. N. (1980). Scaling of subunit structures in booklungs of spiders (Araneae). *J. Morph.* **165**, 167-74.

Anderson, J. F. & Prestwich, K. N. (1982). Respiratory gas exchange in spiders. *Physiol. Zoöl.* **55**, 72-90.

Andrews, E. B. (1965). The functional anatomy of the mantle cavity, kidney and blood system of some pilid gastropods (Prosobranchia). *J. Zoöl., Lond.* **146**, 70-94.

Andrews, E. B. (1976). The ultrastructure of the heart and kidney of the pilid gastropod mollusc *Marisa cornuarietis*, with special reference to filtration throughout the Architaenioglossa. *J. Zoöl., Lond.* **179**, 85-106.

Andrews, E. B. (1979). Fine structure in relation to function in the excretory system of two species of *Viviparus*. *J. moll. Stud.* **45**, 186-206.

Andrews, E. B. (1981). Osmoregulation and excretion in prosobranch gastropods. Part 2. Structure in relation to function. *J. moll. Stud.* **47**, 248-89.

Andrews, E. B. & Little, C. (1972). Structure and function in the excretory systems of some terrestrial prosobranch snails (Cyclophoridae). *J. Zool., Lond.* **168**, 395-422.

Andrews, E. B. & Little, C. (1982). Renal structure and function in relation to habitat in some cyclophorid land snails from Papua New Guinea. *J. moll. Stud.* **48**, 124-43.

Andrews, S. M. (1973). Interrelationships of crossopterygians. In *Interrelationships of fishes*, pp. 138-77, ed. P. H. Greenwood, R. S. Miles & C. Patterson. Academic Press, London.

Andrews, S. M. & Westoll, T. S. (1970). The postcranial skeleton of *Eusthenopteron foordi* Whiteaves. *Trans. roy. Soc. Edinb.* **68**, 207-329.

Angersbach, D. (1975). Oxygen pressures in haemolymph and various tissues of the tarantula, *Eurypelma helluo*. *J. comp. Physiol.* (B) **98**, 133-46.

Angersbach, D. (1978). Oxygen transport in the blood of the tarantula *Eurypelma californicum*: pO_2 and pH during rest, activity and recovery. *J. comp. Physiol.* **123**, 113-25.

Appleton, T. C., Newell, P. F. & Machin, J. (1979). Ionic gradients within mantle-collar epithelial cells of the land snail *Otala lactea*. *Cell Tiss. Res.* **199**, 83-97.

Arendse, M. C. & Barendregt, A. (1981). Magnetic orientation in the semi-terrestrial amphipod, *Orchestia cavimana*, and its interrelationship with photo-orientation and water loss. *Physiol. Ent.* **6**, 333-42.

Arey, L. B. (1937). Observations on two types of respiration in *Onchidium*. *Biol. Bull. mar. biol. Lab., Woods Hole* **72**, 41-6.

Arey, L. B. & Crozier, W. J. (1921). On the natural history of *Onchidium*. *J. exp. Zool.* **32**, 443-502.

Arlian, L. G. (1979). Significance of passive sorption of atmospheric water vapor and feeding in water balance of the rice weevil, *Sitophilus oryzae*. *Comp. Biochem. Physiol.* **62A**, 725-33.

Arlian, L. G. & Veselica, M. M. (1979). Water balance in insects and mites. *Comp. Biochem. Physiol.* **64A**, 191-200.

Arlian, L. G. & Veselica, M. M. (1981). Effect of temperature on the equilibrium body water mass in the mite *Dermatophagoides farinae*. *Physiol. Zool.* **54**, 393-9.

Avens, A. C. & Sleigh, M. A. (1965). Osmotic balance in gastropod molluscs. 1. Some marine and littoral gastropods. *Comp. Biochem. Physiol.* **16**, 121-41.

Babiker, M. M. (1979). Respiratory behaviour, oxygen consumption and relative dependence on aerial respiration in the African lungfish (*Protopterus annectens* Owen) and an air-breathing teleost (*Clarias lazera* C.). *Hydrobiologia* **65**, 177-87.

Baccetti, B. & Bedini, C. (1964). Research on the structure and physiology of the eyes of a lycosid spider. I. Microscopic and ultramicroscopic structure. *Archs ital. Biol.* **102**, 97-122.

Badman, D. G. (1971). Nitrogen excretion in two species of pulmonate land snails. *Comp. Biochem. Physiol.* **38A**, 663-73.

Bahamonde, N. & Lopez, M. T. (1961). Estudios biologicos en la poblacion de *Aegla laevis laevis* (Latreille) de El Monte. *Invest. zool. chilenas* **7**, 19-58.

Bahl, K. N. (1927). On the reproductive processes of earthworms. Part I. The process of copulation and exchange of sperms in *Eutyphoeus waltoni* Mich. *Q. Jl microsc. Sci.* **71**, 479-502.

Bahl, K. N. (1946). Studies on the structure, development, and physiology of the nephridia of Oligochaeta. Part VIII. Biochemical estimations of nutritive and excretory substances in the blood and coelomic fluid of the earth-worm and their bearing on the role of the two fluids in metabolism. *Q. Jl microsc. Sci.* **87**, 357-71.

Bahl, K. N. (1947). Excretion in the Oligochaeta. *Biol. Rev.* **22**, 109-47.

Bähr, R. R. (1974). Contribution to the morphology of chilopod eyes. *Symp. zool. Soc. Lond.* **32**, 383-404.

Bailey, T. G. (1971). Osmotic pressure and pH of slug haemolymph. *Comp. Biochem. Physiol.* **40A**, 83-8.

Baker, G. H. (1980). The water and temperature relationships of *Ommatoiulus moreletii* (Diplopoda: Iulidae). *J. Zool., Lond.* **190**, 97-108.

Baldwin, E. (1964). *An introduction to comparative biochemistry* (4th edition). Cambridge University Press, Cambridge, England.

Balinsky, J. B. (1981). Adaptation of nitrogen metabolism to hyperosmotic environment in Amphibia. *J. exp. Zool.* **215**, 335-50.

Ball, I. R. (1981). The phyletic status of the Paludicola. In *The biology of the Turbellaria*, pp. 7-12, ed. E. R. Schockaert & I. R. Ball. W. Junk, The Hague.

Bareth, C. (1968). Biologie sexuelle et formations endocrines de *Campodea remyi* Denis (diploures campodéides). *Rev. Ecol. Biol. Sol.* **5**, 303-426.

Bareth, C. (1980). Mise à jour des connaissances sur les spermatophores des diploures campodéides (Insecta Apterygota). In *Proceedings of the 1st international seminary on Apterygota, Siena, September 1978*, pp. 41-51, ed. R. Dallai. Academia delle Scienze di Siena detta de Fisiocritia.

Barnes, H. (1954). Some tables for the ionic composition of sea water. *J. exp. Biol.* **31**, 582-8.

Barnes, H., Finlayson, D. M. & Piatigorsky, J. (1963). The effects of desiccation and anaerobic conditions on the behaviour, survival and general metabolism of three common cirripedes. *J. anim. Ecol.* **32**, 233-52.

Barnes, R. D. (1980). *Invertebrate zoology* (4th edition). Saunders, Philadelphia.

Barnes, R. S. K. (1967). The osmotic behaviour of a number of grapsoid crabs with respect to their differential penetration of an estuarine system. *J. exp. Biol.* **47**, 535-51.

Barnes, R. S. K. (1968a). On the evolution of elongate ocular peduncles by the Brachyura. *Syst. Zool.* **17**, 182-7.

Barnes, R. S. K. (1968b). Individual variation in osmotic pressure of an ocypodid crab. *Comp. Biochem. Physiol.* **27**, 447-50.

Barnes, R. S. K. (1974). *Estuarine biology*. Arnold, London.

Barrett, R. T. (1972). Unpublished report. University of Bristol, Woodland Road, Bristol BS8 1UG, England.

Barth, F. G. & Wadepuhl, M. (1975). Slit sense organs on the scorpion leg (*Androctonus australis* L., Buthidae). *J. Morph.* **145**, 209-28.

Bartholomew, G. A. & Heinrich, B. (1978). Endothermy in African dung beetles during flight, ball making, and ball rolling. *J. exp. Biol.* **73**, 65-83.

Barton-Browne, L. B. (1964). Water regulation in insects. *Ann. Rev. Ent.* **9**, 63-82.

Barton-Browne, L. & Dudzinski, A. (1968). Some changes resulting from water deprivation in the blowfly, *Lucilia cuprina*. *J. Insect Physiol.* **14**, 1423-34.

Basch, P. F. (1959). The anatomy of *Laevapex fuscus*, a freshwater limpet. *Misc. Publ. Univ. Michigan.* **108**, 1-56.

Bautz, A. (1977). Structure fine de l'épiderme chez des planaires triclades terrestre et paludicoles. *Archs Zool. exp. gén.* **118**, 155-72.

Beadle, L. C. (1934). Osmotic regulation in *Gunda ulvae*. *J. exp. Biol.* **11**, 382-96.

Beadle, L. C. (1957). Respiration in the African swampworm *Alma emini* Mich. *J. exp. Biol.* **34**, 1-10.

Beament, J. W. L. (1946). The waterproofing process in eggs of *Rhodnius prolixus* Stähl. *Proc. roy. Soc. Lond.* (B) **133**, 407-18.

Beament, J. W. L. (1949). The penetration of insect egg shells. II. The properties and permeability of the sub-chorial membranes during development of *Rhodnius prolixus* Stähl. *Bull. ent. Res.* **39**, 467-88.

Beament, J. W. L. (1959). The waterproofing mechanism of arthropods. I. The effect of temperature on cuticle permeability in terrestrial insects and ticks. *J. exp. Biol.* **36**, 391–422.

Beament, J. W. L. (1964). The active transport and passive movement of water in insects. *Adv. Insect Physiol.* **2**, 67–129.

Beament, J. W. L. (1965). The active transport of water: evidence, models and mechanisms. In *The state and movement of water in living organisms*, pp. 273–306. Cambridge University Press, Cambridge, England.

Beament, J. W. L., Noble-Nesbitt, J. & Watson, J. A. L. (1964). The waterproofing mechanism of arthropods. III. Cuticular permeability in the firebrat, *Thermobia domestica* (Packard). *J. exp. Biol.* **41**, 323–30.

Beauchamp, P. de. (1959). Archiannélides. In *Traité de zoologie*, Vol. V, Fasc. I, pp. 197–223, ed. P.-P. Grassé. 'Masson et Cie, Paris.

Bedford, J. J. & Leader, J. P. (1975). The composition of the haemolymph of the New Zealand centipede, *Cormocephalus rubriceps* (Newport). *Comp. Biochem. Physiol.* **50A**, 561–4.

Bedini, C. (1967). The fine structure of the eyes of *Euscorpius carpathicus* L. (Arachnida Scorpiones). *Archs ital. Biol.* **105**, 361–78.

Bedini, C. (1968). The ultrastructure of the eye of a centipede *Polybothrus fasciatus* (Newport). *Monit. Zool. ital.* **2**, 31–47.

Bedini, C. (1970). The fine structure of the eye in *Glomeris* (Diplopoda). *Monit. Zool. ital.* **4**, 201–19.

Bedini, C. & Mirolli, M. (1967). The fine structure of the temporal organs of a pill millipede, *Glomeris romana* Verhoeff. *Monit. Zool. ital.* **1**, 41–63.

Bedini, C. & Papi, F. (1974). Fine structure of the turbellarian epidermis. In *Biology of the Turbellaria*, pp. 108–47, ed. N. W. Riser & M. P. Morse. McGraw-Hill, New York.

Bedini, C. & Tongiorgi, P. (1971). The fine structure of the pseudoculus of acerentomids Protura (Insecta Apterygota). *Monit. Zool. ital.* **5**, 25–38.

Beever, J. W., Simberloff, D. & King, L. (1979). Herbivory and predation by the mangrove tree crab *Aratus pisoni.* *Oecologia* **43**, 317–28.

Bennett, A. F. & Licht, P. (1975). Evaporative water loss in scaleless snakes. *Comp. Biochem. Physiol.* **52A**, 213–15.

Bennett, D. S. (1971). Nitrogen excretion in the diplopod *Cylindroiulus londinensis. Comp. Biochem. Physiol.* **39A**, 611–24.

Bennett, D. S. & Manton, S. M. (1962). Arthropod segmental organs and Malpighian tubules, with particular reference to their function in the Chilopoda. *Ann. Mag. nat. Hist.* **5**, 545–56.

Benson, L. (1979). *Plant classification* (2nd edition). D. C. Heath & Co., Lexington, Mass.

Bentley, P. J. & Schmidt-Nielsen, K. (1966). Cutaneous water loss in reptiles. *Science, N.Y.* **151**, 1547–9.

Bentley, P. J. & Yorio, T. (1979). Evaporative water loss in anuran Amphibia: a comparative study. *Comp. Biochem. Physiol.* **62A**, 1005–9.

Berg, T. & Steen, J. B. (1965). Physiological mechanisms for aerial respiration in the eel. *Comp. Biochem. Physiol.* **15**, 469–84.

Bergström, J. (1979). Morphology of fossil arthropods as a guide to phylogenetic relationships. In *Arthropod phylogeny*, pp. 3–56, ed. A. P. Gupta. Van Nostrand Reinhold Co., New York.

Berneys, E. A. & Chapman, R. F. (1974). Changes in haemolymph osmotic pressure in *Locusta migratoria* larvae in relation to feeding. *J. Ent. (A)* **48**, 149–55.

Berntsson, K.-E., Haglund, B. & Løvtrup, S. (1965). Osmoregulation in the amphibian egg. The influence of calcium. *J. cell. comp. Physiol.* **65**, 101–12.

Berridge, M. J. (1965a). The physiology of excretion in the cotton stainer, *Dysdercus fasciatus* Signoret. 1. Anatomy, water excretion and osmoregulation. *J. exp. Biol.* **43**, 511–21.

Berridge, M. J. (1965b). The physiology of excretion in the cotton stainer, *Dysdercus fasciatus* Signoret. 3. Nitrogen excretion and excretory metabolism. *J. exp. Biol.* **43**, 535–52.

Berridge, M. J. (1968). Urine formation by the Malpighian tubules of *Calliphora. J. exp. Biol.* **48**, 159–74.

Berridge, M. J. (1970). Osmoregulation in terrestrial arthropods. In *Chemical zoology*, vol. VA, pp. 287–319, ed. M. Florkin & B. T. Scheer. Academic Press, London.

Berridge, M. J. & Gupta, B. L. (1967). Fine-structural changes in relation to ion and water transport in the rectal papillae of the blowfly, *Calliphora. J. Cell Sci.* **2**, 89–112.

Berrill, N. J. (1955). *The origin of vertebrates.* Oxford University Press, Oxford.

Berry, A. J. (1962). The growth of *Opisthostoma (Plectostoma) retrovertens* Tomlin, a minute cyclophorid from a Malayan limestone hill. *Proc. malac. Soc. Lond.* **35**, 46–9.

Berry, A. J. (1964). Faunal zonation in mangrove swamps. *Bull. natn. Mus. Singapore* **32**, 90–8.

Berry, A. J., Lim, R. & Sasekumar, A. (1973). Reproductive systems and breeding condition in *Nerita birmanica* (Archaeogastropoda: Neritacea) from Malayan mangrove swamps. *J. Zool., Lond.* **170**, 189–200.

Bertheau, P. (1971). Histologie comparée des tubes de Malpighi de quelques chilopodes (Myriapods). *C.r. hebd. Séanc. Acad. Sci., Paris* (Ser. D) **272**, 2913–15.

Bertmar, G. (1968). Lungfish phylogeny. In *Current problems of lower vertebrate phylogeny*, pp. 259–83, ed. T. Ørvig. Wiley Interscience, London.

Betsch, J.-M. & Vannier, G. (1977). Caractérisation des deux phases juvéniles d'*Allacma fusca* (Collembola, Symphypleona) par leur morphologie et leur écophysiologie. *Z. zool. Syst. Evolutionsforsch.* **15**, 124–41.

Betsch-Pinot, M.-C. (1976). Le comportement réproducteur de *Sminthurus viridis* (L.) (Collembola, Symphypleona). *Z. Tierpsychol.* **40**, 427–39.

Betsch-Pinot, M.-C. (1980). Relation entre le comportement réproducteur et le biotope chez quelques collemboles symphypleones. In *Proceedings of the 1st international seminary on Apterygota, Siena, September 1978*, pp. 13–17, ed. R. Dallai. Academia delle Scienze di Siena detta de Fisiocritia.

Bhatt, B. D. (1963). On the excretion of some terrestrial and freshwater leeches. *Zool. Beitr.* **8**, 167–72.

Binyon, J. (1979). *Branchiostoma lanceolatum* – a freshwater reject? *J. mar. Biol. Ass. UK* **59**, 61–7.

Binyon, J. & Lewis, J. G. E. (1963). Physiological adaptations of two species of centipede (Chilopoda: Geophilomorpha) to life on the shore. *J. mar. biol. Ass. UK* **43**, 49–55.

Birch, L. C. & Clark, D. P. (1953). Forest soil as an ecological community with special reference to the fauna. *Q. Rev. Biol.* **28**, 13–36.

Bitsch, J. & Palévody, C. (1973). L'epithelium absorbant des vésicules coxales des machilides (Insecta Thysanura). *Z. Zellforsch. mikrosk. Anat.* **143**, 169–82.

Black, R. (1979). Competition between intertidal limpets: an intrusive niche on a steep resource gradient. *J. anim. Ecol.* **48**, 401–11.

Blaxter, J. H. S. (1969). Development: eggs and larvae. In *Fish physiology*, vol. III, pp. 177–252, ed. W. S. Hoar & D. J. Randall, Academic Press, London.

Blinn, W. C. (1964). Water in the mantle cavity of land snails. *Physiol. Zoöl.* **37**, 329–37.

Bliss, D. E. (1968). Transition from water to land in decapod crustaceans. *Am. Zool.* **8**, 355–92.

Bliss, D. E. (1979). From sea to tree: saga of a land crab. *Am. Zool.* **19**, 385–410.

Bliss, D. E., Wang, S. M. & Martinez, E. A. (1966). Water balance in the land crab, *Gecarcinus lateralis*, during the intermolt cycle. *Am. Zool.* **6**, 197–212.

Blower, G. (1951). A comparative study of the chilopod and diplopod cuticle. *Q. Jl microsc. Sci.* **92**, 141–61.

Blower, J. G. (1955). Millipedes and centipedes as soil animals. In *Soil zoology*, pp. 138–51, ed. D. K. McE. Kevan. Butterworths, London.

Blumenthal, H. (1935). Untersuchungen uber das 'Tarsalorgan' der Spinnen. *Z. Morph. Okol. Tiere* **29**, 667–719.

Board, R. G. & Scott, V. D. (1980). Porosity of the avian egg-shell. *Am. Zool.* **20**, 339–349.

Bock, K.-D. (1967). Experimente zur Okologie von *Orchestia platensis* Kröyer. *Z. Morph. Okol. Tiere* **58**, 405–28.

Bolt, J. R. (1977). Dissorophoid relationships and ontogeny, and the origin of the Lissamphibia. *J. Paleont.* **51**, 235–49.

Bolwig, N. (1953). On the variation of the osmotic pressure of the haemolymph in flies. *S. Afr. ind. Chemist* **7**, 113–15.

Boné, G. (1943). Recherches sur les glandes coxales et la régula-tion de milieu interne chez l'*Ornithodorus moubata*. *Annls Soc. roy. zool. Belg.* **74**, 16–31.

Bone, Q. (1960). The origin of chordates. *J. Linn. Soc. (Zool.)* **44**, 252–69.

Bone, Q. (1972). *The origin of chordates*. Oxford University Press, Oxford.

Boroffka, I. (1965). Elektrolyttransport im Nephridium von *Lumbricus terrestris*. *Z. vergl. Physiol.* **51**, 25–48.

Boroffka, I., Altner, H. & Haupt, J. (1970). Funktion und Ultrastruktur des Nephridiums von *Hirudo medicinalis*. 1. Ort und Mechanismus der Primärharnbildung. *Z. vergl. Physiol.* **66**, 421–38.

Boss, K. J. (1978). On the evolution of gastropods in ancient lakes. In *Pulmonates*, vol. 2A, pp. 385–428, ed. V. Fretter & J. Peake. Academic Press, London.

Bott, R. (1970). Betrachtungen über die Entwicklungsgeschichte und Verbreitung der Susswasser-Krabben nach der Sammlung des Naturhistorischen Museums in Genf/ Schweiz. *Rev. Suisse Zool.* **77**, 327–44.

Boudreaux, H. B. (1979*a*). Significance of intersegmental tendon system in arthropod phylogeny and a monophyletic classification of Arthropoda. In *Arthropod phylogeny*, pp. 551–586, ed. A. P. Gupta. Van Nostrand Reinhold Co., New York.

Boudreaux, H. B. (1979*b*). *Arthropod phylogeny with special reference to insects*. Wiley Interscience, New York.

Bouillon, J. (1960). Ultrastructure des cellules rénales des mollusques. I. Gastéropodes pulmonés terrestres. *Annls Sci. nat. (Zool.)* **2**, 719–49.

Bourne, G. C. (1908). Contributions to the morphology of the group Neritacea of aspidobranch gastropods. I. The Neritidae. *Proc. zool. Soc. Lond.* 1908, 810–87.

Bourne, G. C. (1911). Contributions to the morphology of the group Neritacea of the aspidobranch gastropods. II. The Helicinidae. *Proc. zool. Soc. Lond.* 1911, 759–809.

Bousfield, E. L. (1968). Discussion: transition to land. *Am. Zool.* **8**, 393–5.

Boutilier, R. G., McDonald, D. G. & Toews, D. P. (1980). The effects of enforced activity on ventilation, circulation and blood acid–base balance in the aquatic gill-less urodele *Cryptobranchus alleganiensis*; a comparison with the semi-terrestrial anuran, *Bufo marinus*. *J. exp. Biol.* **84**, 289–302.

Boutilier, R. G., Randall, D. J., Shelton, G. & Toews, D. P. (1979*a*). Acid base relationships in the blood of the toad, *Bufo marinus*. I. The effects of environmental CO_2. *J. exp. Biol.* **82**, 331–44.

Boutilier, R. G., Randall, D. J., Shelton, G. & Toews, D. P. (1979*b*). Acid base relationships in the blood of the toad, *Bufo marinus*. II. The effects of dehydration. *J. exp. Biol.* **82**, 345–55.

Boutilier, R. G., Randall, D. J., Shelton, G. & Toews, D. P. (1979*c*). Acid base relationships in the blood of the toad, *Bufo marinus*. III. The effects of burrowing. *J. exp. Biol.* **82**, 357–65.

Bowerman, R. F. (1972). A muscle receptor organ in the scorpion post-abdomen. 1. The sensory system. *J. comp. Physiol.* **81**, 133–46.

Bowerman, R. F. (1976). Ion concentrations and pH of the haemolymph of the scorpions *Hadrurus arizonensis* and *Paruroctonus mesaensis*. *Comp. Biochem. Physiol.* **54A**, 331–3.

Bowers, D. E. (1964). Natural history of two beach hoppers of the genus *Orchestoidea* (Crustacea: Amphipoda) with reference to their complemental distribution. *Ecology* **45**, 677–96.

Boyden, C. R. (1972). The behaviour, survival and respiration of the cockles *Cerastoderma edule* and *C. glaucum* in air. *J. mar. Biol. Ass. UK* **52**, 661–80.

Brandenburg, J. (1975). The morphology of the protonephridia. *Fortschr. Zool.* **23**, 1–17.

Braun, G., Kümmel, G. & Mangos, J. A. (1966). Studies on the ultrastructure and function of a primitive excretory organ, the protonephridium of the rotifer *Asplanchna priodonta*. *Pflügers Arch. ges. Physiol.* **289**, 141–54.

Brenner, G. J. (1976). Middle Cretaceous floral provinces and early migrations of angiosperms. In *Origin and early evolution of angiosperms*, pp. 23–47, ed. C. B. Beck. Columbia University Press, New York.

Bresler, E. H. (1978). A model for transepithelial fluid transport. *Am. J. Physiol.* **235**, F626–37.

Bretfeld, G. (1970). Grundzüge des Paarungsverhaltens euro-päischer Bourletiellini (Collembola, Sminthuridae) und daraus abgeleitete taxonomische-nomenclatorische Folgerungen. *Z. zool. Syst. Evol.-Forsch.* **8**, 259–73.

Bricteux-Grégoire, S., Duchâteau-Bosson, G., Jeuniaux, C., Schoffeniels, E. & Florkin, M. (1963). Constituants osmo-tiquement actifs du sang et des muscles du scorpion *Androctonus australis* L. *Archs int. Physiol. Biochim.* **71**, 393–400.

Brien, P. (1962). Formation du cloaque urinaire et origine des sacs pulmonaires chez *Protopterus*. *Ann. Reeks Zool. Wetenschap.* **108**, 1–51.

Brinkhurst, R. O. (1982). Evolution in the Annelida. *Can. J. Zool.* **60**, 1043–59.

Brinkhurst, R. O. & Jamieson, B. G. M. (1971). *Aquatic Oligo-chaeta of the world*. Oliver & Boyd, Edinburgh.

Bristowe, W. S. (1971). *The world of spiders* (2nd edition). Collins, London.

Brocas, J. & Cherruault, Y. (1973). Association des systèmes trachéens et circulatoires dans le transport des gas respiratoires chez l'insecte aerien. *Bio-Medical Computing* **4**, 173–204.

Broch, E. S. (1969). The osmotic adaptation of the fairy shrimp *Branchinecta campestris* Lynch to saline astatic waters. *Limnol. Oceanogr.* **14**, 485–92.

Brown, D. & Ilic, V. (1979). Freeze-fracture differences in plasma membranes of the stratum corneum and replacement layer cells of amphibian epidermis. *J. Ultrastruct. Res.* **67**, 55–64.

Brownell, P. H. & Farley, R. D. (1974). The organization of the malleolar sensory system in the solpugid, *Chanbria* sp. *Tiss. Cell* **6**, 471–85.

Brownell, P. & Farley, R. D. (1979*a*). Detection of vibrations in sand by tarsal sense organs of the nocturnal scorpion *Paruroctonus mesaensis*. *J. comp. Physiol.* **131**, 23–30.

Brownell, P. & Farley, R. D. (1979*b*). Orientation to vibrations in sand by the nocturnal scorpion *Paruroctonus mesaensis*: mechanism of target localization. *J. comp. Physiol.* **131**, 31–8.

Browning, T. O. (1954). On the structure of the spiracle of the tick *Ornithodoros moubata* Murray. *Parasitology* **44**, 310–12.

Brunhuber, B. S. (1969). The mode of sperm transfer in the scolopendromorph centipede: *Cormocephalus anceps anceps* Porat. *Zool. J. Linn. Soc.* **48**, 409–20.

Brunhuber, B. S. (1970). Egg laying, maternal care and develop-ment of young in the scolopendromorph centipede, *Cormocephalus anceps anceps* Porat. *Zool. J. Linn. Soc.* **49**, 225–34.

Buck, J. (1958). Cyclic CO_2 release in insects. IV. A theory of mechanism. *Biol. Bull. mar. biol. Lab., Woods Hole* **114**, 118–40.

Buck, J. (1962). Some physical aspects of insect respiration. *Ann. Rev. Ent.* **7**, 27–56.

Budgett, J. A. (1901). On the breeding-habits of some west-African fishes, with an account of the external features in the development of *Protopterus annectens*, and a description of the larva of *Polypterus lapradei. Trans. zool. Soc. Lond.* **16**, 115–36.

Burggren, W. W. & McMahon, B. R. (1981*a*). Oxygen uptake during environmental temperature change in hermit crabs: adaptation to subtidal, intertidal and supratidal habitats. *Physiol. Zoöl.* **54**, 325–33.

Burggren, W. W. & McMahon, B. R. (1981*b*). Haemolymph oxygen transport, acid–base status, and hydromineral regulation during dehydration in three terrestrial crabs, *Cardisoma, Birgus* and *Coenobita. J. exp. Zool.* **218**, 53–64.

Burky, A. J., Pacheco, J. & Pereyra, E. (1972). Temperature, water, and respiratory regimes of an amphibious snail, *Pomacea urceus* (Müller), from the Venezuelan savannah. *Biol. Bull. mar. biol. Lab., Woods Hole* **143**, 304–16.

Burrows, M. & Horridge, G. A. (1968). The action of the eyecup muscles of the crab, *Carcinus*, during optokinetic movements. *J. exp. Biol.* **49**, 223–50.

Burrows, M. & Hoyle, G. (1973). The mechanism of rapid running in the ghost crab, *Ocypode ceratophthalma. J. exp. Biol.* **58**, 327–49.

Bursell, E. (1955). The transpiration of terrestrial isopods. *J. exp. Biol.* **32**, 238–55.

Bursell, E. (1957). Spiracular control of water loss in the tsetse fly. *Proc. roy. ent. Soc. Lond.* (*A*) **32**, 21–9.

Bursell, E. (1958). The water balance of tsetse pupae. *Phil. Trans. roy. Soc. Lond.* (*B*) **241**, 179–210.

Bursell, E. (1967). The excretion of nitrogen in insects. *Adv. Insect Physiol.* **4**, 33–67.

Bursell, E. & Ewer, D. W. (1950). On the reactions to humidity of *Peripatopsis moseleyi* (Wood-Mason). *J. exp. Biol.* **26**, 335–53.

Burton, R. F. (1965). Sodium, potassium, and magnesium in the blood of the snail, *Helix pomatia* L. *Physiol. Zoöl.* **38**, 335–42.

Burton, R. F. (1968). Ionic balance in the blood of Pulmonata. *Comp. Biochem. Physiol.* **25**, 509–16.

Burton, R. F. (1969). Buffers in the blood of the snail, *Helix pomatia. Comp. Biochem. Physiol.* **29**, 919–30.

Burton, R. F. (1973). The significance of ionic concentrations in the internal media of animals. *Biol. Rev.* **48**, 195–231.

Burton, R. F. (1983). Ionic regulation and water balance. In *Biology of Mollusca*, vol. 4, ed. A. S. M. Saleuddin & K. M. Wilbur. Academic Press, New York. (In Press.)

Buxton, P. A. (1926). The colonization of the sea by insects: with an account of the habits of *Pontomyia*, the only known submarine insect. *Proc. zool. Soc. Lond.* **2**, 807–814.

Bystrow, A. P. (1947). Hydrophilous and xerophilous labyrinthodonts. *Acta Zool., Stockholm* **28**, 137–64.

Calman, W. T. (1911). *The life of Crustacea.* Methuen, London.

Cameron, J. N. (1975). Aerial gas exchange in the terrestrial Brachyura *Gecarcinus lateralis* and *Cardisoma guanhumi. Comp. Biochem. Physiol.* **50***A*, 129–34.

Cameron, J. N. (1978). Regulation of blood pH in teleost fish. *Resp. Physiol.* **33**, 129–44.

Cameron, J. N. (1981*a*). Brief introduction to the land crabs of the Palau Islands: stages in the transition to air breathing. *J. exp. Zool.* **218**, 1–5.

Cameron, J. N. (1981*b*). Acid–base responses to changes in CO_2 in two Pacific crabs: the coconut crab, *Birgus latro*, and a mangrove crab, *Cardisoma carnifex. J. exp. Zool.* **218**, 65–73.

Cameron, J. N. & Mecklenburg, T. A. (1973). Aerial gas exchange in the coconut crab, *Birgus latro*, with some notes on *Gecarcoidea lalandii. Resp. Physiol.* **19**, 245–61.

Cameron, J. N. & Wood, C. M. (1978). Renal function and acid–base regulation in two Amazonian erythrinid fishes: *Hoplias malabaricus*, a water breather, and *Hoplerythrinus*

unitaeniatus, a facultative air breather. *Can. J. Zool.* **56**, 917–30.

Campbell, J. W. (1965). Arginine and urea biosynthesis in the land planaria: its significance in biochemical evolution. *Nature, Lond.* **208**, 1299–301.

Campbell, J. W., Drotman, R. B., McDonald, J. A. & Tramell, P. R. (1972). Nitrogen metabolism in terrestrial invertebrates. In *Nitrogen metabolism and the environment*, pp. 1–54, ed. J. W. Campbell & L. Goldstein. Academic Press, London.

Campbell, K. S. W. & Bell, M. W. (1977). A primitive amphibian from the late Devonian of New South Wales. *Alcheringa* **1**, 369–81.

Campiglia, S. S. (1976). The blood of *Peripatus acacioi* Marcus & Marcus (Onychophora). III. The ionic composition of the haemolymph. *Comp. Biochem. Physiol.* **54***A*, 129–33.

Canziani, G. A. & Cannata, M. A. (1980). Water balance in *Ceratophrys ornata* from two different environments. *Comp. Biochem. Physiol.* **66***A*, 599–603.

Carley, W. A. (1978). Water economy of the earthworm *Lumbricus terrestris* L.: coping with the terrestrial environment. *J. exp. Zool.* **205**, 71–8.

Carlisle, D. B. (1968). *Triops* (Entomostraca) eggs killed only by boiling. *Science, N.Y.* **161**, 279–80.

Carroll, R. L. (1970). Quantitative aspects of the amphibian-reptilian transition. *Forma et Functio* **3**, 165–78.

Carroll, R. L. (1980). The hyomandibular as a supporting element in the skull of primitive tetrapods. In *The terrestrial environment and the origin of land vertebrates*, pp. 293–317, ed. A. L. Panchen. Academic Press, London.

Carroll, R. L. & Currie, P. J. (1975). Microsaurs as possible apodan ancestors. *Zool. J. Linn. Soc.* **57**, 229–47.

Carroll, R. L. & Gaskill, P. (1978). The order Microsauria. *Mem. Am. philos. Soc.* **126**, 1–211.

Carroll, R. L. & Holmes, R. (1980). The skull and jaw musculature as guides to the ancestry of salamanders. *Zool. J. Linn. Soc.* **68**, 1–40.

Carter, G. S. (1951). *A general zoology of the invertebrates.* Sidgwick & Jackson Ltd, London.

Carter, G. S. (1957). Air breathing. In *Physiology of fishes*, vol. 2, pp. 65–79, ed. M. Brown. Academic Press, New York.

Carter, G. S. (1967). *Structure and habit in vertebrate evolution.* Sidgwick & Jackson Ltd, London.

Carter, G. S. & Beadle, L. C. (1930*a*). The fauna of the swamps of the Paraguayan Chaco in relation to its environment. I. Physico-chemical nature of the environment. *J. Linn. Soc. (Zool.)* **37**, 205–58.

Carter, G. S. & Beadle, L. C. (1930*b*). Notes on the habits and development of *Lepidosiren paradoxa. J. Linn. Soc. (Zool.)* **37**, 167–203.

Carter, G. S. & Beadle, L. C. (1931). The fauna of the swamps of the Paraguayan Chaco in relation to its environment. II. Respiratory adaptations in the fishes. *J. Linn. Soc. (Zool.)* **37**, 327–68.

Case, J. & Gwilliam, G. F. (1961). Amino acid sensitivity of the dactyl chemoreceptors of *Carcinides maenas. Biol. Bull. mar. biol. Lab., Woods Hole* **121**, 449–55.

Casey, T. M. (1976*a*). Activity patterns, body temperature and thermal ecology in two desert caterpillars (Lepidoptera: Sphingidae). *Ecology* **57**, 485–97.

Casey, T. M. (1976*b*). Flight energetics in sphinx moths: heat production and heat loss in *Hyles lineata* during free flight. *J. exp. Biol.* **64**, 545–60.

Casey, T. M. (1981). Energetics and thermoregulation of *Malacosoma americanum* (Lepidoptera: Lasiocampidae) during hovering flight. *Physiol. Zoöl.* **54**, 362–71.

Cave, L. D. & Simonetta, A. M. (1975). Notes on the morphology and taxonomic position of *Aysheaia* (Onychophora?) and of *Skania* (undetermined phylum). *Monit. Zool. ital.* **9**, 67–81.

Chaloner, W. G. & Lacey, W. S. (1973). The distribution of late Palaezoic floras. *Spec. Pap. Palaeont.* **12**, 271–89.

Charles, G. H. (1966). Sense organs (less cephalopods). In *Physiology of Mollusca*, vol. II, pp. 455–521, ed. K. M. Wilbur & C. M. Yonge. Academic Press, London.

Chauvin, G., Gueguen, A. & Vannier, G. (1981). Croissance, bilan d'énergie et production d'eau metabolique chez la larve de *Tineola bisselliella. Can. J. Zool.* 59, 297–304.

Chauvin, G., Vannier, G. & Gueguen, A. (1979). Larval case and water balance in *Tinea pellionella. J. Insect Physiol.* 25, 615–19.

Chemical Rubber Co. (1966). *Handbook of chemistry and physics* (47th edition). Chemical Rubber Co., Cleveland, Ohio.

Cheng, L. (1976). Insects in marine environments. In *Marine insects*, pp. 1–4, ed. L. Cheng. North-Holland Publishing Co., Amsterdam.

Chernov, Ju. I., Striganova, B. R. & Ananjeva, S. I. (1977). Soil fauna of the polar desert of Cape Cheluskin, Taimyr Peninsula, USSR. *Oikos* 29, 175–9.

Cherry, L. M. (1969). The production of cuticle wax by engorged females of the cattle tick, *Boophilus microplus* (Canestrini). *J. exp. Biol.* 50, 705–9.

Chesters, K. I. M., Gnauck, F. R. & Hughes, N. F. (1967). Angiospermae. In *The fossil record*, pp. 269–88, ed. W. B. Harland *et al.* Geological Society of London, London.

Choudhuri, D. K. & Bhattacharyya, B. (1975). The effects of temperature and humidity on the development and hatching of eggs of *Lobella* (*Lobella*) *maxillaris* Yosii, 1966. *Rev. Ecol. Biol. Sol.* 12, 643–7.

Christian, E. (1979). Der Sprung der Collembolen. *Zool. Jb. (Physiol.)* 83, 457–90.

Church, N. S. (1960a) Heat loss and the body temperatures of flying insects. I. Heat loss by evaporation of water from the body. *J. exp. Biol.* 37, 171–85.

Church, N. S. (1960b). Heat loss and the body temperatures of flying insects. II. Heat conduction within the body and its loss by radiation and convection. *J. exp. Biol.* 37, 186–212.

Clark, E. (1936). The freshwater and land crayfishes of Australia. *Mem. Natn. Museum Melbourne* 10, 5–58.

Clark, H. (1953). Metabolism of the black snake embryo. I. Nitrogen excretion. *J. exp. Biol.* 30, 492–501.

Clark, L. B. (1935). The visual acuity of the fiddler-crab, *Uca pugnax. J. gen. Physiol.* 19, 311–19.

Clark, R. B. (1964). *Dynamics in metazoan evolution.* Clarendon Press, Oxford.

Clark, R. B. (1969). Systematics and phylogeny: Annelida, Echiura, Sipuncula. In *Chemical zoology*, vol. IV, pp. 1–68, ed. M. Florkin & B. T. Scheer. Academic Press, London.

Clark, R. B. (1978). Composition and relationships. In *Physiology of annelids*, pp. 1–32, ed. P. J. Mill. Academic Press, London.

Clarke, K. U. (1979). Visceral anatomy and arthropod phylogeny. In *Arthropod phylogeny*, pp. 467–549, ed. A. P. Gupta. Van Nostrand Reinhold Co., New York.

Clegg, J. S. (1967). Metabolic studies of cryptobiosis in encysted embryos of *Artemia salina. Comp. Biochem. Physiol.* 20, 801–9.

Clegg, J. S. (1981). Metabolic consequences of the extent and disposition of the aqueous intracellular environment. *J. exp. Zool.* 215, 303–13.

Cloudsley-Thompson, J. L. (1950). The water relations and cuticle of *Paradesmus gracilis* (Diplopoda, Strongylidae). *Q. Jl. microsc. Sci.* 91, 453–64.

Cloudsley-Thompson, J. L. (1951). Studies in diurnal rhythms. I. Rhythmic behaviour in millipedes. *J. exp. Biol.* 28, 165–72.

Cloudsley-Thompson, J. L. (1956). Studies in diurnal rhythms. VI. Bioclimatic observations in Tunisia and their significance in relation to the physiology of the fauna, especially woodlice, centipedes, scorpions and beetles. *Ann. Mag. nat. Hist.* 9, 305–29.

Cloudsley-Thompson, J. L. (1957). Studies in diurnal rhythms. V. Nocturnal ecology and water-relations of the British cribellate spiders of the genus *Ciniflo* Bl. *J. Linn. Soc., Lond.* 43, 134–52.

Cloudsley-Thompson, J. L. (1959). Studies in diurnal rhythms. IX. The water-relations of some nocturnal tropical arthropods. *Ent. exp. appl.* 2, 248–56.

Cloudsley-Thompson, J. L. (1960). A new sound-producing mechanism in centipedes. *Ent. mon. Mag.* 96, 110–13.

Cloudsley-Thompson, J. L. (1961). Some aspects of the physiology and behaviour of *Galeodes arabs. Ent. exp. appl.* 4, 257–63.

Cloudsley-Thompson, J. L. (1962). Microclimates and the distribution of terrestrial arthropods. *Ann. Rev. Ent.* 7, 199–222.

Cloudsley-Thompson, J. L. (1964). On the function of the subelytral cavity in desert Tenebrionidae (Col.). *Ent. mon. Mag.* 100, 148–51.

Cloudsley-Thompson, J. L. (1967). The water-relations of scorpions and tarantulas from the Sonoran desert. *Ent. mon. Mag.* 103, 217–20.

Cloudsley-Thompson, J. L. (1968). *Spiders, scorpions, centipedes and mites.* Pergamon Press, Oxford.

Cloudsley-Thompson, J. L. (1977). *The water and temperature relations of woodlice.* Meadowfield, Durham, North Carolina.

Cloudsley-Thompson, J. L. & Crawford, C. S. (1970). Water and temperature relations, and diurnal rhythms of scolopendromorph centipedes. *Ent. exp. appl.* 13, 187–93.

Coe, W. R. (1930). Unusual types of nephridia in nemerteans. *Biol. Bull. mar. biol. Lab., Woods Hole* 58, 203–16.

Coenen-Stass, D. (1981). Some aspects of the water balance of two desert woodlice, *Hemilepistus aphganicus* and *Hemilepistus reaumuri* (Crustacea, Isopoda Oniscoidea). *Comp. Biochem. Physiol.* 70A, 405–19.

Cohee, G. V., Glaessner, M. F. & Hedberg, H. D. (1978). *Contributions to the geologic time scale.* American Association of Petroleum Geologists, Tulsa, Oklahoma.

Cohen, A. C., March, R. B. & Pinto, J. D. (1981). Water relations of the desert blister beetle *Cysteodemus armatus* (Lecoute) (Coleoptera: Meloidae). *Physiol. Zoöl.* 54, 179–87.

Cohen, S. & Lewis, H. B. (1949). The nitrogen metabolism of the earthworm. *J. biol. Chem.* 180, 79–92.

Coles, G. C. (1968). The termination of aestivation in the large freshwater snail *Pila ovata* (Ampullariidae). I. Changes in oxygen uptake. *Comp. Biochem. Physiol.* 25, 517–22.

Coles, G. C. (1969). The termination of aestivation in the large freshwater snail *Pila ovata.* II. *In vitro* studies. *Comp. Biochem. Physiol.* 29, 373–81.

Cook, S. B. & Cook, C. B. (1978). Tidal amplitude and activity in the pulmonate limpets *Siphonaria normalis* (Gould) and *S. alternata* (Say). *J. exp. mar. Biol. Ecol.* 35, 119–36.

Cooke, A. H. (1895). Molluscs. In *The Cambridge natural history*, vol. III, pp. 1–459, ed. S. F. Harmer & A. E. Shipley. Macmillan & Co., London.

Coons, L. B. & Axtell, R. C. (1971). Ultrastructure of the excretory tubes of the mite *Macrocheles muscaedomesticae* (Mesostigmata, Macrochelidae) with notes on altered mitochondria. *J. Morph.* 133, 319–38.

Copeland, D. E. (1968). Fine structure of salt and water uptake in the land-crab, *Gecarcinus lateralis. Am. Zool.* 8, 417–32.

Coughtrey, P. J., Martin, M. H., Chard, D. & Shales, S. W. (1980). Micro-organisms and metal retention in the woodlouse *Oniscus asellus. Soil Biol. Biochem.* 12, 23–7.

Coutchié, P. A. & Crowe, J. H. (1979a). Transport of water vapor by tenebrionid beetles. I. Kinetics. *Physiol. Zoöl.* 52, 67–87.

Coutchié, P. A. & Crowe, J. H. (1979b). Transport of water vapor by tenebrionid beetles. II. Regulation of the osmolarity and composition of the haemolymph. *Physiol. Zoöl.* 52, 88–100.

Cox, C. B. (1967). Cutaneous respiration and the origin of the modern Amphibia. *Proc. Linn. Soc., Lond.* 178, 37–47.

Cox, R. A. (1965). The physical properties of sea water. In *Chemical oceanography*, vol. I, pp. 73–120, ed. J. P. Riley & G. Skirrow. Academic Press, London.

Crane, J. (1941). Crabs of the genus *Uca* from the west coast of Central America. *Zoologica, N.Y.* **26**, 145-208.

Crane, J. (1949). Comparative biology of salticid spiders at Rancho Grande, Venezuela. Part IV. An analysis of display. *Zoologica, N.Y.* **34**, 159-214.

Crawford, C. S. (1972). Water relations in a desert millepede, *Orthoporus ornatus* (Girard) (Spirostreptidae). *Comp. Biochem. Physiol.* **42A**, 521-35.

Crawford, C. S. (1978). Seasonal water balance in *Orthoporus ornatus*, a desert millipede. *Ecology* **59**, 996-1004.

Crawford, C. S. (1980). Desert millipedes: a rationale for their distribution. In *Myriapod biology*, pp. 171-81, ed. M. Camatini. Academic Press, London.

Crawford, C. S. & Cloudsley-Thompson, J. L. (1971). Water relations and desiccation-avoiding behaviour in the vinegaroon *Mastigoproctus giganteus* (Arachnida: Uropygi). *Ent. exp. appl.* **14**, 99-106.

Crawford, C. S. & Wooten, R. C. (1973). Water relations in *Diplocentrus spitzeri*, a semimontane scorpion from the southwestern United States. *Physiol. Zoöl.* **46**, 218-29.

Creek, G. A. (1951). The reproductive system and embryology of the snail *Pomatias elegans* (Müller). *Proc. zool. Soc. Lond.* **121**, 599-640.

Creek, G. A. (1953). The morphology of *Acme fusca* (Montagu) with special reference to the genital system. *Proc. malac. Soc. Lond.* **29**, 228-40.

Crofton, H. D. (1966). *Nematodes.* Hutchinson, London.

Croghan, P. C. (1958). The osmotic and ionic regulation of *Artemia salina* (L.). *J. exp. Biol.* **35**, 219-33.

Croghan, P. C. (1959). The interstitial soil-water habitat and the evolution of terrestrial arthropods. *Proc. roy. Soc. Edinb.* **27**, 103-4.

Croghan, P. C. & Lockwood, A. P. M. (1968). Ionic regulation of the Baltic and fresh-water races of the isopod *Mesidotea (Saduria) entomon* (L.). *J. exp. Biol.* **48**, 141-58.

Croll, N. A. & Matthews, B. E. (1977). *Biology of nematodes.* Blackie, Glasgow.

Crowe, J. H. (1972). Evaporative water loss by tardigrades under controlled relative humidities. *Biol. Bull. mar. biol. Lab., Woods Hole* **142**, 407-16.

Crowe, J. H., Madin, K. A. C. & Loomis, S. H. (1977). Anhydrobiosis in nematodes: metabolism during resumption of activity. *J. exp. Zool.* **201**, 57-63.

Crowe, J. H. & Magnus, K. A. (1974). Studies in acarine cuticles. II. Plastron respiration and levitation in a water mite. *Comp. Biochem. Physiol.* **49A**, 301-9.

Crowe, J. H., O'Dell, S. J. & Armstrong, D. A. (1979). Anhydrobiosis in nematodes: permeability during rehydration. *J. exp. Zool.* **207**, 431-8.

Curran, P. F. (1960). Na, Cl and water transport by rat ileum in vitro. *J. gen. Physiol.* **43**, 1137-48.

Curry, A. (1974). The spiracle structure and resistance to desiccation of centipedes. *Symp. zool. Soc. Lond.* **32**, 365-82.

Daguzan, J. (1971). Contribution à l'étude de l'excrétion azotée chez *Littorina littorea* (L.) adulte (Mollusque Mésogastéropode Littorinidae). *Archs Sci. physiol.* **25**, 293-302.

Dainton, B. H. (1954). The activity of slugs. I. The induction of activity by changing temperatures. *J. exp. Biol.* **31**, 165-87.

Dakin, W. J. (1921). The eye of *Peripatus*. *Q. Jl. microsc. Sci.* **65**, 163-72.

Dales, R. P. (1963). *Annelids.* Hutchinson, London.

Dallai, R. (1976). Fine structure of the pyloric region and Malpighian papillae of Protura (Insecta Apterygota). *J. Morph.* **150**, 727-62.

Dallinger, R. & Wieser, W. (1977). The flow of copper through a terrestrial food chain. I. Copper and nutrition in isopods. *Oecologia* **30**, 253-64.

Dalton, T. & Windmill, D. M. (1980). Fluid secretion by isolated Malpighian tubules of the housefly *Musca domestica*. *J. Insect Physiol.* **26**, 281-6.

Dandy, J. W. T. & Ewer, D. W. (1961). The water economy of three species of the amphibious crab, *Potamon*. *Trans. roy. Soc. S. Afr.* **36**, 137-62.

Danielopol, D. L. & Betsch, J. M. (1980). Ostracodes terrestres de Madagascar: systématique, origine, adaptations. *Rev. Ecol. Biol. Sol.* **17**, 87-123.

Darwin, C. R. (1881). *The formation of vegetable mould through the action of worms, with observations on their habits.* John Murray, London.

Das, B. K. (1927). The bionomics of certain air-breathing fishes of India, together with an account of the development of their air-breathing organs. *Phil. Trans. roy. Soc. (B)* **216**, 183-219.

Daumer, K., Jander, R. & Waterman, T. H. (1963). Orientation of the ghost-crab *Ocypode* in polarized light. *Z. vergl. Physiol.* **47**, 56-76.

Davies, L. & Richardson, J. (1970). Distribution in Britain and habitat requirements of *Petrobius maritimus* (Leach) and *P. brevistylis* Carpenter (Thysanura). *Entomologist* **103**, 97-114.

Davies, M. E. & Edney, E. B. (1952). The evaporation of water from spiders. *J. exp. Biol.* **29**, 571-82.

Davies, M.-T. B. (1974a). Changes in critical temperature during nymphal and adult development in the rabbit tick, *Haemaphysalis leporispalustris* (Acari: Ixodides: Ixodidae). *J. exp. Biol.* **60**, 85-94.

Davies, M.-T. B. (1974b). Critical temperature and changes in cuticular lipids in the rabbit tick, *Haemaphysalis leporispalustris. J. Insect Physiol.* **20**, 1087-100.

Dawes, C. M. (1975). Acid–base relationships within the avian egg. *Biol. Rev.* **50**, 351-71.

Daxboeck, C., Barnard, D. K. & Randall, D. J. (1981). Functional morphology of the gills of the bowfin, *Amia calva* L., with special reference to their significance during air exposure. *Resp. Physiol.* **43**, 349-64.

Decary, R. (1950). *La faune malgache.* Payot, Paris.

Dehadrai, P. V. & Tripathi, S. D. (1976). Environment and ecology of freshwater air-breathing teleosts. In *Respiration of amphibious vertebrates*, pp. 39-72, ed. G. M. Hughes. Academic Press, London.

Dejours, P. (1978). Carbon dioxide in water- and air-breathers. *Resp. Physiol.* **33**, 121-8.

Delamare Deboutteville, C. (1960). *Biologie des eaux souterraines littorales et continentales.* Hermann, Paris.

Delamarc Deboutteville, C., Gerlach, S. & Siewing, R. (1954). Recherches sur la faune des eaux souterraines littorales du Golfe de Gascogne. Littoral des landes. *Vie et Milieu* **5**, 373-407.

Delamare Deboutteville, C. & Massoud, Z. (1967). Un groupe panchronique: les collemboles. Essai critique sur *Rhyniella praecursor. Ann. Soc. ent. Fr. (N.S.)* **3**, 625-9.

Delany, M. J. (1959). The life histories and ecology of two species of *Petrobius* Leach, *P. brevistylis* and *P. maritimus. Trans. roy. Soc. Edinb.* **63**, 501-33.

Delevoryas, T. (1977). *Plant diversification* (2nd edition). Holt, Rhinehart & Winston, New York.

Delhaye, W. (1974a). Histophysiologie comparée des organes excréteurs chez quelques Neritacea (Mollusca - Prosobranchia. *Archs Biol., Bruxelles* **85**, 235-62.

Delhaye, W. (1974b). Histophysiologie comparée du rein chez les mésogastéropodes Archaeotaenioglossa et Littorinoidea (Molluscs - Prosobranchia). *Archs Biol., Bruxelles* **85**, 461-507.

Delhaye, W. (1974c). Contribution à l'étude des glandes pédieuses de *Pomatias elegans* (Mollusque, Gastéropode, Prosobranche). *Annls Sci. nat. (Zool.)* **16**, 97-110.

Delhaye, W. (1974d). Recherches sur les glandes pédieuses des gastéropodes prosobranches, principalement les formes terrestres et leur rôle possible dans l'osmorégulation chez les Pomatiasidae et les Chondropomidae. *Forma et Functio* **7**, 181-200.

Delhaye, W. & Bouillon, J. (1972a). L'évolution et l'adaptation de l'organe excréteur chez les mollusques gastéropodes pulmonés. I. Introduction générale et histophysiologie comparée du rein chez les basommatophores. *Bull. biol. Fr. Belg.* **106**, 45-77.

Delhaye, W. & Bouillon, J. (1972b). L'évolution et l'adaptation de l'organe excréteur chez les mollusques gastéropodes pulmonés. II. Histophysiologie comparée du rein chez les stylommatophores. Bull. biol. Fr. Belg. 106, 123-42.

Delhaye, W. & Bouillon, J. (1972c). L'évolution et l'adaptation de l'organe excréteur chez les mollusques gastéropodes pulmonés. III. Histophysiologie comparée du rein chez les soleolifères et conclusions générales pour tous les pulmonés. Bull. biol. Fr. Belg. 106, 295-314.

Délye, G. (1969). Permeabilité du tégument et résistance aux températures élevées de quelques arthropodes sahariens. Bull. Soc. ent. Fr. 74, 51-55.

Demange, J.-M. (1967). Recherches sur la segmentation du tronc des chilopodes et les diplopodes chilognathes (Myriapodes). Mem. Mus. natn. Hist. nat., Paris (A) 44, 1-188.

Dembowski, J. (1913). Uber den Bau der Augen von Ocypoda ceratophthalma Fabr. Zool. Jb. (Anat.) 36, 513-24.

Den Boer, P. J. (1961). The ecological significance of activity patterns in the woodlouse, Porcellio scaber Latr. (Isopoda). Archs neerl. Zool. 14, 283-409.

Denis, R. (1949). Sousclasse des Apterygotes. In Traité de zoologie, vol. 9, pp. 111-275, ed. P.-P. Grassé. Masson et Cie, Paris.

Denison, R. H. (1956). A review of the habitat of the earliest vertebrates. Fieldiana, Geol. 11, 359-457.

Dépêche, J., Gilles, R., Daufresne, S. & Chiapello, H. (1979). Urea content and urea production via the ornithine-urea cycle pathway during the ontogenetic development of two teleost fishes. Comp. Biochem. Physiol. 63A, 51-6.

Deshpande, U. D., Nagabhushanam, R. & Hanumante, M. M. (1980). Reproductive ecology of the marine pulmonate, Onchidium verrucalatum. Hydrobiologia 71, 83-5.

Dethier, V. G. (1963). The physiology of insect senses. Methuen, London.

DeVoe, R. D. (1975). Ultraviolet and green receptors in principal eyes of jumping spiders. J. gen. Physiol. 66, 193-207.

Deyrup, I. J. (1964). Water balance and kidney. In Physiology of the Amphibia, pp. 251-328, ed. J. A. Moore. Academic Press, London.

Diamond, J. M. & Bossert, W. H. (1967). Standing gradient osmotic flow. A mechanism for coupling of water and solute transport in epithelia. J. gen. Physiol. 50, 2061-83.

Díaz, H. & Rodríguez, G. (1977). The branchial chamber in terrestrial crabs: a comparative study. Biol. Bull. mar. biol. Lab., Woods Hole 153, 485-504.

Dietz, R. S. & Holden, J. C. (1976). The breakup of Pangaea. In Continents adrift and continents aground, pp. 126-37, ed. J. T. Wilson. W. H. Freeman, San Francisco.

Dietz, T. H. (1974). Active chloride transport across the skin of the earthworm, Lumbricus terrestris L. Comp. Biochem. Physiol. 49A, 251-8.

Dietz, T. H. & Alvarado, R. H. (1970). Osmotic and ionic regulation in Lumbricus terrestris L. Biol. Bull. mar. biol. Lab., Woods Hole 138, 247-61.

Digby, P. S. B. (1955). Factors affecting the temperature excess of insects in sunshine. J. exp. Biol. 32, 279-98.

Dohle, W. (1974). The origin and inter-relations of the myriapod groups. Symp. zool. Soc. Lond. 32, 191-8.

Dott, R. H. & Batten, R. L. (1981). Evolution of the earth (3rd edition). McGraw-Hill, New York.

Douce, G. K. (1976). Biomass of soil mites (Acari) in Arctic coastal tundra. Oikos 27, 324-30.

Dreisig, H. (1980). Daily activity, thermoregulation and water loss in the tiger beetle Cicindela hybrida. Oecologia 44, 376-89.

Dresel, E. I. B. & Moyle, V. (1950). Nitrogenous excretion of amphipods and isopods. J. exp. Biol. 27, 210-25.

Drummond, F. H. (1953). The eversible vesicles of Campodea. Proc. roy. ent. Soc. Lond. (A) 28, 145-8.

Dubuisson, M. (1928). Recherches sur la ventilation trachéenne chez les chilopodes et sur la circulation sanguine chez les scutigéres. Arch. zool. exp. gén. 67, 49-63.

Duelli, P. (1978). Movement detection in posterolateral eyes of jumping spiders (Evarcha arcuata, Salticidae). J. comp. Physiol. 124, 15-26.

Dunbar, B. S. & Winston, P. W. (1975). The site of active uptake of atmospheric water in larvae of Tenebrio molitor. J. Insect Physiol. 21, 495-500.

Duncan, C. J. (1975). Reproduction. In Pulmonates, vol. 1, pp. 309-65, ed. V. Fretter & J. F. Peake. Academic Press, London.

Duncan, K. W. (1969). The ecology of two species of terrestrial Amphipoda (Crustacea: family Talitridae) living in waste grassland. Pedobiologia 9, 323-41.

Dunger, W. & Steinmetzger, K. (1981). Okologische Unter-suchungen an Diplopoden einer Rasen-Wald-Catena im Thüringer Kalkgebiet. Zool. J. (Syst.) 108, 519-53.

Dwarakanath, S. K. & Job, S. V. (1965). Studies in transpiration in millipedes. I. Spirostreptus asthenes Poc. from a tropical jungle near Madurai. Proc. Indian Acad. Sci. 61B, 142-6.

Eakin, R. M. (1964). Electron microscopy of the nephridium of Perpatonder novaezealandiae (Phylum Onychophora). Am. Zool. 4, 433.

Eakin, R. M. & Westfall, J. A. (1965). Fine structure of the eye of Peripatus (Onychophora). Z. Zellforsch. 68, 278-300.

Ebeling, W. (1974). Permeability of insect cuticle. In The physio-logy of Insecta (2nd edition), vol. 6, pp. 271-343, ed. M. Rockstein. Academic Press, London.

Edney, E. B. (1951). The evaporation of water from woodlice and the millipede Glomeris. J. exp. Biol. 28, 91-115.

Edney, E. B. (1953). The temperature of woodlice in the sun. J. exp. Biol. 30, 331-49.

Edney, E. B. (1957). The water relations of terrestrial arthropods. Cambridge University Press, Cambridge, England.

Edney, E. B. (1960). Terrestrial adaptations. In The physiology of Crustacea, vol. I, pp. 367-93, ed. T. H. Waterman. Academic Press, New York & London.

Edney, E. B. (1961). The water and heat relationships of fiddler crabs (Uca spp.). Trans. roy. Soc. S. Afr. 36, 71-91.

Edney, E. B. (1962). Some aspects of the temperature relations of fiddler crabs (Uca spp.). In Biometeorology, pp. 79-85, ed. S. W. Tromp. Pergamon Press, Oxford.

Edney, E. B. (1968). Transition from water to land in isopod crustaceans. Am. Zool. 8, 309-26.

Edney, E. B. (1971a). Some aspects of water balance in tene-brionid beetles and a thysanuran from the Namib Desert of southern Africa. Physiol. Zoöl. 44, 61-76.

Edney, E. B. (1971b). The body temperature of tenebrionid beetles in the Namib Desert of southern Africa. J. exp. Biol. 55, 253-72.

Edney, E. B. (1977). Water balance in land arthropods. Springer-Verlag, New York.

Edney, E. B. & Barrass, R. (1962). The body temperature of the tsetse fly, Glossina morsitans Westwood (Diptera, Muscidae). J. Insect Physiol. 8, 469-81.

Edney, E. B. & McFarlane, J. (1974). The effect of temperature on transpiration in the desert cockroach, Arenivaga investigata, and in Periplaneta americana. Physiol. Zool. 47, 1-12.

Edney, E. B. & Spencer, J. (1955). Cutaneous respiration in woodlice. J. exp. Biol. 32, 256-69.

Edwards, C. A. & Lofty, J. R. (1972). Biology of earthworms. Chapman & Hall, London.

Edwards, C. A., Reichle, D. E. & Crossley, D. A. (1970). The role of soil invertebrates in turnover of organic matter and nutrients. In Analysis of temperate forest ecosystems, pp. 147-72, ed. D. E. Reichle. Chapman & Hall, London.

Edwards, J. L. (1976). The evolution of terrestrial locomotion. In Major patterns in vertebrate evolution, pp. 553-77, ed. M. K. Hecht, P. C. Goody & B. M. Hecht. Plenum Press, London.

Edwards, R. R. C. (1978). Ecology of a coastal lagoon complex in Mexico. Est. coast. mar. Sci. 6, 75-92.

Ehrlich, P. R. & Raven, P. H. (1964). Butterflies and plants: a study in coevolution. Evolution 18, 586-608.

Eisenbeis, G. (1974). Licht- und elektronenmikroskopische Untersuchungen zur Ultrastruktur des Transportepithels am Ventraltubus arthropleoner Collembolen (Insecta). *Cytobiologie* 9, 180-202.

Eisenbeis, G. (1976a). Zur Feinstruktur und Histochemie des Transport-Epithels abdominaler Koxalblasen der Doppelschwanz-Art *Campodea staphylinus* (Diplura: Campodeidae). *Ent. Germ.* 3, 185-201.

Eisenbeis, G. (1976b). Zur Morphologie des Ventraltubus von *Tomocerus* spp. (Collembola: Tomoceridae) unter besonderer Berucksichtigung der Muskulatur, der cuticularen Strukturen und der Ventralrinne. *Int. J. Insect Morph. Embryol.* 5, 357-79.

Eisenbeis, G. (1982). Physiological absorption of liquid water by Collembola: absorption by the ventral tube at different salinities. *J. Insect Physiol.* 28, 11-20.

Eisenbeis, G. & Wichard, W. (1975a). Histochemischer Chlorid-nachweis im Transportepithel am Ventraltubus arthropleoner Collembolen. *J. Insect. Physiol.* 21, 231-6.

Eisenbeis, G. & Wichard, W. (1975b). Feinstruktureller und histochemischer Nachweis des Transportepithels am Ventraltubus symphypleoner Collembolen (Insecta, Collembola). *Z. Morph. Tiere* 81, 103-10.

Eisenbeis, G. & Wichard, W. (1977). Zur feinstrukturellen Anpassung des Transportepithels am Ventraltubus von Collembolen bei unterschiedlicher Salinitat. *Zoomorphol.* 88, 175-88.

Eldredge, N. (1970). Observations on burrowing behaviour in *Limulus polyphemus* (Chelicerata, Merostomata), with implications on the functional anatomy of trilobites. *Am. Mus. Novitates* 2436, 1-17.

Elgmork, K. (1967). Ecological aspects of diapause in copepods. Proc. Symp. on Crustacea, Ernakulam. Part III. *Symp. mar. biol. Ass. India* (Ser. 2) 947-54.

El-Hifnawi, E.-S. (1973). Topographie und Ultrastruktur der Maxillarnephridien von Diplopoden. *Z. wiss. Zool.* 186, 118-48.

El-Hifnawi, E.-S. & Seifert, G. (1971). Uber den Feinbau der Maxillarnephridien von *Polyxenus lagurus* (L.) (Diplopoda, Penicillata). *Z. Zellforsch.* 113, 518-30.

Emilio, M. G. & Shelton, G. (1980). Carbon dioxide exchange and its effects on pH and bicarbonate equilibria in the blood of the amphibian, *Xenopus laevis. J. exp. Biol.* 85, 253-62.

Engelhardt, W. (1964). Die mitteleuropäischen Arten der Gattung *Trochosa* C. L. Kock 1848 (Araneae, Lycosidae). Morphologie, Chemotaxonomie, Biologie, Autökologie. *Z. Morph. Okol. Tiere* 54, 219-392.

Erséus, C. (1980). Specific and generic criteria in marine Oligochaeta, with special emphasis on Tubificidae. In *Aquatic oligochaete biology*, pp. 9-24, ed. R. O. Brinkhurst & D. G. Cook. Plenum Press, London.

Evans, A. A. F. & Perry, R. N. (1976). Survival strategies in nematodes. In *The organization of nematodes*, pp. 383-424, ed. N. A. Croll. Academic Press, London.

Evans, A. A. F. & Womersley, C. (1980). Longevity and survival in nematodes: models and mechanisms. In *Nematodes as biological models*, vol. 2, pp. 193-211, ed. B. M. Zuckerman. Academic Press, New York.

Evans, M. E. G. (1975). The jump of *Petrobius* (Thysanura, Machilidae). *J. Zool., Lond.* 176, 49-65.

Evoy, W. H. & Fourtner, C. R. (1973). Nervous control of walking in the crab, *Cardisoma guanhumi*. III. Proprioceptive influences on intra- and intersegmental coordination. *J. comp. Physiol.* 83, 303-18.

Eyre, S. R. (1968). *Vegetation and soils*. Arnold, London.

Fage, L. (1949). Classe des mérostomacés. In *Traité de zoologie*, vol. VI, pp. 219-62, ed. P.-P. Grassé. Masson et Cie, Paris.

Fahlander, K. (1938). Anatomie und systematische Einteilung der Chilopoden. *Zool. Bidr. Uppsala* 17, 1-148.

Fahlander, K. J. (1939). Die Segmentalorgane der Diplopoda,

Symphyla und Insecta Apterygota. *Zool. Bidr. Uppsala* 18, 243-51.

Fahrenbach, W. H. (1975). The visual system of the horseshoe crab *Limulus polyphemus. Int. Rev. Cytol.* 41, 285-349.

Fain-Maurel, M. A. & Cassier, P. (1971). Différenciations cytoplasmiques en relation avec la fonction excrétrice dans les reins céphaliques de *Petrobius maritimus* Leach (Insecte, Aptérygote). *J. Microsc.* 10, 163-78.

Fain-Maurel, M. A., Cassier, P. & Alibert, J. (1973). Etude infra-structurale et cytochimique de l'intestin moyen de *Petrobius maritimus* Leach en rapport avec ses fonctions excrétrices et digestives. *Tiss. Cell* 5, 603-31.

Falkowski, P. (1973). The respiratory physiology of hemocyanin in *Limulus polyphemus. J. exp. Zool.* 186, 1-6.

Fänge, R. (1976). Gas exchange in the swimbladder. In *Respiration of amphibious vertebrates*, pp. 189-211, ed. G. M. Hughes. Academic Press, London.

Farber, J. & Rahn, H. (1970). Gas exchange between air and water and the ventilation pattern of the electric eel. *Resp. Physiol.* 9, 151-61.

Farquharson, P. A. (1974a). A study of the Malpighian tubules of the pill millipede, *Glomeris marginata* (Villers). I. The isolation of the tubules in a Ringer solution. *J. exp. Biol.* 60, 13-28.

Farquharson, P. A. (1974b). A study of the Malpighian tubules of the pill millipede, *Glomeris marginata* (Villers). II. The effect of variation in osmotic pressure and sodium and potassium concentration on fluid production. *J. exp. Biol.* 60, 29-39.

Farquharson, P. A. (1974c). A study of the Malpighian tubules of the pill millipede, *Glomeris marginata* (Villers). III. The permeability characteristics of the tubule. *J. exp. Biol.* 60, 41-51.

Farrell, A. P. & Randall, D. J. (1978). Air-breathing mechanics in two Amazonian teleosts, *Arapaima gigas* and *Hoplerythrinus unitaeniatus. Can. J. Zool.* 56, 939-45.

Feeny, P. (1975). Biochemical coevolution between plants and their insect herbivores. In *Coevolution of animals and plants*, pp. 3-19, ed. L. E. Gilbert & P. H. Raven. University of Texas Press, Austin.

Fenchel, T., Jansson, B.-O. & von Thun, W. (1967). Vertical and horizontal distribution of the metazoan microfauna and of some physical factors in a sandy beach in the northern part of the Øresund. *Ophelia* 4, 227-43.

Fenchel, T. M. & Jørgensen, B. B. (1977). Detritus food chains of aquatic ecosystems: the role of bacteria. *Adv. microb. Ecol.* 1, 1-58.

Fernando, C. H. (1960). The Ceylonese freshwater crabs (Potamonidae). *Ceylon J. Sci. (Bio. Sci.)* 3, 191-222.

Feustel, H. (1958). Untersuchungen über die Exkretion bei Collembolen. *Z. wiss. Zool.* 161, 209-38.

Fielder, D. R. (1971). Some aspects of distribution and population structure in the sand bubbler crab *Scopimera inflata* Milne Edwards 1873 (Decapoda, Ocypodidae). *Aust. J. mar. Freshwat. Res.* 22, 41-7.

Fielding, P. J. & Nicolson, S. W. (1980). Regulation of haemolymph osmolarity and ions in the green protea beetle, *Trichostetha fascicularis*, during dehydration and rehydration. *Comp. Biochem. Physiol.* 674, 691-4.

Fimpel, E. (1975). Phänomene der Landadaptation bei terrestrischen und semiterrischen Brachyura der brasilianischen Küste (Malacostraca, Decapoda). *Zool. Jb. (Syst.)* 102, 173-214.

Fischer, P.-H. & Brunel, A. (1953). Relation entre l'habitat des gastéropodes et leur processus d'excrétion. *C.r. Séanc. Soc. Biogéogr.* 30, 36-7.

Fisk, A. & Tribe, M. (1949). The development of the amnion and chorion of reptiles. *Proc. zool. Soc. Lond.* 119, 83-114.

Fitzgerrel, W. W. & Vanatta, J. C. (1980). Factors affecting bicarbonate excretion in the urinary bladder of *Bufo marinus. Comp. Biochem. Physiol.* 66A, 277-81.

Flemister, L. J. (1958). Salt and water anatomy, constancy and regulation in related crabs from marine and terrestrial habitats. *Biol. Bull. mar. biol. Lab., Woods Hole* 115, 180–200.

Florkin, M. (1949). *Biochemical evolution*, trans. S. Morgulis. Academic Press, New York.

Florkin, M. (1966). Nitrogen metabolism. In *Physiology of Mollusca*, vol. II, pp. 309–51, ed. K. M. Wilbur & C. M. Yonge. Academic Press, London.

Florkin, M. & Jeuniaux, C. (1974). Hemolymph: composition. In *The physiology of insecta*, vol. 5, pp. 255–307, ed. M. Rockstein. Academic Press, London.

Foelix, R. F. & Chu-Wang, I.-W. (1973). The morphology of spider sensillae. II. Chemoreceptors. *Tiss. Cell* 5, 461–78.

Foelix, R. F., Chu-Wang, I.-W. & Beck, L. (1975). Fine structure of tarsal sensory organs in the whip spider *Admetus pumilio* (Amblypygi, Arachnida). *Tiss. Cell* 7, 331–46.

Foster, B. A. (1971). Desiccation as a factor in the zonation of barnacles. *Mar. Biol.* 8, 12–29.

Foster, W. A. & Moreton, R. B. (1981). Synchronization of activity rhythms with the tide in a saltmarsh collembolan, *Anurida maritima. Oecologia* 50, 265–70.

Foster, W. A. & Treherne, J. E. (1976). Insects of marine salt-marshes: problems and adaptations. In *Marine insects*, pp. 5–42, ed. L. Cheng. North-Holland Publishing Co., Amsterdam.

Foster, W. A. & Treherne, J. E. (1980). Feeding, predation and aggregation behaviour in a marine insect, *Halobates robustus* Barber (Hemiptera: Gerridae), in the Galapagos Islands. *Proc. roy. Soc. Lond. (B)* 209, 539–53.

Fotheringham, N. (1975). Structure of seasonal migrations of the littoral hermit crab, *Clibanarius vittatus* (Bosc.) *J. exp. mar. Biol. Ecol.* 18, 47–53.

Foxon, G. E. H. (1964). Blood and respiration. In *Physiology of Amphibia*, pp. 151–209, ed. J. A. Moore. Academic Press, London.

Franc, A. (1968). Classe des Gastéropodes. In *Traité de zoologie*, vol. V, Fasc. III, pp. 1–893, ed. P.-P. Grassé. Masson et Cie, Paris.

François, J. (1969). Anatomie et morphologie céphalique des protoures (Insecta Apterygota). *Mem. Mus. natn. Hist. nat. (A)* 59, 1–144.

François, J. (1972). Ultrastructure du rein labiale céphalique de *Campodea chardardi* Condé (Diplura, Insecta). *Z. Zellforsch.* 127, 34–49.

Frazier, L. W. & Vanatta, J. C. (1980). Evidence that the frog skin excretes ammonia. *Comp. Biochem. Physiol.* 66A, 525–7.

Freeman, B. M. (1974). Excretion and water balance. In *Development of the avian embryo*, pp. 191–207, by B. M. Freeman & M. A. Vince. Chapman & Hall, London.

Fretter, V. (1943). Studies in the functional morphology and embryology of *Onchidella celtica. J. mar. biol. Ass. UK* 25, 685–720.

Fretter, V. (1965). Functional studies of the anatomy of some neritid prosobranchs. *J. Zool., Lond.* 147, 46–74.

Fretter, V. (1975). Introduction. In *Pulmonates*, vol. 1, pp. xi–xxix, ed. V. Fretter & J. Peake. Academic Press, London.

Fretter, V. & Graham, A. (1962). *British prosobranch molluscs.* Ray Society, London.

Fretter, V. & Graham, A. (1980). The prosobranch molluscs of Britain and Denmark. Part 5. Marine Littorinacea. *J. moll. Stud.* (suppl.) 7, 243–84.

Fretter, V. & Peake, J. F. (eds.) (1978). *Pulmonates*, vol. 2A. (Appendix.) Academic Press, London.

Frings, H. & Frings, M. (1966). Reactions of orb-weaving spiders (Argiopidae) to airborne sounds. *Ecology* 47, 578–88.

Froehlich, C. G. (1954). Sôbre morfologia e taxonomia das Geoplanidae. *Bol. Fac. Philos. Sci. S. Paulo (Zool.)* 19, 195–251.

Froehlich, C. G. (1955). On the biology of land planarians. *Zoologia* 20, 263–71.

Füller, H. (1960). Untersuchungen über den Bau der Stigmen bei Chilopoden. *Zool. Jb.* (Anat.) 78, 129–44.

Füller, H. (1966). Elektronenmikroskopische Untersuchungen der Malpighischen Gefasse von *Lithobius forficatus* (L.). *Z. wiss. Zool.* 173, 191–217.

Fyhn, H. J., Petersen, J. A. & Johansen, K. (1972). Eco-physiological studies of an intertidal crustacean, *Pollicipes polymerus* (Cirripedia, Lepadomorpha). 1. Tolerance to body temperature change, desiccation and osmotic stress. *J. exp. Biol.* 57, 83–102.

Gabe, M. (1957). Données histologiques sur les organes segmentaires des Peripatopsidae (Onychophores). *Arch. Anat. microsc. Morph. exp.* 46, 283–306.

Gabe, M. (1967a). Données histologiques sur le rein céphalique des thysanoures (Insectes apterygotes). *Ann. Soc. ent. Fr.* 3, 681–713.

Gabe, M. (1967b). Caractères cytologiques et histochimiques du rein maxillaire des chilopodes. *C. r. Acad. Sci., Paris* 264, 726–9.

Gabe, M., Cassier, P. & Fain-Maurel, M. A. (1973). Données morphologiques sur les organes excréteurs abdominaux de *Petrobius maritimus* Leach (Insecte apterygote). *Arch. Anat. microsc. Morph. exp.* 62, 101–43.

Gans, C. (1970a). Respiration in early tetrapods: the frog is a red herring. *Evolution* 24, 723–34.

Gans, C. (1970b). Strategy and sequence in the evolution of the external gas exchangers of ectothermal vertebrates. *Forma et Functio* 3, 66–104.

van Gansen, P. S. (1959). Structure des glandes calciques d'*Eisenia foetida* Sav. *Bull. biol. Fr. Belg.* 93, 38–63.

Gardiner, B. (1980). Tetrapod ancestry: a reappraisal. In *The terrestrial environment and the origin of land vertebrates*, pp. 177–85, ed. A. L. Panchen. Academic Press, London.

Gee, J. H. & Graham, J. B. (1978). Respiratory and hydrostatic functions of the intestine of the catfishes *Hoplosternum thoracatum* and *Brochis splendens* (Callichthyidae). *J. exp. Biol.* 74, 1–16.

Geethabali & Rao, K. P. (1973). A metasomatic neural photoreceptor in the scorpion. *J. exp. Biol.* 58, 189–96.

Geiger, R. (1961). *Das Klima der bodennahen Luftschicht.* F. Vieweg & Sohn, Brunswick, Germany. Translated as *The climate near the ground*, Harvard University Press, Cambridge, Mass.

Gerard, B. M. (1967). Factors affecting earthworms in pastures. *J. anim. Ecol.* 36, 235–52.

Ghabbour, S. I. & Shakir, S. H. (1980). Ecology of soil fauna of Mediterranean desert ecosystems in Egypt. III. Analysis of *Thymelaea* mesofauna populations at the Mariut frontal plain. *Rev. Ecol. Biol. Sol.* 17, 327–52.

Ghilarov, M. S. (1956). Soil as the environment of the invertebrate transition from the aquatic to the terrestrial life. *Rapp. VIe Congr. int. Sci. Sol., Paris* 3, 307–13.

Ghilarov, M. S. (1959). Adaptations of insects to soil dwelling. In *Proceedings of the XVth international congress of zoology, London*, pp. 354–7.

Ghiradella, H. T., Case, J. F. & Cronshaw, J. (1968). Structure of aesthetascs in selected marine and terrestrial decapods: chemoreceptor morphology and environment. *Am. Zool.* 8, 603–21.

Ghiretti, F. (1966). Molluscan hemocyanins. In *Physiology of Mollusca*, vol. 2, pp. 233–48, ed. K. M. Wilbur & C. M. Yonge. Academic Press, London.

Ghiretti, F. & Ghiretti-Magaldi, A. (1975). Respiration. In *Pulmonates*, vol. 1, pp. 33–52, ed. V. Fretter & J. F. Peake. Academic Press, London.

Gibson, R. (1972). *Nemerteans.* Hutchinson, London.

Gibson, R. & Moore, J. (1976). Freshwater nemerteans. *Zool. J. Linn. Soc.* 58, 177–218.

Gibson, R., Moore, J. & Crandall, F. B. (1982). A new semi-terrestrial nemertean from California. *J. Zool., Lond.* 196, 463–74.

Gibson-Hill, C. A. (1947). Field notes on the terrestrial crabs. *Bull. Raffles Mus.* 18, 43–52.

Gifford, C. A. (1962a). Some aspects of osmotic and ionic regulation in the blue crab, *Callinectes sapidus*, and the ghost crab, *Ocypode albicans*. *Publ. Inst. mar. Sci., Univ. Texas* **8**, 97-125.

Gifford, C. A. (1962b). Some observations on the general biology of the land crab, *Cardisoma guanhumi* (Latreille), in south Florida. *Biol. Bull. mar. biol. Lab., Woods Hole* **132**, 207-23.

Gifford, C. A. (1968). Accumulation of uric acid in the land crab, *Cardisoma guanhumi*. *Am. Zool.* **8**, 521-8.

Gilai, A. & Parnas, I. (1970). Neuromuscular physiology of the closer muscles in the pedipalp of the scorpion *Leiurus quinquestriatus*. *J. exp. Biol.* **52**, 325-44.

Gilbert, L. E. (1975). Ecological consequence of a coevolved mutualism between butterflies and plants. In *Coevolution of animals and plants*, pp. 210-40, ed. L. E. Gilbert & P. H. Raven. University of Texas Press, Austin.

Gislén, T. (1930). Affinities between the Echinodermata, Enteropneusta and Chordonia. *Zool. Bidr. Uppsala* **12**, 199-304.

Gislén, T. (1944). Zur Verbreitung und Okologie von *Rhynchodemus terrestris* (O. F. Müll.) mit Bemerkungen über Bitemporalität. *Arch. Hydrobiol.* **40**, 667-86.

Gislén, T. (1947). Conquering terra firma. The transition from water to land-life. *K. fysiogr. Sällsk. Lund Förh.* **17**, 216-35.

Glaessner, M. F. (1969). Decapoda. In *Treatise on invertebrate paleontology*, part R, Arthropoda 4 (vol. 2), pp. 399-566, ed. R. C. Moore. Geological Society of America and Kansas University Press.

Goin, C. J. & Goin, O. B. (1974). *Journey onto land*. Macmillan, New York.

Goldstein, L. (1972). Adaptation of urea metabolism in aquatic vertebrates. In *Nitrogen metabolism and the environment*, pp. 55-77, ed. J. W. Campbell & L. Goldstein. Academic Press, London.

Goodrich, E. S. (1930). *Studies on the structure and development of vertebrates*. Macmillan, London.

Goodrich, E. S. (1945). The study of nephridia and genital ducts since 1895. *Q. Jl microsc. Sci.* **86**, 113-392.

Gordon, H. R. S. (1958). Synchronous claw-waving of fiddler crabs. *Anim. Behav.* **6**, 238-41.

Gordon, M. S. (1962). Osmotic regulation in the green toad (*Bufo viridis*). *J. exp. Biol.* **39**, 261-70.

Gordon, M. S., Boetius, I., Evans, D. H., McCarthy, R. & Oglesby, L. D. (1969). Aspects of the physiology of terrestrial life in amphibious fishes. I. The mudskipper, *Periophthalmus sobrinus*. *J. exp. Biol.* **50**, 141-9.

Gordon, M. S., Boetius, J., Evans, D. H. & Oglesby, L. C. (1968). Additional observations on the natural history of the mudskipper, *Periophthalmus sobrinus*. *Copeia* 1968, 853-7.

Gordon, M. S., Fischer, S. & Tarifeno, E. (1970). Aspects of the physiology of terrestrial life in amphibious fishes. II. The Chilean clingfish, *Sicyases sanguineus*. *J. exp. Biol.* **53**, 559-72.

Gordon, M. S., Ng, W. W. & Yip, A. Y. (1978). Aspects of the physiology of terrestrial life in amphibious fishes. III. The Chinese mudskipper *Periophthalmus cantonensis*. *J. exp. Biol.* **72**, 57-75.

Gordon, M. S., Schmidt-Nielsen, K. & Kelly, H. M. (1961). Osmotic regulation in the crab-eating frog (*Rana cancrivora*). *J. exp. Biol.* **38**, 659-78.

Gordon, M. S. & Tucker, V. A. (1965). Osmotic regulation in the tadpoles of the crab-eating frog (*Rana cancrivora*). *J. exp. Biol.* **42**, 437-45.

Görner, P. (1959). Optische Orientierungsreaktionen bei Chilopoden. *Z. vergl. Physiol.* **42**, 1-5.

Gorvett, H. (1956). Tegumental glands and terrestrial life in woodlice. *Proc. zool. Soc. Lond.* **126**, 291-314.

Graff, L. von (1912-1917). Tricladida. In *Bronn's Klassen und Ordnung des Tier-Reichs*, Bd. IV, Vermes, 1c. Abt. II. Akademische Verlagsgesellschaft M.B.H., Leipzig.

Graham, J. B. (1973). Terrestrial life of the amphibious fish *Mnierpes macrocephalus*. *Mar. Biol.* **23**, 83-91.

Graham, J. B. (1976). Respiratory adaptations of marine air-breathing fishes. In *Respiration of amphibious vertebrates*, pp. 165-87, ed. G. M. Hughes. Academic Press, London.

Graham, J. B. & Baird, T. A. (1982). The transition to air breathing in fishes. I. Environmental effects on the facultative air breathing of *Ancistrus chagresi* and *Hypostomus plecostomus* (Loricariidae). *J. exp. Biol.* **96**, 53-67.

Graham, J. B., Rosenblatt, R. H. & Gans, C. (1978). Vertebrate air breathing arose in fresh waters and not in the ocean. *Evolution* **32**, 459-63.

Grainger, F. & Newell, G. E. (1965). Aerial respiration in *Balanus balanoides*. *J. mar. biol. Ass. UK* **45**, 469-79.

Graszynski, K. (1963). Die Feinstruktur des Nephridialkanals von *Lumbricus terrestris* L. Eine elektronen mikroskopische Untersuchung. *Zool. Beitr.* **8**, 189-296.

Greenaway, P. (1970). Sodium regulation in the freshwater mollusc *Limnaea stagnalis* (L.) (Gastropoda: Pulmonata). *J. exp. Biol.* **53**, 147-63.

Greenaway, P. (1980). Water balance and urine production in the Australian arid-zone crab *Holthuisana transversa*. *J. exp. Biol.* **87**, 237-46.

Greenaway, P. (1981). Sodium regulation in the freshwater/land crab *Holthuisana transversa*. *J. comp. Physiol.* **142**, 451 6.

Greenaway, P. & MacMillen, R. E. (1978). Salt and water balance in the terrestrial phase of the inland crab *Holthuisana* (*Austrothelphusa*) *transversa* Martens (Parathelphusoidea: Sundathelphusidae). *Physiol. Zoöl.* **51**, 217-29.

Greenaway, P. & Taylor, H. H. (1976). Aerial gas exchange in Australian arid-zone crab, *Parathelphusa transversa* von Martens. *Nature, Lond.* **262**, 711-13.

Greenwood, P. H. (1958). Reproduction in the east African lung-fish *Protopterus aethiopicus* Heckel. *Proc. zool. Soc. Lond.* **130**, 547-67.

Gregory, J. T. (1965). Microsaurs and the origin of captorhinomorph reptiles. *Am. Zool.* **5**, 277-86.

Gregory, R. B. (1977). Synthesis and total excretion of waste nitrogen by fish of the *Periophthalmus* (mudskipper) and *Scartelaos* families. *Comp. Biochem. Physiol.* **57A**, 33-6.

Grey, R. D., Working, P. K. & Hedrick, J. L. (1977). Alteration of structure and penetrability of the vitelline envelope after passage of eggs from coelom to oviduct in *Xenopus laevis*. *J. exp. Zool.* **201**, 73-84.

Griffin, D. J. G. (1971). The ecological distribution of grapsid and ocypodid shore crabs (Crustacea: Brachyura) in Tasmania. *J. anim. Ecol.* **40**, 597-621.

Grigg, G. C. (1965a). Studies on the Queensland lungfish, *Neoceratodus forsteri* (Krefft). I. Anatomy, histology, and functioning of the lung. *Aust. J. Zool.* **13**, 243-53.

Grigg, G. C. (1965b). Studies on the Queensland lungfish, *Neoceratodus forsteri* (Krefft). III. Aerial respiration in relation to habits. *Aust. J. Zool.* **13**, 413-21.

Grimstone, A. V., Mullinger, A. M. & Ramsay, J. A. (1968). Further studies on the rectal complex of the mealworm, *Tenebrio molitor* L. (Coleoptera, Tenebrionidae). *Phil. Trans. roy. Soc. Lond.* **253**, 343-82.

Groepler, W. (1969). Feinstruktur der Coxalorgane bei der Gattung *Ornithodorus* (Acari, Argasidae). *Z. wiss. Zool.* **178**, 235-75.

Gross, W. J. (1955). Aspects of osmotic regulation in crabs showing the terrestrial habit. *Am. Nat.* **89**, 205-22.

Gross, W. J. (1957). A behavioural mechanism for osmotic regulation in a semi-terrestrial crab. *Biol. Bull. mar. biol. Lab., Woods Hole* **113**, 268-74.

Gross, W. J. (1961). Osmotic tolerance and regulation in crabs from a hypersaline lagoon. *Biol. Bull. mar. biol. Lab., Woods Hole* **121**, 290-301.

Gross, W. J. (1963). Cation and water balance in crabs showing the terrestrial habit. *Physiol. Zoöl.* **36**, 312-24.

Gross, W. J. (1964a). Water balance in anomuran land crabs on a dry atoll. *Biol. Bull. mar. biol. Lab., Woods Hole* **126**, 54–68.

Gross, W. J. (1964b). Trends in water and salt regulation among aquatic and amphibious crabs. *Biol. Bull. mar. biol. Lab., Woods Hole* **127**, 447–66.

Gross, W. J. & Holland, P. V. (1960). Water and ionic regulation in a terrestrial hermit crab. *Physiol. Zoöl.* **33**, 21–8.

Gross, W. J., Lasiewski, R. C., Dennis, M. & Rudy, P. (1966). Salt and water balance in selected crabs of Madagascar. *Comp. Biochem. Physiol.* **17**, 641–60.

Grove, A. J. & Cowley, L. F. (1926). On the reproductive processes of the brandling worm, *Eisenia foetida* (Sav.). *Q. Jl microsc. Sci.* **70**, 559–81.

Gunderson, H. L. (1976). *Mammalogy.* McGraw-Hill, New York.

Gupta, A. P. (1979). Arthropod hemocytes and phylogeny. In *Arthropod phylogeny*, pp. 669–735, ed. A. P. Gupta. Van Nostrand Reinhold Co., New York.

Gupta, A. P. (1980). Origin and affinities of Myriapoda. In *Myriapod biology*, pp. 373–90, ed. M. Camatini. Academic Press, London.

Gupta, B. L. & Berridge, M. J. (1966). Fine structural organisation of the rectum in the blowfly, *Calliphora erythrocephala* (Meig.) with special reference to connective tissue, tracheae and neurosecretory innervation in the rectal papillae. *J. Morph.* **120**, 23–82.

Gupta, B. L., Hall, T. A. & Moreton, R. B. (1977). Electron probe X-ray microanalysis. In *Transport of ions and water in animals*, pp. 83–143, ed. B. L. Gupta, R. B. Moreton, J. L. Oschman & B. J. Wall. Academic Press, London.

Haacker, U. (1974). Patterns of communication in courtship and mating behaviour of millipedes (Diplopoda). *Symp. zool. Soc. Lond.* **32**, 317–28.

Hackman, R. H. (1974). Chemistry of the insect cuticle. In *The physiology of Insecta* (2nd edition), vol. 6, pp. 215–70, ed. M. Rockstein. Academic Press, London.

Hadley, N. F. (1970a). Micrometeorology and energy exchange in two desert arthropods. *Ecology* **51**, 434–44.

Hadley, N. F. (1970b). Water relations of the desert scorpion, *Hadrurus arizonensis. J. exp. Biol.* **53**, 547–58.

Hadley, N. F. (1981). Cuticular lipids of terrestrial plants and arthropods: a comparison of their structure, composition, and waterproofing mechanism. *Biol. Rev.* **56**, 23–47.

Hagen, H.-O. (1970a). Anpassungen an das spezielle Gezeitenzonen-Niveau bei Ocypodiden (Decapoda, Brachyura). *Forma et Functio* **2**, 361–413.

Hagen, H.-O. (1970b). Zur Deutung langstieliger und gehornter Augen bei Ocypodiden (Decapoda, Brachyura). *Forma et Functio* **2**, 13–57.

Haggag, G. & Fouad, Y. (1965). Nitrogenous excretion in arachnids. *Nature, Lond.* **207**, 1003–4.

Halstead, L. B. (1973). The heterostracan fishes. *Biol. Rev.* **48**, 279–332.

Hamilton, A. G. (1964). The occurrence of periodic or conti-nuous discharge of carbon dioxide by male desert locusts (*Schistocerca gregaria* Forskål) measured by an infra-red gas analyser. *Proc. roy. Soc. Lond.* (B) **160**, 373–95.

Hamilton, W. J. (1975). Coloration and its thermal consequences for diurnal desert insects. In *Environmental physiology of desert organisms*, pp. 67–89, ed. N. F. Hadley. Dowden, Hutchinson & Ross Inc., Stroudsburg, Pennsylvania.

Hamilton, W. J. & Seely, M. K. (1976). Fog basking by the Namib Desert beetle, *Onymacris unguicularis. Nature, Lond.* **262**, 284–5.

Haniffa, M. A. (1978). Energy loss in an aestivating population of the tropical snail *Pila globosa. Hydrobiologia* **61**, 169–82.

Hanström, B. (1934). Bemerkungen uber das Komplex-Auge der Scutigeriden. *Lunds. Univ. Arsskr.* (N.F. Avd. 2) **30**, 1–14.

Harant, H. & Grassé, P.-P. (1959). Classe des annélides achètes ou hirudinées ou sangsues. In *Traité de zoologie*, vol. 5, Fasc. I, pp. 471–593, ed. P.-P. Grassé. Masson et Cie, Paris.

Harder, W. (1975). *Anatomy of fishes.* E. Schweizerbart'sche Verlagsbuchhandlung, Stuttgart.

Harding, J. P. (1953). The first known example of a terrestrial ostracod, *Mesocypris terrestris* sp. nov. *Ann. Natal Mus.* **12**, 359–65.

Harland, W. B., Holland, C. H., House, M. R., Hughes, N. F., Reynolds, A. B., Rudwick, M. J. S., Satterthwaite, G. E., Tarlo, L. B. H. & Willey, E. C. (eds.) (1967). *The fossil record.* Geological Society, London.

Harms, J. W. (1938). Lebenslauf und Stammesgeschichte des *Birgus latro* L. von der Weihnachtsinsel. *Jena Zeits. f. Naturwissenschaft.* **71**, 1–34.

Harris, D. J. & Mill, P. J. (1973). The ultrastructure of chemo-receptor sensillae in *Ciniflo* (Araneida, Arachnida). *Tiss. Cell* **5**, 679–89.

Harris, R. R. (1977). Urine production rate and water balance in the terrestrial crabs *Gecarcinus lateralis* and *Cardisoma guanhumi. J. exp. Biol.* **68**, 57–67.

Harris, R. R. & Kormanik, G. A. (1981). Salt and water balance and antennal gland function in three Pacific species of terrestrial crab (*Gecarcoidea lalandii, Cardisoma carnifex, Birgus latro*). II. The effects of desiccation. *J. exp. Zool.* **218**, 107–116.

Harris, V. A. (1960). On the locomotion of the mud-skipper *Periophthalmus koelreuteri* (Pallas): (Gobiidae). *Proc. zool. Soc. Lond.* **134**, 107–35.

Harrison, J. L. (1953). Sexual behaviour of land leeches. *J. Bombay nat. Hist. Soc.* **51**, 959–60.

Hartman, O. (1959). Capitellidae and Nereidae (marine annelids) from the Gulf side of Florida, with a review of freshwater Nereidae. *Bull. mar. Sci. Gulf Caribb.* **9**, 153–68.

Hartnoll, R. G. (1963). The freshwater grapsid crabs of Jamaica. *Proc. Linn. Soc., Lond.* **175**, 145–69.

Hartnoll, R. G. (1973). Factors affecting the distribution and behaviour of the crab *Dotilla fenestrata* on east African shores. *Est. coast. mar. Sci.* **1**, 137–52.

den Hartog, C. (1974). Salt-marsh Turbellaria. In *Biology of the Turbellaria*, pp. 229–47, ed. N. W. Riser & M. P. Morse. McGraw-Hill, New York.

Haswell, M. S. & Randall, D. J. (1978). The pattern of carbon dioxide excretion in the rainbow trout *Salmo gairdneri. J. exp. Biol.* **72**, 17–24.

Hattingh, J. (1972). A comparative study of the transepidermal water loss through the skin of various animals. *Comp. Biochem. Physiol.* **43A**, 715–18.

Haupt, J. (1969a). Zur Feinstruktur der Labialniere des Silber-fischens *Lepisma saccharina* L. (Thysanura, Insecta). *Zool. Beitr.* **15**, 139–70.

Haupt, J. (1969b). Zur Feinstruktur der Maxillarnephridien von *Scutigerella immaculata* Newport (Symphyla, Myriapoda). *Z. Zellforsch.* **101**, 401–7.

Haupt, J. (1971). Beitrag zur Kenntnis der Sinnesorgane von Symphylen (Myriapoda). II. Feinstruktur des Toemo-esvaryschen Organs von *Scutigerella immaculata* Newport. *Z. Zellforsch.* **122**, 172–89.

Haupt, J. (1972). Ultrastruktur des Pseudoculus von *Eosentomon* (Protura, Insecta). *Z. Zellforsch.* **135**, 539–51.

Haupt, J. (1973). Die Ultrastruktur des Pseudoculus von *Allopaurus* (Pauropoda) und die Homologie der Schlafenorgane. *Z. Morph. Tiere* **76**, 173–91.

Haupt, J. (1976). Die segmentalen Kopfdrüsen von *Scutigerella* (Symphyla, Myriapoda). *Zool. Beitr.* **22**, 19–37.

Haupt, J. (1980). Phylogenetic aspects of recent studies on myriapod sense organs. In *Myriapod biology*, pp. 391–406, ed. M. Camatini. Academic Press, London.

Haupt, J. & Coineau, Y. (1975). Trichobothrien und Tasborsten der Milbe *Microcaeculus* (Acari, Prostigmata, Caeculidae). *Z. Morph. Okol. Tiere* **81**, 305–22.

Hayes, W. F. (1971). Fine structure of the chemoreceptor sensillum in *Limulus. J. Morph.* **133**, 205–40.

Hecker, H., Diehl, P. A. & Aeschlimann, A. (1969). Recherches sur l'ultrastructure et l'histochimie de l'organe coxal d'*Ornithodoros moubata* (Murray). *Acta trop.* **26**, 346–59.

Heeg, J. (1967a). Studies on Thysanura. I. The water economy of *Machiloides delanyi* Wygodzinsky and *Ctenolepisma longicaudata* Escherisch. *Zool. Africana* 3, 21–41.

Heeg, J. (1967*b*). Studies on Thysanura. II. Orientation reactions of *Machiloides delanyi* Wygodzinsky and *Ctenolepisma longicaudata* Escherisch to temperature, light and atmospheric humidity. *Zool. Africana* 3, 43–57.

Heeg, J. (1969). Studies on Thysanura. III. Some factors affecting the distribution of South African Thysanura. *Zool. Africana* 4, 135–43.

Heeg, J. & Cannone, A. J. (1966). Osmoregulation by means of a hitherto unsuspected osmoregulatory organ in two grapsid crabs. *Zool. Africana* 2, 127–9.

Heinrich, B. (1975). Thermoregulation and flight energetics of desert insects. In *Environmental physiology of desert organisms*, pp. 90–105, ed. N. F. Hadley. Dowden, Hutchinson & Ross Inc., Stroudsburg, Pennsylvania.

Heinrich, B. (1976). Heat exchange in relation to blood flow between thorax and abdomen in bumblebees. *J. exp. Biol.* 64, 561–85.

Heinrich, B. (1980*a*). Mechanisms of body-temperature regulation in honeybees, *Apis mellifera*. I. Regulation of head temperature. *J. exp. Biol.* 85, 61–72.

Heinrich, B. (1980*b*). Mechanisms of body-temperature regulation in honeybees, *Apis mellifera*. II. Regulation of thoracic temperature at high air temperatures. *J. exp. Biol.* 85, 73–87.

Heinrich, B. & Bartholomew, G. A. (1971). An analysis of preflight warm-up in the sphinx moth, *Manduca sexta. J. exp. Biol.* 55, 223–39.

Heinrich, B. & Casey, T. M. (1978). Heat transfer in dragonflies: 'fliers' and 'perchers'. *J. exp. Biol.* 74, 17–36.

Held, E. E. (1963). Moulting behaviour of *Birgus latro. Nature, Lond.* 200, 799–800.

Heller, J. (1975). The taxonomy of some British *Littorina* species with notes on their reproduction (Mollusca: Prosobranchia). *Zool. J. Linn. Soc.* 56, 131–51.

Heller, J. & Magaritz, M. (1983). From where do land snails obtain the chemicals to build their shells? *J. moll. Stud.* 49. (In press.)

Hendelberg, J. (1974). Spermiogenesis, sperm morphology, and the biology of fertilization in the Turbellaria. In *Biology of the Turbellaria*, pp. 148–64, ed. N. W. Riser & M. P. Morse. McGraw-Hill, New York.

Hennig, W. (1965). Phylogenetic systematics. *Ann. Rev. Ent.* 10, 97–116.

Hennig, W. (1969). *Die Stammesgeschichte der Insekten.* Kramer, Frankfurt.

Hennig, W. (1981). *Insect phylogeny*, trans. and ed. A. C. Pont. Revisionary notes by D. Schlee. J. Wiley, Chichester.

Henry, R. P. & Cameron, J. N. (1981). A survey of blood and tissue nitrogen compounds in terrestrial decapods of Palau. *J. exp. Zool.* 218, 83–8.

Henry, R. P., Kormanik, G. A., Smatresk, N. J. & Cameron, J. N. (1981). The role of $CaCO_3$ dissolution as a source of HCO_3^- for the buffering of hypercapnic acidosis in aquatic and terrestrial decapod crustaceans. *J. exp. Biol.* 94, 269–74.

Henwood, K. (1975). A field-tested thermoregulation model for two diurnal Namib Desert tenebrionid beetles. *Ecology* 56, 1329–42.

Herlant-Meewis, H. (1950). Cyst formation in *Aeolosoma hemprichi* (Ehr.). *Biol. Bull. mar. biol. Lab., Woods Hole* 99, 173–80.

Hermans, C. O. (1969). The systematic position of the Archiannelida. *Syst. Zool.* 18, 85–102.

Herreid, C. F. (1969*a*). Water loss of crabs from different habitats. *Comp. Biochem. Physiol.* 28, 829–39.

Herreid, C. F. (1969*b*). Integument permeability of crabs and adaptation to land. *Comp. Biochem. Physiol.* 29, 423–9.

Herreid, C. F. (1978). Metabolism of land snails (*Otala lactea*) during dormancy, arousal, and activity. *Comp. Biochem. Physiol.* 56A, 211–15.

Herreid, C. F. & Gifford, C. A. (1963). The burrow habitat of

the land crab *Cardisoma guanhumi* Latrille. *Ecology* 44, 773–5.

Herreid, C. F., Lee, L. W. & Shah, G. M. (1979). Respiration and heart rate in exercising land crabs. *Resp. Physiol.* 37, 109–20.

Herreid, C. F., O'Mahoney, P. M. & Shah, G. M. (1979). Cardiac and respiratory response to hypoxia in the land crab, *Cardisoma guanhumi* (Latreille). *Comp. Biochem. Physiol.* 63A, 145–51.

Herrnkind, W. F. (1968). Adaptive visually-directed orientation in *Uca pugilator. Am. Zool.* 8, 585–98.

Herrnkind, W. F. (1972). Orientation in shore-living arthropods, especially the sand fiddler crab. In *Behavior of marine animals*, vol. 1, pp. 1–59, ed. H. E. Winn & B. L. Olla. Plenum Press, New York.

Hessler, R. R. (1969). Peracarida. In *Treatise on invertebrate paleontology*, part R, Arthropoda 4 (vol. 1), pp. 360–93, ed. R. C. Moore. Geological Society of America and Kansas University Press.

Heurtault, J. (1973). Contribution à la connaissance biologique et anatomophysiologique des pseudoscorpions. *Bull. Mus. Hist. nat., Paris* 124, 571–670.

Hiatt, R. W. (1948). The biology of the lined shore crab, *Pachygrapsus crassipes* Randall. *Pacific Sci.* 2, 135–213.

Hill, A. E. (1975). Solute-solvent coupling in epithelia: an eléctro-osmotic theory of fluid transfer. *Proc. roy. Soc. Lond.* (B) 190, 115–34.

Hinton, H. E. (1963). The respiratory system of the egg-shell of the blowfly, *Calliphora erythrocephala* Meigen, as seen with the electron microscope. *J. Insect Physiol.* 9, 121–9.

Hinton, H. E. (1967). The structure of the spiracles of the cattle tick, *Boophilus microplus. Aust. J. Zool.* 15, 941–5.

Hinton, H. E. (1968). Reversible suspension of metabolism and the origin of life. *Proc. roy. Soc. Lond.* (B) 171, 43–57.

Hinton, H. E. (1969). Respiratory systems of insect egg-shells. *Ann. Rev. Ent.* 14, 343–68.

Hinton, H. E. (1971). Plastron respiration in the mite, *Platyseius italicus. J. Insect Physiol.* 17, 1185–99.

Hinton, H. E. (1976). Respiratory adaptations of marine insects. In *Marine insects*, pp. 43–78, ed. L. Cheng. North-Holland Publishing Co., Amsterdam.

Hinton, H. E. (1977*a*). Enabling mechanisms. *Proc. Int. Congr. Ent.* 15, 71–83.

Hinton, H. E. (1977*b*). Function of shell structures of pig louse and how the egg maintains a low equilibrium temperature in direct sunlight. *J. Insect Physiol.* 23, 785–800.

Hinton, H. E. & Blum, M. S. (1965). Suspended animation and the origin of life. *New Scient.* 28(467), 270–71.

Hinton, H. E. & Wilson, R. S. (1970). Stridulatory organs in spiny orb-weaver spiders. *J. Zool., Lond.* 162, 481–4.

Hoese, B. (1981). Morphologie und Funktion des Wasserleitungssystems der terrestrischen Isopoden (Crustacea, Isopoda, Oniscoidea). *Zoomorphol.* 98, 135–67.

Hoffman, C. (1965). Bau und Funktion der Trichobothrien von *Euscorpius carpathicus* L. *Z. vergl. Physiol.* 54, 290–352.

Hoffman, R. L. (1969). Myriapoda, exclusive of Insecta. In *Treatise on invertebrate palaeontology*, part R, Arthropoda 4, pp. 572–606, ed. R. C. Moore. Geological Society of America and University of Kansas Press.

Hoffmann, H. (1929). Zur Kenntnis der Onchiiden (Gastrop. Pulmon.). II Teil. Phylogenie und Verbreitung. *Zool. Jb. (Syst.)* 57, 253–302.

Hogben, L. & Kirk, R. L. (1944). Studies on temperature regulation. I. The Pulmonata and Oligochaeta. *Proc. roy. Soc. Lond.* (B) 132, 68–82.

Holdgate, M. W. & Seal, M. (1956). The epicuticular wax layers of the pupa of *Tenebrio molitor* L. *J. exp. Biol.* 33, 82–106.

Holdich, D. M. & Mayes, K. R. (1976). Blood volume and total water content of the woodlouse, *Oniscus asellus*, in conditions of hydration and desiccation. *J. Insect Physiol.* 22, 547–53.

Holmes, W. N. & Donaldson, E. M. (1969). The body compartments and the distribution of electrolytes. In *Fish physiology*, vol. I, pp. 1–89, ed. W. S. Hoar & D. J. Randall. Academic Press, London.

Holtfreter, J. (1943). Properties and functions of the surface coat in amphibian embryos. *J. exp. Zool.* **93**, 251–323.

Horch, K. (1971). An organ for hearing and vibration sense in the ghost crab *Ocypode*. *Z. vergl. Physiol.* **73**, 1–21.

Horch, K. W. & Salmon, M. (1969). Production, perception and reception of acoustic stimuli by semiterrestrial crabs (Genus *Ocypode* and *Uca*, family Ocypodidae). *Forma et Functio* **1**, 1–25.

Horne, F. R. (1968*a*). Survival and ionic regulation of *Triops longicaudatus* in various salinities. *Physiol. Zoöl.* **41**, 180–6.

Horne, F. R. (1968*b*). Nitrogen excretion in Crustacea. I. The herbivorous land crab *Cardisoma guanhumi* Latreille. *Comp. Biochem. Physiol.* **26**, 687–95.

Horne, F. R. (1969). Purine excretion in five scorpions, a uropygid and a centipede. *Biol. Bull. mar. biol. Lab., Woods Hole* **137**, 155–60.

Horne, F. R. (1971). Accumulation of urea by a pulmonate snail during aestivation. *Comp. Biochem. Physiol.* **38A**, 565–70.

Horne, F. R. (1977). Ureotelism in the slug, *Limax flavus* Linné. *J. exp. Zool.* **199**, 227–31.

Horne, F. R. (1979). Comparative aspects of estivating metabolism in the gastropod, *Marisa. Comp. Biochem. Physiol.* **64A**, 309–11.

Horne, F. R. & Barnes, G. (1970). Re-evaluation of urea biosynthesis in prosobranch and pulmonate snails. *Z. vergl. Physiol.* **69**, 452–7.

Horowitz, M. (1970). The water balance of the terrestrial isopod *Porcellio scaber. Ent. exp. appl.* **13**, 173–8.

Houlihan, D. F. (1976). Water transport by the eversible abdominal vesicles of *Petrobius brevistylis. J. Insect Physiol.* **22**, 1683–95.

Houlihan, D. F. (1977). Increased oxygen consumption during the uptake of water by the eversible vesicles of *Petrobius brevistylis. J. Insect Physiol.* **23**, 1285–94.

Houlihan, D. F. (1979). Respiration in air and water of three mangrove snails. *J. exp. mar. Biol. Ecol.* **41**, 143–61.

Howell, B. F. (1962). Worms. In *Treatise on invertebrate paleontology*, part W, Miscellanea, pp. 144–77, ed. R. C. Moore. Geological Society of America and Kansas University Press.

Howes, N. H. & Wells, G. P. (1934). The water relations of snails and slugs. I. Weight rhythms in *Helix pomatia* L. *J. exp. Biol.* **11**, 327–43.

Hoyle, G. & Castilo, J. (1979). Neuromuscular transmission in *Peripatus. J. exp. Biol.* **83**, 13–29.

Hoyle, G. & Williams, M. (1980). The musculature of *Peripatus* and its innervation. *Phil. Trans. roy. Soc. Lond.* **288**, 481–510.

Hsu, M.-H. & Sauer, J. R. (1975). Ion and water balance in the feeding lone star tick. *Comp. Biochem. Physiol.* **52A**, 269–76.

Hubendick, B. (1978). Systematics and comparative morphology of the Basommatophora. In *Pulmonates*, vol. 2A, pp. 1–47, ed. V. Fretter & J. Peake. Academic Press, London.

Hubert, M. (1970). Etude des organes excréteurs chez les diplopodes: morphologie des tubes de Malpighi et des reins labiaux ou maxillaires. *Bull. Soc. zool. Fr.* **95**, 847–61.

Hubert, M. (1979). Données histophysiologiques complémentaires sur les bioaccumulations minérales et puriques chez *Cylindroiulus londinensis* (Leach, 1814) (Diplopode, Iuloidea). *Archs Zool. exp. gén.* **119**, 669–83.

Hubert, M. (1980). Localization and identification of mineral elements and nitrogenous waste in Diplopoda. In *Myriapod biology*, pp. 127–34, ed. M. Camatini. Academic Press, London.

Hubert, M. & Razet, P. (1965). Sur les principaux éléments du catabolisme azoté chez les myriapodes. *C. r. Acad. Sci., Paris* **261**, 797–800.

Huebner, E. & Chee, G. (1978). Histological and ultrastructural specialization of the digestive tract of the intestinal air breather *Hoplosternum thoracatum* (Teleost). *J. Morph.* **157**, 301–28.

Huggins, A. K., Skutsch, G. & Baldwin, E. (1969). Ornithineurea cycle enzymes in teleostean fish. *Comp. Biochem. Physiol.* **28**, 587–602.

Hughes, D. A. (1966). Behavioural and ecological investigations of the crab *Ocypode ceratophthalmus* (Crustacea: Ocypodidae). *J. Zool., Lond.* **150**, 129–43.

Hughes, D. A. (1973). On mating and the 'copulation burrows' of crabs of the genus *Ocypode* (Decapoda, Brachyura). *Crustaceana* **24**, 72–6.

Hughes, G. M. (1967). Evolution between air and water. In *Development of the lung*, pp. 64–84, ed. A. V. S. de Reuck & R. Porter. J. & A. Churchill Ltd, London.

Hughes, G. M. (1976). On the respiration of *Latimeria chalumnae*. *Zool. J. Linn. Soc.* **59**, 195–208.

Hughes, G. M., Knights, B. & Scammell, C. A. (1969). The distribution of pO_2 and hydrostatic pressure changes within the branchial chambers in relation to gill ventilation of the shore crab *Carcinus maenas* L. *J. exp. Biol.* **51**, 203–20.

Hughes, G. M. & Mill, P. J. (1974). Locomotion: terrestrial. In *The physiology of insects* (2nd edition), vol. III, pp. 335–79, ed. M. Rockstein. Academic Press, New York & London.

Hughes, G. M. & Singh, B. N. (1970). Respiration in an airbreathing fish, the climbing perch *Anabas testudineus* Bloch. I. Oxygen uptake and carbon dioxide release into air and water. *J. exp. Biol.* **53**, 265–80.

Hughes, N. F. (1976). *Palaeobiology of angiosperm origins*. Cambridge University Press, Cambridge, England.

Hughes, N. F. & Smart, J. (1967). Plant–insect relationships in Palaeozoic and later time. In *The fossil record*, pp. 107–17, ed. W. B. Harland *et al.* Geological Society, London.

Hughes, T. E. (1959). *Mites, or the Acari*. The Athlone Press, University of London, London.

Humbert, W. (1974). Localisation, structure et genèse des concrétions minérales dans le mésentéron des collemboles Tomoceridae (Insecta, Collembola). *Z. Morph. Tiere* **78**, 93–109.

Humphreys, W. F. (1974). Behavioural thermoregulation in a wolf spider. *Nature, Lond.* **251**, 502–3.

Humphreys, W. F. (1975). The influence of burrowing and thermoregulatory behaviour on the water relations of *Geolycosa godeffroyi* (Araneae: Lycosidae), an Australian wolf spider. *Oecologia* **21**, 291–311.

Hunter, P. J. (1978). Slugs – a study in applied ecology. In *Pulmonates*, vol. 2A, pp. 271–86, ed. V. Fretter & J. F. Peake. Academic Press, London.

Hunter, W. R. (1953). The condition of the mantle cavity in two pulmonate snails living in Loch Lomond. *Proc. roy. Soc. Edinb.* (*B*) **65**, 143–65.

Hunter, W. R. (1964). Physiological aspects of ecology in nonmarine molluscs. In *Physiology of Mollusca*, vol. I, pp. 83–126, ed. K. M. Wilbur & C. M. Yonge. Academic Press, London.

Hurley, D. E. (1968). Transition from water to land in amphipod crustaceans. *Am. Zool.* **8**, 327–53.

Hutchinson, G. E. (1931). Restudy of some Burgess Shale fossils. *Proc. US natn. Mus.* **78** (art. 11), 1–24.

Hutchinson, G. E. (1967). *A treatise on limnology*, vol. II, *Introduction to lake biology and the limnoplankton*. John Wiley & Sons Inc, London.

Huxley, T. H. (1881). *The crayfish. An introduction to the study of zoology*. C. Kegan Paul, London.

Hyatt, A. D. & Marshall, A. T. (1977). Sequestration of haemolymph sodium and potassium by fat body in the water-stressed cockroach *Periplaneta americana. J. Insect Physiol.* **23**, 1437–41.

Hyman, L. H. (1951*a*). *The invertebrates*, vol. II, *Platyhelminthes and Rhynchocoela.* McGraw-Hill, New York.

Hyman, L. H. (1951*b*). *The invertebrates*, vol. III, *Acanthocephala, Aschelminthes, and Entoprocta.* McGraw-Hill, New York.

Hyman, L. H. (1967). *The invertebrates*, vol. VI, *Mollusca I.* McGraw-Hill, New York.

Icely, J. D. & Jones, D. A. (1978). Factors affecting the distribution of the genus *Uca* (Crustacea: Ocypodidae) on an east African shore. *Est. coast. mar. Sci.* 6, 315-25.

Imms, A. D. (1936). The ancestry of insects. *Trans. Soc. Brit. Ent.* 3, 1-32.

Imms, A. D. (1947). *Insect natural history.* Collins, London.

Inger, R. F. (1957). Ecological aspects of the origin of the tetrapods. *Evolution* 11, 373-6.

Irvine, H. B. (1969). Sodium and potassium excretion by isolated insect Malpighian tubules. *Am. J. Physiol.* 217, 1520-7.

Jans, D. & Ross, K. F. A. (1963). A histological study of the peripheral receptors in the thorax of land isopods, with special reference to the location of possible hygroreceptors. *Q. Jl microsc. Sci.* 104, 337-50.

Jarvik, E. (1968*a*). Aspects of vertebrate phylogeny. In *Current problems of lower vertebrate phylogeny*, pp. 497-527, ed. T. Ørvig. Wiley Interscience, London.

Jarvik, E. (1968*b*). The systematic position of the Dipnoi. In *Current problems of lower vertebrate phylogeny*, pp. 223-45, ed. T. Ørvig. Wiley Interscience, London.

Jarvik, E. (1980). *Basic structure and evolution of vertebrates.* 2 vols. Academic Press, London.

Jarvik, E. (1981). Review of 'Lungfishes, tetrapods, paleontology, and plesiomorphy'. *Syst. Zool.* 30, 378-84.

Jarvis, J. H. & King, P. E. (1973). Ultrastructure of the photoreceptors in the pycnogonid species, *Nymphon gracile* (Leach) and *Pycnogonum littorale* (Strom). *Mar. Behav. Physiol.* 2, 1-13.

Jefferies, R. P. S. (1975). Fossil evidence concerning the origin of the chordates. *Symp. zool. Soc. Lond.* 36, 253-318.

Jefferies, R. P. S. (1981). In defence of the calcichordates. *Zool. J. Linn. Soc.* 73, 351-96.

Johannes, R. E. (1981). *Words of the lagoon. Fishing and marine lore in the Palau district of Micronesia.* University of California Press, Berkeley.

Johansen, K. (1966). Air breathing in the teleost *Symbranchus marmoratus. Comp. Biochem. Physiol.* 18, 383-95.

Johansen, K. (1970). Air breathing in fishes. In *Fish physiology*, vol. IV, pp. 361-411, ed. W. S. Hoar & D. J. Randall. Academic Press, London.

Johansen, K. & Hanson, D. (1968). Functional anatomy of the hearts of lungfishes and amphibians. *Am. Zool.* 8, 191-210.

Johansen, K., Hanson, D. & Lenfant, C. (1970). Respiration in a primitive air breather, *Amia calva. Resp. Physiol.* 9, 162-74.

Johansen, K. & Lenfant, C. (1967). Respiratory function in the South American lungfish, *Lepidosiren paradoxa. J. exp. Biol.* 46, 205-18.

Johansen, K. & Lenfant, C. (1968). Respiration in the African lungfish *Protopterus aethiopicus.* II. Control of breathing. *J. exp. Biol.* 49, 453-68.

Johansen, K., Lenfant, C. & Grigg, G. C. (1967). Respiratory control in the lungfish, *Neoceratodus forsteri* (Krefft). *Comp. Biochem. Physiol.* 20, 835-54.

Johansen, K., Lenfant, C. & Hanson, D. (1968). Cardiovascular dynamics in the lungfish. *Z. vergl. Physiol.* 59, 157-86.

Johansen, K., Lenfant, C., Schmidt-Nielsen, K. & Petersen, J. A. (1968). Gas exchange and control of breathing in the electric eel, *Electrophorus electricus. Z. vergl. Physiol.* 61, 137-63.

Johansen, K., Lomholt, J. P. & Maloiy, G. M. O. (1976). Importance of air and water breathing in relation to size of the African lungfish *Protopterus amphibius* Peters. *J. exp. Biol.* 65, 395-9.

Johansen, K. & Martin, A. W. (1966). Circulation in a giant earthworm, *Glossoscolex giganteus.* II. Respiratory properties

of the blood and some patterns of gas exchange. *J. exp. Biol.* 45, 165-72.

Johnels, A. G. & Svensson, G. S. O. (1954). On the biology of *Protopterus annectens* (Owen). *Ark. Zool.* 7, 131-64.

Johnson, I. T. & Riegel, J. A. (1977*a*). Ultrastructural studies on the Malpighian tubule of the pill millipede. *Glomeris marginata* (Villers). *Cell Tiss. Res.* 180, 357-66.

Johnson, I. T. & Riegel, J. A. (1977*b*). Ultrastructural tracer studies on the permeability of the Malpighian tubule of the pill millipede, *Glomeris marginata* (Villers). *Cell Tiss. Res.* 182, 549-56.

Johnson, W. H., Delanney, L. E., Williams, E. C. & Cole, T. A. (1977). *Principles of zoology* (2nd edition). Holt, Rinehart & Winston, New York.

Jollie, M. (1982). What are the 'Calcichordata'? and the larger question of the origin of chordates. *Zool. J. Linn. Soc.* 75, 167-88.

Jones, H. D. (1978*a*). Observations on the locomotion of two British terrestrial planarians (Platyhelminthes, Tricladida). *J. Zool., Lond.* 186, 407-16.

Jones, H. D. (1978*b*). Fluid skeletons in aquatic and terrestrial animals. In *Comparative physiology: water, ions and fluid mechanics*, pp. 267-81, ed. K. Schmidt-Nielsen, L. Bolis & S. H. P. Maddrell. Cambridge University Press, Cambridge, England.

Jones, J. D. (1961). Aspects of respiration in *Planorbis corneus* L. and *Lymnaea stagnalis* L. (Gastropoda: Pulmonata). *Comp. Biochem. Physiol.* 4, 1-29.

Jones, L. L. (1941). Osmotic regulation in several crabs of the Pacific coast of North America. *J. cell. comp. Physiol.* 18, 79-92.

de Jorge, F. B., Petersen, J. A. & Ditadi, A. S. F. (1969). Variations in nitrogenous compounds in the urine of *Strophocheilus* (Pulmonata, Mollusca) with different diets. *Experientia* 25, 614-15.

de Jorge, F. B., Petersen, J. A. & Ditadi, A. S. F. (1970). Influence of dormancy on the chemical composition of *Strophocheilus* (Strophocheilidae, Pulmonata, Mollusca). *Experientia* 26, 41-3.

Juberthie-Jupeau, L. (1976). Fine structure of postgonopodial glands of a myriapod *Glomeris marginata* (Villers). *Tiss. Cell* 8, 293-304.

Jungreis, A. M. (1976). Partition of excretory nitrogen in Amphibia. *Comp. Biochem. Physiol.* 53A, 133-41.

Junqua, C. (1966). Recherches biologiques et histophysiologiques sur un solifuge saharien *Othoes saharae* Panouse. *Mem. Mus. Hist. nat., Paris* 43, 1-124.

Jurgens, J. D. (1971). The morphology of the nasal region of Amphibia and its bearing on the phylogeny of the group. *Ann. Univ. Stellenbosch* 46A, 1-146.

Kaczmarek, M. (1977). Comparison of the role of Collembola in different habitats. *Ecol. Bull., Stockholm* 25, 64-74.

Kaestner, A. (1968). *Invertebrate zoology*, vol. II, *Arthropod relatives, Chelicerata, Myriapoda*, trans. H. W. & L. H. Levi. Interscience Publishers, London.

Kaestner, A. (1970). *Invertebrate zoology*, vol. III, *Crustacea*, trans. H. W. Levi & L. R. Levi. Interscience Publishers, New York & London.

Kamemoto, F. I., Spalding, A. E. & Keister, S. M. (1962). Ionic balance in blood and coelomic fluid of earthworms. *Biol. Bull. mar. biol. Lab., Woods Hole* 122, 228-31.

Kammer, A. E. & Bracchi, J. (1973). Role of the wings in the absorption of radiant energy by a butterfly. *Comp. Biochem. Physiol.* 45A, 1057-63.

Kammer, A. E. & Heinrich, B. (1978). Insect flight metabolism. *Adv. Insect Physiol.* 13, 133-228.

Kanungo, M. S. (1955). Physiology of the heart of a scorpion. *Nature, Lond.* 176, 980-1.

Kanungo, M. S., Bohidar, S. C. & Patnaik, B. K. (1962). Excretion in the scorpion, *Palamnaeus bengalensis* C. Koch. *Physiol. Zoöl.* 35, 201-3.

Kanwisher, J. W. (1966). Tracheal gas dynamics in pupae of the cecropia silkworm. *Biol. Bull. mar. biol. Lab., Woods Hole* **130**, 96-105.

Karling, T. G. (1958). Zur Kenntnis von *Stygocapitella subterranea* Knöllner und *Parergodrilus heideri* Reisinger (Annelida). *Ark. Zool.* **11**, 307-42.

Kasinathan, R. (1975). Some studies of five species of cyclophorid snails from peninsular India. *Proc. malac. Soc. Lond.* **41**, 379-94.

Kästner, A. (1929). Bau und Funktion der Fächertracheen einiger Spinnen. *Z. Morph. Okol. Tiere* **13**, 463-558.

Katz, U. & Graham, R. (1980). Water relations in the toad (*Bufo viridis*) and a comparison with the frog (*Rana ridibunda*). *Comp. Biochem. Physiol.* **67A**, 245-51.

Kaufman, S. E., Kaufman, W. R. & Phillips, J. E. (1981). Fluid balance in the argasid tick, *Ornithodorus moubata*, fed on modified blood meals. *J. exp. Biol.* **93**, 225-42.

Kaufman, S. E., Kaufman, W. R. & Phillips, J. E. (1982). Mechanism and characteristics of coxal fluid excretion in the argasid tick *Ornithodorus moubata. J. exp. Biol.* **98**, 343-52.

Kaufman, W. R. & Phillips, J. E. (1973). Ion and water balance in the ixodid tick *Dermacentor andersoni* I. Routes of ion and water excretion. *J. exp. Biol.* **58**, 523-36.

Kawaguti, S. (1932). On the physiology of land planarians. III. The problems of desiccation. *Mem. Fac. Sci. Agr., Taihoku Imp. Univ., Formosa*, 7, 39-55.

Keil, T. (1976). Sinnesorganes auf den Antennen von *Lithobius forficatus* L. (Myriapoda, Chilopoda). 1. Die Funktionsmorphologie der 'Sensilla trichoidea'. *Zoomorphol.* **84**, 77-102.

Keilin, D. (1959). The problem of anabiosis or latent life: history and current concept. *Proc. roy. Soc. Lond.* (*B*) **150**, 149-91.

Kemp, T. S. (1980). Origin of the mammal-like reptiles. *Nature, Lond.* **283**, 378-80.

Kerkut, G. A. & Walker, R. J. (1975). Nervous system, eye and statocyst. In *Pulmonates*, vol. 1, pp. 165-244, ed. V. Fretter & J. Peake. Academic Press, London.

Kerney, M. P. & Cameron, R. A. D. (1979). *A field guide to the land snails of Britain and north-west Europe*. Collins, London.

Kevan, D. K. McE. (1962). *Soil animals*. H. F. & G. Witherby Ltd, London.

Kilian, E. F. (1951). Untersuchungen zur Biologie von *Pomatias elegans* (Müller) und ihrer 'Konkrementdrüse'. *Arch. Molluskenk.* **80**, 1-16.

King, P. E. (1973). *Pycnogonids*. Hutchinson, London.

Kirby, P. K. & Harbaugh, R. D. (1974). Diurnal patterns of ammonia release in marine and terrestrial isopods. *Comp. Biochem. Physiol.* **47A**, 1313-21.

Kirsch, R. & Nonnotte, G. (1977). Cutaneous respiration in three freshwater teleosts. *Resp. Physiol.* **29**, 339-54.

Kirsteuer, E. (1971). The interstitial nemertean fauna of marine sand. In *Proceedings of the first international conference on meiofauna*, pp. 17-19, ed. N. C. Hulings. Smithsonian Contributions to Zoology, No. 76, Washington.

Kitazawa, Y. (1971). Biological regionality of the soil fauna and its function in forest ecosystem types. In *Productivity of forest ecosystems*, pp. 485-98, ed. P. Duvigneaud. Unesco, Paris.

Klaassen, F. (1973). Stridulation und Kommunikation durch Substratschall bei *Gecarcinus lateralis* (Crustacea Decapoda). *J. comp. Physiol.* **83**, 73-9.

Klaassen, F. (1975). Okologische und ethologische Untersuchungen zur Fortpflanzungsbiologie von *Gecarcinus lateralis* (Decapoda, Brachyura). *Forma et Functio* **8**, 101-74.

Klaver, E. (1975). Some aspects of the reproductive biology of *Bourletiella* (*Cassagnaudiella*) *pruinosa* (Tullberg, 1871) (Collembola: Sminthuridae). *Bull. zool. Mus. Univ. Amsterdam* **4**, 179-86.

Klekowski, R. Z. (1961). Die Resistenz gegan Austrocknung bei einigen Wirbellosen aus astatischen Gewässern. *Verh. Int. Ver. Limnol.* **14**, 1023-8.

Klekowski, R. Z. (1963). Water balance and osmoregulation in the snail *Coretus corneus* (L.) under conditions of desiccation and in diluted sea water. *Polskie Archwm Hydrobiol.* **11**, 219-40.

Klika, E. & Lelek, A. (1967). A contribution to the study of the lungs of the *Protopterus annectens* and *Polypterus senegalensis. Folia Morph., Prague* **15**, 168-75.

Klingel, H. (1962). Das Paarungsverhalten des malaischen Höhlentausendfusses *Thereuopoda decipiens cavernicola* Verhoeff (Scutigeromorpha, Chilopoda). *Zool. Anz.* **169**, 458-60.

Knight, J. B., Cox, L. R., Keen, A. M., Batten, R. L., Yochelson, E. L. & Robertson, R. (1960). Systematic descriptions. In *Treatise on invertebrate palaeontology*, part I, Mollusca (vol. I), pp. 169-331, ed. R. C. Moore. Geological Society of America, Inc. & University of Kansas Press.

Knowles, G. (1975). The reduced glucose permeability of the isolated Malpighian tubules of the blowfly *Calliphora vomitoria. J. exp. Biol.* **62**, 327-40.

Knülle, W. (1965). Die Sorption und Transpiration des Wasserdampfes bei der mehlmilbe (*Acarus siro* L.). *Z. vergl. Physiol.* **49**, 586-604.

Koechlin, N. (1979). In situ perfusion study of sodium transport across the nephridia of a marine invertebrate *Sabella pavonia* S. (Polychaeta; Annelida). *Comp. Biochem. Physiol.* **62A**, 797-800.

Koefoed-Johnsen, V. (1979). Control mechanisms in amphibians. In *Mechanisms of osmoregulation in animals*, pp. 223-72, ed. R. Gilles. Wiley, Chichester.

Koepcke, H.-W. & Koepcke, M. (1953). Contribution al conocimiento de la forma de vida de *Ocypode gaudichaudii* Milne Edwards et Lucas (Decapoda, Crustacea). *Publ. Mus. Hist. nat. 'Javier Prado' (Lima)* (Ser. A., Zool.) **13**, 1-46.

Kormanik, G. A. & Harris, R. R. (1981). Salt and water balance and antennal gland function in three Pacific species of terrestrial crab (*Gecarcoidea lalandii, Cardisoma carnifex, Birgus latro*). I. Urine production and salt exchanges in hydrated crabs. *J. exp. Zool.* **218**, 97-105.

Korte, R. (1965). Durch polarisiertes Licht hervorgernfene Optomotorik bei *Uca tangeri. Experientia* **21**, 98.

Korte, R. (1966). Untersuchungen zum Sehrvermogen einiger Dekapoden, insbesondere von *Uca tangeri. Z. Morph. Okol. Tiere* **58**, 1-37.

Krafsur, E. S. (1971). Behaviour of thoracic spiracles of *Aedes* mosquitoes in controlled relative humidities. *Ann. ent. Soc. Am.* **64**, 93-7.

Kraus, O. (1974). On the morphology of Palaeozic diplopods. *Symp. zool. Soc. Lond.* **32**, 13-22.

Kraus, O. (1976). Zur phylogenetischen Stellung und Evolution der Chelicerata. *Ent. Germ.* **3**, 1-12.

Krishnamoorthy, R. V. & Srihari, K. (1973). Changes in the excretory patterns of the fresh-water field crab *Paratelphusa hydrodromous* upon adaptations to higher salinities. *Mar. Biol.* **21**, 341-8.

Krishnan, G. (1952). On the nephridia of Nereidae in relation to habitat. *Proc. natn. Inst. Sci. India.* **18**, 241-55.

Krishnan, G. & Sundara Rajulu, G. (1964). Study of the cuticle of the annectant symphylid *Polyxenella krishnani* together with observations on its phylogenetic significance. *Z. Naturforsch.* **19b**, 640-5.

Kristensen, N. P. (1975). The phylogeny of hexapod 'orders'. A critical review of recent accounts. *Z. zool. Syst. Evol.-Forsch.* **13**, 1-44.

Kristensen, N. P. (1981). Phylogeny of insect orders. *Ann. Rev. Ent.* **26**, 135-57.

Krogh, A. (1939). *Osmotic regulation in aquatic animals*. Cambridge University Press, Cambridge, England.

Krogh, A., Schmidt-Nielsen, K. & Zeuthen, E. (1938). The osmotic behaviour of frogs eggs and young tadpoles. *Z. vergl. Physiol.* **26**, 230-8.

Kromhout, G. A. (1943). A comparison of the protonephridia of fresh-water, brackish-water and marine specimens of *Gyratrix hermaphroditus. J. Morph.* **72**, ...

Kuenen, D. J. (1959). Excretion and water balance in some land isopods. *Ent. exp. appl.* **2**, 287–94.

Kühnelt, W. (1976). *Soil biology with special reference to the animal kingdom* (2nd edition), trans. N. Walker, with contributions by J. W. Butcher & C. Laughlin. Faber & Faber, London.

Kümmel, G. (1975). The physiology of protonephridia. *Fortschr. Zool.* **23**, 18–32.

Kümmel, G. (1981). Fine structural indications of an osmoregulatory function of the 'gills' in terrestrial isopods (Crustacea, Oniscoidea). *Cell Tiss. Res.* **214**, 663–66.

Lagarrigue, J.-G. (1969). Composition ionique de l'hémolymphe des oniscoïdes. *Bull. Soc. zool. Fr.* **94**, 137–44.

Lagerspetz, K. & Jäynäs, E. (1958). The behavioural regulation of the water content in *Linyphia montana* (Aran., Linyphiidae) and some other spiders. *Ann. Ent. Fenn.* **25**, 210–33.

Lall, A. B. & Chapman, R. M. (1973). Phototaxis in *Limulus* under natural conditions: evidence for reception of near-ultraviolet light in the median dorsal ocellus. *J. exp. Biol.* **58**, 213–24.

Land, M. F. (1969a). Structure of the retinae of the principal eyes of jumping spiders (Salticidae: Dendryphantinae) in relation to visual optics. *J. exp. Biol.* **51**, 443–70.

Land, M. F. (1969b). Movements of the retinae of jumping spiders (Salticidae: Dendryphantinae) in response to visual stimuli. *J. exp. Biol.* **51**, 471–93.

Land, M. F. (1972). Stepping movements made by jumping spiders during turns mediated by the lateral eyes. *J. exp. Biol.* **57**, 15–40.

Lanzavecchia, G. & Camatini, M. (1980). Phylogenetic problems and muscle cell ultrastructure in Onychophora. In *Myriapod biology*, pp. 407–17, ed. M. Camatini. Academic Press, London.

Lauterbach, K.-E. (1972). Die morphologischen Grundlagen für die Entstehung der Entognathie bei den apterygoten Insekten in phylogenetische Sicht. *Zool. Beitr.* (N.F.) **18**, 25–69.

Lavallard, R. (1965). Etude au microscope électronique de l'épithélium tégumentaire chez *Peripatus acacioi* Marcus et Marcus. *C. r. Acad. Sci., Paris* **260**, 965–8.

Lavallard, R. & Campiglia, S. S. (1975). Contribution to the biology of *Peripatus acacioi* Marcus and Marcus (Onychophora). V. Studies of the breeding in a laboratory culture. *Zool. Anz.* **195**, 338–50.

Lavallard, R., Campiglia, S., Parisi Alvares, E. & Valle, C. M. C. (1975). Contribution à la biologie de *Peripatus acacioi* Marcus et Marcus. III. Etude descriptive de l'habitat. *Vie et Milieu* **25**, 87–118.

Laverack, M. S. (1963). *The physiology of earthworms*. Pergamon, London.

Lawrence, R. F. (1953). *The biology of the cryptic fauna of forests, with special reference to the indigenous forests of South Africa*. A. A. Balkema, Cape Town & Amsterdam.

Lebour, M. V. (1945). The eggs and larvae of some prosobranchs from Bermuda. *Proc. zool. Soc. Lond.* **114**, 462–89.

Lee, D. L. (1970). The fine structure of the excretory system in adult *Nippostrongylus brasiliensis* (Nematoda) and a suggested function for the 'excretory glands'. *Tiss. Cell* **2**, 225–31.

Lee, D. L. & Atkinson, H. J. (1976). *Physiology of nematodes* (2nd edition). Macmillan, London.

de Leersnyder, N. & Hoestlandt, H. (1963). Premières données sur la régulation osmotique et la régulation ionique du crabe terrestre, *Cardisoma armatum* Herklots. *Cah. Biol. mar.* **4**, 211–18.

Lees, A. D. (1946a). The water balance in *Ixodes ricinus* L. and certain other species of ticks. *Parasitology* **37**, 1–20.

Lees, A. D. (1946b). Chloride regulation and the function of the coxal gland in ticks. *Parasitology* **37**, 172–84.

Lees, A. D. (1947). Transpiration and the structure of the epicuticle in ticks. *J. exp. Biol.* **23**, 379–410.

Lenfant, C. & Johansen, K. (1968). Respiration in the African lungfish *Protopterus aethiopicus*. I. Respiratory properties of blood and normal patterns of breathing and gas exchange. *J. exp. Biol.* **49**, 437–52.

Lenfant, C., Johansen, K. & Grigg, G. C. (1966). Respiratory properties of blood and pattern of gas exchange in the lungfish *Neoceratodus forsteri* (Krefft). *Resp. Physiol.* **2**, 1–21.

Leslie, C. J. (1951). Mating behaviour of leeches. *J. Bombay nat. Hist. Soc.* **50**, 422–3.

Levi, H. W. (1967). Adaptations of respiratory systems of spiders. *Evolution* **21**, 571–83.

Lewis, J. G. E. (1961). The life history and ecology of the littoral centipede *Strigamia* (=*Scolioplanes*) *maritima* (Leach). *Proc. zool. Soc. Lond.* **137**, 221–48.

Lewis, J. G. E. (1963). On the spiracle structure and resistance to desiccation of four species of geophilomorph centipede. *Ent. exp. appl.* **6**, 89–94.

Lewis, J. G. E. (1965). The food and reproductive cycles of the centipedes *Lithobius variegatus* and *Lithobius forficatus* in a Yorkshire woodland. *Proc. zool. Soc. Lond.* **144**, 269–83.

Lewis, J. G. E. (1971a). The life history and ecology of the millipede *Tymbodesmus falcatus* (Polydesmida: Gomphodesmidae) in northern Nigeria with notes on *Sphenodesmus sheribongensis*. *J. Zool., Lond.* **164**, 551–63.

Lewis, J. G. E. (1971b). The life history and ecology of three paradoxosomatid millipedes (Diplopoda: Polydesmida) in northern Nigeria. *J. Zool., Lond.* **165**, 431–52.

Lewis, J. G. E. (1981). *The biology of centipedes*. Cambridge University Press, Cambridge, England.

Lewis, J. R. (1964). *The ecology of rocky shores*. English Universities Press Ltd, London.

Lieber, A. (1931). Beitrag zur Kenntnis eines arboricolen Feuchtland-Nereiden aus Amboina. *Zool. Anz.* **96**, 255–65.

Liesenfeld, F. J. (1961). Über Leistung und Sitz des Erschutterungssinnes von Netzspinnen. *Biol. Zbl.* **80**, 466–75.

Lindqvist, O. V. (1971). Evaporation in terrestrial isopods is determined by oral and anal discharge. *Experientia* **27**, 1496–8.

Lindqvist, O. V. (1972). Components of water loss in terrestrial isopods. *Physiol. Zoöl.* **45**, 316–24.

Lindqvist, O. V. & Fitzgerald, G. (1976). Osmotic interrelationship between blood and gut fluid in the isopod *Porcellio scaber* Latr. (Crustacea). *Comp. Biochem. Physiol.* **53A**, 57–9.

Lindqvist, O. V., Salminen, I. & Winston, P. W. (1972). Water content and water activity in the cuticle of terrestrial isopods. *J. exp. Biol.* **56**, 49–55.

Linke, O. (1935). Zur Morphologie und Physiologie des Genitalapparates der süsswasserlittorinide *Cremnoconchus syhadrensis* Blanford. *Arch. Naturgesch.* (N.F.) **4** 72–87.

Linsenmair, K. E. (1967). Konstruktion und Signalfunktion der Sandpyramide der Reiterkrabbe *Ocypode saratan* Forsk. (Decapoda Brachyura Ocypodidae). *Z. Tierpsychol.* **24**, 403–56.

Lissmann, H. W. (1945a). The mechanism of locomotion in gastropod molluscs. I. Kinematics. *J. exp. Biol.* **21**, 58–69.

Lissmann, H. W. (1945b). The mechanism of locomotion in gastropod molluscs. II. Kinetics. *J. exp. Biol.* **22**, 37–50.

Little, C. (1965a). Osmotic and ionic regulation in the prosobranch gastropod mollusc, *Viviparus viviparus* Linn. *J. exp. Biol.* **43**, 23–37.

Little, C. (1965b). The formation of urine by the prosobranch gastropod mollusc *Viviparus viviparus* Linn. *J. exp. Biol.* **43**, 39–54.

Little, C. (1967). Ionic regulation in the queen conch, *Strombus gigas* (Gastropoda, Prosobranchia). *J. exp. Biol.* **46**, 459–74.

Little, C. (1968). Aestivation and ionic regulation in two species of *Pomacea* (Gastropoda, Prosobranchia). *J. exp. Biol.* **48**, 569–85.

Little, C. (1972). The evolution of kidney function in the Neritacea (Gastropoda, Prosobranchia). *J. exp. Biol.* **56**, 249–61.

Little, C. (1981). Osmoregulation and excretion in prosobranch gastropods. Part 1. Physiology and biochemistry. *J. moll. Stud.* **47**, 221–47.

Little, C. & Andrews, E. B. (1977). Some aspects of excretion and osmoregulation in assimineid snails. *J. moll. Stud.* **43**, 265–85.

Locke, M. (1974). The structure and formation of the integument in insects. In *The physiology of Insecta* (2nd edition), vol. 6, pp. 123–213, ed. M. Rockstein. Academic Press, London.

Lockwood, A. P. M. (1959a). The osmotic and ionic regulation of *Asellus aquaticus* (L.). *J. exp. Biol.* **36**, 546–55.

Lockwood, A. P. M. (1959b). The extra-haemolymph sodium of *Asellus aquaticus* (L.). *J. exp. Biol.* **36**, 562–5.

Lockwood, A. P. M. (1961). The urine of *Gammarus duebeni* and *G. pulex. J. exp. Biol.* **38**, 647–58.

Lockwood, A. P. M. & Croghan, P. C. (1957). The chloride regulation of the brackish and fresh-water races of *Mesidotea entomon* (L.). *J. exp. Biol.* **34**, 253–8.

Lockwood, A. P. M. & Croghan, P. C. (1959). Composition of the haemolymph of *Petrobius maritimus* Leach. *Nature, Lond.* **184**, 370–1.

Lockwood, A. P. M. & Riegel, J. A. (1969). The excretion of magnesium by *Carcinus maenas. J. exp. Biol.* **51**, 575–89.

Loest, R. A. (1979a). Ammonia volatilization and absorption by terrestrial gastropods: a comparison between shelled and shell-less species. *Physiol. Zoöl.* **52**, 461–9.

Loest, (1979b). Ammonia-forming enzymes and calcium-carbonate deposition in terrestrial pulmonates. *Physiol. Zoöl.* **52**, 470–83.

Loewe, R. & de Eggert, H. B. (1979). Blood gas analysis and acid–base status in the haemolymph of a spider (*Eurypelma californicum*) – influence of temperature. *J. comp. Physiol.* **134**, 331–8.

Loewe, R. & Linzen, B. (1975). Haemocyanins in spiders. II. Automatic recording of oxygen binding curves, and the effect of Mg^{++} on oxygen affinity, cooperativity and subunit association of *Cupiennius salei* haemocyanin. *J. comp. Physiol.* (*B*) **98**, 147–56.

Loewe, R, Linzen, B. & von Stackelberg, W. (1970). Die gelösten Stoffe in der Hämolymphe einer Spinne, *Cupiennius salei. Z. vergl. Physiol.* **66**, 27–34.

Lofts, B. (1974). Reproduction. In *Physiology of Amphibia*, vol. 2, pp. 107–218, ed. B. Lofts. Academic Press, London.

Lombard, R. E. & Bolt, J. R. (1979). Evolution of the tetrapod ear: an analysis and reinterpretation. *Biol. J. Linn. Soc.* **11**, 19–76.

Loveridge, J. P. (1968). The control of water loss in *Locusta migratoria migratorioides* R & F. II. Water loss through the spiracles. *J. exp. Biol.* **49**, 15–29.

Loveridge, J. P. (1974). Studies on the water relations of adult locusts. II. Water gain in the food and loss in the faeces. *Trans. Rhodesia Sci. Ass.* **56**, 1–30.

Løvtrup, S. (1960). Water permeation in the amphibian embryo. *J. exp. Zool.* **145**, 139–50.

Luckett, W. P. (1976). Ontogeny of amniote fetal membranes and their application to phylogeny. In *Major patterns in vertebrate evolution*, pp. 439–516, ed. M. K. Hecht, P. C. Goody & B. M. Hecht. Plenum Press, London.

Lutz, P. L. (1969). Salt and water balance in the West African freshwater/land crab *Sudanonautes africanus africanus* and the effects of desiccation. *Comp. Biochem. Physiol.* **30**, 469–80.

Lutz, P. L., Bentley, T. B., Harrison, K. E. & Marszalek, D. S. (1980). Oxygen and water vapour conductance in the shell and shell membrane of the American crocodile egg. *Comp. Biochem. Physiol.* **66A**, 335–8.

Macallum, A. B. (1926). The paleochemistry of the body fluids and tissues. *Physiol. Rev.* **6**, 316–57.

McCann, C. (1937). Notes on the common land crab *Paratelphusa* (*Barytelphusa*) *guerini* (M.-Eds.) of Salsette Island. *J. Bombay nat. Hist. Soc.* **39**, 531–42.

McClanahan, L. (1967). Adaptation of the spadefoot toad, *Scaphiopus couchii*, to desert environments. *Comp. Biochem. Physiol.* **20**, 73–99.

McClanahan, L. (1972). Changes in body fluids of burrowed spadefoot toads as a function of soil water potential. *Copeia* 1972, 209–16.

McClary, A. (1964). Surface inspiration and ciliary feeding in *Pomacea paludosa* (Prosobranchia: Mesogastropoda: Ampullariidae). *Malacologia* **2**, 87–104.

McColl, H. P. (1974). The arthropods of the floors of six forest types on the west coast, South Island: a preliminary report. *Proc. NZ ecol. Soc.* **21**, 11–16.

MacDonald, J. D., Pike, R. B. & Williamson, D. I. (1957). Larvae of the British species of *Diogenes, Pagurus, Anapagurus* and *Lithodes* (Crustacea, Decapoda). *Proc. zool. Soc. Lond.* **128**, 209–57.

McEnroe, W. D. (1961). Guanine excretion by the two-spotted spider mite (*Tetranychus telarius* (L.)). *Ann. ent. Soc. Am.* **54**, 925–6.

Machan, L. (1968). Spectral sensitivity of scorpion eyes and the possible role of shielding pigment effect. *J. exp. Biol.* **49**, 95–105.

Machin, J. (1964). The evaporation of water from *Helix aspersa.* I. The nature of the evaporating surface. *J. exp. Biol.* **41**, 759–69.

Machin, J. (1966). The evaporation of water from *Helix aspersa.* IV. Loss from the mantle of the inactive snail. *J. exp. Biol.* **45**, 269–78.

Machin, J. (1972). Water exchange in the mantle of a terrestrial snail during periods of reduced evaporative loss. *J. exp. Biol.* **57**, 103–11.

Machin, J. (1975). Water relationships. In *Pulmonates*, vol. 1, pp. 105–163; ed. V. Fretter & J. Peake. Academic Press, London.

Machin, J. (1976). Passive exchange during water vapour absorption in mealworms (*Tenebrio molitor*): a new approach to studying the phenomenon. *J. exp. Biol.* **65**, 603–15.

Machin, J. (1979a). Compartmental osmotic pressures in the rectal complex of *Tenebrio* larvae: evidence for a single tubular pumping site. *J. exp. Biol.* **82**, 123–37.

Machin, J. (1979b). Atmospheric water absorption in arthropods. *Adv. Insect Physiol.* **14**, 1–48.

McIntyre, A. D. (1969). Ecology of marine meiobenthos. *Biol. Rev.* **44**, 245–90.

McKanna, J. A. (1968). Fine structure of the protonephridial system in planaria. I. Flame cells. *Z. Zellforsch. mikrosk. Anat.* **92**, 509–23.

Mackenzie, F. T. (1975). Sedimentary cycling and the evolution of sea water. In *Chemical oceanography* (2nd edition), vol. I, pp. 309–64, ed. J. P. Riley & G. Skirrow. Academic Press, London.

MacLean, S. F., Douce, G. K., Morgan, E. A. & Skeel, M. A. (1977). Community organization in the soil invertebrates of Alaskan arctic tundra. *Ecol. Bull., Stockholm* **25**, 90–101.

McMahon, B. R. (1969). A functional analysis of the aquatic and aerial respiratory movements of an African lungfish, *Protopterus aethiopicus*, with reference to the evolution of the lung-ventilation mechanism in vertebrates. *J. exp. Biol.* **51**, 407–30.

McMahon, B. R. (1970). The relative efficiency of gaseous exchange across the lungs and gills of an African lungfish *Protopterus aethiopicus. J. exp. Biol.* **52**, 1–15.

McMahon, B. R. & Burggren, W. W. (1979). Respiration and adaptation to the terrestrial habitat in the land hermit crab, *Coenobita clypeatus. J. exp. Biol.* **79**, 265–81.

McMahon, B. R. & Burggren, W. W. (1981). Acid–base balance following temperature acclimation in land crabs. *J. exp. Zool.* **218**, 45–52.

McMahon, R. F. & Russell-Hunter, W. D. (1977). Temperature relations of aerial and aquatic respiration in six littoral snails in relation to their vertical zonation. *Biol. Bull. mar. biol. Lab., Woods Hole* **152**, 182–98.

McMahon, R. F. & Russell-Hunter, W. D. (1981). The effects of physical variables and acclimation on survival and oxygen consumption in the high littoral salt-marsh snail *Melampus bidentatus* Say. *Biol. Bull. mar. biol. Lab., Woods Hole* 161, 246-69.

MacMillen, R. E. & Greenaway, P. (1978). Adjustments of energy and water metabolism to drought in an Australian arid-zone crab. *Physiol. Zoöl.* 51, 231-40.

McMullen, H. L., Sauer, J. R. & Burton, R. L. (1976). Possible role in uptake of water vapour by ixodid tick salivary glands. *J. Insect Physiol.* 22, 1281-5.

McNabb, R. A. & McNabb, F. M. A. (1975). Urate excretion by the avian kidney. *Comp. Biochem. Physiol.* 51A, 253-8.

Macnae, W. (1968). A general account of the fauna and flora of mangrove swamps and forests in the Indo-West-Pacific region. *Adv. mar. Biol.* 6, 73-270.

Maddrell, S. H. P. (1971). The mechanisms of insect excretory systems. *Adv. Insect Physiol.* 8, 199-331.

Maddrell, S. H. P. (1977). Insect Malpighian tubules. In *Transport of ions and water in animals*, pp. 541-69, ed. B. L. Gupta, R. B. Moreton, J. L. Oschman & B. J. Wall. Academic Press, London.

Maddrell, S. H. P. (1981). The functional design of the insect excretory system. *J. exp. Biol.* 90, 1-15.

Maetz, J., Payan, P. & de Renzis, G. (1976). Controversial aspects of ionic uptake in freshwater animals. In *Perspectives in experimental biology*, vol. I, pp. 77-92, ed. P. Spencer Davies. Pergamon Press, Oxford.

Magni, F., Papi, F., Savely, H. E. & Tongiorgi, P. (1965). Research on the structure and physiology of the eyes of a lycosid spider. III. Electroretinographic responses to polarised light. *Archs ital. Biol.* 103, 146-58.

Magnus, D. (1960). Zur Okologie des Landeisiedlers *Coenobita jousseaumei* Bouvier und der Krabbe *Ocypode aegyptiaca* Gerstaecker am Roten Meer. *Verh. dt. zool. Ges.*, S. 1960, 316-29.

Main, A. R. (1968). Ecology, systematics and evolution of Australian frogs. *Adv. ecol. Res.* 5, 37-86.

Mangum, C. P., Booth, C. E., DeFur, P. L., Heckel, N. A., Henry, R. P., Oglesby, L. C. & Polites, G. (1976). The ionic environment of hemocyanin in *Limulus polyphemus*. *Biol. Bull. mar. biol. Lab., Woods Hole* 150, 453-67.

Mangum, C. P., Freadman, M. A. & Johansen, K. (1975). The quantitative role of hemocyanin in aerobic respiration of *Limulus polyphemus*. *J. exp. Zool.* 191, 279-85.

Mangum, C. P., Lykkeboe, G. & Johansen, K. (1975). Oxygen uptake and the role of hemoglobin in the east African swampworm *Alma emini*. *Comp. Biochem. Physiol.* 52A, 477-82.

Mann, K. H. (1962). *Leeches (Hirudinea)*. Pergamon Press, Oxford.

Mann, R. H. K. & Mills, C. A. (1979). Demographic aspects of fish fecundity. *Symp. zool. Soc. Lond.* 44, 161-77.

Mantel, L. H. (1968). The foregut of *Gecarcinus lateralis* as an organ of salt and water balance. *Am. Zool.* 8, 433-42.

Manton, S. M. (1938). Studies on the Onychophora. VI. The life history of *Peripatopsis*. *Ann. Mag. nat. Hist.* 1, 515-29.

Manton, S. M. (1950). The evolution of arthropodan locomotory mechanisms. Part 1. The locomotion of *Peripatus*. *J. Linn. Soc. (Zool.)* 41, 529-70.

Manton, S. M. (1952a). The evolution of arthropodan locomotory mechanisms. Part 2. General introduction to the locomotory mechanisms of the Arthropoda. *J. Linn. Soc. (Zool.)* 42, 93-117.

Manton, S. M. (1952b). The evolution of arthropodan locomotory mechanisms. Part 3. The locomotion of the Chilopoda and Pauropoda. *J. Linn. Soc., Lond.* 42, 118-67.

Manton, S. M. (1954). The evolution of arthropodan locomotory mechanisms. Part 4. The structure, habits and evolution of the Diplopoda. *J. Linn. Soc., Lond.* 42, 299-368.

Manton, S. M. (1957). The evolution of arthropodan locomotory mechanisms. Part 5. The structure, habits and evolution of the Pselaphognatha (Diplopoda). *J. Linn. Soc., Lond.* 43, 153-87.

Manton, S. M. (1958a). Hydrostatic pressure and leg extension in arthropods, with special reference to arachnids. *Ann. Mag. nat. Hist.* (Ser. 13) 1, 161-82.

Manton, S. M. (1958b). Habits of life and evolution of body design in Arthropoda. *J. Linn. Soc. (Zool.)* 44, 58-72.

Manton, S. M. (1958c). The evolution of arthropodan locomotory mechanisms. Part 6. Habits and evolution of the Lysiopetaloidea (Diplopoda), some principles of leg design in Diplopoda and Chilopoda, and limb structure of Diplopoda. *J. Linn. Soc., Lond.* 43, 487-556.

Manton, S. M. (1961). The evolution of arthropodan locomotory mechanisms. Part 7. Functional requirements and body design in Colobognatha (Diplopoda), together with a comparative account of diplopod burrowing techniques, trunk musculature, and segmentation. *J. Linn. Soc., Lond.* 44, 383-461.

Manton, S. M. (1964). Mandibular mechanisms and the evolution of arthropods. *Phil. Trans. roy. Soc. Lond.* (B) 247, 1-183.

Manton, S. M. (1965). The evolution of arthropodan locomotory mechanisms. Part 8. Functional requirements and body design in Chilopoda, together with a comparative account of their skeleto-muscular systems. *J. Linn. Soc., Lond.* 45, 251-484.

Manton, S. M. (1966). The evolution of arthropodan locomotory mechanisms. Part 9. Functional requirements and body design in Symphyla and Pauropoda and the relationships between Myriapoda and pterygote insects. *J. Linn. Soc. (Zool.)* 46, 103-41.

Manton, S. M. (1970). Arthropods: introduction. In *Chemical zoology*, vol. V, Arthropoda, part A, pp. 1-34, ed. M. Florkin & B. T. Scheer. Academic Press, New York & London.

Manton, S. M. (1972). The evolution of arthropodan locomotory mechanisms. Part 10. Locomotory habits, morphology and evolution of the hexapod classes. *Zool. J. Linn. Soc.* 51, 203-400.

Manton, S. M. (1973). The evolution of arthropodan locomotory mechanisms. Part 11. Habits, morphology and evolution of the Uniramia (Onychophora, Myriapoda, Hexapoda) and comparisons with the Arachnida, together with a functional review of uniramian musculature. *Zool. J. Linn. Soc.* 53, 257-75.

Manton, S. M. (1974). Segmentation in Symphyla, Chilopoda and Pauropoda in relation to phylogeny. *Symp. zool. Soc. Lond.* 32, 163-90.

Manton, S. M. (1977). *The Arthropoda. Habits, functional morphology and evolution.* Oxford University Press, Oxford.

Manton, S. M. (1978). Habits, functional morphology and the evolution of pycnogonids. *Zool. J. Linn. Soc.* 63, 1-21.

Manton, S. M. (1979). Functional morphology and the evolution of the hexapod classes. In *Arthropod phylogeny*, pp. 387-465, ed. A. P. Gupta. Van Nostrand Reinhold Co., New York.

Manton, S. M. (1980). Uniramian evolution with particular reference to the pleuron. In *Myriapod biology*, pp. 317-43, ed. M. Camatini. Academic Press, London.

Manton, S. M. & Heatley, N. G. (1937). Studies on the Onychophora. II. The feeding, digestion, excretion, and food storage of *Peripatopsis*, with biochemical estimations and analyses. *Phil. Trans. roy. Soc. Lond.* (B) 227, 411-64.

Manton, S. M. & Ramsay, J. A. (1937). Studies on the Onychophora. III. The control of water loss in *Peripatopsis*. *J. exp. Biol.* 14, 470-2.

Marcus, E. (1944). Sobre Oligochaeta limnicos do Brasil. *Bolm Univ. S. Paulo (Zool.)* 43, 5-135.

Marsh, B. A. & Branch, G. M. (1979). Circadian and circatidal rhythms of oxygen consumption in the sandy-beach isopod *Tylos granulatus* Krauss. *J. exp. mar. Biol. Ecol.* 37, 77-89.

Marshall, N. B. (1965). *The life of fishes*. Weidenfeld & Nicolson, London.

Martin, A. W., Stewart, D. M. & Harrison, F. M. (1965). Urine formation in a pulmonate land snail, *Achatina fulica*. *J. exp. Biol.* **42**, 99–123.

Martoja, M. (1975). Le rein de *Pomatias* (=*Cyclostoma*) *elegans* (Gastéropode, Prosobranche): données structurales et analytiques. *Annls Sci. nat. (Zool.)* **17**, 535–58.

Mason, C. A. (1970). Function of the pericardial sacs during the molt cycle in the land crab *Gecarcinus lateralis*. *J. exp. Zool.* **174**, 381–90.

Mathews, R. S. (1954). Land leeches. *J. Bombay nat. Hist. Soc.* **52**, 655–6.

Matthews, E. G. (1976). *Insect ecology*. University of Queensland Press, St Lucia, Queensland.

Mattox, N. T. (1949). Effects of drying on certain marine snails from Puerto Rico. *Ecology* **30**, 242–4.

Mautz, W. (1980). Factors influencing evaporative water loss in lizards. *Comp. Biochem. Physiol.* **67A**, 429–37.

May, M. L. (1977). Thermoregulation and reproductive activity in tropical dragonflies in the genus *Micrathyria*. *Ecology* **58**, 787–98.

May, R. M. (1949). Cytologie de la reviviscence. *Proc. 6th Int. Congr. exp. Cytol.* (suppl.) **1**, 390.

Mayes, K. R. & Holdich, D. M. (1975). Water exchange between woodlice and moist environments with particular reference to *Oniscus asellus*. *Comp. Biochem. Physiol.* **51A**, 295–300.

Mazokhin-Porshnyakov, G. A. (1969). *Insect vision*. Plenum Press, New York.

Mead, F., Gabouriaut, D. & Corbière-Tichané, G. (1976). Structure de l'organe sensoriel apical de l'antenne chez l'isopode terrestre *Metoponorthus sexfasiatus* Budde-Lund (Crustacea, Isopoda). *Zoomorphol.* **83**, 253–69.

Mead-Briggs, A. R. (1956). The effect of temperature upon the permeability to water of arthropod cuticles. *J. exp. Biol.* **33**, 737–49.

Meenakshi, V. R. (1956). Physiology of hibernation of the apple-snail *Pila virens* (Lamarck). *Curr. Sci.* **25**, 321–2.

Meenakshi, V. R. (1964). Aestivation in the Indian apple snail *Pila*. I. Adaptation in natural and experimental conditions. *Comp. Biochem. Physiol.* **11**, 379–86.

Meincke, K.-F. (1972). Osmotische Druck und ionale Zusammensetzung der Hämolymphe winterschlafender *Helix pomatia* bei konstanter und sich zyklisch ändernder Temperatur. *Z. vergl. Physiol.* **76**, 226–32.

Melamed, J. & Trujillo-Cenóz, O. (1966). The fine structure of the visual system of *Lycosa* (Araneae: Lycosidae). Part I. Retina and optic nerve. *Z. Zellforsch. mikrosk. Anat.* **74**, 12–31.

Meske, C. (1960). Schallreaktionen von *Lithobius forficatus* L. (Chilopoden). *Z. vergl. Physiol.* **43**, 526–30.

Meske, C. (1961). Untersuchungen zur Sinnesphysiologie von Diplopoden und Chilopoden. *Z. vergl. Physiol.* **45**, 61–77.

Meyer, K. F. (1925). The bacterial symbiosis in the concretion deposits of certain operculate land molluscs of the families Cyclostomatidae and Annulariidae. *J. infect. Dis.* **36**, 1–107.

Miall, L. C. (1895). *The natural history of aquatic insects*. Macmillan, London.

Michel, C. & DeVillez, E. J. (1978). Digestion. In *Physiology of annelids*, pp. 509–54, ed. P. J. Mill. Academic Press, London.

Michelbacher, A. E. (1949). The ecology of Symphyla. *Pan-Pacific Ent.* **25**, 1–12.

Michon, J. (1949). Influence de la dessiccation sur la diapause des lombriciens. *C. r. hebd. Séanc. Acad. Sci., Paris*. **228**, 1455–6.

Miles, R. S. (1977). Dipnoan (lungfish) skulls and the relationships of the group: a study based on new species from the Devonian of Australia. *Zool. J. Linn. Soc.* **61**, 1–328.

Miley, H. H. (1930). Internal anatomy of *Euryurus erythropygus* (Brandt). (Diplopoda.) *Ohio J. Sci.* **30**, 229–49.

Miller, P. L. (1960). Respiration in the desert locust. III. Ventilation and the spiracles during flight. *J. exp. Biol.* **37**, 264–78.

Miller, P. L. (1964*a*). Factors affecting spiracle control in adult dragonflies: water balance. *J. exp. Biol.* **41**, 331–43.

Miller, P. L. (1964*b*). Factors affecting spiracle control in adult dragonflies: hypoxia and temperature. *J. exp. Biol.* **41**, 345–57.

Miller, P. L. (1974). Respiration – aerial gas transport. In *The physiology of Insecta* (2nd edition), vol. 6, pp. 345–402, ed. M. Rockstein. Academic Press, London.

Miller, P. J. (1979). Adaptiveness and implications of small size in teleosts. *Symp. zool. Soc. Lond.* **44**, 263–306.

Milner, A. R. (1980). The tetrapod assemblage from Nýrany, Czechoslovakia. In *The terrestrial environment and the origin of land vertebrates*, pp. 439–96, ed. A. L. Panchen. Academic Press, London.

Minichev, Y. S. & Slavoshevskaia, L. V. (1971). Peculiarities in the evolution of the renopericardial complex of terrestrial pulmonate molluscs. *Zool. Zh.* **50**, 350–60. In Russian. Translated as RTS 7038, National Lending Library.

Mitchell, R. (1972). The tracheae of water mites. *J. Morph.* **136**, 327–36.

Moffett, D. F. (1975). Sodium and potassium transport across the isolated hindgut of the desert millipede *Orthoporus ornatus* (Girard). *Comp. Biochem. Physiol.* **50A**, 57–63.

Monk, C. & Stewart, D. M. (1966). Urine formation in a freshwater snail, *Viviparus malleatus*. *Am. J. Physiol.* **210**, 647–51.

Monterosso, B. (1930). Studi cirripedologici. VI. Sul comportamento di *Chthamalus stellatus* in diverse condizioni sperimentali. *R. C. Accad. Lincei.* II, 501–5. (English abstract in *Biol. Abstr.* **5**, Entry 15577.)

Moore, J. (1973). Land nemertines of New Zealand. *Zool. J. Linn. Soc.* **52**, 293–313.

Moore, J. & Gibson, R. (1973). A new genus of freshwater hoplonemerteans from New Zealand. *Zool. J. Linn. Soc.* **52**, 141–57.

Moore, J. & Gibson, R. (1981). The *Geonemertes* problem (Nemertea). *J. Zool., Lond.* **194**, 175–201.

Moore, R. C., Lalicker, C. G. & Fischer, A. G. (1952). *Invertebrate fossils*. McGraw-Hill, New York.

Morgan, E. (1971). The swimming of *Nymphon gracile* (Pycnogonida). The mechanics of the leg-beat cycle. *J. exp. Biol.* **55**, 273–87.

Morii, H., Nishikata, K. & Tamura, O. (1978). Nitrogen excretion of mudskipper fish *Periophthalmus cantonensis* and *Boleophthalmus pectinirostris* in water and on land. *Comp. Biochem. Physiol.* **60A**, 189–93.

Morris, J. E. & Afzelius, B. A. (1967). The structure of the shell and outer membranes in encysted *Artemia salina* embryos during cryptobiosis and development. *J. Ultrastruct. Res.* **20**, 244–59.

Morris, R. (1960). General problems of osmoregulation with special reference to cyclostomes. *Symp. zool. Soc. Lond.* **1**, 1–16.

Morrison, P. R. (1946). Physiological observations on water loss and oxygen consumption in *Peripatus*. *Biol. Bull. mar. biol. Lab., Woods Hole* **91**, 181–8.

Morton, J. E. (1952). A preliminary study of the land operculate *Murdochia pallidum* (Cyclophoridae, Mesogastropoda). *Trans. roy. Soc. NZ* **80**, 69–79.

Morton, J. E. (1954). Notes on the ecology and annual cycle of *Carychium tridentatum* at Box Hill. *Proc. malac. Soc.* **31**, 30–46.

Morton, J. E. (1955*a*). The evolution of the Ellobiidae with a discussion on the origin of the Pulmonata. *Proc. zool. Soc. Lond.* **125**, 127–68.

Morton, J. E. (1955*b*). The functional morphology of the British Ellobiidae (Gastropoda Pulmonata) with special reference to the digestive and reproductive systems. *Phil. Trans. roy. Soc. (B)* **239**, 89–160.

Morton, J. E. (1979). *Molluscs* (5th edition). Hutchinson, London.

Morton, J. E. & Miller, M. (1973). *The New Zealand sea shore* (2nd edition). Collins, London.

Morton, J. E. & Yonge, C. M. (1964). Classification and structure of the Mollusca. In *Physiology of Mollusca*, vol. I. pp. 1–58, ed. K. M. Wilbur & C. M. Yonge. Academic Press, London.

Moy-Thomas, J. A. (1971). *Palaeozoic fishes* (2nd edition), revised by R. S. Miles. Chapman & Hall, London.

Mullins, D. E. & Cochran, D. G. (1973). Nitrogenous excretory materials from the American cockroach. *J. Insect Physiol.* **19**, 1007–18.

Munshi, J. S. D. (1976). Gross and fine structure of the respiratory organs of air-breathing fishes. In *Respiration of amphibious vertebrates*, pp. 73–104, ed. G. M. Hughes. Academic Press, London.

Mykles, D. L. (1977). The ultrastructure of the posterior midgut caecum of *Pachygrapsus crassipes* (Decapoda, Brachyura), adapted to low salinity. *Tiss. Cell* **9**, 681–91.

Nagabhushanam, R., Deshpande, U. D. & Hanumante, M. M. (1981). Hormonal control of egg-laying in the marine pulmonate, *Onchidium verruculatum*. *Hydrobiologia* **80**, 277–81.

Naire, V. S. K. & Prabhu, V. K. K. (1971). On the free amino acids in the haemolymph of a millipede. *Comp. Biochem. Physiol.* **388**, 1–4.

Nakahara, H. & Bevelander, G. (1969). An electron microscope and autoradiographic study of the calciferous glands of the earthworm, *Lumbricus terrestris*. *Calcif. Tiss. Res.* **4**, 193–201.

Needham, A. E. (1957). Components of nitrogenous excreta in the earthworms *Lumbricus terrestris* L. and *Eisenia foetida* (Savigny). *J. exp. Biol.* **34**, 425–46.

Needham, A. E. (1970). Nitrogen metabolism in Annelida. In *Comparative biochemistry of nitrogen metabolism*, vol. I, *The invertebrates*, pp. 207–97, ed. J. W. Campbell. Academic Press, London.

Needham, J. (1938). Contributions of chemical physiology to the problem of reversibility in evolution. *Biol. Rev.* **13**, 225–51.

Needham, J. (1963). *Chemical embryology*. Hafner, New York.

Neumann, D. (1960). Osmotische Resistenz und Osmoregulation der Flussdeckelschnecke *Theodoxus fluviatilis* L. *Biol. Zbl.* **79**, 585–605.

Neville, A. C. (1975). *Biology of the arthropod cuticle*. Springer-Verlag, New York.

Newell, P. F. & Skelding, J. M. (1973). Studies on the permeability of the septate junction in the kidney of *Helix pomatia* L. *Malacologia* **14**, 89–91.

Newell, R. C. (1979). *Biology of intertidal animals* (3rd edition). Marine Ecological Surveys Ltd, Faversham, Kent.

Newell, R. C., Ahsanullah, M. & Pye, V. I. (1972). Aerial and aquatic respiration in the shore crab *Carcinus maenas* (L.). *Comp. Biochem. Physiol.* **43A**, 239–52.

Newman, W. A. (1967). On physiology and behaviour of estuarine barnacles. Proc. Symp. on Crustacea, Ernakulam, Part III. *Symp. mar. biol. Ass. India* (Ser. 2) 1038–66.

Nicholas, W. L. (1975). *The biology of free-living nematodes*. Clarendon Press, Oxford.

Nicolson, S. W. (1980). Water balance and osmoregulation in *Onymacris plana*, a tenebrionid beetle from the Namib Desert. *J. Insect Physiol.* **26**, 315–20.

Nielsen, C. (1976). Notes on *Littorina* and *Murex* from the mangrove at Ao Nam-Bor, Phuket, Thailand. *Res. Bull. Phuket mar. biol. Center* **11**, 1–4.

Niggemann, R. (1968). Zur Biologie und Okologie des Landeinsiedlerkrebses *Coenobita scaevola* Forskal am Roten Meer. *Oecologia* **1**, 236–64.

Noble-Nesbitt, J. (1963a). Transpiration in *Podura aquatica* L. (Collembola, Isotomidae) and the wetting properties of its cuticle. *J. exp. Biol.* **40**, 681–700.

Noble-Nesbitt, J. (1963b). The fully formed intermoult cuticle and associated structures of *Podura aquatica* (Collembola). *Q. Jl microsc. Sci.* **104**, 253–70.

Noble-Nesbitt, J. (1963c). A site of water and ionic exchange with the medium in *Podura aquatica* L. (Collembola, Isotomidae). *J. exp. Biol.* **40**, 701–11.

Noble-Nesbitt, J. (1970). Water balance in the firebrat, *Thermobia domestica* (Packard). The site of uptake of water from the atmosphere. *J. exp. Biol.* **52**, 193–200.

Noble-Nesbitt, J. (1975). Reversible arrest of uptake of water from subsaturated atmospheres by the firebrat, *Thermobia domestica* (Packard). *J. exp. Biol.* **62**, 657–69.

Noble-Nesbitt, J. (1978). Absorption of water vapour by *Thermobia domestica* and other insects. In *Comparative physiology: water, ions and fluid mechanics*, pp. 53–66, ed. K. Schmidt-Nielsen, L. Bolis & S. H. P. Maddrell. Cambridge University Press, Cambridge, England.

Noirot, C. & Noirot-Timothée, C. (1971a). Ultrastructure du proctodeum chez le thysanoure *Lepismodes inquilinus* Newman (=*Thermobia domestica* Packard). II. Le sac anal. antérieure (iléon et rectum). *J. Ultrastruct. Res.* **37**, 119–37.

Noirot, C. & Noirot-Timothée, C. (1971b). Ultrastructure du proctodeum chez le thysanoure *Lepismodes inquilinus* Newman (=*Thermobia domestica* Pakcard). II. Le sac anal. *J. Ultrastruct. Res.* **37**, 335–50.

Nordström, S. (1975). Seasonal activity of lumbricids in southern Sweden. *Oikos* **26**, 307–15.

Nørgaard, E. (1951). On the ecology of two lycosid spiders (*Pirata piraticus* and *Lycosa pullata*) from a Danish *Sphagnum* bog. *Oikos* **3**, 1–21.

Nørgaard, E. (1956). Environment and behaviour of *Theridion saxatile*. *Oikos* **7**, 159–92.

Nosek, J. (1973). *The European Protura*. Muséum d'Histoire Naturelle, Geneva.

Nosek, J. (1975). Niches of Protura in biogeocoenoses. *Pedobiologia* **15**, 290–8.

Nunnemacher, R. F. (1966). The fine structure of optic tracts of Decapoda. In *The functional organization of the compound eye*, pp. 363–75, ed. C. G. Bernard. Pergamon Press, Oxford.

Nutman, S. R. (1941). Function of the ventral tube in *Onychiurus armatus* (Collembola). *Nature, Lond.* **148**, 168–9.

O'Donnell, M. J. (1978). The site of water vapour absorption in *Arenivaga investigata*. In *Comparative physiology: water, ions and fluid mechanics*, pp. 115–21, ed. K. Schmidt-Nielsen, L. Bolis & S. H. P. Maddrell. Cambridge University Press, Cambridge, England.

O'Donnell, M. J. (1981). Fluid movements during water-vapour absorption by the desert burrowing cockroach, *Arenivaga investigata*. *J. Insect Physiol.* **27**, 877–87.

O'Donnell, M. J. (1982). Water vapour absorption by the desert burrowing cockroach, *Arenivaga investigata*: evidence against a solute dependent mechanism. *J. exp. Biol.* **96**, 251–62.

Oduleye, S. O. (1977). Unidirectional water and sodium fluxes and respiratory metabolism in the African lungfish, *Protopterus annectens*. *J. comp. Physiol.* **119**, 127–39.

Oglesby, L. C. (1969). Inorganic components and metabolism; ionic and osmotic regulation: Annelida, Sipuncula, and Echiura. In *Chemical zoology*, vol. IV, pp. 555–658, ed. M. Florkin & B. T. Scheer. Academic Press, London.

Oglesby, L. C. (1978). Salt and water balance. In *Physiology of annelids*, pp. 555–658, ed. P. J. Mill. Academic Press, London.

Okasha, A. Y. K. (1971). Water relations in an insect, *Thermobia domestica*. I. Water uptake from subsaturated atmospheres as a means of volume regulation. *J. exp. Biol.* **55**, 435–48.

Okasha, A. Y. K. (1972). Water relations in an insect, *Thermobia domestica*. II. Relationships between water content, water uptake from subsaturated atmospheres and water loss. *J. exp. Biol.* **57**, 285–96.

Okasha, A. Y. K. (1973). Water relations in an insect, *Thermobia domestica*. III. Effects of desiccation and rehydration on the haemolymph. *J. exp. Biol.* **58**, 385–400.

de Oliveira, L. P. H. (1946). Estudos ecologicos dos crustaceos commestiveis uca e guaiamu, *Cardisoma guanhumi* Latreille e *Ucides cordatus* (L.). Gecarcinidae, Brachyura. *Mem. Inst. Oswaldo Cruz, Brazil* **44**, 295–322.

Olivier, L. & Barbosa, F. S. (1956). Observations on vectors of *Schistosomiasis mansoni* kept out of water in the laboratory, II. *J. Parasit.* **42**, 277–86.

Olson, E. C. (1965*a*). Evolution and relationships of the Amphibia. Introductory remarks. *Am. Zool.* **5**, 263–5.

Olson, E. C. (1965*b*). Relationships of *Seymouria, Diadectes*, and *Chelonia. Am. Zool.* **5**, 295–307.

Olthof, H. J. (1936). Uber die Luftatmung von *Eriocheir sinensis* H. Milne-Edwards. *Z. vergl. Physiol.* **23**, 293–300.

O'Neill, R. V. (1969*a*). Adaptive responses to desiccation in the millipede, *Narceus americanus* (Beauvois). *Am. Mid. Nat.* **81**, 578–83.

O'Neill, R. V. (1969*b*). Comparative desiccation tolerance in seven species of millipedes. *Am. Mid. Nat.* **82**, 182–7.

Oschman, J. L. & Wall, B. J. (1969). The structure of the rectal pads of *Periplaneta americana* L. with regard to fluid transport. *J. Morph.* **127**, 475–510.

Packard, A. (1972). Cephalopods and fish: the limits of convergence. *Biol. Rev.* **47**, 241–307.

Packard, G. C. (1974). The evolution of air-breathing in paleozoic gnathostome fishes. *Evolution* **28**, 320–5.

Packard, G. C. & Packard, M. J. (1980). Evolution of the cleidoic egg among reptilian antecedents of birds. *Am. Zool.* **20**, 351–62.

Packard, G. C., Taigen, T. L., Packard, M. J. & Shuman, R. D. (1979). Water-vapor conductance of testudinian and crocodilian eggs (class Reptilia). *Resp. Physiol.* **38**, 1–10.

Packard, G. C., Tracy, C. R. & Roth, J. J. (1977). The physiological ecology of reptilian eggs and embryos, and the evolution of viviparity within the class Reptilia. *Biol. Rev.* **52**, 71–105.

Packard, M. J., Packard, G. C. & Boardman, T. J. (1982). Structure of eggshells and water relations of reptilian eggs. *Herpetologica* **38**, 136–55.

Padmanabhanaidu, B. (1962). Ionic composition of the blood of scorpion. *Curr. Sci.* **31**, 21.

Padmanabhanaidu, B. (1966). Physiological properties of the blood and haemocyanin of the scorpion *Heterometrus fulvipes. Comp. Biochem. Physiol.* **17**, 167–81.

Padmanabhanaidu, B. (1967). Perfusion fluid for the scorpion *Heterometrus fulvipes. Nature, Lond.* **213**, 410.

Padmanabhanaidu, B. & Ramamurthy, R. (1961). The effect of sex and size on osmotic pressure, the chloride and the free amino acids of the blood of the freshwater field crab, *Paratelphusa* sp. and the freshwater mussel *Lamellidens marginalis. J. exp. Biol.* **38**, 35–41.

Pagés, J. (1967). La notion de territoire chez les diploures japygidés. *Ann. Soc. ent. Fr.* (N.S.) **3**, 715–19.

Palmer, M. F. (1966). Investigations of the blood capillary system of *Tubifex tubifex. J. Zool., Lond.* **148**, 449–52.

Palmer, M. F. (1968). Aspects of the respiratory physiology of *Tubifex tubifex* in relation to its ecology. *J. Zool., Lond.* **154**, 463–73.

Pampapathi Rao, K. (1963). Physiology of low temperature acclimation in tropical poikilotherms. I. Ionic changes in the blood of the freshwater mussel, *Lamellidens marginalis*, and the earthworm, *Lampito mauritii. Proc. Indian Acad. Sci.* **57B**, 290–6.

Panchen, A. L. (1967). The nostrils of choanate fishes and early tetrapods. *Biol. Rev.* **42**, 374–420.

Panchen, A. L. (1972). The interrelationships of the earliest tetrapods. In *Studies in vertebrate evolution*, pp. 65–87, ed. K. A. Joysey & T. S. Kemp. Oliver & Boyd, Edinburgh.

Panchen, A. L. (1975). A new genus and species of anthracosaur amphibian from the Lower Carboniferous of Scotland and the status of *Pholidogaster pisciformis* Huxley. *Phil. Trans. roy. Soc.* (*B*) **269**, 581–640.

Panchen, A. L. (1980). The origin and relationships of the anthracosaur Amphibia from the late Palaeozoic. In *The terrestrial environment and the origin of land vertebrates*, pp. 319–50, ed. A. L. Panchen. Academic Press, London.

Pantin, C. F. A. (1931*a*). The origin of the composition of the body fluids in animals. *Biol. Rev.* **6**, 459–82.

Pantin, C. F. A. (1931*b*). The adaptation of *Gunda ulvae* to salinity. I. The environment. *J. exp. Biol.* **8**, 63–72.

Pantin, C. F. A. (1944). Terrestrial nemertines and planarians in Britain. *Nature, Lond.* **154**, 80.

Pantin, C. F. A. (1947). The nephridia of *Geonemertes dendyi. Q. Jl microsc. Sci.* **88**, 15–25.

Pantin, C. F. A. (1950). Locomotion in British terrestrial nemertines and planarians: with a discussion on the identity of *Rhynchodemus bilineatus* (Mecznikow) in Britain, and on the name *Fasciola terrestris* O. F. Müller. *Proc. Linn. Soc., Lond.* **162**, 23–37.

Pantin, C. F. A. (1961). *Geonemertes* – a study in island life. *Proc. Linn. Soc., Lond.* **172**, 137–52.

Pantin, C. F. A. (1969). The genus *Geonemertes. Bull. Brit. Mus. nat. Hist.* **18**, 263–310.

Parakkal, P. & Matoltsy, A. G. (1964). A study of the fine structure of the epidermis of *Rana pipiens. J. Cell Biol.* **20**, 85–94.

Paris, O. H. (1963). The ecology of *Armadillidium vulgare* (Isopoda: Oniscoidea) in California grassland: food, enemies and weather. *Ecol. Monogr.* **33**, 1–22.

Paris, O. H. (1965). Vagility of P^{32} labeled isopods in grassland. *Ecology* **46**, 635–48.

Parry, D. A. (1951). Factors determining the temperature of terrestrial arthropods in sunlight. *J. exp. Biol.* **28**, 445–62.

Parry, D. A. (1954). On the drinking of soil capillary water by spiders. *J. exp. Biol.* **31**, 218–27.

Parry, D. A. (1957). Spider leg muscles and the autotomy mechanism. *Q. Jl microsc. Sci.* **98**, 331–4.

Parry, D. A. (1965). The signal generated by an insect in a spider's web. *J. exp. Biol.* **43**, 185–92.

Parry, D. A. & Brown, R. H. J. (1959*a*). The hydraulic mechanism of the spider leg. *J. exp. Biol.* **36**, 422–33.

Parry, D. A. & Brown, R. H. J. (1959*b*). The jumping mechanisms of salticid spiders. *J. exp. Biol.* **36**, 654–64.

Parry, G. (1953). Osmotic and ionic regulation in the isopod crustacean *Ligia oceanica. J. exp. Biol.* **30**, 567–74.

Parry, G. (1960). Excretion. In *The physiology of Crustacea*, vol. I, pp. 341–66, ed. T. H. Waterman. Academic Press, London.

Parsons, T. S. & Williams, E. E. (1963). The relationships of the modern Amphibia: a re-examination. *Q. Rev. Biol.* **38**, 26–35.

Patten, D. T. & Smith, E. M. (1975). Heat flux and the thermal regime of desert plants. In *Environmental physiology of desert organisms*, pp. 1–19, ed. N. F. Hadley. Dowden, Hutchinson & Ross, Inc., Stroudsberg, Pennsylvania.

Patterson, C. (1978). Arthropods and ancestors. *Bull. roy. ent. Soc., Lond.* **2**, 99–103.

Patterson, C. (1980). Origin of tetrapods: historical introduction to the problem. In *The terrestrial environment and the origin of land vertebrates*, pp. 159–75, ed. A. L. Panchen. Academic Press, London.

Pattle, R. E. (1976). The lung surfactant in the evolutionary tree. In *Respiration of amphibious vertebrates*, pp. 233–55, ed. G. M. Hughes. Academic Press, London.

Paulpandian, A. (1965). A preliminary report on the diurnal rhythm in the locomotor activity of pill-millipede, *Arthrosphaera dalyi* (Pocock). *Proc. Indian Acad. Sci.* **62B**, 235–41.

Paulus, H. F. (1974). Die phylogenetische Bedeutung der Ommatidien der apterygoten Insekten (Collembola, Archaeognatha und Zygentoma). *Pedobiologia* **14**, 123–33.

Paulus, H. F. (1979). Eye structure and the monophyly of the Arthropoda. In *Arthropod phylogeny*, pp. 299–383, ed. A. P. Gupta. Van Nostrand Reinhold Co., New York.

Peake, J. F. (1978). Distribution and ecology of the Stylommatophora. In *Pulmonates*, vol. 2A, pp. 429–526, ed. V. Fretter & J. F. Peake. Academic Press, London.

Peaker, M. & Linzell, J. L. (1975). *Salt glands in birds and reptiles*. Cambridge University Press, Cambridge, England.

Pearse, A. S. (1911). On the habits of *Thalassina anomala* (Herbst). *Philipp. J. Sci.* 6, 213–15.

Pearse, A. S. (1929a). Observations on certain littoral and terrestrial animals at Tortugas, Florida, with special reference to migrations from marine to terrestrial habitats. *Pap. Tortugas Lab.* 26, 205–23.

Pearse, A. S. (1929b). The ecology of certain estuarine crabs at Beaufort, N.C. *J. Elisha Mitchell scient. Soc.* 44, 230–7.

Pearse, A. S. (1936). *The migrations of animals from sea to land*. Duke University Press, Durham, North Carolina.

Pearson, R. (1978). *Climate and evolution*. Academic Press, London.

Pennak, R. W. (1951). Comparative ecology of the interstitial fauna of fresh-water and marine beaches. *Année Biologique* 27, 449–80.

Perttunen, V. (1953). Reactions of diplopods to the relative humidity of the air. Investigations of *Orthomorpha gracilis*, *Iulus terrestris* and *Schizophyllum sabulosum*. *Ann. Soc. zool. Fenn. Vanamo.* 16, 1–69.

Perttunen, V. (1955). Antennectomy and the humidity reactions of normal and desiccated specimens of *Schizophyllum sabulosum* (Diplopoda, Iulidae). *Ann. Ent. Fenn.* 21, 157–62.

Petersen, J. A., Fyhn, H. J. & Johansen, K. (1974). Eco-physiological studies of an intertidal crustacean, *Pollicipes polymerus* (Cirripedia, Lepadomorpha): aquatic and aerial respiration. *J. exp. Biol.* 61, 309–20.

Pflugfelder, O. (1933). Landpolychäten aus Niederländisch-Indien. *Zool. Anz.* 105, 65–76.

Philip, G. M. (1979). Carpoids – echinoderms or chordates? *Biol. Rev.* 54, 439–71.

Phillips, J. E. (1961). Studies on the rectal absorption of water and salts in the locust, *Schistocerca gregaria*, and the blowfly, *Calliphora erythrocephala*. PhD thesis, University of Cambridge.

Phillips, J. E. (1964). Rectal absorption in the desert locust. *Schistocerca gregaria* Forskal. III. The nature of the excretory process. *J. exp. Biol.* 41, 69–80.

Phillips, J. E. (1969). Osmotic regulation and rectal absorption in the blowfly *Calliphora erythrocephala*. *Can. J. Zool.* 47, 851–63.

Phleger, C. F. & Saunders, B. S. (1978). Swim-bladder surfactants of Amazon air-breathing fishes. *Can. J. Zool.* 56, 946–52.

Piavis, G. W. (1971). Embryology. In *The biology of lampreys*, vol. I. pp. 361–400, ed. M. W. Hardisty & I. C. Potter. Academic Press, London.

Picken, L. E. R. (1936). The mechanism of urine formation in invertebrates. I. The excretion mechanism in certain Arthropoda. *J. exp. Biol.* 13, 309–28.

Piiper, J., Tazawa, H., Ar, A. & Rahn, H. (1980). Analysis of chorioallantoic gas exchange in the chick embryo. *Resp. Physiol.* 39, 273–84.

Pilkington, M. C. & Pilkington, J. B. (1982). The planktonic veliger of *Amphibola crenata* (Gmelin). *J. moll. Stud.* 48, 24–9.

Plate, L. H. (1894). Studien über opisthopneumone Lungen-schnecken. II. Die Oncidiiden. *Zool. Jb. (Anat.)* 7, 93–234.

Plate, L. H. (1898). Beiträge zur Anatomie und Systematik der Janelliden. *Zool. Jb. (Anat.)* 11, 193–280.

Platt, H. M. & Warwick, R. M. (1980). The significance of free-living nematodes to the littoral ecosystem. In *The shore environment*, vol. 2, pp. 729–59, ed. J. H. Price, D. E. G. Irvine & W. F. Farnham. Academic Press, London.

Pohunková, H. (1969). Lung ultrastructure of the Arachnida-Arachnoidea. *Folia Morph.*, Prague 17, 309–16.

Poinsot-Balaguer, N. (1976). Dynamique des communautes de collemboles en milieu xérique méditerranéen. *Pedobiologia* 16, 1–17.

Poll, M. (1962). Etude sur la structure adulte et la formation des sacs pulmonaires des protoptères. *Ann. Reeks Zool. Wetenschap.* 108, 129–72.

Pond, C. M. (1975). The role of the 'walking legs' in aquatic and terrestrial locomotion of the crayfish *Austropota-mobius pallipes* (Lereboullet). *J. exp. Biol.* 62, 447–54.

Pontin, R. M. (1966). The osmoregulatory function of the vibratile flames and the contractile vesicle of *Asplanchna* (Rotifera). *Comp. Biochem. Physiol.* 17, 1111–26.

Porter, K. R. (1972). *Herpetology*. Saunders, London.

Potswald, H. E. (1971). A fine-structural analysis of the epidermis and cuticle of the oligochaete *Aeolosoma bengalense* Stephenson. *J. Morph.* 135, 185–212.

Potts, W. T. W. (1954a). The inorganic composition of the blood of *Mytilus edulis* and *Anodonta cygnea*. *J. exp. Biol.* 31, 376–85.

Potts, W. T. W. (1954b). The energetics of osmotic regulation in brackish- and fresh-water animals. *J. exp. Biol.* 31, 618–30.

Potts, W. T. W. & Parry, G. (1964). *Osmotic and ionic regulation in animals*. Pergamon Press, Oxford.

Potts, W. T. W. & Rudy, P. P. (1969). Water balance in the eggs of the Atlantic salmon *Salmo salar*. *J. exp. Biol.* 50, 223–37.

Powers, L. W. & Cole, J. F. (1976). Temperature variation in fiddler crab micro-habitats. *J. exp. mar. Biol. Ecol.* 21, 141–57.

Prashad, B. (1925). Respiration of gastropod molluscs. *Proc. Indian Sci. Congr.* 12, 126–43.

Prell, H. (1910). Beiträge zur Kenntniss der Lebensweise einiger Pantopoden. *Bergens Mus. Aarborg* 10, 1–30.

Price, C. H. (1980). Water relations and physiological ecology of the salt marsh snail, *Melampus bidentatus* Say. *J. exp. mar. Biol. Ecol.* 45, 51–67.

Price, J. B. & Holdich, D. M. (1980). Changes in osmotic pressure and sodium concentration of the haemolymph of wood-lice with progressive desiccation. *Comp. Biochem. Physiol.* 66A, 297–305.

von Prince, G. (1967). Uber Lebensweise, Fortpflanzung und Genitalorgane des terrestrischen Prosobranchiers *Cochlo-stoma septemspirale*. *Arch. Molluskenk.* 96, 1–18.

Pringle, J. W. S. (1955). The function of the lyriform organs of arachnids. *J. exp. Biol.* 32, 270–8.

Prosser, C. L. (1973). *Comparative animal physiology*. Saunders, Philadelphia.

Prosser, C. L., Green, J. W. & Chow, T. J. (1955). Ionic and osmotic concentrations in blood and urine of *Pachygrapsus crassipes* acclimated to different salinities. *Biol. Bull. mar. biol. Lab., Woods Hole* 109, 99–107.

Prusch, R. D. (1971). The site of ammonia excretion in the blowfly larva, *Sarcophaga bullata*. *Comp. Biochem. Physiol.* 39A, 761–7.

Prusch, R. D. (1972). Secretion of NH_4Cl by the hindgut of *Sarcophaga bullata* larva. *Comp. Biochem. Physiol.* 41A, 215–23.

Prusch, R. D. (1976). Unidirectional ion movements in the hindgut of larval *Sarcophaga bullata* (Diptera: Sarco-phagidae). *J. exp. Biol.* 64, 89–100.

Pugach, S. & Crawford, C. S. (1978). Seasonal changes in hemolymph amino acids, proteins, and inorganic ions of a desert millipede *Orthoporus ornatus* (Girard) (Diplopoda: Spirostreptidae). *Can. J. Zool.* 56, 1460–5.

Pullin, R. S. V. (1971). Composition of the haemolymph of *Lymnaea truncatula*, the snail host of *Fasciola hepatica*. *Comp. Biochem. Physiol.* 40A, 617–26.

von Raben, K. (1934). Veranderungen im Kiemendeckel und in den Kiemen einiger Brachyuren (Decapoden) im Verlauf der Anpassung an die Feuchtluftatmung. *Z. wiss. Zool.* 145, 425–61.

Rackoff, J. S. (1980). The origin of the tetrapod limb and the ancestry of tetrapods. In *The terrestrial environment and the origin of land vertebrates*, pp. 255–92, ed. A. L. Panchen. Academic Press, London.

Rafaeli-Bernstein, A. & Mordue, W. (1979). The effects of phlorizin, phloretin and ouabain on the reabsorption of glucose by the Malpighian tubules of *Locusta migratoria migratorioides*. *J. Insect Physiol.* **25**, 241–7.

Rahn, H. (1967). Gas transport from the external environment to the cell. In *Development of the lung*, pp. 3–29, ed. A. V. S. de Reuck & R. Porter. J. & A. Churchill Ltd, London.

Rahn, H. & Howell, B. J. (1976). Bimodal gas exchange. In *Respiration of amphibious vertebrates*, pp. 271–85, ed. G. M. Hughes. Academic Press, London.

Ramamurthi, P. (1967). Oxygen consumption of a freshwater crab, *Paratelphusa hydrodromous*, in relation to salinity stress. *Comp. Biochem. Physiol.* **23**, 599–605.

Ramsay, J. A. (1935). The evaporation of water from the cockroach. *J. exp. Biol.* **12**, 373–83.

Ramsay, J. A. (1949a). The osmotic relations of the earthworm. *J. exp. Biol.* **26**, 46–56.

Ramsay, J. A. (1949b). The site of formation of hypotonic urine in the nephridium of *Lumbricus*. *J. exp. Biol.* **26**, 65–75.

Ramsay, J. A. (1953). Active transport of potassium by the Malpighian tubules of insects. *J. exp. Biol.* **30**, 358–69.

Ramsay, J. A. (1955a). The excretory system of the stick insect, *Dixippus morosus* (Orthoptera, Phasmidae). *J. exp. Biol.* **32**, 183–99.

Ramsay, J. A. (1955b). The excretion of sodium, potassium and water by the Malpighian tubules of the stick insect, *Dixippus morosus* (Orthoptera, Phasmidae). *J. exp. Biol.* **32**, 200–16.

Ramsay, J. A. (1964). The rectal complex of the mealworm *Tenebrio molitor*, L. (Coleoptera, Tenebrionidae). *Phil. Trans. roy. Soc. Lond.* (*B*) **248**, 279–314.

Ramsay, J. A. (1971). Insect rectum. *Phil. Trans. roy. Soc. Lond.* (*B*) **262**, 251–60.

Ramsay, J. A. (1976). The rectal complex in the larvae of Lepidoptera. *Phil. Trans. roy. Soc. Lond.* (*B*) **274**, 203–26.

Randall, D. J., Burggren, W. W., Farrell, A. P. & Haswell, M. S. (1981). *The evolution of air breathing in vertebrates*. Cambridge University Press, Cambridge, England.

Randall, D. J., Cameron, J. N., Daxboeck, C. & Smatresk, N. (1981). Aspects of bimodal gas exchange in the bow fin, *Amia calva* L. (Actinopterygii: Amiiformes). *Resp. Physiol.* **43**, 339–48.

Randall, D. J. & Wood, C. M. (1981). Carbon dioxide excretion in the land crab (*Cardisoma carnifex*). *J. exp. Zool.* **218**, 37–44.

Ranzi, S. (1980). On the origin of the Arthropoda. In *Myriapod biology*, pp. 345–51, ed. M. Camatini. Academic Press, London.

Rao, K. P. (1968). The pericardial sacs of *Ocypode* in relation to the conservation of water, molting, and behaviour. *Am. Zool.* **8**, 561–7.

Rao, K. P. & Gopalakrishnareddy, T. (1962). Nitrogen excretion in arachnids. *Comp. Biochem. Physiol.* **7**, 175–8.

Rasmont, R. (1960). Structure et ultrastructure de la glande coxale d'un scorpion. *Annls Soc. roy. zool. Belg.* **89**, 239–68.

Rasmont, R. & Rabaey, M. (1958). Comparaison, par micro-electrophorèse, de la composition protidique de l'hémolymphe et du contenu des glandes coxales du scorpion *Euscorpius carpathicus* L. *C. r. Séanc. Soc. Biol.* **152**, 1020–2.

Rathmeyer, W. (1965). Neuromuscular transmission in a spider and the effect of calcium. *Comp. Biochem. Physiol.* **14**, 673–87.

Raynal, G. (1973). Le comportement reproducteur dans le genre *Bourletiella s. str.* (Colemboles Sminthuridae). *Rev. Ecol. Biol. Sol.* **10**, 317–25.

Read, L. J. (1971). The presence of high ornithine-urea cycle enzyme activity in the teleost *Opsanus tau*. *Comp. Biochem. Physiol.* **39B**, 409–13.

Redfield, A. C. & Ingalls, E. N. (1933). The oxygen dissociation curves of some bloods containing hemocyanin. *J. cell. comp. Physiol.* **3**, 169–202.

Rees, W. J. (1964). A review of breathing devices in land operculate snails. *Proc. malac. Soc. Lond.* **36**, 55–67.

Reese, E. S. (1968). Shell use: an adaptation for emigration from the sea by the coconut crab. *Science, N.Y.* **161**, 385–6.

Reid, D. M. (1947). *Talitridae (Crustacea Amphipoda)*. (Synopses of the British Fauna No. 7, Linnean Society of London.) Academic Press, London.

Reisinger, E. (1925). Ein landbewohnender Archiannelide. *Z. Morph. Okol. Tiere.* **111**, 199–254.

Reisinger, E. (1960). Die Lösung des *Parergodrilus*-Problems. *Z. Morph. Okol. Tiere.* **48**, 517–44.

Remane, A. (1951). Die Besiedlung des Sandbodens im Meere und die Bedeutung der Lebensformtypen für die Okologie. *Verh. dt. zool. Ges.* 1965, 327–59.

Remmert, H. (1968). Die *Littorina*-Arten: Kein Modell für die Entstehung der Landschnecken. *Oecologia* **2**, 1–6.

Reynoldson, T. B. (1974). Ecological separation in British triclads (Turbellaria) with a comment on two American species. In *Biology of Turbellaria*, pp. 213–28, ed. N. W. Riser & M. P. Morse. McGraw-Hill, New York.

Richards, F. A. (1965). Dissolved gases other than carbon dioxide. In *Chemical oceanography*, vol. I. pp. 197–225, ed. J. P. Riley & G. Skirrow. Academic Press, London.

Richards, O. W. (1940). The capsular fluid of *Amblystoma punctatum* eggs compared with Holtfreter's and Ringer's solutions. *J. exp. Zool.* **83**, 401–6.

Richards, O. W. & Davies, R. G. (1977). *Imms' general textbook of entomology* (10th edition). Chapman & Hall, London.

Richardson, L. R. (1968). Observations on the Australian land-leech, *Chtonobdella limbata* (Grube 1886) (Hirudinea: Haemadipsidae). *Aust. Zool.* **14**, 294–305.

Riddle, W. A. (1975). Water relations and humidity-related metabolism of the desert snail *Rabdotus schiedeanus* (Pfeiffer) (Helicidae). *Comp. Biochem. Physiol.* **51A**, 579–83.

Riddle, W. A. (1978). Comparative respiratory physiology of a desert snail *Rabdotus schiedeanus*, and a garden snail, *Helix aspersa. Comp. Biochem. Physiol.* **56A**, 369–73.

Riddle, W. A., Crawford, C. S. & Zeitone, A. M. (1976). Patterns of hemolymph osmoregulation in three desert arthropods. *J. comp. Physiol.* **112**, 295–305.

Riegel, J. A. (1963). Micropuncture studies of chloride concentration and osmotic pressure in the crayfish antennal gland. *J. exp. Biol.* **40**, 487–92.

Riegel, J. A. (1972). *Comparative physiology of renal excretion*. Oliver & Boyd, Edinburgh.

Riegel, J. A. & Lockwood, A. P. M. (1961). The role of the antennal gland in the osmotic and ionic regulation of *Carcinus maenas. J. exp. Biol.* **38**, 491–9.

Riemann, F. (1977). Causal aspects of nematode evolution: relations between structure, function, habitat and evolution. *Mikrofauna Meeresboden* **61**, 217–30.

Rigby, J. E. (1965). *Succinea putris*: a terrestrial opisthobranc mollusc. *Proc. zool. Soc. Lond.* **144**, 445–86.

Riggs, A. (ed.) (1979). The Alpha Helix expedition to the Amazon for the study of fish bloods and hemoglobins. *Comp. Biochem. Physiol.* **62A**, 1–272.

Robertson, H. G., Nicolson, S. W. & Louw, G. N. (1982). Osmoregulation and temperature effects on water loss and oxygen consumption in two species of African scorpion. *Comp. Biochem. Physiol.* **71A**, 605–9.

Robertson, J. D. (1936). The function of the calciferous glands of earthworms. *J. exp. Biol.* **13**, 279–97.

Robertson, J. D. (1941). The function and metabolism of calcium in the Invertebrata. *Biol. Rev.* **16**, 106–33.

Robertson, J. D. (1957). The habitat of the early vertebrates. *Biol. Rev.* **32**, 156–87.

Robertson, J. D. (1960). Osmotic and ionic regulation. In *The physiology of Crustacea*, vol. 1, pp. 317–39, ed. T. H. Waterman. Academic Press, New York & London.

Robertson, J. D. (1970). Osmotic and ionic regulation in the horseshoe crab *Limulus polyphemus* (Linnaeus). *Biol. Bull. mar. biol. Lab., Woods Hole* 138, 157-83.

Robinson, M. H. & Robinson, B. C. (1978). Thermoregulation in orb-web spiders: new descriptions of thermoregulatory postures and experiments on the effects of posture and coloration. *Zool. J. Linn. Soc.* 64, 87-102.

Robson, E. A. (1964). The cuticle of *Peripatopsis moseleyi*. *Q. Jl microsc. Sci.* 105, 281-99.

Robson, E. A., Lockwood, A. P. M. & Ralph, R. (1966). Composition of the blood in Onychophora. *Nature, Lond.* 209, 533.

Romanoff, A. L. (1960). *The avian embryo.* Macmillan, New York.

Romer, A. S. (1949). *The vertebrate body.* W. B. Saunders, Philadelphia.

Romer, A. S. (1955). Herpetichthyes, Amphibioidei, Choanichthyes or Sarcopterygii? *Nature, Lond.* 176, 126-7.

Romer, A. S. (1958). Tetrapod limbs and early tetrapod life. *Evolution* 12, 365-9.

Romer, A. S. (1972). Skin-breathing – primary or secondary? *Resp. Physiol.* 14, 183-92.

Romer, A. S. & Grove, B. H. (1935). Environment of the early vertebrates. *Am. Mid. Nat.* 16, 805-56.

Roots, B. I. (1955). The water relations of earthworms. I. The activity of the nephridiostome cilia of *Lumbricus terrestris* L. and *Allolobophora chlorotica* Savigny, in relation to the concentration of the bathing medium. *J. exp. Biol.* 32, 765-74.

Roots, B. I. (1956). The water relations of earthworms. II. Resistance to desiccation and immersion, and behaviour when submerged and when allowed a choice of environment. *J. exp. Biol.* 33, 29-44.

Rosen, D. E., Forey, P. L., Gardiner, B. G. & Patterson, C. (1981). Lungfishes, tetrapods, palaeontology, and plesiomorphy. *Bull. Am. Mus. nat. Hist.* 167, 159-276.

Rosenberg, J. (1979). Topographie und Feinstruktur des Maxillarnephridium von *Scutigera coleoptrata* (L.) (Chilopoda, Notostigmophora). *Zoomorphol.* 92, 141-59.

Rosenberg, J. (1982). Coxal organs in Geophilomorpha (Chilopoda). Organization and fine structure of the transporting epithelium. *Zoomorphol.* 100, 107-20.

Rosenberg, J. & Seifert, G. (1977). The coxal glands of the Geophilomorpha (Chilopoda): organs of osmoregulation. *Cell Tiss. Res.* 182, 247-51.

Roth, V. D. & Brown, W. L. (1976). Other intertidal air-breathing arthropods. In *Marine insects*, pp. 119-50, ed. L. Cheng. North-Holland Publishing Co., Amsterdam.

Rovner, J. S. (1967). Acoustic communication in a lycosid spider (*Lycosa rabida* Walckenaer). *Anim. Behav.* 15, 273-81.

Rudolph, D. & Knülle, W. (1974). Site and mechanism of water vapour uptake from the atmosphere in ixodid ticks. *Nature, Lond.* 249, 84-5.

Rudolph, D. & Knülle, W. (1978). Uptake of water vapour from the air: process, site and mechanism in ticks. In *Comparative physiology: water, ions and fluid mechanics*, pp. 97-113, ed. K. Schmidt-Nielsen, L. Bolis & S. H. P. Maddrell. Cambridge University Press, Cambridge, England.

Rudy, P. P. (1966). Sodium balance in *Pachygrapsus crassipes*. *Comp. Biochem. Physiol.* 18, 881-907.

Rumsey, T. J. (1971). Adaptations of the superfamily Littorinacea to marine and terrestrial environments. Ph.D. thesis, University of Bristol.

Rumsey, T. J. (1972). Osmotic and ionic regulation in a terrestrial snail, *Pomatias elegans* (Gastropoda, Prosobranchia) with a note on some tropical Pomatiasidae. *J. exp. Biol.* 57, 205-15.

Rumsey, T. J. (1973). Some aspects of osmotic and ionic regulation in *Littorina littorea* (L.) (Gastropoda, Prosobranchia). *Comp. Biochem. Physiol.* 45A, 327-44.

Rundgren, S. (1975). Vertical distribution of lumbricids in southern Sweden. *Oikos* 26, 299-306.

Runham, N. W. & Hunter, P. J. (1970). *Terrestrial Slugs.* Hutchinson, London.

Russell, E. J. (1957). *The world of the soil.* Collins, London.

Russell-Hunter, W. (1964). Physiological aspects of ecology in nonmarine molluscs. In *Physiology of Mollusca*, vol. I, pp. 83-126, ed. K. M. Wilbur & C. M. Yonge. Academic Press, London.

Russell-Hunter, W. D. (1978). Ecology of freshwater pulmonates. In *Pulmonates*, vol. 2A, pp. 335-83, ed. V. Fretter & J. Peake. Academic Press, London.

Russell-Hunter, W. D., Apley, M. L. & Hunter, R. D. (1972). Early life-history of *Melampus* and the significance of semilunar synchrony. *Biol. Bull. mar. biol. Lab., Woods Hole* 143, 623-56.

Sahrhage, D. (1954). Okologische Untersuchungen an *Thermobia domestica* (Packard) und *Lepisma saccharina* L. *Z. wiss. Zool.* 157, 77-168.

Saigusa, M. (1981). Adaptive significance of a semilunar rhythm in the terrestrial crab *Sesarma*. *Biol. Bull. mar. biol. Lab., Woods Hole* 160, 311-21.

Saigusa, M. & Hidaka, T. (1978). Semilunar rhythm in the zoea-release activity of the land crabs *Sesarma*. *Oecologia* 37, 163-76.

Saini, R. S. (1962). Histology and physiology of the crypto-nephridial system in insects. Ph. D. Thesis, Cambridge University.

Salmon, M. (1971). Signal characteristics and acoustic detection by the fiddler crabs, *Uca rapax* and *Uca pugilator*. *Physiol. Zoöl.* 44, 210-24.

Salmon, M. & Horch, K. W. (1972). Acoustic signalling and detection by semiterrestrial crabs of the family Ocypodidae. In *Behavior of marine animals*, vol. 1, pp. 60-96, ed. H. E. Winn & B. L. Olla. Plenum Press, New York.

Salpeter, M. & Walcott, C. (1960). The anatomy of the spider vibration receptor. *Exp. Neurol.* 2, 232-50.

Salthe, S. N. (1965). Increase in volume of the perivitelline chamber during development of *Rana pipiens* Schreber. *Physiol. Zoöl.* 38, 80-98.

Salthe, S. N. & Mecham, J. S. (1974). Reproductive and courtship patterns. In *Physiology of the Amphibia*, vol. 2, pp. 310-521, ed. B. Lofts. Academic Press, London.

Sandeman, D. C. (1978). Eye-scanning during walking in the crab *Leptograpsus variegatus*. *J. comp. Physiol.* 124, 249-57.

Sankey, J. H. P. & Savory, T. H. (1974). *British Harvestmen*. (Linnean Society of London, Synopses of the British Fauna No. 4.) Academic Press, London.

Satchell, G. H. (1976). The circulatory system of air-breathing fish. In *Respiration of amphibious vertebrates*, pp. 105-23, ed. G. M. Hughes. Academic Press, London.

Saussey, M. (1963). Précisions sur le déterminisme de la diapause d'*Allolobophora*. *C.r. hebd. Séanc. Acad. Sci., Paris* 257, 2527-9.

Savory, T. H. (1964). *Arachnida*. Academic Press, London.

Savory, T. H. (1971). *Evolution in the Arachnida*. Merrow Publishing Co., Watford, England.

Schaeffer, B. (1968). The origin and basic radiation of the Osteichthyes. In *Current problems of lower vertebrate phylogeny*, pp. 207-22, ed. T. Ørvig. Wiley Interscience, London.

Schaller, F. (1971). Indirect sperm transfer by soil arthropods. *Ann. Rev. Ent.* 16, 407-46.

Schaller, F. (1979). Significance of sperm transfer and formation of spermatophores in arthropod phylogeny. In *Arthropod phylogeny*, pp. 587-608, ed. A. P. Gupta. Van Nostrand Reinhold Co., New York.

Schlüter, U. (1980a). Struktur und Funktion des Enddarms chilognather Diplopoden. *Zool. Jb. (Anat.)* 103, 607-39.

Schlüter, U. (1980b). Plasmalemma–mitochondrial complexes involved in water transport in the hindgut of a millipede, *Scaphiostreptus* sp. *Cell Tiss. Res.* 205, 333-6.

Schmalfuss, H. (1975). Morphologie, Funktion und Evolution der Tergithöcker bei Landisopoden (Oniscoidea, Isopoda, Crustacea). *Z. Morph. Tiere* 80, 287-316.

Schmalhausen, I. I. (1968). *The origin of terrestrial vertebrates*. Academic Press, London.

Schmid, W. D. (1965). Some aspects of the water economies of nine species of amphibians. *Ecology* 46, 261-9.

Schmidt-Nielsen, B., Gertz, K. H. & Davis, L. (1968). Excretion and ultrastructure of the antennal gland of the fiddler crab *Uca mordax. J. Morph.* 125, 472-96.

Schmidt-Nielsen, K. (1964). *Desert animals. Physiological problems of heat and water.* Clarendon Press, Oxford.

Schmidt-Nielsen, K., Taylor, C. R. & Shkolnik, A. (1971). Desert snails: problems of heat, water and food. *J. exp. Biol.* 55, 385-98.

Schoffeniels, E. & Terkafs, R. R. (1965). Osmotic and ionic balance in brackish water amphibians. *Ann. roy. Soc. Belg.* 96, 23-9.

Schömann, K. (1956). Zur Biologie von *Polyxenus lagurus* (L. 1758). *Zool. Jb. (Syst.)* 84, 195-256.

Schöne, H. (1968). Agonistic and sexual display in aquatic and semi-terrestrial brachyuran crabs. *Am. Zool.* 8, 641-54.

Schöttle, E. (1932). Morphologie und Physiologie der Atmung bei wasser-, schlamm- und landlebenden Gobiiformes. *Z. wiss. Zool.* 140, 1-114.

Schröder, O. (1918). Beiträge zur Kenntniss von *Geonemertes palaensis* Semper. *Senckenbergiana* 35, 153-77.

Schultze, H.-P. (1972). Early growth stages in coelacanth fishes. *Nature New Biol.* 236, 90-1.

Schuster, R. M. (1976). Plate tectonics and its bearing on the geographical origin and dispersal of angiosperms. In *Origin and early evolution of angiosperms*, pp. 48-138, ed. C. B. Beck. Columbia University Press, New York.

Seelemann, U. (1968a). Zur Uberwindung der biologischen Grenze Meer-Land durch Mollusken. Untersuchungen an *Alderia modesta* (Opisth.) und *Ovatella myosotis* (Pulmonat.). *Oecologia* 1, 130-54.

Seelemann, U. (1968b). Zur Uberwindung der biologischen Grenze Meer-Land durch Mollusken. II. Untersuchungen an *Limapontia capitata, Limapontia depressa* und *Assiminea grayana. Oecologia* 1, 356-68.

Seely, M. K. (1979). Irregular fog as a water source for desert dune beetles. *Oecologia* 42, 213-27.

Seifert, G. (1967). Die Cuticula von *Polyxenus lagurus* L. (Diplopoda, Pselaphognatha). *Z. Morph. Okol. Tiere* 59, 42-53.

Seifert, G. (1980). Considerations about the evolution of excretory organs in terrestrial arthropods. In *Myriapod biology*, pp. 353-72, ed. M. Camatini. Academic Press, London.

Seifert, G. & Rosenberg, J. (1976). Fine structure of the sacculus of the nephridia of *Peripatoides leuckarti* (Onychophora: Peripatopsidae). *Ent. Germ.* 3, 202-11.

Sevilla, C. & Lagarrigue, J.-G. (1979). Oxygen binding characteristics of Oniscoidea hemocyanins (Crustacea: terrestrial isopods). *Comp. Biochem. Physiol.* 64A, 531-6.

Seymour, R. S. (1974). Convective and evaporative cooling in sawfly larvae. *J. Insect Physiol.* 20, 2447-57.

Shachak, M. (1980). Energy allocation and life history strategy of the desert isopod *Hemilepistus reaumuri. Oecologia* 45, 404-13.

Shachak, M. & Steinberger, Y. (1980). An algae–desert snail food chain: energy flow and soil turnover. *Oecologia* 46, 402-11.

Shachak, M., Steinberger, Y. & Orr, Y. (1979). Phenology, activity and regulation of radiation load in the desert isopod *Hemilepistus reaumuri. Oecologia* 40, 133-40.

Sharov, A. G. (1966). *Basic arthropodan stock, with special reference to insects.* Pergamon Press, Oxford.

Sharp, M. S. & Neff, J. M. (1980). Steady state hemolymph osmotic and chloride ion regulation and percent body water in *Clibanarius vittatus* (Decapoda: Anomura) from the Texas gulf coast. *Comp. Biochem. Physiol.* 66A, 455-60.

Shaw, J. (1955). Ionic regulation and water balance in the aquatic larva of *Sialis lutaria. J. exp. Biol.* 32, 353-82.

Shaw, J. (1959). Salt and water balance in the east African fresh-water crab, *Potamon niloticus* (M. Edw.). *J. exp. Biol.* 36, 157-76.

Shaw, J. (1961). Sodium balance in *Eriocheir sinensis* (M. Edw.). The adaptation of the Crustacea in fresh water. *J. exp. Biol.* 38, 153-62.

Shear, W. A. (1970). Stridulation in *Acanthophrynus coronatus* (Butler) (Amblypygi, Tarantulidae). *Psyche* 77, 181-3.

Shoemaker, V. H., Balding, D. & Ruibal, R. (1972). Uricotelism and low evaporative water loss in a South American frog. *Science, N.Y.* 175, 1018-20.

Shoemaker, V. H., McClanahan, L. & Ruibal, R. (1969). Seasonal changes in body fluids in a field population of spadefoot toads. *Copeia* 1969, 585-91.

Shumway, S. E. (1978). Osmotic balance and respiration in the hermit crab, *Pagurus bernhardus*, exposed to fluctuating salinities. *J. mar. biol. Ass. UK* 58, 869-76.

Shumway, S. E. & Marsden, I. D. (1982). The combined effects of temperature, salinity, and declining oxygen tension on oxygen consumption in the marine pulmonate *Amphibola crenata* (Gmelin, 1791). *J. exp. mar. Biol. Ecol.* 61, 133-46.

Silveira, M. & Corinna, A. (1976). Fine structural observations on the protonephridium of the terrestrial triclad *Geoplana pasipha. Cell Tiss. Res.* 168, 455-63.

Simkiss, K. (1980). Water and ionic fluxes inside the egg. *Am. Zool.* 20, 385-93.

Simmons, J. E. (1970). Nitrogen metabolism in Platyhelminthes. In *Comparative biochemistry of nitrogen metabolism*, vol. I. *The invertebrates*, pp. 67-90, ed. J. W. Campbell. Academic Press, London.

Sinclair, F. G. (1895). Myriapods. In *The Cambridge natural history*, vol. V, Peripatus *Myriapods Insects*, Part I, pp. 29-80. Macmillan, London.

Singh, B. N. (1976). Balance between aquatic and aerial respiration. In *Respiration of amphibious vertebrates*, pp. 125-64, ed. G. M. Hughes. Academic Press, London.

Singh, B. N. & Hughes, G. M. (1971). Respiration of an air-breathing catfish *Clarias batrachus* (Linn.). *J. exp. Biol.* 55, 421-34.

Singh, B. N. & Hughes, G. M. (1973). Cardiac and respiratory responses in the climbing perch *Anabas testudineus. J. comp. Physiol.* 84, 205-26.

Singh, B. N. & Munshi, J. S. D. (1968). On the respiratory organs and mechanics of breathing in *Periophthalmus vulgaris. Zool. Anz.* 183, 92-110.

Skaer, H. Le B. (1974a). The water balance of a serpulid polychaete, *Mercierella enigmatica* (Fauvel). I. Osmotic concentration and volume regulation. *J. exp. Biol.* 60, 321-30.

Skaer, H.Le. B. (1974b). The water balance of a serpulid polychaete, *Mercierella enigmatica* (Fauvel). IV. The excitability of the longitudinal muscle cells. *J. exp. Biol.* 60, 351-70.

Skelding, J. M. (1973a). The fine structure of the kidney of *Achatina achatina* (L.). *Z. Zellforsch.* 147, 1-29.

Skelding, J. M. (1973b). Studies on the renal physiology of *Achatina achatina* (L.). *Malacologia* 14, 93-6.

Skirrow, G. (1965). The dissolved gases – carbon dioxide. In *Chemical oceanography*, vol. I, pp. 227-322, ed. J. P. Riley & G. Skirrow. Academic Press, London.

Slifer, E. H. & Sekhon, S. S. (1970). Sense-organs of a thysanuran, *Ctenolepisma lineata pilifera*, with special reference to those on the antennal flagellum (Lepismatidae). *J. Morph.* 132, 1-26.

Slobodkin, L. B. & Rapoport, A. (1974). An optimal strategy of evolution. *Q. Rev. Biol.* 49, 181-200.

Smart, J. & Hughes, N. F. (1973). The insect and the plant: progressive palaeoecological integration. *Symp. roy. ent. Soc. Lond.* 6, 143-55.

Smatresk. N. J. & Cameron, J. N. (1981). Post-exercise acid–base balance and ventilatory control in *Birgus latro*, the coconut crab. *J. exp. Zool.* 218, 75-82.

Smith, D. G. & Campbell, G. (1976). The anatomy of the pulmonary vascular bed in the toad *Bufo marinus. Cell Tiss. Res.* 165, 199-213.

Smith, G. (1909). Crustacea. In *Cambridge natural history*, vol. 4, pp. 55–217, ed. S. F. Harmer & A. E. Shipley. Macmillan, London.

Smith, H. W. (1930). Metabolism of the lung-fish *Protopterus aethiopicus. J. biol. Chem.* **88**, 97–130.

Smith, H. W. (1931). Observations on the African lung-fish, *Protopterus aethiopicus*, and on evolution from water to land environments. *Ecology* **12**, 164–81.

Smith, H. W. (1935). The metabolism of the lung-fish. I. General considerations of the fasting metabolism in active fish. *J. cell. comp. Physiol.* **6**, 43–67.

Smith, H. W. (1959). *From fish to philosopher.* Little, Brown & Co., Boston.

Smith, L. H. (1981). Quantified aspects of pallial fluid and its effect on the duration of locomotor activity in the terrestrial gastropod *Triodopsis albolabris. Physiol. Zoöl.* **54**, 407–14.

Smith, S. (1957). Early development and hatching. In *The physiology of fishes*, vol. I, pp. 323–59, ed. M. E. Brown. Academic Press, New York.

Smith, W. K. & Miller, P. C. (1973). The thermal ecology of two south Florida fiddler crabs: *Uca rapax* Smith and *U. pugilator* Bosc. *Physiol. Zoöl.* **46**, 186–207.

Snodgrass, R. E. (1938). Evolution of the Annelida, Onychophora, and Arthropoda. *Smithson. misc. Collns* **97**, 1–159.

Sohlenius, B. (1980). Abundance, biomass and contribution to energy flow by soil nematodes in terrestrial ecosystems. *Oikos* **34**, 186–94.

Solem, A. (1974). *The shell makers. Introducing mollusks.* Wiley Interscience, New York.

Solem, A. (1978). Classification of the land mollusca. In *Pulmonates*, vol. 2A, pp. 49–97, ed. V. Fretter & J. Peake. Academic Press, London.

Southwood, T. R. E. (1973). The insect/plant relationship – an evolutionary perspective. *Symp. roy. ent. Soc. Lond.* **6**, 3–30.

Southwood, T. R. E. (1977). The stability of the trophic milieu, its influence on the evolution of behaviour and of responsiveness to trophic signals. *Colloques internationaux, CNRS* **265**, 471–93.

Spaargaren, D. H. (1978). A comparison of the blood osmotic composition of various marine and brackish-water animals. *Comp. Biochem. Physiol.* **60**, 327–33.

Spearman, R. I. C. (1966). The keratinization of epidermal scales, feathers and hairs. *Biol. Rev.* **41**, 59–96.

Speeg, K. V. & Campbell, J. W. (1968a). Formation and volatilization of ammonia gas by terrestrial snails. *Am. J. Physiol.* **214**, 1392–402.

Speeg, K. V. & Campbell, J. W. (1968b). Purine biosynthesis and excretion in *Otala* (=*Helix*) *lactea*: an evaluation of the nitrogen excretory potential. *Comp. Biochem. Physiol.* **26**, 579–95.

Spencer, J. O. & Edney, E. B. (1954). The absorption of water by woodlice. *J. exp. Biol.* **31**, 491–6.

Spies, T. (1981). Structure and phylogenetic interpretation of diplopod eyes (Diplopoda). *Zoomorphol.* **98**, 241–60.

Spitzer, J. M. (1937). Physiologische-ökologische Untersuchungen über den Excretstoffwechsel der Mollusken. *Zool. Jb.* **57**, 457–96.

Sponder, D. L. & Lauder, G. V. (1981). Terrestrial feeding in the mudskipper *Periophthalmus* (Pisces: Teleostei): a cineradiographic analysis. *J. Zool., Lond.* **193**, 517–30.

Spotila, J. R. & Berman, E. N. (1976). Determination of skin resistance and the role of the skin in controlling water loss in amphibians and reptiles. *Comp. Biochem. Physiol.* **55A**, 407–11.

Srivastava, P. N. & Gupta, P. D. (1961). Excretion of uric acid in *Periplaneta americana* L. *J. Insect Physiol.* **6**, 163–7.

Staddon, B. W. (1959). Nitrogen excretion in nymphs of *Aeshna cyanea* (Mull.) (Odonata, Anisoptera). *J. exp. Biol.* **36**, 566–74.

Staddon, B. W. (1963). Water balance in the aquatic bugs *Notonecta glauca* L. and *Notonecta marmorea* Fabr. (Hemiptera: Heteroptera). *J. exp. Biol.* **40**, 563–71.

Stahl, B. J. (1974). *Vertebrate history: problems in evolution.* McGraw-Hill, New York.

Stammers, F. M. G. (1950). Observations on the behaviour of land-leeches (Genus *Haemadipsa*). *Parasitology* **40**, 237–45.

Starling, J. H. (1944). Ecological studies of the Pauropoda of the Duke Forest. *Ecol. Monogr.* **14**, 291–310.

Stebbing, T. R. R. (1893). *A history of Crustacea. Recent Malacostraca.* Kegan Paul, London.

Stebbins, R. C. & Kalk, M. (1961). Observations on the natural history of the mudskipper, *Periophthalmus sobrinus. Copeia* 1961, 18–27.

Steinberger, Y., Grossman, S. & Dubinsky, Z. (1981). Some aspects of the ecology of the desert snail *Sphincterochila prophetarum* in relation to energy and water flow. *Oecologia* **50**, 103–8.

Steinberger, Y., Grossman, S. & Dubinsky, Z. (1982). Changes in organic storage compounds during the active and inactive periods in a desert snail, *Sphincterochila prophetarum. Comp. Biochem. Physiol.* **71A**, 41–6.

Stephenson, J. (1930). *The Oligochaeta.* Clarendon Press, Oxford.

Stephenson, J. & Prashad, B. (1919). The calciferous glands of earthworms. *Trans. roy. Soc. Edinb.* **52**, 455–85.

Stephenson, W. (1942). The effect of variations in osmotic pressure upon a free-living soil nematode. *Parasitology* **34**, 253–65.

Stewart, D. M. & Martin, A. W. (1970). Blood and fluid balance of the common tarantula, *Dugesiella hentzi. Z. vergl. Physiol.* **70**, 233–46.

Stewart, D. M. & Martin, A. W. (1974). Blood pressure in the tarantula, *Dugesiella hentzi. J. comp. Physiol.* **88**, 141–72.

Stewart, T. C. & Woodring, J. P. (1973). Anatomical and physiological studies of water balance in the millipedes *Pachydesmus crassicutis* (Polydesmida) and *Orthoporus texicolens* (Spirobolida). *Comp. Biochem. Physiol.* **44A**, 735–50.

Stobbart, R. H. & Shaw, J. (1974). Salt and water balance; excretion. In *The physiology of Insecta* (2nd edition), vol. 5, pp. 362–446, ed. M. Rockstein. Academic Press, London.

Storch, V., Cases, E. & Rosito, R. (1979). Recent findings on the coconut crab, *Birgus latro* (L.). *Philippine Sci.* **16**, 57–67.

Storch, V. & Ruhberg, H. (1977). Fine structure of the sensilla of *Peripatopsis moseleyi* (Onychophora). *Cell Tiss. Res.* **177**, 539–53.

Storch, V., Ruhberg, H. & Alberti, G. (1978). Zur ultrastruktur der Segmentalorgane der Peripatopsidae (Onychophora). *Zool. Jb. (Anat.)* **100**, 47–63.

Storch, V. & Welsch, U. (1972). Ultrastructure and histochemistry of the integument of air-breathing polychaetes from mangrove swamps of Sumatra. *Mar. Biol.* **17**, 137–44.

Storch, V. & Welsch, U. (1975). Uber Bau und Funktion der Kiemen und Lungen von *Ocypode ceratophthalma* (Decapoda: Crustacea). *Mar. Biol.* **29**, 363–71.

Størmer, L. (1955). Arthropoda 2. Chelicerata. In *Treatise on invertebrate palaeontology*, pp. 1–41, ed. R. C. Moore. Geological Society of America and University of Kansas Press.

Størmer, L. (1976). Arthropods from the Lower Devonian (lower Emsian) of Alken an der Mosel, Germany. Part 5. Myriapoda and additional forms, with general remarks on fauna and problems regarding invasion of land by arthropods. *Senckenberg. lethaea* **57**, 87–183.

Stout, J. D. (1952). The occurrence of aquatic oligochaetes in soil. *Trans. roy. Soc. NZ* **80**, 97–101.

Stout, J. D. (1956). Aquatic oligochaetes occurring in forest litter. I. *Trans. roy. Soc. NZ* **84**, 97–102.

Sturm, H. (1955). Beitrage zur Ethologie einiger mitteldeutscher Machiliden. *Z. Tierpsychol.* **12**, 337–63.

Sundara Rajulu, G. (1969). Presence of haemocyanin in the blood of a centipede *Scutigera longicornis* (Chilopoda: Myriapoda). *Curr. Sci.* **38**, 168–9.

Sundara Rajulu, G. (1970*a*). A comparative study of the free amino acids in the haemolymph of a millipede, *Spirostreptus asthenes*, and a centipede *Ethmostigmus spinosus* (Myriapoda). *Comp. Biochem. Physiol.* **37**, 339-44.

Sundara Rajula, G. (1970*b*). Tracheal pulsation in a marine centipede *Mixophilus indicus*. *Curr. Sci.* **39**, 397-8.

Sundara Rajula, G. (1971). Cardiac physiology of a millipede *Cingalobolus bugnioni* Carl (Myriapoda Diplopoda). *Monit. Zool. ital.* **5**, 39-52.

Sundara Rajulu, G. (1974). A comparative study of the organic components of the haemolymph of a millipede, *Cingalobolus bugnioni*, and a centipede, *Scutigera longicornis* (Myriapoda). *Symp. zool. Soc. Lond.* **32**, 347-64.

Sundara Rajulu, G. & Manavalaramanujam, R. (1972). Free amino acids in the haemolymph of *Eoperipatus weldoni* (Onychophora). *Experientia* **28**, 87.

Sundnes, G. & Valen, E. (1969). Respiration of dry cysts of *Artemia salina* L. *J. Cons. perm. int. Explor. Mer* **32**, 413-15.

Sutcliffe, D. W. (1963). The chemical composition of haemolymph in insects and some other arthropods, in relation to their phylogeny. *Comp. Biochem. Physiol.* **9**, 121-35.

Sutcliffe, D. W. (1968). Sodium regulation and adaptation to fresh water in gammarid crustaceans. *J. exp. Biol.* **48**, 359-80.

Sutcliffe, D. W. (1974). Sodium regulation and adaptation to fresh water in the isopod genus *Asellus*. *J. exp. Biol.* **61**, 719-36.

Sutton, S. (1972). *Woodlice*. Ginn & Co. Ltd, London.

Sutton, S. L. (1968). The population dynamics of *Trichoniscus pusillus* and *Philoscia muscorum* (Crustacea, Oniscoidea) in limestone grassland. *J. anim. Ecol.* **37**, 425-44.

Sverdrup, H. U., Johnson, M. W. & Fleming, R. H. (1942). *The oceans*. Prentice-Hall, Englewood Cliffs, New Jersey.

Swedmark, B. (1964). The interstitial fauna of marine sand. *Biol. Rev.* **39**, 1-42.

Sweetman, H. L. (1938). Physiological ecology of the firebrat, *Thermobia domestica* (Packard). *Ecol. Monogr.* **8**, 285-311.

Sykes, A. H. (1971). Formation and composition of urine. In *Physiology and biochemistry of the domestic fowl*, vol. I, pp. 233-78, ed. D. J. Bell & B. M. Freeman. Academic Press, London.

Szarski, H. (1968). The origin of vertebrate foetal membranes. *Evolution* **22**, 211-14.

Szarski, H. (1976). Sarcopterygii and the origin of tetrapods. In *Major patterns in vertebrate evolution*, pp. 517-40, ed. M. K. Hecht, P. C. Goody & B. M. Hecht. Plenum Press, London.

Takeuchi, N. (1980). Control of coelomic fluid concentration and brain neurosecretion in the littoral earthworm *Pontodrilus matsushimensis* Iizuka. *Comp. Biochem. Physiol.* **67A**, 357-9.

Tamura, H. & Koseki, K. (1974). Population study on a terrestrial amphipod, *Orchestia platensis japonica* (Tattelsall), in a temperate forest. *Jap. J. Ecol.* **24**, 123-39.

Tamura, S. O., Morii, H. & Yuzuriha, M. (1976). Respiration of the amphibious fishes *Periophthalmus cantonensis* and *Boleophthalmus chinensis* in water and on land. *J. exp. Biol.* **65**, 97-107.

Tarling, D. & Runcorn, S. (eds.) (1973). *Implications of continental drift to the earth sciences*. 2 vols. Academic Press, London.

Taylor, A. C. & Davies, P. S. (1981). Respiration in the land crab, *Gecarcinus lateralis*. *J. exp. Biol.* **93**, 197-208.

Taylor, D. W. & Sohl, N. F. (1962). An outline of gastropod classification. *Malacologia* **1**, 7-32.

Taylor, E. W. & Butler, P. J. (1978). Aquatic and aerial respiration in the shore crab, *Carcinus maenas* (L.), acclimated to 15 °C. *J. comp. Physiol.* **127**, 315-23.

Taylor, E. W. & Wheatly, M. G. (1981). The effect of long-term aerial exposure on heart rate, ventilation, respiratory gas exchange and acid-base status in the crayfish *Austropotamobius pallipes*. *J. exp. Biol.* **92**, 109-24.

Taylor, H. H. & Greenaway, P. (1979). The structure of the gills and lungs of the arid-zone crab, *Holthuisana (Austrothelphusa) transversa* (Brachyura: Sundathelphusidae) including observations on arterial vessels in the gills. *J. Zool., Lond.* **189**, 359-84.

Tazawa, H. (1980). Oxygen and CO_2 exchange and acid-base regulation in the avian embryo. *Am. Zool.* **20**, 395-404.

Teal, J. M. & Carey, F. G. (1967). Skin respiration and oxygen debt in the mudskipper, *Periophthalmus sobrinus*. *Copeia* 1967, 677-9.

Termier, G. & Termier, H. (1952). Classe des gastéropodes. In *Traité de paléontologie*, vol. II, pp. 365-460, ed. J. Piveteau. Masson et Cie, Paris.

Thiele, H. (1971). Uber die Facettenaugen von land- und wasser-bewohnenden Crustaceen. *Z. Morph. Tiere* **69**, 9-22.

Thiele, J. (1927). Uber die Schneckenfamilie Assimineidae. *Zool. Jb. (Syst.)* **53**, 113-46.

Thiele, J. (1931). *Handbuch der Systematischen Weichtierkunde*. A. Asher & Co., Amsterdam. (Reprinted 1963.)

Thompson, F. G. (1969). Some Mexican and Central American land snails of the family Cyclophoridae. *Zoologica, N.Y.* **54**, 35-77.

Thompson, F. G. (1980). Proserpinoid land snails and their relationships within the Archaeogastropoda. *Malacologia* **20**, 1-33.

Thomson, K. S. (1967). Mechanisms of intracranial kinetics in fossil rhipidistian fishes (Crossopterygii) and their relatives. *J. Linn. Soc. (Zool.)* **46**, 223-53.

Thomson, K. S. (1968). A critical review of the diphyletic theory of rhipidistian-amphibian relationships. In *Current problems of lower vertebrate phylogeny*, pp. 285-306, ed. T. Ørvig. Interscience, London.

Thomson, K. S. (1969). The biology of the lobe-finned fishes. *Biol. Rev.* **44**, 91-154.

Thomson, K. S. (1980). The ecology of the Devonian lobe-finned fish. In *The terrestrial environment and the origin of land vertebrates*, pp. 187-222, ed. A. L. Panchen. Academic Press, London.

Thomson, K. S. & Bossy, K. H. (1970). Adaptive trends and relationships in early Amphibia. *Forma et Functio* **3**, 7-31.

Tichy, H. (1973). Untersuchungen uber die Feinstruktur des Tömösvaryschen Sinnesorgane von *Lithobius forficatus* L. (Chilopoda) und zur Frage seiner Funktion. *Zool. Jb. (Anat.)* **91**, 93-139.

Tichy, H. (1975). Unusual fine structure of sensory hair triad of the millipede, *Polyxenus*. *Cell Tiss. Res.* **156**, 229-38.

Tiegs, O. W. (1940). The embryology and affinities of the Symphyla, based on a study of *Hanseniella agilis*. *Q. Jl microsc. Sci.* **82**, 1-225.

Tiegs, O. W. & Manton, S. M. (1958). The evolution of the Arthropoda. *Biol. Rev.* **33**, 255-337.

Tielecke, H. (1940). Anatomie, Phylogenie und Tiergeographie der Cyclophoriden. *Arch. Naturgesch.* (N.F.) **9**, 317-71.

Tilbrook, P. J. (1967). The terrestrial invertebrate fauna of the maritime Antarctic. *Phil. Trans. roy. Soc.* (B) **252**, 261-78.

Tillinghast, E. K. (1967). Excretory pathways of ammonia and urea in the earthworm *Lumbricus terrestris* L. *J. exp. Zool.* **166**, 295-300.

Timm, T. (1980). Distribution of aquatic oligochaetes. In *Aquatic oligochaete biology*, pp. 55-77, ed. R. O. Brinkhurst & D. G. Cook. Plenum Press, London.

Todd, E. S. & Ebeling, A. W. (1966). Aerial respiration in the longjaw mudsucker *Gillichthys mirabilis* (Teleostei: Gobiidae). *Biol. Bull. mar. biol. Lab., Woods Hole* **130**, 265-88.

Todd, M. E. (1963). Osmoregulation in *Ligia oceanica* and *Idotea granulosa*. *J. exp. Biol.* **40**, 381-92.

Tolbert, W. A. (1979). Thermal stress of the orb-weaving spider *Argiope trifasciata* (Araneae). *Oikos* **32**, 386-92.

Tompa, A. S. (1976). A comparative study of the ultrastructure and mineralogy of calcified land snail eggs (Pulmonata: Stylommatophora). *J. Morph.* **150**, 861–88.

Toolson, E. C. & Hadley, N. F. (1977). Cuticular permeability and epicuticular lipid composition in two Arizona vejovid scorpions. *Physiol. Zoöl.* **50**, 323–30.

Toye, S. A. (1966). The effect of desiccation on the behaviour of three species of Nigerian millipedes: *Spirostreptus assiniensis*, *Oxydesmus* sp. and *Habrodesmus falx*. *Ent. exp. appl.* **9**, 378–84.

Tramell, P. R. & Campbell, J. W. (1970). Nitrogenous excretory products of the giant South American land snail, *Strophocheilus oblongus*. *Comp. Biochem. Physiol.* **32**, 569–71.

Treherne, J. E. & Willmer, P. G. (1975). Hormonal control of integumentary water-loss: evidence for a novel neuroendocrine system in an insect (*Periplaneta americana*). *J. exp. Biol.* **63**, 143–59.

Tucker, L. E. (1977*a*). Effect of dehydration and rehydration on the water content and Na$^+$ and K$^+$ balance in adult male *Periplaneta americana*. *J. exp. Biol.* **71**, 49–66.

Tucker, L. E. (1977*b*). The influence of diet, age and state of hydration on Na$^+$, K$^+$ and urate balance in the fat body of the cockroach *Periplaneta americana*. *J. exp. Biol.* **71**, 67–79.

Tucker, L. E. (1977*c*). The influence of age, diet and lipid content on survival, water balance and Na$^+$ and K$^+$ regulation in dehydrating cockroaches. *J. exp. Biol.* **71**, 81–93.

Tucker, L. E. (1977*d*). Regulation of ions in the haemolymph of the cockroach *Periplaneta americana* during dehydration and rehydration. *J. exp. biol.* **71**, 95–110.

Tuerkay, M. (1970). Die Gecarcinidae Amerikas. *Senkenberg. Biol.* **51**, 333–54.

Tullett, S. G. & Board, R. G. (1977). Determinants of avian eggshell porosity. *J. Zool., Lond.* **183**, 203–11.

Tuxen, S. L. (1970). The systematic position of entognathous apterygotes. *An. Esc. nac. Cienc. biol. Méx.* **17**, 65–79.

Tweedie, M. W. F. (1950). Notes on the grapsoid crabs from the Raffles Museum. 2. On the habits of three ocypodid crabs. *Bull. Raffles Mus.* **23**, 317–24.

Tynen, M. J. (1969). Littoral distribution of *Lumbricillus reynoldsoni* Backlund and other Enchytraeidae (Oligochaeta) in relation to salinity and other factors. *Oikos* **20**, 41–53.

Ubaghs, G. (1979). Classification of echinoderms. In *Treatise on invertebrate paleontology*, vol. T2 (1) (Crinoidea), pp. T359–401, ed. R. A. Robison. Geological Society of America, New York.

Underwood, A. J. (1975). Comparative studies on the biology of *Nerita atramentosa* Reeve, *Bembicium nanum* (Lamarck) and *Cellana tramoserica* (Sowerby) (Gastropoda: Prosobranchia) in S. E. Australia. *J. exp. mar. Biol. Ecol.* **18**, 153–72.

Usinger, R. L. (1957). Marine insects. *Geol. Soc. Am. Mem.* **67**(1), 1177–82.

Valentine, J. W. (1973). Plates and provinciality, a theoretical history of environmental discontinuities. *Spec. Pap. Palaeont.* **12**, 79–92.

Valentine, J. W. & Moores, E. M. (1976). Plate tectonics and the history of life in the oceans. In *Continents adrift and continents aground*, pp. 196–205, ed. J. T. Wilson. W. H. Freeman, San Francisco.

Vandel, A. (1943). Essai sur l'origine, l'évolution et la classification des Oniscoidea (Isopodes terrestres). *Bull. biol. Fr. Belg.* (suppl.) **30**, 1–143.

Vandel, A. (1960). *Isopodes Terrestres. Faune de France*, vol. 64. Librairie de la Faculté des Sciences, Paris.

Vandel, A. (1965). Sur l'existence d'oniscoides très primitifs menant une vie aquatique et sur le polyphylétisme des isopodes terrestres. *Ann. Spéléologie* **20**, 489–518.

Vannier, G. (1973*a*). Régulation du flux d'évaporation corporelle chez un insecte collembole vivant dans une atmosphère totalement deséchée. *C. R. Acad. Sci., Paris* **277**, 85–8.

Vannier, G. (1973*b*). Etude de la transpiration chez un insecte collembole au cours de son exuviation. *C. r. Acad. Sci., Paris* **277**, 2231–4.

Vannier, G. (1973*c*). Originalité des conditions de vie dans le sol due à la présence de l'eau: importance thermodynamique et biologique de la porosphere. *Annls Soc. roy. zool. Belg.* **103**, 157–67.

Vannier, G. (1977). Relations hydriques chez deux espèces de Tomoceridae (insectes collemboles) peuplant des niveaux écologiques séparés. *Bull. Soc. zool. Fr.* **102**, 63–79.

Vannier, G. (1978). La résistance à la dessiccation chez les premiers arthropodes terrestres. *Bull. Soc. ecophysiol.* **3**, 13–42.

Vannier, G. (1983). The importance of ecophysiology for both biotic and abiotic studies of the soil. In *New trends in soil biology, VIIIth international colloquium on soil zoology, 1982, Louvain-la-Neuve*. (In press.)

Vannier, G. & Ghabbour, S. I. (1983). Effect of rising ambient temperature on transpiration in the cockroach *Heterogamia syriaca* Sauss. from the Mediterranean coastal desert of Egypt. In *New trends in soil biology, VIIIth international colloquium on soil biology, 1982, Louvain-la-Neuve*. (In press.)

Vannier, G. & Verhoef, H. A. (1978). Effect of starvation on transpiration and water content in the populations of two co-existing Collembola species. *Comp. Biochem. Physiol.* **60A**, 483–9.

Verhoef, H. A. (1981). Water balance in Collembola and its relation to habitat selection: water content, haemolymph osmotic pressure and transpiration during an instar. *J. Insect Physiol.* **27**, 755–60.

Verhoef, H. A., Bosman, C., Bierenbroodspot, A. & Boer, H. H. (1979). Ultrastructure and function of the labial nephridia and the rectum of *Orchesella cincta* (L.) (Collembola). *Cell Tiss. Res.* **198**, 246.

Verhoef, H. A. & Witteveen, J. (1980). Water balance in Collembola and its relation to habitat selection: cuticular water loss and water uptake. *J. Insect Physiol.* **26**, 201–8.

Verhoeff, K. W. (1917). Zur Kenntnis der Atmung und der Atmungsorgane der Isopoda Oniscoidea. *Biol. Zbl.* **37**, 113–27.

Verhoeff, K. W. (1920). Uber die Atmung der Landasseln, zugleich ein Beitrag zur Kenntnis der Entstehung der Landtiere (Uber Isopoden. 21. Aufsatz). *Z. wiss. Zool.* **118**, 365–447.

Verhoeff, K. W. (1941). Zur Kenntniss der Chilopoden stigmen. *Z. Morph. Okol. Tiere* **38**, 96–117.

Verhoeff, K. W. (1949). *Tylos*, eine terrestrisch-marine Rückwanderergattung der Isopoden. *Arch. Hydrobiol.* **42**, 329–40.

Vernon, G. M., Herold, L. & Witkus, E. R. (1974). Fine structure of the digestive tract epithelium in the terrestrial isopod, *Armadillidium vulgare*. *J. Morph.* **144**, 337–60.

Versluys, J. & Demoll, R. (1920). Das Limulus-Problem. Die Verwandtschaftsbeziehungen der Merostomen und Arachnoideen unter sich und mit anderen Arthropoden. *Ergebn. Fortschr. Zool.* **5**, 67–388.

Verwey, J. (1930). Einiges aus die Biologie ostindischer Mangrovekrabben. *Treubia* **12**, 167–261.

Vielmetter, W. (1958). Physiologie des Verhaltens zur Sonnenstrahlung bei dem Tagfalter *Argynnis paphia* L. I. Untersuchungen im Freiland. *J. Insect Physiol.* **2**, 13–37.

Vollmer, A. T. & MacMahon, J. A. (1974). Comparative water relations of five species of spiders from different habitats. *Comp. Biochem. Physiol.* **47A**, 753–65.

Vorwohl, G. (1961). Zur Funktion der Exkretionsorgane von *Helix pomatia* L. und *Archachatina ventricosa* Gould. *Z. vergl. Physiol.* **45**, 12–49.

Walcott, C. (1963). The effect of the web on vibration sensitivity in the spider *Achaearanea tepidariorum* (Koch). *J. exp. Biol.* **40**, 595–611.

Walcott, C. & van der Kloot, W. G. (1959). The physiology of the spider vibration receptor. *J. exp. Zool.* **141**, 191–244.

Walcott, C. D. (1931). Addenda to descriptions of Burgess shale fossils. *Smithson. misc. Collns* **85**(No. 3), 1–46.

Wald, G. (1968). Single and multiple visual systems in arthropods. *J. gen. Physiol.* **51**, 125–50.

Wall, B. J. (1970). Effects of dehydration and rehydration on *Periplaneta americana*. *J. Insect Physiol.* **16**, 1027–42.

Wall, B. J. (1977). Fluid transport in the cockroach rectum. In *Transport of ions and water in animals*, pp. 599–612, ed. B. L. Gupta, R. B. Moreton, J. L. Oschman & B. L. Wall. Academic Press, London.

Wall, B. J. & Oschman, J. L. (1970). Water and solute uptake by rectal pads of *Periplaneta americana*. *Am. J. Physiol.* **218**, 1208–15.

Wallace, H. R. (1958). Movement of eelworms II. *Ann. appl. Biol.* **46**, 86–94.

Walls, G. L. (1942). *The vertebrate eye and its adaptive radiation*. Cranbrook Institute, Bloomfield Hills, Michigan.

Wallwork, J. A. (1970). *Ecology of soil animals*. McGraw-Hill, London.

Wallwork, J. A. (1976). *The distribution and diversity of soil fauna*. Academic Press, London.

Walsh, G. E. (1974). Mangroves: a review. In *Ecology of halophytes*, pp. 51–174, ed. R. J. Reimold & W. H. Queen. Academic Press, London.

Walvig, F. (1963). The gonads and the formation of the sexual cells. In *The biology of Myxine*, pp. 530–580, ed. A. Brodal & R. Fänge. Universitets-forlaget, Oslo.

Wangensteen, O. D. (1972). Gas exchange by a bird's embryo. *Resp. Physiol.* **14**, 64–74.

Waraska, J. C., Proos, A. A. & Tillinghast, E. K. (1980). The utilization of carbohydrate reserves with the transition to ureotelism in the earthworm *Lumbricus terrestris* L. *Comp. Biochem. Physiol.* **65A**, 255–7.

Warburg, M. R. (1965a). The evaporative water loss of three isopods from semi-arid habitats in south Australia. *Crustaceana* **9**, 302–8.

Warburg, M. R. (1965b). Water relation and internal body temperature of isopods from mesic and xeric habitats. *Physiol. Zoöl.* **38**, 99–109.

Warburg, M. R. (1965c). The microclimate in the habitats of two isopod species in southern Arizona. *Am. Mid. Nat.* **73**, 363–75.

Warburg, M. R. (1968). Behavioural adaptations of terrestrial isopods. *Am. Zool.* **8**, 545–59.

Warburg, M. R. & Degani, G. (1979). Evaporative water loss and uptake in juvenile and adult *Salamandra salamandra* (L.) (Amphibia: Urodela). *Comp. Biochem. Physiol.* **62A**, 1071–5.

Warburg, M. R., Goldenberg, S. & Ben-Horin, A. (1980). Thermal effect on evaporative water loss and haemolymph osmolarity in scorpions at low and high humidities. *Comp. Biochem. Physiol.* **67A**, 47–57.

Ward, D. V. (1969). Leg extension in *Limulus*. *Biol. Bull. mar. biol. Lab., Woods Hole* **136**, 288–300.

Warner, G. F. (1967). The life history of the mangrove tree crab, *Aratus pisoni*. *J. Zool., Lond.* **153**, 321–35.

Warner, G. F. (1969). The occurrence and distribution of crabs in a Jamaican mangove swamp. *J. anim. Ecol.* **38**, 379–89.

Warner, G. F. (1977). *The biology of crabs*. Elek Science, London.

Wasserthal, L. T. (1975). The role of butterfly wings in regulation of body temperature. *J. Insect Physiol.* **21**, 1921–30.

Wasserthal, L. T. (1980). Oscillating haemolymph 'circulation' in the butterfly *Papilio machaon* L. revealed by contact thermography and photocell measurements. *J. comp. Physiol.* **139**, 145–63.

Wasserthal, L. T. (1981). Oscillating haemolymph 'circulation' and discontinuous tracheal ventilation in the giant silkmoth *Attacus atlas* L. *J. comp. Physiol.* **145**, 1–15.

Waterhouse, F. L. (1955). Microclimatological profiles in grass cover in relation to biological problems. *Q. J. roy. met. Soc.* **81**, 63–71.

Waterman, T. H. (1961). Light sensitivity and vision. In *The physiology of Crustacea*, vol. II, pp. 1–64, ed. T. H. Waterman. Academic Press, New York and London.

Waterman, T. H. & Chace, F. A. (1960). General crustacean biology. In *The physiology of Crustacea*, vol. I, pp. 1–33, ed. T. H. Waterman. Academic Press, London.

Watson, D. M. S. (1951). *Paleontology and modern biology*. Yale University Press, New Haven.

Weigmann, G. (1973). Zur Ökologie der Collembolen und Oribatiden im Grenzbereich Land-Meer (Collembola, Insecta – Oribatei, Acari). *Z. wiss. Zool.* **186**, 295–391.

Wells, H. G., Huxley, J. & Wells, G. P. (1931). *The science of life*. Cassell, London.

Wells, M. J. (1962). *Brain and behaviour in cephalopods*. Heinemann, London.

Wells, R. M. G. & Shumway, S. E. (1980). The effects of salts on haemocyanin–oxygen binding in the marine pulmonate snail *Amphibola crenata* (Martyn). *J. exp. mar. Biol. Ecol.* **43**, 11–27.

Wells, R. M. G. & Wong, P. P. S. (1978). Respiratory functions of the blood in the limpet *Siphonaria zelandica* (Gastropoda: Pulmonata). *NZ J. Zool.* **5**, 417–20.

Welsch, U. & Storch, V. (1976). Elektronenmikroskopische Beobachtungen am Kiemenepithel des amphibischen Teleosteers *Periophthalmus vulgaris*. *Z. mikrosk.-anat. Forsch.* **90**, 447–57.

Wenning, A. (1978). Struktur und Funktion des Exkretionssystems von *Lithobius forficatus* L. (Myriapoda, Chilopoda). *Zool. Jb. (Physiol.)* **82**, 419–33.

Wenning, A. (1980). Structure and function of the hind-gut of *Lithobius forficatus* L. (Myriapoda, Chilopoda). In *Myriapod biology*, pp. 135–42, ed. M. Camatini. Academic Press, London.

Weyda, F. (1974). Coxal vesicles of Machilidae (Thysanura). *Pedobiologia* **14**, 138–41.

Weyda, F. (1976). Histology and ultrastructure of the abdominal vesicles of *Campodea franzi* (Diplura, Campodeidae). *Acta ent. bohemoslov.* **73**, 237–42.

Weyda, F. (1980). Diversity of cuticular types in the abdominal vesicles of *Campodea silvestrii* (Diplura, Campodeidae). *Acta ent. bohemoslov.* **77**, 297–302.

Weygoldt, P. (1969). *The biology of pseudoscorpions*. Harvard University Press, Cambridge, Mass.

Weygoldt, P. (1979). Significance of later embryonic stages and head development in arthropod phylogeny. In *Arthropod phylogeny*, pp. 107–35, ed. A. P. Gupta. Van Nostrand Reinhold Co., New York.

Wharton, G. W. (1978). Uptake of water vapour by mites and mechanisms utilized by the Acaridei. In *Comparative physiology: water, ions and fluid mechanics*, pp. 79–95, ed. K. Schmidt-Nielsen, L. Bolis & S. H. P. Maddrell. Cambridge University Press, Cambridge, England.

Wharton, G. W. & Kanungo, K. (1962). Some effects of temperature and relative humidity on water balance in females of the spiny rat mite *Echinolaelaps echidninus* (Acarina: Laelaptidae). *Ann. ent. Soc. Am.* **55**, 483–92.

Wharton, G. W. & Richards, A. G. (1978). Water vapor exchange kinetics in insects and acarines. *Ann. Rev. Ent.* **23**, 309–28.

Whittington, H. B. (1978). The lobopod animal *Aysheaia pedunculata* Walcott, Middle Cambrian, Burgess Shale, British Columbia. *Phil. Trans. roy. Soc. (B)* **284**, 165–97.

Whittington, H. B. (1979). Early arthropods, their appendages and relationships. In *The origin of major invertebrate groups*, pp. 253–68, ed. M. R. House. Academic Press, London.

Wieser, W. (1967). Conquering terra firma: the copper problem from the isopod's point of view. *Helgolander Wiss. Meeresuntersuch.* **15**, 282–93.

Wieser, W. (1968). Aspects of nutrition and the metabolism of copper in isopods. *Am. Zool.* **8**, 495–506.

Wieser, W. (1972). Oxygen consumption and ammonia excretion in *Ligia beaudiana* M.-E. *Comp. Biochem. Physiol.* **43A**, 869–76.

Wieser, W. (1979). The flow of copper through a terrestrial food web. In *Copper in the environment*, pp. 326–55, ed. J. O. Nriagu. John Wiley & Son Ltd, London.